Structural Engineering Reference Manual

Fifth Edition

Alan Williams, PhD, SE, FICE, C ENG

The Power to Pass®
www.ppi2pass.com

Professional Publications, Inc. • Belmont, California

Benefit by Registering This Book with PPI

- Get book updates and corrections.
- Hear the latest exam news.
- Obtain exclusive exam tips and strategies.
- Receive special discounts.

Register your book at **www.ppi2pass.com/register**.

Report Errors and View Corrections for This Book

PPI is grateful to every reader who notifies us of a possible error. Your feedback allows us to improve the quality and accuracy of our products. You can report errata and view corrections at **www.ppi2pass.com/errata**.

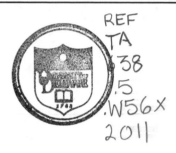
STRUCTURAL ENGINEERING REFERENCE MANUAL
Fifth Edition

Current printing of this edition: 1

Printing History

edition number	printing number	update
4	3	Minor corrections.
4	4	Minor corrections. Copyright update.
5	1	New edition. Code update. Copyright update.

Printed in the United States of America.

PPI
1250 Fifth Avenue, Belmont, CA 94002
(650) 593-9119
www.ppi2pass.com

ISBN: 978-1-59126-335-7

Library of Congress Control Number: 2010939275

Table of Contents

Preface and Acknowledgments v

Introduction vii

Codes and References Used to Prepare
This Book xix

Chapter 1: Reinforced Concrete Design
1. Strength Design Principles 1-1
2. Strength Design of Reinforced
   Concrete Beams 1-3
3. Serviceability Requirements for Beams . . . 1-10
4. Elastic Design Method 1-14
5. Shear and Torsion 1-15
6. Concrete Columns 1-27
7. Development and Splice Length
   of Reinforcement 1-34
8. Two-Way Slab Systems 1-42
   Practice Problems 1-47
   Solutions 1-48

Chapter 2: Foundations and Retaining Structures
1. Strip Footing 2-1
2. Isolated Column with Square Footing 2-5
3. Isolated Column with
   Rectangular Footing 2-10
4. Combined Footing 2-12
5. Strap Footing 2-16
6. Cantilever Retaining Wall 2-20
7. Counterfort Retaining Wall 2-25
   Practice Problems 2-27
   Solutions 2-28

Chapter 3: Prestressed Concrete Design
1. Design Stages 3-1
2. Design for Shear and Torsion 3-12
3. Prestress Losses 3-17
4. Composite Construction 3-21
5. Load Balancing Procedure 3-26
6. Statically Indeterminate Structures 3-27
   Practice Problems 3-30
   Solutions 3-31

Chapter 4: Structural Steel Design
1. Load and Resistance Factor Design 4-1
2. Design for Flexure 4-2
3. Design for Shear 4-8
4. Design of Compression Members 4-10
5. Plastic Design 4-24
6. Design of Tension Members 4-29
7. Design of Bolted Connections 4-33
8. Design of Welded Connections 4-38

9. Plate Girders 4-42
10. Composite Beams 4-49
    Practice Problems 4-54
    Solutions 4-55

Chapter 5: Timber Design
1. Adjustment Factors 5-1
2. Design for Flexure 5-6
3. Design for Shear 5-8
4. Design for Compression 5-11
5. Design for Tension 5-16
6. Design of Connections 5-17
   Practice Problems 5-26
   Solutions 5-27

Chapter 6: Design of Reinforced Masonry
1. Design Principles 6-1
2. Design for Flexure 6-1
3. Design for Shear 6-5
4. Design of Masonry Columns 6-6
5. Design of Shear Walls 6-10
   Practice Problems 6-15
   Solutions 6-16

Chapter 7: Seismic Design: International Building Code Lateral Force Procedure
1. Equivalent Lateral Force Procedure 7-2
2. Vertical Distribution of Seismic Forces . . . 7-10
3. Diaphragm Loads 7-11
4. Story Drift 7-12
5. P-Delta Effects 7-13
6. Simplified Lateral Force Procedure 7-14
7. Seismic Load on an Element of a Structure . 7-18
   Practice Problems 7-20
   Solutions 7-21

Chapter 8: Design of Bridges
1. Design Loads 8-1
2. Reinforced Concrete Design 8-14
3. Prestressed Concrete Design 8-21
4. Structural Steel Design 8-33
5. Wood Structures 8-40
6. Seismic Design 8-43
   Practice Problems 8-53
   Solutions 8-53

Appendices A-1

Index I-1

Index of Codes IC-1

Preface and Acknowledgments

The *Structural Engineering Reference Manual*, 5th edition, is a comprehensive resource to help you prepare for the NCEES 16-hour Structural exam. My goal in writing these eight chapters is to provide you with a guide to the relevant codes, and to demonstrate their use in calculations for structures as necessary to help you pass the exam.

This fifth edition has been revised to reflect the most current NCEES 16-hour Structural exam specifications and design standards. (For a complete list of updated design standards, see Codes and References Used to Prepare This Book.) Nomenclature, equations, examples, and practice problems have been checked and updated so that they are consistent with current codes.

New text has also been added to sections previously dependent on older editions of exam-referenced codes. Chapter 8, Design of Bridges, has been completely revised to conform with *AASHTO LRFD Bridge Design Specifications*, 4th edition, which has been adopted by NCEES for the 16-hour Structural exam.

Thank you to PPI's production and editorial staff, including Kate Hayes (typesetter), Tom Bergstrom (illustrator), Amy Schwertman (cover designer), Cathy Schrott (director of production), Sarah Hubbard (director of new product development), and Jenny Lindeburg King (project editor).

Alan Williams, PhD, SE, FICE, C Eng

Introduction

Part 1: How You Can Use This Book

This *Structural Engineering Reference Manual* is intended to help you prepare for the 16-hour Structural exam administered by the National Council of Examiners for Engineering and Surveying (NCEES). The NCEES 16-hour Structural exam will test your knowledge of structural principles by presenting problems that cover the design of an entire structure or portion of a structure. The exam is given in four modules—two concerning vertical forces and two concerning lateral forces. The eight chapters of this book are organized around the eight areas in which these forces are applied. These eight areas include

- reinforced concrete design
- foundations and retaining structures
- prestressed concrete design
- structural steel design
- timber design
- masonry design
- seismic design
- bridge structures

Each chapter presents structural design principles that build on the ones before, so you should read the chapters in the order in which they are presented. The examples in each chapter should also be read in sequence. Taken together in this way, they constitute the solution to a complete design problem similar to that on the exam.

Your solutions to the NCEES 16-hour Structural exam problems must be based on the NCEES-adopted codes and design standards. Therefore, you should carefully review the appropriate sections of the exam-adopted design standards and codes that are presented, analyzed, and explained in each chapter of this book. The examples in this book also focus on one specific code principle and offer a clear interpretation of that principle.

Table 1 lists the NCEES structural design standards that code-based problems on the exam will reference. You will not receive credit for solutions based on other editions or standards. All problems are in U.S. customary (English) units, and you will not receive credit for solutions using SI units.

Abbreviations are used throughout this book to refer to the codes and design standards referenced by the NCEES 16-hour Structural exam. This book's "Codes and References Used to Prepare This Book" section lists these abbreviations in brackets after their appropriate code or design standard. This book also cites other publications that discuss pertinent structural design procedures, which may also be found in the "Codes" section. Text references to any other publications are numbered and the publications are cited in the "References" section that precedes the practice problems at the end of each chapter. These references are provided for your additional review.

As you prepare for the NCEES 16-hour Structural exam, the following suggestions may also help.

- Become intimately familiar with this book. This means knowing the order of the chapters, the approximate locations of important figures and tables, and so on.

- Use the subject title tabs along the side of each page.

- Skim through a chapter to familiarize yourself with the subjects before starting the practice problems.

- To minimize time spent in searching for often-used formulas and data, prepare a one-page summary of all the important formulas and information in each subject area. You can then refer to this summary during the exam instead of searching in this book.

- Use the subject index extensively. Every significant term, law, theorem, and concept has been indexed. If you don't recognize a term used, look for it in the index. Some subjects appear in more than one chapter. Use the index liberally to learn all there is to know about a particular subject.

Table 1 *NCEES 16-Hour Structural Exam Design Standards*

abbreviation	design standard title
AASHTO	*AASHTO LRFD Bridge Design Specifications*, 4th ed., 2007, with 2008 Interim Revisions, American Association of State Highway and Transportation Officials, Washington, DC.
IBC	*International Building Code*, 2006 ed. (without supplements), International Code Council, Falls Church, VA.
ASCE 7	*Minimum Design Loads for Buildings and Other Structures*, 2005 ed., American Society of Civil Engineers, Reston VA.
ACI 318	*Building Code Requirements for Structural Concrete*, 2005 ed., American Concrete Institute, Farmington Hills, MI.
ACI 530/530.1[a]	*Building Code Requirements and Specifications for Masonry Structures* (and related commentaries), 2005 ed., American Concrete Institute, Detroit, MI; Structural Engineering Institute of the American Society of Civil Engineers, Reston VA; and The Masonry Society, Boulder, CO.
AISC[b]	*Steel Construction Manual*, 13th ed., American Institute of Steel Construction, Inc., Chicago, IL.
AISC	*Seismic Design Manual*, 2nd printing, 2006 ed., American Institute of Steel Construction, Inc., Chicago, IL.
AISI	*North American Specification for the Design of Cold-Formed Steel Structural Members*, 2001 ed., with 2004 supplement, American Iron and Steel Institute, Washington, DC.
NDS	*National Design Specification for Wood Construction ASD/LRFD*, 2005 ed., and *National Design Specification Supplement, Design Values for Wood Construction*, 2005 ed., American Forest and Paper Association (formerly National Forest Products Association), Washington, DC.
PCI	*PCI Design Handbook: Precast and Prestressed Concrete*, 6th ed., 2004, Precast/Prestressed Concrete Institute, Chicago, IL.

[a]Masonry problems must be solved using only the allowable stress design (ASD) method, except for those on walls with out-of-plane loads, which may be solved using the load and resistance factor design (LRFD) method in Sec. 3.3.5.
[b]Steel problems may be solved using either ASD or LRFD methods. (The LRFD method is used in this book.)

Part 2: Everything You Ever Wanted to Know About the 16-Hour Exam

HISTORY OF THE NCEES 16-HOUR STRUCTURAL EXAM

Through October 2010, NCEES offered three different structural engineering exams: the four-hour structural depth section of the Civil PE exam, the eight-hour Structural I exam, and the eight-hour Structural II exam. (In addition, California, Oregon, and Washington administered an eight-hour, state-specific structural exam.) In 2004, NCEES also adopted the Model Law Structural Engineer (MLSE) designation that outlines minimum competency standards for structural engineering licensure. To qualify for the MLSE, you need to meet specific education and experience requirements, as well as pass 16 hours of structural engineering exams. In the past, to satisfy the exam requirement you could choose various combinations of the NCEES and state-specific structural exams. However, the many possible combinations led to questions and concerns about what

exactly was needed to become a licensed structural engineer.

To address these questions and concerns, NCEES appointed a Structural Exam Task Force (SETF) in 2006. Composed of engineers with PE and SE licenses from various states, the SETF evaluated the NCEES Structural I and II exams, as well as any possible additional needs for states with high seismic activity. The NCEES 16-hour Structural exam is the result of their findings.

ABOUT THE EXAM

The 16-hour Structural exam is offered in two components. The first component—vertical forces (gravity/other) and incidental lateral—takes place on a Friday. The second component—lateral forces (wind/earthquake)—takes place on a Saturday. Each component comprises a morning breadth and an afternoon depth module, as outlined in Table 2.

Table 2 NCEES 16-Hour Structural Exam Component Module Specifications

Friday: vertical forces (gravity/other) and incidental lateral	

morning breadth module
4 hours
40 multiple-choice problems
	analysis of structures (30%)
	    loads (10%)
	    methods (20%)
	design and details of structures (65%)
	    general structural considerations (7.5%)
	    structural systems integration (2.5%)
	    structural steel (12.5%)
	    light gage/cold-formed steel (2.5%)
	    concrete (12.5%)
	    wood (10%)
	    masonry (7.5%)
	    foundations and retaining structures (10%)
	construction administration (5%)
	    procedures for mitigating noncomforming work (2.5%)
	    inspection methods (2.5%)

afternoon depth module[a]
4 hours
essay problems
	buildings[b]
	    steel structure (1-hour problem)
	    concrete structure (1-hour problem)
	    wood structure (1-hour problem)
	    masonry structure (1-hour problem)
	bridges
	    concrete superstructure (1-hour problem)
	    steel superstructure (2-hour problem)
	    other elements of bridges (e.g., culverts, abutments, retaining walls)
	        (1-hour problem)

Saturday: lateral forces (wind/earthquake)	

morning breadth module
4 hours
40 multiple-choice problems
	analysis of structures (37%)
	    lateral forces (10%)
	    lateral force distribution (22%)
	    methods (5%)
	design and detailing of structures (60%)
	    general structural considerations (7.5%)
	    structural systems integration (5%)
	    structural steel (10%)
	    light gage/cold-formed steel (2.5%)
	    concrete (12.5%)
	    wood (7.5%)
	    masonry (7.5%)
	    foundations and retaining structures (7.5%)
	construction administration (3%)
	    structural observation (3%)

afternoon depth module[a]
4 hours
essay problems
	buildings[c]
	    steel structure (1-hour problem)
	    concrete structure (1-hour problem)
	    wood and/or masonry structure (1-hour problem)
	    general analysis (e.g., existing structures, secondary structures,
	        nonbuilding structures, and/or computer verification) (1-hour problem)
	bridges
	    columns (1-hour problem)
	    footings (1-hour problem)
	    general analysis (e.g., seismic and/or wind) (2-hour problem)

[a]Afternoon sessions focus on a single area of practice. You must choose *either* the buildings or bridges depth module and you must work the same depth module across both exam components.
[b]At least one problem will contain a multistory building and at least one problem will contain a foundation.
[c]At least two problems will include seismic content with a seismic design category of D or above. At least one problem will include wind content with a base wind speed of at least 110 mph. Problems may include a multistory building and/or a foundation.

The morning breadth modules are each four hours and contain 40 multiple-choice problems that cover a range of structural engineering topics specific to vertical and lateral forces. The afternoon depth modules are also each four hours, but instead of multiple-choice problems, they contain essay problems. You may choose either the bridges or the buildings depth module, but you must work the same depth module across both exam components. That is, if you choose to work buildings for the lateral forces component, you must also work buildings for the vertical forces component.

According to NCEES, the vertical forces (gravity/other) and incidental lateral depth module in buildings covers loads, lateral earth pressures, analysis methods, general structural considerations (e.g., element design), structural systems integration (e.g., connections), and foundations and retaining structures. The depth module in bridges covers gravity loads, superstructures, substructures, and lateral loads other than wind and seismic. It may also require pedestrian bridge and/or vehicular bridge knowledge.

The lateral forces (wind/earthquake) depth module in buildings covers lateral forces, lateral force distribution, analysis methods, general structural considerations (e.g., element design), structural systems integration (e.g., connections), and foundations and retaining structures. The depth module in bridges covers gravity loads, superstructures, substructures, and lateral forces. It may also require pedestrian bridge and/or vehicular bridge knowledge.

WHAT DOES "MOST NEARLY" REALLY MEAN?

One of the more disquieting aspects of multiple-choice problems is that the available answer choices are seldom exact. Answer choices generally have only two or three significant digits. Exam questions ask, "Which answer choice is most nearly the correct value?" or they instruct you to complete the sentence, "The value is approximately..." A lot of self-confidence is required to move on to the next question when you don't find an exact match for the answer you calculated, or if you have had to split the difference because no available answer choice is close.

The NCEES has described it like this:

Many of the questions on NCEES exam require calculations to arrive at a numerical answer. Depending on the method of calculation used, it is very possible that examinees working correctly will arrive at a range of answers. The phrase "most nearly" is used to accommodate answers that have been derived correctly but that may be slightly different from the correct answer choice given on the exam. You should use good engineering judgment when selecting your choice of answer. For example, if the question asks you to calculate an electrical current or determine the load on a beam, you

should literally select the answer option that is most nearly what you calculated, regardless of whether it is more or less than your calculated value. However, if the question asks you to select a fuse or circuit breaker to protect against a calculated current or to size a beam to carry a load, you should select an answer option that will safely carry the current or load. Typically, this requires selecting a value that is closest to but larger than the current or load.

WHY DOES NCEES REUSE SOME PROBLEMS?

NCEES reuses some of the more reliable problems from each exam. The percentage of repeat problems isn't high—no more than 25% of the exam. NCEES repeats problems in order to equate the performance of one group of examinees with the performance of an earlier group. The repeated problems are known as *equaters*.

HOW MUCH MATHEMATICS IS NEEDED FOR THE EXAM?

Generally, only simple algebra, trigonometry, and geometry are needed on the exam. You will need to use the trigonometric, logarithm, square root, and similar buttons on your calculator. There is no need to use any other method for these functions.

There are no pure mathematics problems (algebra, geometry, trigonometry, etc.) on the exam. However, you will need to apply your knowledge of these subjects to the exam problems.

Except for simple quadratic equations, you will probably not need to find the roots of higher-order equations. Occasionally, it will be convenient to use the equation-solving capability of an advanced calculator. However, other solution methods will always exist.

There is little or no use of calculus on the exam. Rarely, you may need to take a simple derivative to find a maximum or minimum of some function. Even more rare is the need to integrate to find an average or moment of inertia.

Basic statistical analysis of observed data may be necessary. Statistical calculations are generally limited to finding means, medians, standard deviations, variances, percentiles, and confidence limits. The only population distribution you need to be familiar with is the normal curve. Probability, reliability, hypothesis testing, and statistical quality control are not explicit exam subjects.

The exam is concerned with numerical answers, not with proofs or derivations. You will not be asked to prove or derive formulas.

Occasionally, a calculation may require an iterative solution method. Generally, there is no need to complete more than two iterations. You will not need to program your calculator to obtain an "exact" answer. Nor will you generally need to use complex numerical methods.

WHAT ABOUT CALCULATORS?

The exam requires use of a scientific calculator. However, it may not be obvious that you should bring a spare calculator with you to the exam. It is always unfortunate when an examinee is not able to finish because his or her calculator was dropped or stolen or stopped working for some unknown reason.

The exam has not been "optimized" for any particular brand or type of calculator. In fact, for most calculations, a $15 scientific calculator will produce results as satisfactory as those from a $200 calculator. There are definite advantages to having built-in statistical functions, graphing, unit-conversion, and equation-solving capabilities. However, these advantages are not so great as to give anyone an unfair advantage.

Calculators with text-editing or communications capabilities are banned. This excludes all but a limited number of models. Details are provided on the PPI website at **www.ppi2pass.com/calculators**. Check with your state board to see if nomographs and specialty slide rules are permitted.

The calculator you use for the exam should have the following essential functions.

- trigonometric and inverse trigonometric functions

- hyperbolic and inverse hyperbolic functions

- π

- $\sqrt{}$ and x^2

- both common and natural logarithms

- y^x and e^x

For maximum speed, your calculator should also have or be programmed for the following functions.

- interpolation

- finding standard deviations and variances

- extracting roots of quadratic and higher-order equations

- calculating determinants of matrices

- linear regression

You may not share calculators with other examinees.

Laptop computers are generally not permitted in the exam. Their use has been considered, and some states may actually permit them. However, considering the nature of the exam problems, it is very unlikely that laptops would provide any advantage.

You may not use a walkie-talkie, cellular telephone, interactive pager, or other communications device during the exam.

Be sure to take your calculator with you whenever you leave the exam room for any length of time.

WHAT REFERENCE MATERIAL IS PERMITTED IN THE EXAM?

The exam is open-book format. Most states do not have any limits on the numbers and types of books you can use. For the exam, the references you bring into the exam room in the morning do not have to be the same as the references you use in the afternoon. You cannot share books with other examinees during the exam.

Loose paper and scratch pads are not permitted in the exam room. Personal notes in a three-ring binder and other semipermanent covers can usually be used. Some states use a "shake test" to eliminate loose papers from binders. Make sure that nothing escapes from your binders when they are inverted and shaken.

HOW MANY BOOKS SHOULD YOU BRING?

You actually won't use many books in the exam. The trouble is, you can't know in advance which ones you will need. That's the reason why many examinees show up with boxes and boxes of books. Without a doubt, there are things that you will need that are not in this book. But there are not so many that you need to bring your entire company's library. The exam is very fast-paced. You will not have time to use books with which you are not thoroughly familiar. The exam doesn't require you to know obscure solution methods or to use difficult-to-find data. A general rule is that you shouldn't bring books that you have not looked at during your review. If you didn't need a book while doing the practice problems in this book, you won't need it during the exam.

So, it really is unnecessary to bring a large quantity of books with you. Essential books are identified in the References section, and you should be able to decide which support you need for the areas in which you intend to work.

MAY I PLACE TABS ON CERTAIN PAGES?

It is common to tab pages in your books in an effort to reduce the time required to locate useful sections. Inasmuch as some states consider Post-it® Notes to be "loose paper," your tabs should be of the "permanent" variety. Although you can purchase tabs with gummed attachment points, it is also possible simply to use transparent tape to attach the Post-its you have already placed in your books.

HOW ARE THE EXAM COMPONENTS GRADED AND SCORED?

For the morning multiple-choice problems, answers are recorded on an answer sheet that is machine graded. The minimum number of points for passing (referred to by NCEES as the "cut score") varies from administration to administration. The cut score is determined through a rational procedure, without the benefit of knowing examinees' performance on the exam. That is, the exam is not graded on a curve. The cut score is selected based on what you are expected to know, not based on allowing a certain percentage of engineers "through."

Grading of multiple-choice problems is straightforward, since a computer grades your score sheet. Either you get the problem right or you don't. There is no deduction for incorrect answers, so guessing is encouraged. However, if you mark two or more answers, no credit is given for the problem.

Solutions for the afternoon essay problems are evaluated for overall compliance with established scoring criteria and for general quality. The scores from each of the morning and afternoon modules are combined for a component's final score.

Exam results are given a pass/fail grade approximately 8–10 weeks after the exam date. You will receive the results of your exam from your state board (not NCEES) by mail. Allow at least four months for notification. You will receive a pass or fail notice only and will not receive a numerical score. A diagnostic report is provided to those who fail.

HOW YOU SHOULD GUESS

NCEES produces defensible licensing exam. As a result, there is no pattern to the placement of correct responses in multiple-choice problems. Therefore, it is not important whether you randomly guess all "A," "B," "C," or "D."

The proper way to guess is as an engineer. You should use your knowledge of the subject to eliminate illogical answer choices. Illogical answer choices are those that violate good engineering principles, that are outside normal operating ranges, or that require extraordinary assumptions. Of course, this requires you to have some basic understanding of the subject in the first place. Otherwise, it's back to random guessing. That's the reason that the minimum passing score is higher than 25%.

You won't get any points using the "test-taking skills" that helped you in college—the skills that helped with tests prepared by amateurs. You won't be able to eliminate any [verb] answer choices from "Which [noun]..." questions. You won't find questions with four answer choices, two of which are of the "more than 50" and "less than 50" variety. You won't find one answer choice among the four that has a different number of significant digits, or whose verb is written in a different tense, or that has some singular/plural discrepancy with the stem. The distractors will always match the stem, and they will be logical.

CHEATING AND EXAM SUBVERSION

There aren't very many ways to cheat on an open-book test. The proctors are well trained on the few ways that do exist. It goes without saying that you should not talk to other examinees in the room, nor should you pass notes back and forth. The number of people who are released to use the restroom may be limited to prevent discussions.

NCEES regularly reuses good problems that have appeared on a previous exam. Therefore, exam security is a serious issue with NCEES, which goes to great lengths to make sure nobody copies the problems. You may not keep your exam booklet, enter text of problems into your calculator, or copy problems into your own material.

The proctors are concerned about exam subversion, which generally means activity that might invalidate the exam or the exam process. The most common form of exam subversion involves trying to copy exam problems for future use.

Part 3: How to Prepare for and Pass the NCEES 16-Hour Exam

HOW LONG SHOULD YOU STUDY?

A thorough review takes approximately 300 hours. Most of this time is spent solving problems. Some examinees spread this time over a year. Others cram it all into two months. The best time to start studying will depend on how much time you can spend per week.

A SIMPLE PLANNING SUGGESTION

Designate some location (a drawer, a corner, a cardboard box, or even a paper shopping bag left on the floor) as your "exam catch-all." Use your catch-all during the months before the exam when you have revelations about things you should bring with you. For example, you might realize that the plastic ruler marked off in tenths of an inch that is normally kept in the kitchen junk drawer can help you with some soil pressure questions. Or, you might decide that a certain book is particularly valuable. Or that it would be nice to have dental floss after lunch. Or that large rubber bands are useful for holding books open.

It isn't actually necessary to put these treasured items in the catch-all during your preparation. You can, of course, if it's convenient. But if these items will have other functions during the time before the exam, at least write yourself a note and put the note into the catch-all. When you go to pack your exam kit a few days before the exam, you can transfer some items immediately, and the notes will be your reminder for the other items that are back in the kitchen drawer.

HOW YOU CAN MAKE YOUR REVIEW REALISTIC

In the exam, you must be able to quickly recall solution procedures, formulas, and important data. When you played a sport back in school, your coach tried to put you in game-related situations. Preparing for the exam isn't much different from preparing for a big game. Some part of your preparation should be realistic and representative of the exam environment.

There are several things you can do to make your review more representative. For example, if you gather most of your review resources (i.e., books) in advance and try to use them exclusively during your review, you will become more familiar with them. (Of course, you can also add to or change your references if you find inadequacies.)

Learning to use your time wisely is one of the most important lessons you can learn during your review. You will undoubtedly encounter problems that end up taking much longer than you expected. In some instances, you will cause your own delays by spending too much time looking through books for things you need, or just by looking for the books themselves. Other times, the problems will entail too much work. Learn to recognize these situations so that you can make an intelligent decision about skipping such problems in the exam.

WHAT TO DO A FEW DAYS BEFORE THE EXAM

There are a few things you should do a week or so before the exam. You should arrange for childcare and transportation. Since the exam does not always start or end at the designated time, make sure that your arrangements are flexible.

Check PPI's website for last-minute updates and errata to any PPI books you might have and are bringing to the exam.

If you haven't already done so, read the "Advice from Examinees" section of PPI's website.

If you haven't been following along on the Engineering Exam Forum on PPI's website, use the search function to locate discussions on this bulletin board.

If it's convenient, visit the exam location in order to find the building, parking areas, exam room, and restrooms. If it's not convenient, you may find driving directions and/or site maps on the web.

Take the battery cover off your calculator and check to make sure you are bringing the correct size replacement batteries. Some calculators require a different kind of battery for their "permanent" memories. Put the cover back on and secure it with a piece of masking tape. Write your name on the tape to identify your calculator.

If your spare calculator is not the same as your primary calculator, spend a few minutes familiarizing yourself with how it works. At the very least, you should verify that your spare calculator is functional.

PREPARE YOUR CAR

[ ] Gather snow chains, shovel, and tarp to lie on while installing chains.

[ ] Check your tire pressures.

[ ] Check your spare tire.

[ ] Check for tire installation tools.

[ ] Verify that you have the vehicle manual.

[ ] Check fluid levels (oil, gas, water, brake fluid, transmission fluid, window-washing solution).

[ ] Fill up with gas.

[ ] Check battery and charge if necessary.

[ ] Know something about your fuse system (where they are, how to replace them, etc.).

[ ] Assemble all required maps.

[ ] Fix anything that might slow you down (missing wiper blades, etc.).

[ ] Check your taillights.

[ ] Affix the recently arrived DMV license sticker.

[ ] Fix anything that might get you pulled over on the way to the exam (burned-out taillight or headlight, broken lenses, bald tires, missing license plate, noisy muffler).

[ ] Treat the inside windows with anti-fog solution.

[ ] Put a roll of paper towels in the back seat.

[ ] Gather exact change for any bridge tolls or toll roads.

[ ] Put $20 in your glove box.

[ ] Check for current registration and proof of insurance.

[ ] Locate a spare key.

[ ] Find your AAA or other roadside-assistance cards and phone numbers.

[ ] Plan out alternate routes.

PREPARE YOUR EXAM KITS

Second in importance to your scholastic preparation is the preparation of your two exam kits. The first kit consists of a bag, box (plastic milk crates hold up better than cardboard in the rain), or wheeled travel suitcase containing items to be brought with you into the exam room. NCEES provides pencils, leads, and erasers, and does not allow you to bring your own.

[ ] letter admitting you to the exam

[ ] photographic identification (e.g., driver's license)

[ ] this book

[ ] other textbooks and reference books

[ ] regular dictionary

[ ] scientific/engineering dictionary

[ ] cardboard boxes or plastic milk crates to use as a bookcase

[ ] primary calculator (non-QWERTY)

[ ] spare calculator

[ ] instruction booklets for your calculators

[ ] extra calculator batteries

[ ] triangles

[ ] scales

[ ] straightedge and rulers

[ ] compass

[ ] protractor

[ ] scissors

[ ] stapler

[ ] transparent tape

[ ] magnifying glass

[ ] small (jeweler's) screwdriver for fixing your glasses or for removing batteries from your calculator

[ ] unobtrusive (quiet) snacks or candies, already unwrapped

[ ] two small plastic bottles of water

[ ] travel pack of tissue (keep in your pocket)

[ ] headache remedy

[ ] personal medication

[ ] $3.00 in miscellaneous change

[ ] light, comfortable sweater

[ ] loose shoes or slippers

[ ] cushion for your chair

[ ] earplugs

[ ] wristwatch with alarm

[ ] several large trash bags ("rain coats" for your boxes of books)

[ ] wire coat hanger (to hang up your jacket or to get back into your car in an emergency)

[ ] extra set of car keys on a string around your neck

The second kit consists of the following items and should be left in a separate bag or box in your car in case they are needed.

[ ] copy of your application

[ ] proof of delivery

[ ] light lunch

[ ] beverage in thermos or cans

[ ] sunglasses

[ ] extra pair of prescription glasses

[ ] raincoat, boots, gloves, hat, and umbrella

[ ] street map of the exam area

[ ] parking permit

[ ] battery-powered desk lamp

[ ] your cellular phone

[ ] piece of rope

The following items cannot be used during the exam and should be left at home.

[ ] fountain pens

[ ] radio or tape/CD player

[ ] battery charger

[ ] extension cords

[ ] scratch paper

[ ] note pads

PREPARE FOR THE WORST

All of the occurrences listed in this section happen to examinees on a regular basis. Granted, you cannot prepare for every eventuality. But, even though each of these occurrences taken individually is a low-probability event, taken together, they are worth considering in advance.

- Imagine getting a flat tire, getting stuck in traffic, or running out of gas on the way to the exam.

- Imagine rain and snow as you are carrying your cardboard boxes of books into the exam room. Would plastic trash bags be helpful?

- Imagine arriving late. Can you get into the exam without having to make two trips from your car?

- Imagine having to park two blocks from the exam site. How are you going to get everything to the exam room? Can you actually carry everything that far? Could you use a furniture dolly, a supermarket basket, or perhaps a helpmate?

- Imagine a Star Trek convention, square-dancing contest, construction, or auction in the next room.

- Imagine a site without any heat, with poor lighting, or with sunlight streaming directly into your eyes.

- Imagine a hard folding chair and a table with one short leg.

- Imagine a site next to an airport with frequent take-offs, or next to a construction site with a pile driver, or next to the NHRA's Drag Racing Championship.

- Imagine a seat where someone nearby chews gum with an open mouth; taps his pencil or drums her fingers; or wheezes, coughs, and sneezes for hours on end.

- Imagine the distraction of someone crying, or of proctors evicting yelling and screaming examinees (who have been found cheating), or of the tragedy of another examinee's serious medical emergency.

- Imagine a delay of an hour while they find someone to unlock the building, turn on the heat, or wait for the head proctor to bring instructions.

- Imagine a power outage occurring sometime during the exam.

- Imagine a proctor who (a) tells you that one of your favorite books can't be used in the exam, (b) accuses you of cheating, or (c) calls "time up" without giving you any warning.

- Imagine not being able to get your lunch out of your car or find a restaurant.

- Imagine getting sick or nervous in the exam.

- Imagine someone stealing your calculator during lunch.

WHAT TO DO THE DAY BEFORE THE EXAM

Take the day before the exam off from work to relax. Do not cram the last night. A good night's sleep is the best way to start the exam. If you live a considerable distance from the exam site, consider getting a hotel room in which to spend the night.

Practice setting up your exam work environment. Carry your boxes to the kitchen table. Arrange your "bookcases" and supplies. Decide what stays on the floor

in boxes and what gets an "honored position" on the tabletop.

Use your checklist to make sure you have everything. Make sure your exam kits are packed and ready to go. Wrap your boxes in plastic bags in case it's raining when you carry them from the car to the exam room.

Calculate your wake-up time and set the alarms on two bedroom clocks. Select and lay out your clothing items. (Dress in layers.) Select and lay out your breakfast items.

If it's going to be hot on exam day, put your (plastic) bottles of water in the freezer.

Make sure you have gas in your car and money in your wallet.

WHAT TO DO THE DAY OF THE EXAM

Turn off the quarterly and hourly alerts on your wristwatch. Change your pager or cell phone to silent mode.

Bring or buy a morning newspaper.

You should arrive at least 30 minutes before the exam starts. This will allow time for finding a convenient parking place, bringing your materials to the exam room, making room and seating changes, and calming down. Be prepared, though, to find that the exam room is not open or ready at the designated time.

Once you have arranged the materials around you on your table, take out your morning newspaper and look cool. (Only nervous people work crossword puzzles.)

WHAT TO DO DURING THE EXAM

All of the procedures typically associated with timed, proctored assessment tests will be in effect when you take the exam.

The proctors will distribute the exam booklet and answer sheet if they are not already on your tables. However, you should not open the booklet until instructed to do so. You may read the information on the front and back covers, and you should write your name in the appropriate blank spaces.

Listen carefully to everything the proctors say. Do not ask your proctors any engineering questions. Even if they are knowledgeable in engineering, they will not be permitted to answer your questions.

Answers to problems are recorded on an answer sheet contained in the test booklet. The proctors will guide you through the process of putting your name and other biographical information on this sheet when the time comes, which will take approximately 15 minutes. You will be given the full four hours to answer problems. Time to initialize the answer sheet is not part of your four hours.

The common suggestions to "completely fill the bubbles and erase completely" apply here when working multiple-choice problems. NCEES provides each examinee with a mechanical pencil with 0.7 mm lead. Use of ballpoint pens and felt-tip markers is prohibited for several reasons.

If you finish an exam early and there are still more than 30 minutes remaining, you will be permitted to leave the room. If you finish less than 30 minutes before the end of the exam, you may be required to remain until the end. This is done to be considerate of the people who are still working.

When you leave, you must return your exam booklet. You may not keep the exam booklet for later review.

If there are any problems that you think were flawed, in error, or unsolvable, ask a proctor for a "reporting form" on which you can submit your comments. Follow your proctor's advice in preparing this document.

WHAT ABOUT EATING AND DRINKING DURING THE EXAM?

The official rule is probably the same in every state: no eating or drinking in the exam. That makes sense, for a number of reasons. Some exam sites don't want (or don't permit) stains and messes. Others don't want crumbs to attract ants and rodents. Your table partners don't want spills or smells. Nobody wants the distractions. Your proctors aren't going to give you a new exam booklet when the first one is ruined with coffee.

How this rule is administered varies from site to site and from proctor to proctor. Some proctors enforce the letter of law, threatening to evict you from the exam room when they see you chewing gum. Others may permit you to have bottled water, as long as you store the bottles on the floor where any spills will not harm what's on the table. No one is going to let you crack peanuts while you work on the exam, but it's unlikely that anyone will complain about a hard candy melting away in your mouth. You'll just have to find out when you get there.

HOW TO SOLVE MULTIPLE-CHOICE PROBLEMS

When approaching multiple-choice problems, observe the following suggestions.

- Do not spend an inordinate amount of time on any single problem. If you have not answered a problem in a reasonable amount of time, make a note of it and move on.

- Set your wristwatch alarm for five minutes before the end of each four-hour session, and use that remaining time to guess at all of the remaining problems. Odds are that you will be successful with about 25% of your guesses, and these points will more than make up for the few points that you might earn by working during the last five minutes.

- Make mental notes about any problems for which you cannot find a correct response, that appears to have two correct responses, or that you believe have some technical flaw. Errors in the exam are rare, but they do occur. Such errors are usually discovered during the scoring process and discounted from the exam, so it is not necessary to tell your proctor, but be sure to mark the one best answer before moving on.

- Make sure all of your responses on the answer sheet are dark and completely fill the bubbles.

SOLVE PROBLEMS CAREFULLY

Many points are lost due to carelessness. Keep the following items in mind when you are solving the end-of-chapter problems. Hopefully, these suggestions will be automatic in the exam.

[ ] Did you recheck your mathematical equations?

[ ] Do the units cancel out in your calculations?

[ ] Did you recheck all data obtained from other sources, tables, and figures?

SHOULD YOU TALK TO OTHER EXAMINEES AFTER THE EXAM?

The jury is out on this question. People react quite differently to the exam experience. Some people are energized. Most are exhausted. Some people need to unwind by talking with other examinees, describing every detail of their experience, and dissecting every exam problem. Other people need lots of quiet space, and prefer to just get into a hot tub to soak and sulk. Most engineers, apparently, are in this latter category.

Since everyone who took an exam has seen it, you will not be violating your "oath of silence" if you talk about the details with other examinees. It's difficult not to ask how someone else approached a problem that had you completely stumped. However, keep in mind that it is very disquieting to think you answered a problem correctly, only to have someone tell you where you went wrong.

AFTER THE EXAM

Yes, there is something to do after the exam. Most people come home, throw their exam "kits" into the corner, and collapse. A week later, when they can bear to think about the experience again, they start integrating their exam kits back into their normal lives. The calculators go back into the desk, the books go back on the shelves, the $3.00 in change goes back into the piggy bank, and all of the miscellaneous stuff you brought with you to the exam is put back wherever it came from.

Here are some suggestions for what to do as soon as you get home, before you collapse.

[ ] Thank your spouse and children for helping you during your preparation.

[ ] Take any paperwork you received on exam day out of your pocket, purse, or wallet. Put this inside your *Structural Engineering Reference Manual*.

[ ] Reflect on any statements regarding exam secrecy to which you signed your agreement in the exam.

[ ] Visit the PPI website and complete the after-exam survey to help PPI improve the quality of its service and products.

[ ] Call your employer and tell him/her that you need to take a mental health day off on Monday.

A few days later, when you can face the world again, do the following.

[ ] Make notes about anything you would do differently if you had to take the exam over again.

[ ] Consolidate all of your application paperwork, correspondence to/from your state, and any paperwork that you received on exam day.

[ ] Visit the Engineering Exam Forum on PPI's website and see what other people are saying about the exam you took.

[ ] Return any books you borrowed.

[ ] Write thank-you notes to all of the people who wrote letters of recommendation or reference for you.

[ ] There were no ethics problems on your exam, but it doesn't make any difference. Ethical behavior is expected of a structural engineer in any case. Spend a few minutes reflecting on how your performance (obligations, attitude, presentation, behavior, appearance, etc.) might be about to change once you are licensed. Consider how you are going to be a role model for others around you.

[ ] Put all of your review books, binders, and notes someplace where they will be out of sight.

AND THEN THERE'S THE WAIT...

It goes without saying that the grading process for essay problems is complex and time consuming. For multiple-choice problems it seems as though grading should be almost instantaneous. Nevertheless, you are going to wait. There are many reasons for the delay.

Although the actual machine grading of multiple-choice problems "only takes seconds," consider the following facts: (a) Multiple exams are prepared for each administration, in case one becomes unusable (i.e., is inappropriately released) before the exam date. (b) Since the actual version of the exam used is not known until after it is finally given, the cut-score determination occurs after the exam date.

It wouldn't be unreasonable to surmise that dozens, if not hundreds, of claims are made by well-meaning examinees who are certain that the exam they took was flawed—that there wasn't a correct answer for such-and-such problem—that there were two answers for such-and-such problem—or even, perhaps, that such-and-such problem was missing from their exam booklet altogether. Each of these claims must be considered as a potential adjustment to the cut-score.

Then, the exam must actually be graded. Since grading a high volume of exam requires specialized equipment, software, and training not normally possessed by the average employee, as well as time to do the work (also not normally possessed by the average employee), grading of the NCEES exam is invariably outsourced.

Outsourced grading cannot begin until all of the states have returned their score sheets to NCEES and NCEES has sorted, separated, organized, and consolidated the score sheets into whatever "secret sauce sequence" is best.

During grading, some of the score sheets "pop out" with any number of a variety of abnormalities that demand manual scoring.

After the individual exam are scored, the results are analyzed in a variety of ways. Some of the analysis looks at passing rates by such delineators as degree, major, university, site, and state. Part of the analysis looks for similarities between physically adjacent examinees (to look for cheating). Part looks for exam sites that have statistically abnormal group performance. And, some of the analysis looks for exam problems that have a disproportionate fraction of successful or unsuccessful examinees. Anyway, you get the idea: It's not merely putting your exam sheet in an electronic reader. All of these steps have to be completed for 100% of the examinees before any results can go out.

Once NCEES has graded your test and notified your state, when you hear about it depends on when the work is done by your state. Some states have to approve the results at a board meeting; others prepare the certificates before sending out notifications. Some states are more computerized than others. Some states have 50 examinees, while others have 10,000. Some states are shut down by blizzards and hurricanes; others are administratively challenged—understaffed, inadequately trained, or over budget.

There is no pattern to the public release of results. None. The exam results are not released to all states simultaneously. (The states with the fewest examinees often receive their results soonest.) They are not released by discipline. They are not released alphabetically by state or examinee name. The people who failed are not uniformly notified first (or last). Your coworker might receive his or her notification today, and you might be waiting another three weeks for yours.

Some states post the names of the successful examinees on their official state websites before the results go out. Others update their websites after the results go out. Some states don't list much of anything on their websites.

Remember, too, that the size or thickness of the envelope you receive from your state does not mean anything. Some states send a big congratulations package and certificate. Others send a big package with a new application to repeat the exam. Some states send a postcard. Some send a one-page letter. Some states simply send you an invoice for your license fees. You just have to open it to find out.

Check the Engineering Exam Forum on the PPI website regularly to find out which states have released their results. You will find many other anxious examinees there, sharing information, humorous conspiracy theories, and rumors.

While you are waiting, we hope you will become a "Forum" regular. Log on often and help other examinees by sharing your knowledge and experiences.

AND WHEN YOU PASS...

[ ] Celebrate. (Don't drink and drive, though.)

[ ] Notify the people who wrote letters of recommendation or reference for you.

[ ] Read "FAQs about What Happens After You Pass the Exam" on PPI's website.

[ ] Ask your employer for a raise.

[ ] Tell the folks at PPI the good news. Nothing makes them happier than hearing your success story.

Codes and References Used to Prepare This Book

1. CODES

American Association of State Highway and Transportation Officials. *LRFD Bridge Design Specifications*, 4th ed. 2007, with 2008 Interim Revisions. [AASHTO]

American Concrete Institute. *Building Code Requirements and Commentary for Structural Concrete*. 2005. [ACI]

American Concrete Institute. *Building Code Requirements for Masonry Structures*. 2005. [BCRMS]

American Forest and Paper Association. *National Design Specification for Wood Construction ASD/LRFD, with Commentary and Supplement*. 2005. [NDS]

American Institute of Steel Construction, Inc. *Steel Construction Manual*, 13th ed. 2005. [LRFD]

American Society of Civil Engineers. *Minimum Design Loads for Buildings and Other Structures*. 2005. [ASCE]

Building Seismic Safety Council. *NEHRP Recommended Provisions for the Development of Seismic Regulations for New Buildings: Part 1, Provisions*. 2003. [NEHRP]

International Code Council. *International Building Code*. 2006. [IBC]

Precast/Prestressed Concrete Institute. *PCI Design Handbook: Precast and Prestressed Concrete*. 6th ed. 2004. [PCI]

2. REFERENCES

Buckner, C.D. *246 Solved Structural Engineering Problems*, 3rd ed. Professional Publications. 2003.

Corps of Engineers. *Seismic Design for Buildings*. U.S. Government Printing Office. 1998. [NAVFAC]

Simpson Strong-Tie Company. *Wood Construction Connectors. Catalog C-2009*. 2009.

Structural Engineers Association of California. *Recommended Lateral Force Requirements and Commentary*. 1999. [SEAOC]

Williams, A. *Structural Steel Design, Volume 1, ASD*. International Code Council. 2010.

1 Reinforced Concrete Design

1. Strength Design Principles 1-1
2. Strength Design of Reinforced
   Concrete Beams 1-3
3. Serviceability Requirements for Beams . . . 1-10
4. Elastic Design Method 1-14
5. Shear and Torsion 1-15
6. Concrete Columns 1-27
7. Development and Splice Length
   of Reinforcement 1-34
8. Two-Way Slab Systems 1-42
   Practice Problems 1-47
   Solutions 1-48

1. STRENGTH DESIGN PRINCIPLES

Nomenclature

D	dead load	kips or lbf
E	earthquake load	kips or lbf
F	load due to weight and pressure of fluids	kips or lbf
H	load due to pressure of soil	kips or lbf
L	live load	kips or lbf
L_r	roof live load	kips or lbf
Q	service level force	kips or lbf
R	load due to rainwater	kips or lbf
S	snow load	kips or lbf
U	required strength to resist factored load	kips or lbf
W	wind load	kips or lbf
w	distributed load	kips/ft

Symbols

γ	load factor	
ϕ	strength-reduction factor	

Required Strength

The required ultimate strength of a member consists of the most critical combination of factored loads applied to the member. Factored loads consist of working, or service, loads Q multiplied by the appropriate load factors γ. The required strength U is defined by ACI[1] Sec. 9.2 as

$$U = 1.4(D + F) \qquad \text{[ACI 9-1]}$$

$$U = 1.2(D + F + T) + 1.6(L + H)$$
$$\qquad + 0.5(L_r \text{ or } S \text{ or } R) \qquad \text{[ACI 9-2]}$$

$$U = 1.2D + 1.6(L_r \text{ or } S \text{ or } R)$$
$$\qquad + (0.5L^* \text{ or } 0.8W) \qquad \text{[ACI 9-3]}$$

$$U = 1.2D + 1.6W^\dagger + 0.5L^*$$
$$\qquad + 0.5(L_r \text{ or } S \text{ or } R) \qquad \text{[ACI 9-4]}$$

$$U = 1.2D + 1.0E^\ddagger + 0.5L^* + 0.2S \qquad \text{[ACI 9-5]}$$

$$U = 0.9D + 1.6W^\dagger + 1.6H \qquad \text{[ACI 9-6]}$$

$$U = 0.9D + 1.0E^\ddagger + 1.6H \qquad \text{[ACI 9-7]}$$

*Replace $0.5L$ with $1.0L$ for garages, places of public assembly, and areas where $L > 100$ lbf/ft.

†Replace $1.6W$ with $1.3W$ where the wind load has not been reduced by a directionality factor.

‡Replace $1.0E$ with $1.4E$ where earthquake load is based on service level seismic forces.

Example 1.1

The illustration on the following page shows a typical frame of a six-story office building. The loading on the frame is

roof dead load, including cladding and columns, w_{Dr}	= 1.2 kips/ft
roof live load, w_{Lr}	= 0.4 kip/ft
floor dead load, including cladding and columns, w_D	= 1.6 kips/ft
floor live load, w_L	= 1.25 kips/ft
horizontal wind pressure, p_h	= 1.0 kip/ft
vertical wind pressure, p_v	= 0.5 kip/ft

Determine the maximum and minimum design loads on the first-floor columns.

Solution

The axial load on one column due to the dead load is

$$D = \frac{l(w_{Dr} + 5w_D)}{2}$$
$$= \frac{(20 \text{ ft})\left(1.2\,\dfrac{\text{kips}}{\text{ft}} + (5 \text{ stories})\left(1.6\,\dfrac{\text{kips}}{\text{ft}}\right)\right)}{2}$$
$$= 92 \text{ kips}$$

The axial load on one column due to the floor live load is

$$L = \frac{l(5w_L)}{2}$$
$$= \frac{(20 \text{ ft})\left((5 \text{ stories})\left(1.25\,\dfrac{\text{kips}}{\text{ft}}\right)\right)}{2}$$
$$= 62.5 \text{ kips}$$

Illustration for Ex. 1.1

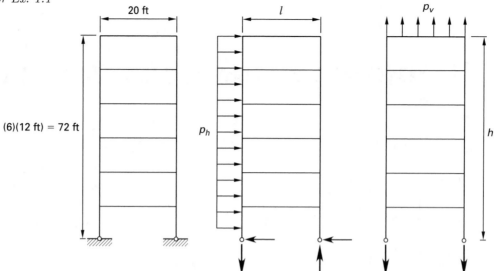

The axial load on one column due to the roof live load is

$$L_r = \frac{lw_{Lr}}{2}$$

$$= \frac{(20 \text{ ft}) \left(0.4 \dfrac{\text{kip}}{\text{ft}}\right)}{2}$$

$$= 4 \text{ kips}$$

The axial load on one column due to horizontal wind pressure is obtained by taking moments about the base of the other column and is given by

$$W_h = \pm\frac{p_h h^2}{2l} = \pm\frac{\left(1 \dfrac{\text{kip}}{\text{ft}}\right)(72 \text{ ft})^2}{(2)(20 \text{ ft})}$$

$$= \pm129.6 \text{ kips}$$

The axial load on one column due to the vertical wind pressure is obtained by resolving forces at the column bases and is given by

$$W_v = -\frac{p_v l}{2} = -\frac{\left(0.5 \dfrac{\text{kip}}{\text{ft}}\right)(20 \text{ ft})}{2}$$

$$= -5 \text{ kips}$$

The maximum strength level design load on a column is

$$\sum \gamma Q = 1.2D + 1.6L + 0.5L_r$$

$$= (1.2)(92 \text{ kips}) + (1.6)(62.5 \text{ kips})$$

$$+ (0.5)(4 \text{ kips})$$

$$= 212 \text{ kips} \quad [\text{compression}]$$

$$\sum \gamma Q = 1.2D + 1.6L_r + 0.5L$$

$$= (1.2)(92 \text{ kips}) + (1.6)(4 \text{ kips})$$

$$+ (0.5)(62.5 \text{ kips})$$

$$= 148 \text{ kips} \quad [\text{compression}]$$

$$\sum \gamma Q = 1.2D + 1.6L_r + 0.8W$$

$$= (1.2)(92 \text{ kips}) + (1.6)(4 \text{ kips})$$

$$+ (0.8)(129.6 \text{ kips} - 5 \text{ kips})$$

$$= 216 \text{ kips} \quad [\text{compression}]$$

$$\sum \gamma Q = 1.2D + 1.6W + 0.5L + 0.5L_r$$

$$= (1.2)(92 \text{ kips})$$

$$+ (1.6)(129.6 \text{ kips} - 5 \text{ kips})$$

$$+ (0.5)(4 \text{ kips}) + (0.5)(62.5 \text{ kips})$$

$$= 343 \text{ kips} \quad [\text{compression; governs}]$$

The minimum strength level design load on a leg is

$$\sum \gamma Q = 0.9D + 1.6W_h + 1.6W_v$$

$$= (0.9)(92 \text{ kips}) + (1.6)(-129.6 \text{ kips})$$

$$+ (1.6)(-5 \text{ kips})$$

$$= -133 \text{ kips} \quad [\text{tension}]$$

Design Strength

The design strength of a member consists of the nominal, or theoretical ultimate, strength of the member multiplied by the appropriate strength reduction factor ϕ. The reduction factor is defined in ACI Sec. 9.3 as

$\phi = 0.90$ [for flexure of tension-controlled sections]

$\phi = 0.75$ [for sheer and torsion]

$\phi = 0.70$ [for columns with spiral reinforcement]

$\phi = 0.65$ [for columns with lateral ties]

$\phi = 0.65$ [for bearing on concrete surfaces]

2. STRENGTH DESIGN OF REINFORCED CONCRETE BEAMS

Nomenclature

a	depth of equivalent rectangular stress block	in
A_{max}	maximum area of tension reinforcement, $\rho_t b_w d$	in²
A_s	area of tension reinforcement	in²
A_s'	area of compression reinforcement	in²
A_{sf}	reinforcement area to develop the outstanding flanges	in²
A_{sw}	reinforcement area to balance the residual moment	in²
b	width of compression face of member	in
b_w	web width	in
c	distance from extreme compression fiber to neutral axis	in
C_u	compressive force in the concrete	kips
d	distance from extreme compression fiber to centroid of tension reinforcement	in
d'	distance from extreme compression fiber to centroid of compression reinforcement	in
f_c'	compressive strength of concrete	lbf/in²
f_s'	stress in compression reinforcement	lbf/in²
f_y	yield strength of reinforcement	lbf/in²
h_f	flange depth	in
K_u	design moment factor, M_u/bd^2	lbf/in²
M_f	design moment strength of the outstanding flanges	in-lbf or ft-kips
M_{max}	maximum design flexural strength for a tension-controlled section	in-lbf or ft-kips
M_n	nominal flexural strength of a member	in-lbf or ft-kips
M_r	residual moment ($M_u - M_{max}$)	in-lbf or ft-kips
M_u	factored moment on the member	in-lbf or ft-kips
T_u	tensile force in the reinforcement	kips

Symbols

β_1	compression zone factor	
ε_c	strain at external compression fiber	
$\varepsilon_{c(max)}$	maximum strain at external compression fiber, 0.003	
ε_t	strain in tension reinforcement	
ε_s'	strain in compression reinforcement	
ρ	ratio of tension reinforcement, A_s/bd	
ρ_b	reinforcement ratio producing balanced strain conditions	
ρ_{max}	maximum tension reinforcement ratio in a rectangular beam with tension reinforcement only	
ρ_{min}	minimum allowable reinforcement ratio	
ρ_t	reinforcement ratio producing a tension-controlled section	
ω	tension reinforcement index, $\rho f_y/f_c'$	
ω_{max}	maximum tension reinforcement index in a rectangular beam with tension reinforcement only	

Beams with Tension Reinforcement Only

In accordance with ACI Sec. 10.2, a rectangular stress block is assumed in the concrete, as shown in Fig. 1.1, and it is also assumed that the tension reinforcement has yielded.[1,2] The nominal flexural strength of a rectangular beam is derived[2] as

$$M_n = A_s f_y d \left(1 - \frac{0.59\rho f_y}{f_c'}\right)$$

Equating the tensile and compressive forces acting on the section gives the depth of the equivalent rectangular stress block as

$$a = \frac{A_s f_y}{0.85 f_c' b}$$
$$M_n = A_s f_y \left(d - \frac{a}{2}\right)$$

The maximum permissible factored moment on the member, or minimum required moment strength, shall not exceed ϕM_n. For a tension-controlled section, $\phi = 0.9$ and, hence,

$$M_u = 0.9 M_n$$

The required reinforcement ratio for a given factored moment is then

$$\rho = \frac{(0.85 f_c')\left(1 - \sqrt{1 - \frac{K_u}{0.383 f_c'}}\right)}{f_y}$$
$$K_u = \frac{M_u}{bd^2}$$

These expressions may be readily programmed using a hand-held calculator.[3] Alternatively, design tables may be used,[4,5,6] and rearranging the expression[3] in terms of the tension reinforcement index ω gives

$$\frac{M_u}{f_c' bd^2} = \omega(0.9 - 0.5294\omega)$$
$$\omega = \frac{\rho f_y}{f_c'}$$

Table A.1 provides a design aid that tabulates ω against $M_u/f_c' bd^2$.

Tension-Controlled and Compression-Controlled Sections

As specified in ACI Sec. 10.2.3 and shown in Fig. 1.1, the nominal flexural strength of a member is reached when the strain in the extreme compression fiber reaches a value of 0.003. Depending on the strain in the tension steel, the section is classified as either tension-controlled or compression-controlled and the strength-reduction factor varies from a value of 0.90 to 0.65.

Figure 1.1 *Rectangular Stress Block[2]*

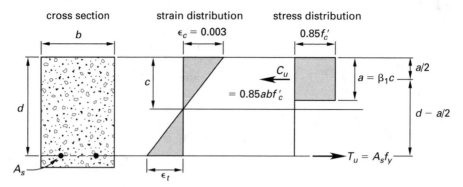

ACI Sec. 10.3.4 defines a tension-controlled section as one in which the strain in the extreme tension steel $\epsilon_t \geq 0.005$ when the concrete reaches its ultimate strain of $\epsilon_c = 0.003$. Hence, from Fig. 1.1, for a value of $\epsilon_t = 0.005$, the neutral axis depth ratio is given by

$$\frac{c}{d} = 0.375$$

The following relationships are obtained.

$$a = 0.375\beta_1 d$$

$$C_u = 0.319\beta_1 f_c' bd$$

$$\rho_t = 0.319\beta_1 \frac{f_c'}{f_y}$$

$$\omega = 0.319\beta_1$$

$$\beta_1 = 0.85 \quad [f_c' \leq 4000 \text{ lbf/in}^2]$$

$$= 0.85 - \frac{f_c' - 4000 \frac{\text{lbf}}{\text{in}^2}}{20,000}$$

$$[4000 \text{ lbf/in}^2 < f_c' \leq 8000 \text{ lbf/in}^2]$$

$$= 0.65 \text{ minimum} \quad [f_c' > 8000 \text{ lbf/in}^2]$$

The strength reduction factor for this condition is given by ACI Sec. 9.3.2.1 as

$$\phi = 0.90$$

In a tension-controlled section at failure, the strength of the reinforcement is fully used and wide cracks and large deflections are produced, giving adequate warning of impending failure.

ACI Sec. 10.3.3 defines a compression-controlled section as that in which the strain in the extreme tension steel $\epsilon_t \leq f_y/E_s$ when the concrete reaches its ultimate strain of $\epsilon_c = 0.003$. For grade 60 reinforcement bars, ACI Sec. 10.3.3 assumes a strain limit of

$$\epsilon_t = 0.002$$

Hence, the neutral axis depth ratio is given by

$$\frac{c}{d} = 0.600$$

The strength reduction factor for this condition, for members with rectangular stirrups, is given by ACI Sec. 9.3.2.2 as

$$\phi = 0.65$$

For sections that lie in the transition region between the tension-controlled and compression-controlled limits, the strength reduction factor is obtained from ACI Fig. R9.3.2 as

$$\phi = 0.48 + 83\epsilon_t$$

$$= 0.23 + \frac{0.25}{\frac{c}{d}}$$

For flexural members with an axial load of $P_u < 0.10 f_c' A_g$, the minimum value of the net tensile strain at nominal strength is limited by ACI Sec. 10.3.5 to

$$\epsilon_t = 0.004$$

Hence, the neutral axis depth ratio is given by

$$\frac{c}{d} = 0.429$$

The corresponding strength reduction factor is

$$\phi = 0.812$$

The maximum allowable reinforcement ratio is derived from Fig. 1.1 as

$$\rho_{\max} = 0.364\beta_1 \frac{f_c'}{f_y}$$

Example 1.2

A reinforced concrete slab is simply supported over a span of 12 ft. The slab has a concrete compressive strength of 3000 lbf/in^2, and the reinforcement consists of no. 4 grade 60 bars at 11 in on center with an effective depth of 6 in. The total dead load, including the self-weight of the slab, is 120 lbf/ft^2.

1. Consider a slab with a 12 in width. The tension reinforcement area provided is most nearly

    (A) 0.15 in^2
    (B) 0.22 in^2
    (C) 0.26 in^2
    (D) 0.35 in^2

2. The depth of the rectangular stress block is most nearly

    (A) 0.24 in
    (B) 0.33 in
    (C) 0.39 in
    (D) 0.43 in

3. The lever-arm of the internal resisting moment is most nearly

    (A) 5.3 in
    (B) 5.8 in
    (C) 6.3 in
    (D) 6.9 in

4. The nominal flexural strength of a 12 in wide slab is most nearly

    (A) 5.75 ft-kips
    (B) 6.35 ft-kips
    (C) 7.10 ft-kips
    (D) 7.95 ft-kips

5. The maximum permissible design factored moment on a 12 in wide slab is most nearly

    (A) 5.75 ft-kips
    (B) 6.25 ft-kips
    (C) 7.40 ft-kips
    (D) 8.10 ft-kips

6. The applied factored dead load moment, in ACI Eq. (9-2), on a 12 in wide slab is most nearly

    (A) 2.0 ft-kips
    (B) 2.5 ft-kips
    (C) 3.0 ft-kips
    (D) 3.5 ft-kips

7. The maximum permissible strength level live load moment on a 12 in wide slab is most nearly

    (A) 1.9 ft-kips
    (B) 2.4 ft-kips
    (C) 2.7 ft-kips
    (D) 3.1 ft-kips

8. The maximum permissible service level live load moment on a 12 in wide slab is most nearly

    (A) 1.1 ft-kips
    (B) 1.4 ft-kips
    (C) 1.7 ft-kips
    (D) 2.0 ft-kips

9. The permissible service level live load is most nearly

    (A) 100 lbf/ft
    (B) 110 lbf/ft
    (C) 118 lbf/ft
    (D) 125 lbf/ft

Solution

Consider a 12 in wide slab.

1. The area of one no. 4 bar is 0.20 in^2. The reinforcement area provided in a 12 in width is

$$A_s = \frac{(0.20 \text{ in}^2)(12 \text{ in})}{11 \text{ in}}$$
$$= 0.22 \text{ in}^2$$

The answer is (B).

2. Equating the tensile and compressive forces acting on the section gives the depth of the equivalent rectangular stress block as

$$a = \frac{A_s f_y}{0.85 f'_c b} = \frac{(0.22 \text{ in}^2)\left(60{,}000 \dfrac{\text{lbf}}{\text{in}^2}\right)}{(0.85)\left(3000 \dfrac{\text{lbf}}{\text{in}^2}\right)(12 \text{ in})}$$
$$= 0.43 \text{ in}$$

The answer is (D).

3. The lever-arm of the internal resisting moment is obtained from Fig. 1.1 as

$$d - \frac{a}{2} = 6 \text{ in} - \frac{0.43 \text{ in}}{2}$$
$$= 5.78 \text{ in}$$

The answer is (B).

4. The nominal moment of resistance is

$$M_n = A_s f_y \left(d - \frac{a}{2}\right)$$
$$= \frac{(0.22 \text{ in}^2)\left(60 \frac{\text{kips}}{\text{in}^2}\right)(5.78 \text{ in})}{12 \frac{\text{in}}{\text{ft}}}$$
$$= 6.36 \text{ ft-kips}$$

The answer is (B).

5. The limiting reinforcement ratio for a tension-controlled section is

$$\rho_t = 0.319 \beta_1 \frac{f'_c}{f_y}$$
$$= (0.319)(0.85) \left(\frac{3000 \frac{\text{lbf}}{\text{in}^2}}{60,000 \frac{\text{lbf}}{\text{in}^2}}\right)$$
$$= 0.014$$

The reinforcement ratio provided is

$$\rho = \frac{A_s}{bd}$$
$$= \frac{0.22 \text{ in}^2}{(12 \text{ in})(6 \text{ in})}$$
$$= 0.003$$
$$< \rho_t$$

Hence, the section is tension-controlled and the strength reduction factor is
$$\phi = 0.9$$

The maximum permissible design factored moment is

$$M_u = \phi M_n$$
$$= (0.9)(6.36 \text{ ft-kips})$$
$$= 5.73 \text{ ft-kips}$$

The answer is (A).

6. The applied factored dead load moment is

$$M_{uD} = \frac{1.2 w_D l^2}{8}$$
$$= \frac{(1.2)\left(0.12 \frac{\text{kip}}{\text{ft}}\right)(12 \text{ ft})^2}{8}$$
$$= 2.59 \text{ ft-kips}$$

The answer is (B).

7. The maximum permissible strength level live load moment is

$$M_{uL} = M_u - M_{uD}$$
$$= 5.73 \text{ ft-kips} - 2.59 \text{ ft-kips}$$
$$= 3.14 \text{ ft-kips}$$

The answer is (D).

8. The maximum permissible service level live load moment is
$$M_L = \frac{M_{uL}}{1.6}$$
$$= \frac{3.14 \text{ ft-kips}}{1.6}$$
$$= 1.96 \text{ ft-kips}$$

The answer is (D).

9. The permissible service level live load is

$$w_L = \frac{8 M_L}{l^2}$$
$$= \frac{(8)(1.96 \text{ ft-kips})\left(1000 \frac{\text{lbf}}{\text{kip}}\right)}{(12 \text{ ft})^2}$$
$$= 109 \text{ lbf/ft}$$

The answer is (B).

Balanced Strain Condition

A balanced strain condition is achieved when the tension reinforcement yields simultaneously with the maximum strain in the concrete reaching a value of 0.003. The reinforcement ratio required to produce a balanced strain condition is given by[2]

$$\rho_b = (0.85) \left(\frac{87,000 \beta_1 f'_c}{f_y (87,000 + f_y)}\right)$$

In accordance with ACI Sec. 10.5, the minimum permissible reinforcement ratio is

$$\rho_{\min} = \frac{3\sqrt{f'_c}}{f_y}$$
$$\geq \frac{200}{f_y}$$

The exception is that the minimum reinforcement need not exceed 33% more than that required by analysis. For slabs and footings, ACI Sec. 7.12 requires a minimum reinforcement ratio for grade 60 deformed bars of

$$\rho_{\min} = 0.0018$$

Example 1.3

A reinforced concrete beam with an effective depth of 16 in and a width of 12 in is reinforced with grade 60 bars and has a concrete compressive strength of 3000 lbf/in^2. Determine the area of tension reinforcement required if the beam supports a total factored moment of 150 ft-kips.

Solution

The design moment factor is

$$K_u = \frac{M_u}{bd^2}$$

$$= \frac{(150 \text{ ft-kips})\left(12\ \frac{\text{in}}{\text{ft}}\right)\left(1000\ \frac{\text{lbf}}{\text{kip}}\right)}{(12 \text{ in})(16 \text{ in})^2}$$

$$= 586 \text{ lbf/in}^2$$

$$\frac{K_u}{f_c'} = \frac{586\ \dfrac{\text{lbf}}{\text{in}^2}}{3000\ \dfrac{\text{lbf}}{\text{in}^2}}$$

$$= 0.195$$

From Table A.1, assuming a tension-controlled section, the corresponding tension reinforcement index is

$$\omega = 0.255$$

The required reinforcement ratio is

$$\rho = \frac{\omega f_c'}{f_y}$$

$$= \frac{(0.255)\left(3000\ \dfrac{\text{lbf}}{\text{in}^2}\right)}{60{,}000\ \dfrac{\text{lbf}}{\text{in}^2}}$$

$$= 0.0128$$

The limiting reinforcement ratio for a tension-controlled section is

$$\rho_t = 0.319\beta_1 \frac{f_c'}{f_y}$$

$$= 0.0136$$

$$> \rho$$

Hence, the section is tension-controlled.

The minimum allowable reinforcement ratio is governed by

$$\rho_{\min} = \frac{200}{f_y}$$

$$= \frac{200\ \dfrac{\text{lbf}}{\text{in}^2}}{60{,}000\ \dfrac{\text{lbf}}{\text{in}^2}}$$

$$= 0.0033$$

$$< \rho \quad [\text{satisfactory}]$$

The reinforcement area required is

$$A_s = \rho b_w d$$

$$= (0.0128)(12 \text{ in})(16 \text{ in})$$

$$= 2.45 \text{ in}^2$$

Beams with Compression Reinforcement

A reinforced concrete beam with compression reinforcement is shown in Fig. 1.2. Compression reinforcement and additional tension reinforcement are required when the factored moment on the member exceeds the design flexural strength of the member with the strain in the tension steel $\epsilon_t = 0.005$. The residual moment is given by

$$M_r = M_u - M_{\max}$$

The areas of compression reinforcement and additional tension reinforcement are

$$A_s' = \frac{M_r}{\phi f_s'(d - d')}$$

$$A_t = \frac{A_s' f_s'}{f_y}$$

$$f_s' = (87{,}000)\left(1 - \frac{d'}{c}\right)$$

$$\leq f_y$$

$$c = 0.375d$$

Figure 1.2 *Beam with Compression Reinforcement*[2]

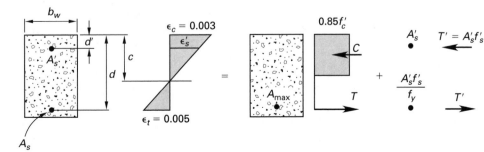

Example 1.4

A reinforced concrete beam with an effective depth of 16 in and a width of 12 in is reinforced with grade 60 bars and has a concrete compressive strength of 3000 lbf/in. The depth to the centroid of the compression reinforcement is 3 in. Determine the areas of tension and compression reinforcement required if the beam supports a total factored moment of 178 ft-kips.

Solution

In Ex. 1.3, it was shown that the maximum allowable tension reinforcement ratio in a tension-controlled beam with a concrete strength of 3000 lbf/in² and grade 60 reinforcement bars is

$$\rho_t = 0.0136$$

The corresponding tension reinforcement area is

$$\begin{aligned}
A_{\max} &= b_w d \rho_t \\
&= (12 \text{ in})(16 \text{ in})(0.0136) \\
&= 2.611 \text{ in}^2
\end{aligned}$$

The corresponding tension reinforcement index is

$$\begin{aligned}
\omega = \frac{\rho_t f_y}{f'_c} &= \frac{(0.0136)\left(60 \dfrac{\text{kips}}{\text{in}^2}\right)}{3 \dfrac{\text{kips}}{\text{in}^2}} \\
&= 0.271
\end{aligned}$$

From Table A.1, the corresponding maximum design flexural strength is

$$\begin{aligned}
M_{\max} &= 0.205 f'_c b_w d^2 \\
&= \frac{(0.205)\left(3000 \dfrac{\text{lbf}}{\text{in}^2}\right)(12 \text{ in})(16 \text{ in})^2}{\left(1000 \dfrac{\text{lbf}}{\text{kip}}\right)\left(12 \dfrac{\text{in}}{\text{ft}}\right)} \\
&= 157.4 \text{ ft-kips}
\end{aligned}$$

The residual moment is

$$\begin{aligned}
M_r &= M_u - M_{\max} \\
&= 178 \text{ ft-kips} - 157.4 \text{ ft-kips} \\
&= 20.6 \text{ ft-kips}
\end{aligned}$$

The additional area of tension reinforcement required is

$$\begin{aligned}
A_t &= \frac{M_r}{\phi f_y (d - d')} \\
&= \frac{(20.6 \text{ ft-kips})\left(12 \dfrac{\text{in}}{\text{ft}}\right)}{(0.9)\left(60 \dfrac{\text{kips}}{\text{in}^2}\right)(16 \text{ in} - 3 \text{ in})} \\
&= 0.352 \text{ in}^2
\end{aligned}$$

The total required area of tension reinforcement is

$$\begin{aligned}
A_s &= A_{\max} + A_t = 2.611 \text{ in}^2 + 0.352 \text{ in}^2 \\
&= 2.963 \text{ in}^2
\end{aligned}$$

The depth of the stress block is

$$\begin{aligned}
a = \frac{f_y A_{\max}}{0.85 f'_c b} &= \frac{\left(60 \dfrac{\text{kips}}{\text{in}^2}\right)(2.611 \text{ in}^2)}{(0.85)\left(3 \dfrac{\text{kips}}{\text{in}^2}\right)(12 \text{ in})} \\
&= 5.12 \text{ in}
\end{aligned}$$

The neutral axis depth is

$$\begin{aligned}
c &= 0.375 d \\
&= (0.375)(16 \text{ in}) \\
&= 6.0 \text{ in}
\end{aligned}$$

The stress in the compression steel is

$$\begin{aligned}
f'_s &= \left(87,000 \dfrac{\text{lbf}}{\text{in}^2}\right)\left(1 - \frac{d'}{c}\right) \\
&= \left(87,000 \dfrac{\text{lbf}}{\text{in}^2}\right)\left(1 - \frac{3 \text{ in}}{6.0 \text{ in}}\right) \\
&= 43,500 \text{ lbf/in}^2
\end{aligned}$$

The required area of compression reinforcement is

$$\begin{aligned}
A'_s = \frac{A_t f_y}{f'_s} &= \frac{(0.352)\left(60 \dfrac{\text{kips}}{\text{in}^2}\right)}{43.50 \dfrac{\text{kips}}{\text{in}^2}} \\
&= 0.486 \text{ in}^2
\end{aligned}$$

Flanged Section with Tension Reinforcement

When the rectangular stress block is wholly contained in the flange, a flanged section may be designed as a rectangular beam.

When the depth of the rectangular stress block exceeds the flange thickness, the flanged beam is designed as shown in Fig. 1.3. The area of reinforcement required to balance the compressive force in the outstanding flanges is

$$A_{sf} = \frac{0.85 f'_c h_f (b - b_w)}{f_y}$$

The corresponding design moment strength is

$$M_f = \phi A_{sf} f_y \left(d - \frac{h_f}{2}\right)$$

Figure 1.3 *Flanged Section with Tension Reinforcement*[2]

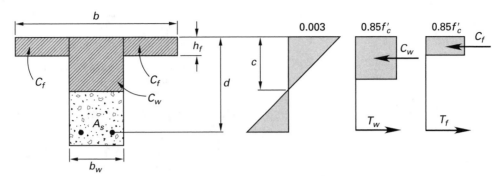

The beam web must develop the residual moment, which is given by

$$M_r = M_u - M_f$$

The value of $M_r/f'_c b_w d^2$ is determined. The corresponding value of ω is obtained from Table A.1, and the additional area of reinforcement required to balance the residual moment is

$$A_{sw} = \frac{\omega b_w d f'_c}{f_y}$$

The total area of reinforcement required is

$$A_s = A_{sf} + A_{sw}$$

Example 1.5

A reinforced concrete flanged beam with a flange width of 24 in, a web width of 12 in, a flange depth of 3 in, and an effective depth of 16 in is reinforced with grade 60 reinforcement. If the concrete compressive strength is 3000 lbf/in², determine the area of tension reinforcement required to support an applied factored moment of 250 ft-kips.

Solution

Assume that the depth of the rectangular stress block exceeds the depth of the flange.

The area of tension reinforcement required to balance the compression force in the flange is

$$A_{sf} = \frac{0.85 f'_c h_f (b - b_w)}{f_y}$$

$$= \frac{(0.85)\left(3\ \dfrac{\text{kips}}{\text{in}^2}\right)(3\text{ in})(24\text{ in} - 12\text{ in})}{60\ \dfrac{\text{kips}}{\text{in}^2}}$$

$$= 1.53\text{ in}^2$$

Assuming the section is tension-controlled, the corresponding design moment strength is

$$M_f = \phi A_{sf} f_y \left(d - \frac{h_f}{2}\right)$$

$$= \frac{(0.9)\,(1.53\text{ in}^2)\left(60\ \dfrac{\text{kips}}{\text{in}^2}\right)(16\text{ in} - 1.5\text{ in})}{12\ \dfrac{\text{in}}{\text{ft}}}$$

$$= 99.83\text{ ft-kips}$$

The residual moment to be developed by the web is

$$M_r = M_u - M_f$$
$$= 250\text{ ft-kips} - 99.83\text{ ft-kips}$$
$$= 150.17\text{ ft-kips}$$

$$\frac{M_r}{f'_c b_w d^2} = \frac{(150.17\text{ ft-kips})\left(12\ \dfrac{\text{in}}{\text{ft}}\right)}{\left(3\ \dfrac{\text{kips}}{\text{in}^2}\right)(12\text{ in})(16\text{ in})^2}$$

$$= 0.196$$

From Table A.1, the corresponding tension reinforcement index is

$$\omega = 0.257$$

The reinforcement required to develop the residual moment is

$$A_{sw} = \frac{\omega b_w d f'_c}{f_y}$$

$$= \frac{(0.257)(12\text{ in})(16\text{ in})\left(3\ \dfrac{\text{kips}}{\text{in}^2}\right)}{60\ \dfrac{\text{kips}}{\text{in}^2}}$$

$$= 2.47\text{ in}^2$$

The total tension reinforcement area required is

$$A_s = A_{sf} + A_{sw}$$
$$= 1.53\text{ in}^2 + 2.47\text{ in}^2$$
$$= 4.00\text{ in}^2$$

The depth of the equivalent rectangular stress block is given by

$$a = \frac{A_{sw} f_y}{0.85 f_c' b}$$

$$= \frac{(2.47 \text{ in}^2)\left(60 \dfrac{\text{kips}}{\text{in}^2}\right)}{(0.85)\left(3 \dfrac{\text{kips}}{\text{in}^2}\right)(12 \text{ in})}$$

$$= 4.84 \text{ in}$$

$$> h_f \quad [\text{as assumed}]$$

For a tension-controlled section, the maximum depth of the equivalent rectangular stress block is given by

$$a_t = 0.375 \beta_1 d$$

$$= (0.375)(0.85)(16 \text{ in})$$

$$= 5.10 \text{ in}$$

$$> a \quad [\text{the section is tension-controlled as assumed}]$$

3. SERVICEABILITY REQUIREMENTS FOR BEAMS

Nomenclature

A_b	area of individual bar	in^2
A_s	area of tension reinforcement	in^2
A_{sk}	area of skin reinforcement per unit height in one side face	in^2
A_{ts}	area of nonprestressed reinforcement in a tie	in^2
c_c	clear cover to tension reinforcement	in
d_b	diameter of bar	in
E_c	modulus of elasticity of concrete, $33\left(w_c^3 f_c\right)^{0.5}$	lbf/in^2
E_s	modulus of elasticity of reinforcement, 29,000	kips/in^2
f_r	modulus of rupture of concrete, $7.5\left(f_c'\right)^{0.5}$	lbf/in^2
f_s	calculated stress in reinforcement at service loads	kips/in^2
h	overall dimension of member	in
I_{cr}	moment of inertia of cracked transformed section, $b_w(kd)^3/3 + nA_s(d-kd)^2$	in^4
I_e	effective moment of inertia	in^4
I_g	moment of inertia of gross concrete section, $b_w h^3/12$	in^4
k	neutral axis depth factor at service load, $(2\rho n + (\rho n)^2)^{0.5} - \rho n$	–
l	span length of beam or one-way slab, projection of cantilever	ft
M_a	maximum moment in member at stage deflection is required	in-lbf or ft-kips
M_{cr}	cracking moment, $2 f_r I_g / h$	in-lbf or ft-kips
n	modular ratio, E_s / E_c	–
s	center-to-center spacing of tension reinforcement	in
w_c	unit weight of concrete	lbf/ft^3

Symbols

ζ	time-dependent factor for sustained load	
λ	multiplier for additional long-time deflection	
ρ'	reinforcement ratio for compression reinforcement, A_s'/bd	

Control of Crack Widths

In accordance with ACI Sec. 10.6, crack width is controlled by limiting the spacing of tension reinforcement to a value given by ACI Equation (10-4), where f_s is in units of kips/in^2.

$$s = \frac{600}{f_s} - 2.5 c_c$$

$$\leq \frac{480}{f_s}$$

As shown in Fig. 1.4, s is the center-to-center spacing, in inches, of the tension reinforcement nearest to the extreme tension face and c_c is the clear concrete cover, in inches, from the nearest surface in tension to the surface of the tension reinforcement. Where there is only one bar nearest to the extreme tension face, s is taken as the width of the extreme tension face. Controlling the spacing of tension reinforcement limits the width of surface cracks to an acceptable level.

Figure 1.4 *Tension Reinforcement Details*

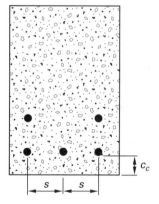

The stress in the reinforcement at service load may be either calculated or assumed equal to $(2/3) f_y$.

When the depth of the beam exceeds 36 in, ACI Sec. 10.6.7 requires that skin reinforcement be placed along both side faces of the web, in the lower half of the beam.

Example 1.6

The beam shown in the following figure is reinforced with eight no. 9 grade 60 bars. Clear cover of $1^1/_2$ in is provided to the no. 4 stirrups. Determine the skin reinforcement required, and check that the spacing of the reinforcement conforms to ACI Sec. 10.6.

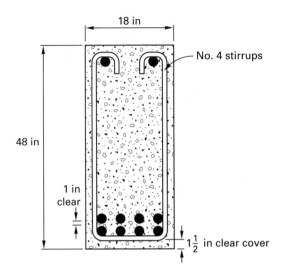

Solution

The clear cover provided to the tension reinforcement is given by

$$c_c = 1.5 \text{ in} + 0.5 \text{ in}$$

$$= 2 \text{ in}$$

The stress in the reinforcement at service load is assumed equal to

$$f_s = \frac{2}{3} f_y$$

$$= \left(\frac{2}{3} \right) \left(60 \, \frac{\text{kips}}{\text{in}^2} \right)$$

$$= 40 \text{ kips/in}^2$$

The maximum allowable bar spacing is given by ACI Eq. (10-4) as

$$s = \frac{600}{f_s} - 2.5 c_c$$

$$= \frac{600}{40 \, \dfrac{\text{kips}}{\text{in}^2}} - (2.5)(2 \text{ in})$$

$$= 10 \text{ in}$$

The actual bar spacing is given by

$$s' = \frac{18 \text{ in} - (2)(1.5 \text{ in}) - (2)(0.5 \text{ in}) - (1.128 \text{ in})}{3}$$

$$= 4.29 \text{ in}$$

Since 4.29 in is less than 10 in, this bar spacing is satisfactory.

The depth of the beam is

$$h = 48 \text{ in}$$

$$> 36 \text{ in}$$

Therefore, skin reinforcement is required.

Using no. 3 bars, the maximum allowable spacing is

$$s_{sk} = 10 \text{ in}$$

The bars shall extend for a distance, $h/2$, from the tension face.

The reinforcement layout is shown in the following illustration.

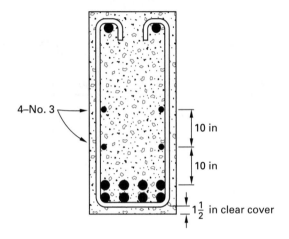

Deflection Limitations

The allowable, immediate deflection of flexural members supporting non-sensitive elements is specified in ACI Table 9.5(b) as $l/180$ for flat roofs and $l/360$ for floors due to the applied live load. The total deflection occurring after the attachment of non-sensitive elements is limited to $l/240$. The total deflection occurring after the attachment of deflection sensitive elements is limited to $l/480$.

For normal weight concrete and grade 60 reinforcement, ACI Table 9.5(a) provides span/depth ratios applicable to members supporting non-sensitive elements. These ratios are shown in Table 1.1.

Table 1.1 Span/Depth Ratios

end conditions	beam	slab
simply supported	$\dfrac{l}{16}$	$\dfrac{l}{20}$
one end continuous	$\dfrac{l}{18.5}$	$\dfrac{l}{24}$
both ends continuous	$\dfrac{l}{21}$	$\dfrac{l}{28}$
cantilever	$\dfrac{l}{8}$	$\dfrac{l}{10}$

For grade 40 reinforcement, the tabulated values are multiplied by the factor 0.8. For lightweight concrete, the tabulated values are multiplied by the factor

$$R = 1.65 - 0.005w_c$$

$$\geq 1.09$$

Deflection Determination

Short-term deflections may be calculated by using the effective moment of inertia given by ACI Sec. 9.5.2.3 and illustrated in Fig. 1.5 as

$$I_e = \left(\frac{M_{cr}}{M_a}\right)^3 I_g + \left(1 - \left(\frac{M_{cr}}{M_a}\right)^3\right) I_{cr} \text{ [ACI 9-8]}$$

Additional long-term deflection is estimated from ACI Sec. 9.5.2.5 by multiplying the short-term deflection by the multiplier

$$\lambda = \frac{\xi}{1 + 50\rho'} \qquad \text{[ACI 9-11]}$$

ξ is the time-dependent factor for sustained load defined in ACI Sec. 9.5.2.5 and shown in Table 1.2.

Table 1.2 Value of ξ

time period (mo)	ξ
60	2.0
12	1.4
6	1.2
3	1.0

The deflection is calculated for each loading case using the appropriate value of the effective moment of inertia. Thus, the short-term deflection δ_D may be calculated for dead load only and the short-term deflection $\delta_{(D+L)}$ calculated for the total applied load. The live load deflection is then given by

$$\delta_L = \delta_{(D+L)} - \delta_D$$

The final total deflection, including additional long-term deflection, is given by

$$\delta_T = \delta_D(1 + \lambda) + \delta_L$$

$$= \delta_{(D+L)} + \lambda\delta_D$$

Example 1.7

A reinforced concrete beam spanning 12 ft has an effective depth of 16 in, an overall depth of 18 in, a compressive strength of 3000 lbf/in², and is reinforced with three no. 8 grade 60 bars. The beam is 12 in wide. The bending moment due to sustained dead load is 60 ft-kips. The weight of the nondeflection sensitive elements, which are attached immediately after removing the falsework, may be neglected.

The transient floor live load moment is 30 ft-kips. Compare the beam deflections with the allowable values and determine the final beam deflection due to long-term effects and transient loads.

Figure 1.5 Service Load Conditions

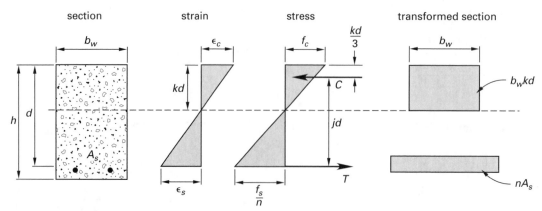

Solution

The allowable live load deflection for floors is given by ACI Table 9.5(b) as

$$\delta_L = \frac{l}{360} = \frac{(12 \text{ ft}) \left(12 \frac{\text{in}}{\text{ft}}\right)}{360}$$

$$= 0.40 \text{ in}$$

The allowable deflection after attachment of nonsensitive elements is

$$\delta_{(\lambda D + L)} = \frac{l}{240} = \frac{(12 \text{ ft}) \left(12 \frac{\text{in}}{\text{ft}}\right)}{240}$$

$$= 0.60 \text{ in}$$

From ACI Sec. 8.5,

$$E_c = 33\sqrt{w_c^3 f_c'}$$

$$= 33\sqrt{\left(150 \frac{\text{lbf}}{\text{ft}^3}\right)^3 \left(3000 \frac{\text{lbf}}{\text{in}^2}\right)}$$

$$= 3320 \text{ kips/in}^2$$

$$E_s = 29{,}000 \text{ kips/in}^2$$

$$n = \frac{E_s}{E_c} = \frac{29{,}000 \dfrac{\text{kips}}{\text{in}^2}}{3320 \dfrac{\text{kips}}{\text{in}^2}}$$

$$= 8.73$$

$$\rho = \frac{A_s}{b_w d} = \frac{2.37 \text{ in}^2}{(12 \text{ in})(16 \text{ in})}$$

$$= 0.0123$$

$$\rho n = (0.0123)(8.73)$$

$$= 0.108$$

From Table A.2, the corresponding neutral axis depth factor is

$$k = 0.3691$$

The moment of inertia of the cracked transformed section is

$$I_{cr} = \frac{b_w (kd)^3}{3} + nA_s(d - kd)^2$$

$$= \frac{(12 \text{ in})\big((0.3691)(16 \text{ in})\big)^3}{3}$$

$$\quad + (8.73)\left(2.37 \text{ in}^2\right)\left(16 \text{ in} - (0.3691)(16 \text{ in})\right)^2$$

$$= 2932 \text{ in}^4$$

$$I_g = \frac{(12 \text{ in})(18 \text{ in})^3}{12}$$

$$= 5832 \text{ in}^4$$

The modulus of rupture is given by ACI Eq. (9-10) as

$$f_r = 7.5\sqrt{f_c'}$$

$$= 7.5\sqrt{3000 \frac{\text{lbf}}{\text{in}^2}}$$

$$= 411 \text{ lbf/in}^2$$

The cracking moment is given by ACI Eq. (9-9) as

$$M_{cr} = \frac{2 f_r I_g}{h}$$

$$= \frac{(2)\left(411 \dfrac{\text{lbf}}{\text{in}^2}\right)\left(5832 \text{ in}^4\right)}{(18 \text{ in})\left(12 \dfrac{\text{in}}{\text{ft}}\right)(1000 \text{ lbf})}$$

$$= 22.18 \text{ ft-kips}$$

$$< 60 \text{ ft-kips} \quad \text{[Section is cracked.]}$$

The effective moment of inertia for the dead load bending moment is given by ACI Eq. (9-8) as

$$I_e = \left(\frac{M_{cr}}{M_D}\right)^3 I_g + \left(1 - \left(\frac{M_{cr}}{M_D}\right)^3\right) I_{cr}$$

$$= \left(\frac{22.18 \text{ ft-kips}}{60 \text{ ft-kips}}\right)^3 \left(5832 \text{ in}^4\right)$$

$$\quad + \left(1 - \left(\frac{22.18 \text{ ft-kips}}{60 \text{ ft-kips}}\right)^3\right)\left(2932 \text{ in}^4\right)$$

$$= 3078 \text{ in}^4$$

The corresponding short-term deflection due to dead-load is

$$\delta_D = \frac{180 M_D L^2}{E_c I_e}$$

$$= \frac{(180)(60 \text{ ft-kips})(12 \text{ ft})^2}{\left(3320 \dfrac{\text{kips}}{\text{in}^2}\right)\left(3078 \text{ in}^4\right)}$$

$$= 0.152 \text{ in}$$

The effective moment of inertia for the dead load plus live load is

$$I_e = \left(\frac{M_{cr}}{M_{(D+L)}}\right)^3 I_g + \left(1 - \left(\frac{M_{cr}}{M_{(D+L)}}\right)^3\right) I_{cr}$$

$$= \left(\frac{22.18 \text{ ft-kips}}{90 \text{ ft-kips}}\right)^3 \left(5832 \text{ in}^4\right)$$

$$\quad + \left(1 - \left(\frac{22.18 \text{ ft-kips}}{90 \text{ ft-kips}}\right)^3\right)\left(2932 \text{ in}^4\right)$$

$$= 2975 \text{ in}^4$$

The corresponding short-term deflection due to the dead load plus live load is

$$\delta_{(D+L)} = \frac{(180)(90 \text{ ft-kips})(12 \text{ ft})^2}{\left(3320 \, \dfrac{\text{kips}}{\text{in}^2}\right)(2975 \text{ in}^4)}$$

$$= 0.236 \text{ in}$$

The short-term deflection due to transient live load is

$$\delta_L = \delta_{(D+L)} - \delta_D$$

$$= 0.236 \text{ in} - 0.152 \text{ in}$$

$$= 0.084 \text{ in}$$

$$< 0.40 \text{ in} \quad [\text{satisfactory}]$$

The multiplier for additional long-term deflection is given by ACI Eq. (9-11) as

$$\lambda = \frac{\xi}{1 + 50\rho'} = \frac{2}{1 + 0}$$

$$= 2$$

The deflection due to short-term live loads and long-term dead loads is

$$\delta_{(\lambda D + L)} = \lambda \delta_D + \delta_L$$

$$= (2)(0.152 \text{ in}) + 0.084 \text{ in}$$

$$= 0.388 \text{ in}$$

$$< 0.60 \text{ in} \quad [\text{satisfactory}]$$

The final deflection due to long-term and short-term effects is

$$\delta_T = \delta_D(1 + \lambda) + \delta_L$$

$$= (0.152 \text{ in})(3) + 0.084 \text{ in}$$

$$= 0.540 \text{ in}$$

4. ELASTIC DESIGN METHOD

Nomenclature

f_c	actual stress in concrete	lbf/in^2
f_s	actual tensile stress in reinforcement	lbf/in^2
j	lever-arm factor	–
j_{bal}	balanced lever-arm factor	–
k_{bal}	balanced neutral axis depth factor	–
M	service design moment	ft-kips
M_{bal}	balanced service design moment	ft-kips
p_{cb}	permissible concrete stress	lbf/in^2
p_{st}	permissible steel stress	lbf/in^2
ρ_{bal}	balanced tension reinforcement ratio, $p_{cb}k_{\text{bal}}/2p_{st}$	–

Determination of Working Stress Values

The elastic design method is referred to in ACI Sec. R1.1 as the alternate design method. The straight-line theory, illustrated in Fig. 1.5, is used to calculate the stresses in a member under the action of the applied service loads and to ensure that these stresses do not exceed permissible values. The permissible stresses are

$$p_{cb} = \text{maximum permissible stress in the concrete}$$

$$= 0.45 f_c'$$

$$p_{st} = \text{maximum permissible stress in the reinforcement}$$

$$= 20 \text{ kips/in}^2 \quad [\text{grade 40 reinforcement}]$$

$$= 24 \text{ kips/in}^2 \quad [\text{grade 60 reinforcement}]$$

From Fig. 1.5, the neutral axis depth factor is derived as

$$k = \sqrt{2\rho n + (\rho n)^2} - \rho n$$

Table A.2 tabulates values of k against ρn. In addition, the lever-arm factor is derived as

$$j = 1 - \frac{k}{3}$$

The stress in the reinforcement due to an applied service moment M is

$$f_s = \frac{M}{A_s j d}$$

The stress in the concrete is

$$f_c = \frac{2M}{jk b_w d^2}$$

For a balanced design, the stress in the reinforcement and the maximum stress in the concrete should simultaneously reach their permissible values. Then, the corresponding design values will be

$$k_{\text{bal}} = \frac{n p_{cb}}{p_{st} + n p_{cb}}$$

$$j_{\text{bal}} = 1 - \frac{k_{\text{bal}}}{3}$$

$$\rho_{\text{bal}} = \frac{p_{cb} k_{\text{bal}}}{2 p_{st}}$$

$$M_{\text{bal}} = A_{s(\text{bal})} p_{st} j_{\text{bal}} d$$

Example 1.8

A reinforced concrete beam with an effective depth of 16 in and a width of 12 in is reinforced with grade 60 bars and has a concrete cylinder strength of 3000 lbf/in^2. Using the elastic design method, determine the area of tension reinforcement required if the beam supports a total service moment of 50 ft-kips.

Solution

From Ex. 1.7, the modular ratio is given as

$$n = 8.73$$

The permissible concrete and reinforcement stresses are

$$p_{cb} = 1350 \text{ lbf/in}^2$$
$$p_{st} = 24{,}000 \text{ lbf/in}^2$$

Sufficient accuracy is obtained by assuming that the neutral axis depth factor equals the balanced value. Then,

$$k = \frac{np_{cb}}{p_{st} + np_{cb}}$$

$$= \frac{(8.73)\left(1350 \dfrac{\text{lbf}}{\text{in}^2}\right)}{\left(24{,}000 \dfrac{\text{lbf}}{\text{in}^2}\right) + (8.73)\left(1350 \dfrac{\text{lbf}}{\text{in}^2}\right)}$$

$$= 0.329$$

$$j = 1 - \frac{k}{3}$$

$$= 0.89$$

$$A_s = \frac{M}{p_{st}jd}$$

$$= \frac{(50 \text{ ft-kips})\left(12 \dfrac{\text{in}}{\text{ft}}\right)}{\left(24 \dfrac{\text{kips}}{\text{in}^2}\right)(0.89)(16 \text{ in})}$$

$$= 1.76 \text{ in}^2$$

$$f_c = \frac{2A_sp_{st}}{b_wkd}$$

$$= \frac{(2)\left(1.76 \text{ in}^2\right)\left(24{,}000 \dfrac{\text{lbf}}{\text{in}^2}\right)}{(12 \text{ in})(0.329)(16 \text{ in})}$$

$$= 1334 \text{ lbf/in}^2$$

$$< p_{cb} \quad \text{[satisfactory]}$$

5. SHEAR AND TORSION

Nomenclature

a_v	shear span, distance between concentrated load and face of supports	ft
A_{cs}	effective cross-sectional area of a strut in a strut-and-tie model taken perpendicular to the axis of the strut	in^2
A_{cp}	area enclosed by outside perimeter of concrete cross section	in^2
A_f	area of reinforcement in bracket or corbel resisting factored moment	in^2
A_h	area of shear reinforcement parallel to flexural tension reinforcement	in^2

A_l	total area of longitudinal reinforcement to resist torsion	in^2
A_n	area of reinforcement in bracket or corbel resisting tensile force N_{uc}	in^2
A_n	effective cross-sectional area of the face of a nodal zone	in^2
A_o	gross area enclosed by shear flow, $0.85A_{oh}$	in^2
A_{oh}	area enclosed by centerline of the outermost closed transverse torsional reinforcement	in^2
A_s	area of nonprestressed tension reinforcement	in^2
A_t	area of one leg of a closed stirrup resisting torsion within a distance s	in^2
A_v	area of shear reinforcement perpendicular to flexural tension reinforcement	in^2
A_{vf}	area of shear-friction reinforcement	in^2
A_{vh}	area of shear reinforcement parallel to flexural tension	in^2
b	width of a deep beam	in
b	width of compression face of member	in
b_w	web width or diameter of circular section	in
C	compressive force acting on a nodal zone	kips
f_{ce}	effective compressive strength of concrete in a strut or node	lbf/in^2
f_y	yield strength of reinforcement	lbf/in^2
f_{yt}	yield strength of transverse reinforcement	lbf/in^2
h	overall thickness of member	in
l_a	anchorage length of a reinforcing bar	in
l_b	width of bearing plate	in
l_n	clear span measured face-to-face of supports	ft
M_u	factored moment at section	ft-kips
N_{uc}	factored tensile force applied at top of corbel	kips
p_{cp}	outside perimeter of the concrete cross section	in
p_h	perimeter of centerline of outermost closed transverse torsional reinforcement	in
R	support reaction acting on a nodal zone	kips
s	spacing of shear or torsion reinforcement in direction parallel to longitudinal reinforcement	in
s_2	spacing of horizontal reinforcement	in
T	tension force acting on a nodal zone	kips
T_n	nominal torsional moment strength	ft-kips
T_u	factored torsional moment at section	ft-kips
V_c	nominal shear strength provided by concrete	kips
V_s	nominal shear strength provided by shear reinforcement	kips
V_u	factored shear force at section	kips
w_s	effective width of strut perpendicular to the axis of the strut	in
w_t	effective width of concrete concentric with a tie	in

Symbols

α	angle between inclined stirrups and longitudinal axis of member	degrees
β_n	factor to account for the effect of the anchorage of ties on the effective compressive strength of a nodal zone	–

β_s factor to account for the effect of cracking and confining reinforcement on the effective compressive strength of the concrete in a strut   –

λ correction factor related to unit weight of concrete   –

μ coefficient of friction   –

ρ_w $A_s/b_w d$ –

Design for Shear

The nominal shear capacity of the inclined stirrups shown in Fig. 1.6 is given by ACI Sec. 11.5.7.4 as

$$V_s = \frac{A_v f_{yt} d(\sin \alpha + \cos \alpha)}{s} \qquad \text{[ACI 11-16]}$$

Figure 1.6 Beam with Inclined Stirrups

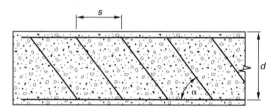

When the shear reinforcement is vertical, ACI Sec. 11.5.7.2 gives the nominal shear capacity as

$$V_s = \frac{A_v f_{yt} d}{s} \qquad \text{[ACI 11-15]}$$

The nominal shear strength of the shear reinforcement is limited by ACI Sec. 11.5.7.9 to a value of

$$V_s = 8 b_w d \sqrt{f_c'}$$

If additional shear capacity is required, the size of the concrete section must be increased. ACI Sec. 11.5.5.1 limits the spacing of the stirrups to a maximum value of $d/2$ or 24 in, and when the value of V_s exceeds $4 b_w d \sqrt{f_c'}$, the spacing is reduced to a maximum value of $d/4$ or 12 in.

The nominal shear capacity of the concrete section is given by ACI Sec. 11.3.1.1 as

$$V_c = 2 b_w d \sqrt{f_c'} \qquad \text{[ACI 11-3]}$$

This value is conservative and is usually sufficiently accurate. A more precise value is provided by ACI Sec. 11.3.2.1 as

$$V_c = \left(1.9 \sqrt{f_c'} + \frac{2500 \rho_w V_u d}{M_u} \right) b_w d \qquad \text{[ACI 11-5]}$$

$$\leq 3.5 b_w d \sqrt{f_c'}$$

$$\frac{V_u d}{M_u} \leq 1.0$$

$$\sqrt{f_c'} \leq 100 \text{ lbf/in}^2$$

M_u is the factored moment occurring simultaneously with V_u at the section being analyzed.

In accordance with ACI Eqs. (11-1) and (11-2), the combined shear capacity of the concrete section and the shear reinforcement is $(\phi V_c + \phi V_s)$, and this must exceed V_u, which is the factored shear force applied to the section. When V_u is less than $\phi V_c/2$, the concrete section is adequate to carry the shear without any shear reinforcement. Within the range $\phi V_c/2 \leq V_u \leq \phi V_c$, a minimum area of shear reinforcement is specified by ACI Sec. 11.5.6.3 as

$$A_{v(\min)} = \frac{0.75 b_w s \sqrt{f_c'}}{f_{yt}} \qquad \text{[ACI 11-13]}$$

$$\geq \frac{50 b_w s}{f_{yt}}$$

When $f_c' > 4.44$ kips/in^2, ACI Eq. (11-13) governs.

When the support reaction produces a compressive stress in the member, as indicated in Fig. 1.7, ACI Sec. 11.1.3 specifies that the critical section for shear is located at a distance from the support equal to the effective depth. In addition, for this condition to apply, loads must be applied near or at the top of the beam and no concentrated load may occur within a distance from the support equal to the effective depth.

Figure 1.7 Critical Section for Shear

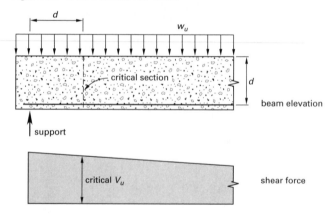

Example 1.9

A reinforced concrete beam with an effective depth of 16 in and a width of 12 in is reinforced with grade 60 bars and has a concrete compressive strength of 3000 lbf/in^2. Determine the shear reinforcement required when

    (a) the factored shear force $V_u = 9$ kips, the factored moment $M_u = 20$ ft-kips, and the reinforcement ratio $\rho_u = 0.015$

    (b) the factored shear force $V_u = 14$ kips

    (c) the factored shear force $V_u = 44$ kips

    (d) the factored shear force $V_u = 71$ kips

    (e) the factored shear force $V_u = 120$ kips

Solution

(a) The shear strength provided by the concrete is given by ACI Eq. (11-3) as

$$\phi V_c = 2\phi b_w d\sqrt{f_c'}$$

$$= \frac{(2)(0.75)(12 \text{ in})(16 \text{ in})\sqrt{3000 \dfrac{\text{lbf}}{\text{in}^2}}}{1000 \dfrac{\text{lbf}}{\text{kip}}}$$

$$= 15.8 \text{ kips}$$

$$< 2V_u$$

Using ACI Eq. (11-5) to verify the section gives

$$\frac{V_u d}{M_u} = \frac{(9 \text{ kips})(16 \text{ in})}{(20 \text{ ft-kips})\left(12 \dfrac{\text{in}}{\text{ft}}\right)}$$

$$= 0.60$$

$$< 1.0 \quad \text{[satisfactory]}$$

$$\phi V_c = \phi\left(1.9\sqrt{f_c'} + \frac{2500\rho_w V_u d}{M_u}\right)b_w d$$

$$= \frac{(0.75)\left(\begin{array}{c}1.9\sqrt{3000 \dfrac{\text{lbf}}{\text{in}^2}} \\ + (2500)(0.015)(0.60)\end{array}\right)(12 \text{ in})(16 \text{ in})}{1000 \dfrac{\text{lbf}}{\text{kip}}}$$

$$= 18.2 \text{ kips}$$

$$> 2V_u$$

Hence, in accordance with ACI Sec. 11.5.6.1, shear reinforcement is not required.

(b) Because $\phi V_c/2 < V_u < \phi V_c$, the minimum shear reinforcement specified by ACI Sec. 11.5.6.3 is required, and this is given by

$$\frac{A_{v(\min)}}{s} = \frac{50 b_w}{f_y} \quad \text{[governs]}$$

$$= \frac{\left(50 \dfrac{\text{lbf}}{\text{in}^2}\right)(12 \text{ in})\left(12 \dfrac{\text{in}}{\text{ft}}\right)}{60{,}000 \dfrac{\text{lbf}}{\text{in}^2}}$$

$$= 0.12 \text{ in}^2/\text{ft}$$

Shear reinforcement consisting of two arms of no. 3 bars at 8 in spacing provides a reinforcement area of

$$\frac{A_v}{s} = 0.33 \text{ in}^2/\text{ft}$$

The spacing of 8 in does not exceed $d/2$ and is satisfactory.

(c) The factored shear force exceeds the shear strength of the concrete, and the shear strength required from shear reinforcement is given by ACI Eqs. (11-1) and (11-2) as

$$\phi V_s = V_u - \phi V_c$$

$$= 44 \text{ kips} - 15.8 \text{ kips}$$

$$= 28.2 \text{ kips}$$

$$< 2\phi V_c$$

Hence, in accordance with ACI Sec. 11.5.5.1, stirrups are required at a maximum spacing of $d/2 = 8$ in. The area of shear reinforcement required is given by ACI Eq. (11-15) as

$$\frac{A_v}{s} = \frac{\phi V_s}{\phi d f_{yt}} = \frac{(28.2 \text{ kips})\left(12 \dfrac{\text{in}}{\text{ft}}\right)}{(0.75)(16 \text{ in})\left(60 \dfrac{\text{kips}}{\text{in}^2}\right)}$$

$$= 0.47 \text{ in}^2/\text{ft}$$

Shear reinforcement consisting of two arms of no. 4 bars at 8 in spacing provides a reinforcement area of

$$\frac{A_v}{s} = 0.60 \text{ in}^2/\text{ft}$$

$$> 0.47 \text{ in}^2/\text{ft} \quad \text{[satisfactory]}$$

(d) The shear strength required from the shear reinforcement is given by ACI Eqs. (11-1) and (11-2) as

$$\phi V_s = V_u - \phi V_c$$

$$= 71 \text{ kips} - 15.8 \text{ kips}$$

$$= 55.2 \text{ kips}$$

$$> 2\phi V_c$$

Hence, in accordance with ACI Sec. 11.5.5.3, stirrups are required at a maximum spacing of $d/4 = 4$ in. The area of shear reinforcement required is given by ACI Eq. (11-15) as

$$\frac{A_v}{s} = \frac{\phi V_s}{\phi d f_{yt}} = \frac{(55.2 \text{ kips})\left(12 \dfrac{\text{in}}{\text{ft}}\right)}{(0.75)(16 \text{ in})\left(60 \dfrac{\text{kips}}{\text{in}^2}\right)}$$

$$= 0.92 \text{ in}^2/\text{ft}$$

Shear reinforcement consisting of two arms of no. 4 bars at 4 in spacing provides a reinforcement area of

$$\frac{A_v}{s} = 1.2 \text{ in}^2/\text{ft}$$

$$> 0.92 \quad \text{[satisfactory]}$$

(e) The shear strength required from the shear reinforcement is given by ACI Eqs. (11-1) and (11-2) as

$$\phi V_s = V_u - \phi V_c$$
$$= 120 \text{ kips} - 15.8 \text{ kips}$$
$$= 104.2 \text{ kips}$$
$$> 4\phi V_c$$

Hence, in accordance with ACI Sec. 11.5.7.9, the section size is inadequate.

Inclined Bars

When a single, bent-up bar or group of bars equidistant from the support is used as shear reinforcement, the nominal shear capacity is given by ACI Sec. 11.5.7.5 as

$$V_s = A_v f_y \sin \alpha \qquad \textit{[ACI 11-17]}$$
$$\leq 3b_w d \sqrt{f_c'}$$

When a series of equally spaced bent-up bars is used, as shown in Fig. 1.8, the nominal shear capacity is given by ACI Sec. 11.5.7.4 as

$$V_s = \frac{A_v f_y (\sin \alpha + \cos \alpha)(d)}{s} \qquad \textit{[ACI 11-16]}$$

Figure 1.8 Beam with Inclined Bars

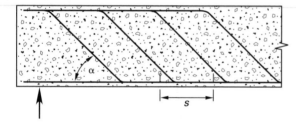

Only the center three-fourths of the inclined bar is considered effective; this limits the spacing, measured in a direction parallel to the longitudinal reinforcement, to a maximum value of

$$s_{\max} = 0.375d(1 + \cot \alpha)$$

This value is halved, in accordance with ACI Sec. 11.5.5.3, when V_s exceeds $4b_w d \sqrt{f_c'}$.

In accordance with ACI Sec. 11.5.1.2, the minimum permitted angle of inclination of the inclined bars is 30°. When shear reinforcement consists of both stirrups and inclined bars, the total combined shear resistance is given by the sum of the shear resistances of each type. The nominal combined shear resistance shall not exceed $8b_w d \sqrt{f_c'}$.

Example 1.10

The reinforced concrete beam shown in the illustration has an effective depth of 16 in, a width of 12 in, a concrete compressive strength of 3000 lbf/in², and is reinforced with grade 60 bars. Determine the design shear capacity provided at section A-A.

Solution

The nominal shear strength of the concrete is given by ACI Eq. (11-3) as

$$V_c = 2b_w d \sqrt{f_c'}$$
$$= \frac{(2)(12 \text{ in})(16 \text{ in})\sqrt{3000 \frac{\text{lbf}}{\text{in}^2}}}{1000 \frac{\text{lbf}}{\text{kip}}}$$
$$= 21.0 \text{ kips}$$

Illustration for Ex. 1.10

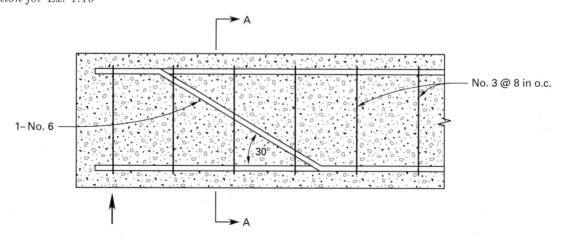

The nominal shear strength of the vertical stirrups is given by ACI Eq. (11-15) as

$$V_{s(\text{str})} = \frac{A_v f_{yt} d}{s} = \frac{\left(0.33 \; \frac{\text{in}^2}{\text{ft}}\right)\left(60 \; \frac{\text{kips}}{\text{in}^2}\right)(16 \; \text{in})}{12 \; \frac{\text{in}}{\text{ft}}}$$

$$= 26.4 \; \text{kips}$$

The nominal shear strength of the inclined bar is given by ACI Eq. (11-17) as

$$V_{s(\text{bar})} = A_v f_y \sin\alpha = \left(0.44 \; \text{in}^2\right)\left(60 \; \frac{\text{kips}}{\text{in}^2}\right)(\sin 30°)$$

$$= 13.2 \; \text{kips}$$

The combined shear strength of the shear reinforcement is

$$V_s = V_{s(\text{str})} + V_{s(\text{bar})}$$

$$= 39.6 \; \text{kips}$$

$$< 2V_c$$

Hence, in accordance with ACI Sec. 11.5.5.1, stirrup spacing shall not exceed $d/2 = 8$ in and the spacing provided is satisfactory. The minimum shear reinforcement required is specified by ACI Eq. (11-13) as

$$\frac{A_v}{s} = \frac{50 b_w}{f_{yt}} \quad [\text{governs}]$$

$$= \frac{\left(50 \; \frac{\text{lbf}}{\text{in}^2}\right)(12 \; \text{in})\left(12 \; \frac{\text{in}}{\text{ft}}\right)}{60{,}000 \; \frac{\text{lbf}}{\text{in}^2}}$$

$$= 0.12 \; \text{in}^2/\text{ft}$$

$$< 0.33 \quad [\text{satisfactory}]$$

The total design shear capacity at section A-A is

$$\phi V_n = \phi\left(V_c + V_s\right)$$

$$= (0.75)(21 \; \text{kips} + 39.6 \; \text{kips})$$

$$= 45.5 \; \text{kips}$$

Figure 1.9 *Minimum Shear Reinforcement for a Deep Beam*

Deep Beams

As shown in Fig. 1.9, a *deep beam*, as defined in ACI Secs. 10.7.1 and 11.8.1, is a beam in which the ratio of clear span to overall depth does not exceed 4. Deep beam conditions also apply to regions of beams loaded with concentrated loads within twice the beam depth from a support. The nominal shear strength of a deep beam is limited by ACI Sec. 11.8.3 to a maximum of

$$V_n = 10 b_w d \sqrt{f_c'}$$

As indicated in Fig. 1.9, minimum areas of vertical and horizontal reinforcement are specified in ACI Secs. 11.8.4 and 11.8.5 in order to restrain cracking. Alternatively, reinforcement may be provided as specified in ACI Sec. A.3.3.

In accordance with ACI Sec. 11.8.2, deep beams shall be designed using either nonlinear analysis or by the *strut-and-tie* method given in ACI App. A. In the strut-and-tie method, a member is divided, as shown in Fig. 1.10, into discontinuity, or *D*-regions, in which the beam theory of ACI Sec. 10.2 does not apply, and *B*-regions in which beam theory does apply. In addition, for the strut-and-tie method to apply, the deep beam must be loaded so that compression struts can develop between the loads and the supports.

Figure 1.10 *B- and D-Regions*

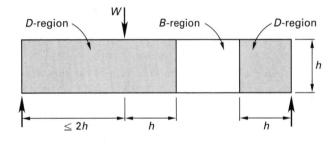

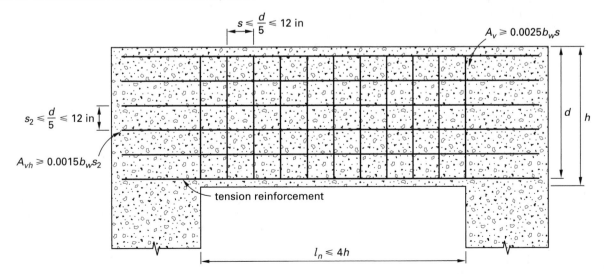

As shown in Fig. 1.11, a strut-and-tie model may be constructed to represent the internal forces in a deep beam. Compression struts are formed in the concrete to resist compressive forces. The strength of these struts is governed by the transverse tension developed by the lateral spread of the applied compression force. Using crack control reinforcement, as specified in ACI Sec. A.3.3, to resist the transverse tension increases the strength of the strut.

Figure 1.11 *Strut-and-Tie Model*

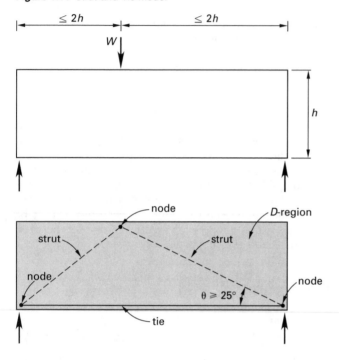

Ties consist of tension reinforcement and the surrounding concrete that is concentric with the axis of the tie. The concrete does not contribute to the strength of the tie.

Nodes occur where the axes of struts, ties, concentrated loads, and support reactions acting on the joint intersect. The angle between the axes of a strut and a tie at a node is limited by ACI Sec. A.2.5 to a minimum of $\theta = 25°$ in order to mitigate cracking.

The effective compressive strength of the concrete in a strut is specified in ACI Sec. A.3.2 as

$$f_{ce} = 0.85\beta_s f_c' \qquad \text{[ACI A-3]}$$

$$\beta_s = 1.0 \quad \begin{bmatrix} \text{for a strut of uniform cross-section,} \\ \text{as in the compression zone of a beam} \end{bmatrix}$$

$$\beta_s = 0.60\lambda \quad \begin{bmatrix} \text{for an unreinforced,} \\ \text{bottle-shaped strut} \end{bmatrix}$$

$$\beta_s = 0.75 \quad \begin{bmatrix} \text{for a bottle-shaped strut with reinforce-} \\ \text{ment as specified in ACI Sec. A.3.3} \end{bmatrix}$$

$$\beta_s = 0.40 \quad \begin{bmatrix} \text{for struts in a tension member} \\ \text{or the tension flange of a member} \end{bmatrix}$$

$$\beta_s = 0.60 \quad \text{[for all other cases]}$$

$\lambda = 1.0$ [for normal weight concrete]

$\lambda = 0.85$ [for sand-lightweight concrete]

$\lambda = 0.75$ [for all lightweight concrete]

In accordance with ACI Sec. A.3.1, the nominal compressive strength of a strut is

$$F_{ns} = f_{ce}A_{cs} \qquad \text{[ACI A-2]}$$
$$= f_{ce}w_s b$$

The strength reduction factor for strut-and-tie models is given by ACI Sec. 9.3.2.6 as

$$\phi = 0.75$$

In a reinforced concrete beam, the nominal strength of a reinforcing bar acting as a tie is given by ACI Sec. A.4.1 as

$$F_{nt} = A_{ts}f_y$$

If the bars in a tie are in one layer, as shown in Fig. 1.12, the width of the tie may be taken as the diameter of the bars in the tie plus twice the cover to the surface of the bars.

Figure 1.12 *Nodal Zone*

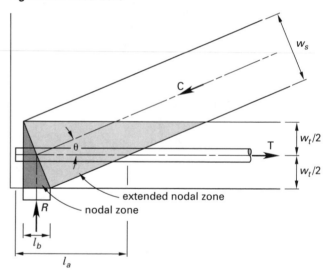

As defined in ACI Sec. A.1 and shown in Fig. 1.12, a *nodal zone* is the volume of concrete surrounding a node that is assumed to transfer strut-and-tie forces through the node. The node illustrated in Fig. 1.12 is classified as C-C-T, with two of the members acting on the node in compression and the third member in tension. Similarly, when all three members acting on the node are in compression, the node is classified as C-C-C. The effective compressive strength of the concrete in a node is specified in ACI Sec. A.5.2 as

$$f_{ce} = 0.85\beta_n f_c' \qquad \text{[ACI A-8]}$$

$$\beta_n = 1.0 \dots \quad \begin{bmatrix} \text{for a nodal zone bounded on all sides} \\ \text{by struts or bearing areas or both} \end{bmatrix}$$

$$= 0.80 \dots \quad \begin{bmatrix} \text{for a nodal zone anchoring} \\ \text{one tie} \end{bmatrix}$$

$$= 0.60 \dots \quad \begin{bmatrix} \text{for a nodal zone anchoring} \\ \text{two or more ties} \end{bmatrix}$$

In accordance with ACI Sec. A.5.1, the nominal compressive strength of a nodal zone is

$$F_{nn} = f_{ce} A_{nz} \qquad \text{[ACI A-7]}$$

$$= f_{ce} w_s b$$

The faces of the nodal zone shown in Fig. 1.12 are perpendicular to the axes of the strut, tie, and bearing plate, and the lengths of the faces are in direct proportion to the forces acting. Hence, the node has equal stresses on all faces and is termed a *hydrostatic*

nodal zone. The effective width of the strut shown in Fig. 1.12 is

$$w_s = w_t \cos\theta + l_b \sin\theta$$

The *extended nodal zone* shown in Fig. 1.12 is that portion of the member bounded by the intersection of the effective strut width and the effective tie width. As specified in ACI Sec. A.4.3.2, the anchorage length of the reinforcement is measured from the point of intersection of the bar and the extended nodal zone. The reinforcement may be anchored by a plate, by hooks, or by a straight development length.

Example 1.11

A reinforced concrete beam, with a clear span of 6 ft, an effective depth of 28 in, and a width of 12 in, as shown in the following illustration, has a concrete compressive strength of 4500 lbf/in². The factored applied force of 80 kips includes an allowance for the self weight of the beam. Determine the number of grade 60, no. 8 bars required for tension reinforcement and check that the equivalent concrete strut and nodal zone at the left support comply with the requirements of ACI App. A.

Illustration for Ex. 1.11

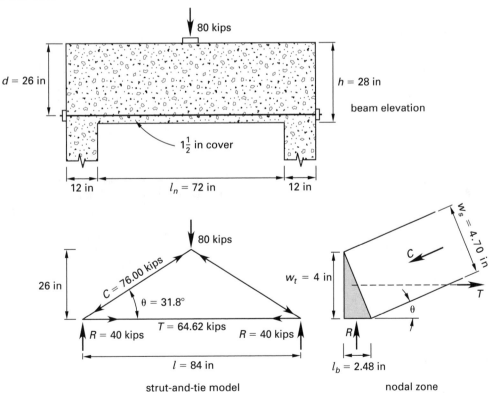

Solution

The clear span-to-depth ratio is given by

$$\frac{l_n}{h} = \frac{72 \text{ in}}{28 \text{ in}}$$
$$= 2.6$$
$$< 4 \quad \text{[satisfies ACI Sec. 11.8.1]}$$

The idealized strut-and-tie model is shown in the illustration and the angle between the struts and the tie is

$$\theta = \tan^{-1} \frac{26 \text{ in}}{42 \text{ in}}$$
$$= 31.8°$$
$$> 25° \quad \text{[satisfies ACI Sec. A.2.5]}$$

The equivalent tie force is determined from the strut-and-tie model as

$$T = \frac{(40 \text{ kips})(42 \text{ in})}{26 \text{ in}}$$
$$= 64.62 \text{ kips}$$

The strength reduction factor is given by ACI Sec. 9.3.2.6 as

$$\phi = 0.75$$

The necessary reinforcement area is given by

$$A_{ts} = \frac{T}{\phi f_y}$$
$$= \frac{64.62}{(0.75)\left(60 \ \dfrac{\text{kips}}{\text{in}^2}\right)}$$
$$= 1.44 \text{ in}^2$$

Use two no. 8 bars which gives an area of

$$A = 1.58 \text{ in}^2$$
$$> A_{ts} \quad \text{[satisfactory]}$$

As shown in the illustration, the dimensions of the nodal zone are

$$
\begin{aligned}
w_t &= \text{equivalent tie width}\\
&= d_b + 2c\\
&= 1 \text{ in} + (2)(1.5 \text{ in})\\
&= 4 \text{ in}\\
l_b &= \text{width of equivalent support strut}\\
&= w_t \tan \theta\\
&= 4 \tan 31.8°\\
&= 2.48 \text{ in}
\end{aligned}
$$

$$
\begin{aligned}
w_s &= \text{width of equivalent concrete strut}\\
&= \frac{w_t}{\cos \theta}\\
&= \frac{4}{\cos 31.8°}\\
&= 4.70 \text{ in}
\end{aligned}
$$

The stress in the equivalent tie is

$$
\begin{aligned}
f_T &= \frac{T}{bw_t}\\
&= \frac{64.62}{(12 \text{ in})(4 \text{ in})}\\
&= 1.35 \text{ kips/in}^2
\end{aligned}
$$

For a hydrostatic nodal zone

$$
\begin{aligned}
f_C &= \text{stress in the equivalent concrete strut}\\
&= f_T\\
&= 1.35 \text{ kips/in}^2\\
f_R &= \text{stress in the equivalent support strut}\\
&= f_T\\
&= 1.35 \text{ kips/in}^2
\end{aligned}
$$

For normal weight concrete, the design compressive strength of the concrete in the strut is given by ACI Eq. (A-3) as

$$
\begin{aligned}
\phi f_{ce} &= 0.85\phi\beta_s f_c'\\
&= (0.85)(0.75)(0.6)(1.0)(4.5 \text{ kips/in}^2)\\
&\quad \begin{bmatrix} \text{for an unreinforced} \\ \text{bottle-shaped strut} \end{bmatrix}\\
&= 1.72 \text{ kips/in}^2\\
&> f_C \quad \text{[satisfactory]}
\end{aligned}
$$

The design compressive strength of a nodal zone anchoring one layer of reinforcing bars without confining reinforcement is given by ACI Eq. (A-8) as

$$
\begin{aligned}
\phi f_{ce} &= \phi 0.85 \beta_n f_c'\\
&= (0.75)(0.85)(0.8)\left(4.5 \ \frac{\text{kips}}{\text{in}^2}\right)\\
&= 2.30 \text{ kips/in}^2\\
&> f_C \quad \text{[satisfactory]}
\end{aligned}
$$

The anchorage length available for the tie reinforcement, using 2 in end cover is

$$
\begin{aligned}
l_a &= \frac{h - d}{\tan \theta} + \frac{l_b}{2} + 6 \text{ in} - 2 \text{ in}\\
&= \frac{2}{\tan 31.8°} + 1.24 \text{ in} + 6 \text{ in} - 2 \text{ in}\\
&= 8.47 \text{ in}
\end{aligned}
$$

The development length for a grade 60, no. 8 bar, with 2.5 in side cover and 2 in end cover and with a standard 90° hook is given by ACI Sec. 12.5.2 as

$$l_{dh} = \frac{(0.7)(1200)d_b}{\sqrt{f'_c}}$$

$$= \frac{(0.7)(1200)(1.0)}{\sqrt{4500}}$$

$$= 12.5 \text{ in}$$

$$> l_a \quad \text{[anchorage length is inadequate]}$$

Hence, use an end plate to anchor the bars.

Corbels

As shown in Fig. 1.13 and specified in ACI Sec. 11.9, the shear span-to-depth ratio a/d and the ratio of horizontal tensile force to vertical force N_{uc}/V_u are limited to a maximum value of unity. The depth of the corbel at the outside edge of bearing area shall not be less than $d/2$.

Figure 1.13 Corbel Details

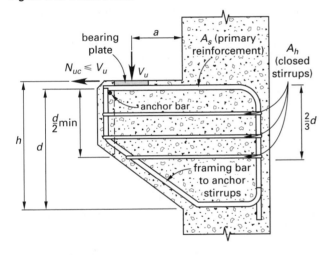

At the face of the support, the forces acting on the corbel are a shear force V_u, a moment $(V_u a + N_{uc}(h - d))$, and a tensile force N_{uc}. These require reinforcement areas of A_{vf}, A_f, and A_n, respectively. The shear friction reinforcement area is specified by ACI Sec. 11.9.3.2 and derived from ACI Eq. (11-25) as

$$A_{vf} = \frac{V_u}{\phi f_y \mu}$$

Also, from ACI Sec. 11.9.3.2.1, the factored shear force on the section is

$$V_u \le 0.2\phi f'_c b_w d$$

$$\le 0.8\phi b_w d$$

μ = coefficient of friction at face of support, as given by ACI Sec. 11.7.4.3

$$= 1.4\lambda \quad \text{[for concrete placed monolithically]}$$

The correction factor related to the unit weight of concrete is defined by ACI Sec. 11.7.4.3 as

$$\lambda = 1.0 \quad \text{[for normal weight concrete]}$$

$$= 0.75 \quad \text{[for all lightweight concrete]}$$

The tensile force N_{uc} may not be less than $0.2V_u$, and the corresponding area of reinforcement required is given by ACI Sec. 11.9.3.4 as

$$A_n = \frac{N_{uc}}{\phi f_y}$$

The required area of primary tension reinforcement is given by ACI Sec. 11.9.3.5 and Sec. 11.9.5 as

$$A_{sc} = A_f + A_n$$

$$\ge \frac{2A_{vf}}{3} + A_n$$

$$\frac{A_{sc}}{bd} \ge \frac{0.04f'_c}{f_y}$$

The minimum required area of closed ties distributed over a depth of $2d/3$ is given by ACI Sec. 11.9.4 as

$$A_h = \frac{A_{sc} - A_n}{2}$$

Example 1.12

The reinforced concrete corbel shown in the illustration, with a width of 15 in, is reinforced with grade 60 bars and has a concrete compressive strength of 3000 lbf/in². Determine whether the corbel is adequate for the applied factored loads indicated.

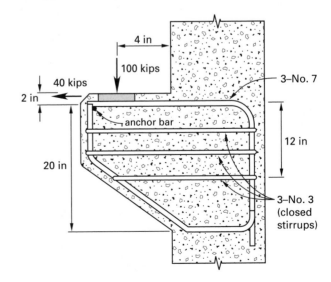

Solution

$$0.2\phi f_c' b_w d = (0.2)(0.75)\left(3 \; \frac{\text{kips}}{\text{in}^2}\right)(15 \text{ in})(20 \text{ in})$$

$$= 135 \text{ kips}$$

$$> V_u$$

$$0.8\phi b_w d = (0.8)(0.75)(15 \text{ in})(20 \text{ in})$$

$$= 180 \text{ kips}$$

$$> V_u$$

Hence, the corbel conforms to ACI Sec. 11.9.3.2.

The shear friction reinforcement area is given by ACI Sec. 11.7.4 as

$$A_{vf} = \frac{V_u}{\phi f_y \mu} = \frac{100 \text{ kips}}{(0.75)\left(60 \; \dfrac{\text{kips}}{\text{in}^2}\right)(1.4)}$$

$$= 1.59 \text{ in}^2$$

The tension reinforcement area is given by ACI Sec. 11.9.3.4 as

$$A_n = \frac{N_{uc}}{\phi f_y} = \frac{40}{(0.75)\left(60 \; \dfrac{\text{kips}}{\text{in}^2}\right)}$$

$$= 0.889 \text{ in}^2$$

The factored moment acting on the corbel is

$$M_u = V_u a + N_{uc}(h - d)$$

$$= (100 \text{ kips})(4 \text{ in}) + (40)(2 \text{ in})$$

$$= 480 \text{ in-kips}$$

The area of flexural reinforcement required for $\phi = 0.75$ as given by ACI Sec. R11.9.3.1 is

$$A_f = \frac{0.85 b d f_c' \left(1 - \sqrt{1 - \dfrac{M_u}{0.319 b_w d^2 f_c'}}\right)}{f_y}$$

$$= 0.545 \text{ in}^2$$

The primary reinforcement area required is given by ACI Sec. 11.9.3.5 as

$$A_{sc} = A_f + A_n = 0.545 \text{ in}^2 + 0.889 \text{ in}^2$$

$$= 1.434 \text{ in}^2$$

Three no. 7 bars are provided, giving an area of

$$A_s' = 1.80 \text{ in}^2$$

$$> 1.434 \text{ in}^2 \quad \text{[satisfactory]}$$

Also, from ACI Sec. 11.9.3.5, the area of primary reinforcement must not be less than

$$\frac{2A_{vf}}{3} + A_n = \frac{(2)\left(1.59 \text{ in}^2\right)}{3} + 0.889 \text{ in}^2$$

$$= 1.95 \text{ in}^2$$

$$> A_s' \quad \text{[unsatisfactory]}$$

The area of closed stirrups required is given by ACI Sec. 11.9.4 as

$$A_h = \frac{A_{sc} - A_n}{2} = \frac{1.95 \text{ in}^2 - 0.889 \text{ in}^2}{2}$$

$$= 0.53 \text{ in}^2$$

Three no. 3 closed stirrups are provided, giving an area of

$$A_h = 0.66 \text{ in}^2$$

$$> 0.53 \text{ in}^2 \quad \text{[satisfactory]}$$

Torsion

The terminology used in torsion design is illustrated in Figs. 1.14 and 1.15.

In accordance with ACI Sec. 11.6.1, for a statically determinate member the equilibrium torsional effects may be neglected, and closed stirrups and longitudinal torsional reinforcement are not required when the factored torque does not exceed

$$T_u = \phi \sqrt{f_c'} \left(\frac{A_{cp}^2}{p_{cp}}\right)$$

When this value is exceeded, reinforcement must be provided to resist the full torsion. When both shear and torsion reinforcements are required, the sum of the individual areas must be provided.

ACI Sec. 11.6.3.6 specifies the required area of one leg of a closed stirrup as

$$\frac{A_t}{s} = \frac{T_u}{2\phi A_o f_{yt}} \quad \begin{bmatrix} \text{for compression} \\ \text{diagonals at } 45° \end{bmatrix}$$

$$= \frac{T_u}{1.7\phi A_{oh} f_{yt}} \quad \text{[ACI 11-21]}$$

The corresponding area of longitudinal reinforcement required is specified in ACI Sec. 11.6.3.7 as

$$A_l = \frac{A_t p_h f_{yt}}{f_y s} \quad \text{[ACI 11-22]}$$

The minimum area of longitudinal reinforcement required is specified in ACI Sec. 11.6.5.3 as

$$A_l = \frac{5 A_{cp} \sqrt{f_c'}}{f_y} - \frac{A_t p_h f_{yt}}{f_y s} \quad \text{[ACI 11-24]}$$

$$\frac{A_t}{s} \geq \frac{25 b_w}{f_{yt}}$$

Figure 1.14 *Torsion in Rectangular Section*

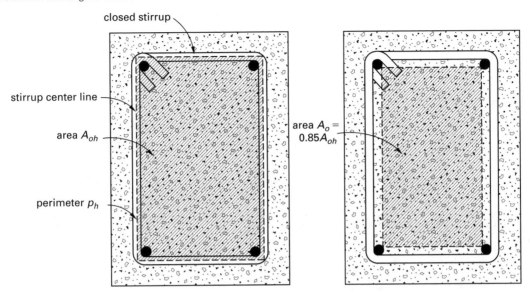

Figure 1.15 *Torsion in Flanged Section*

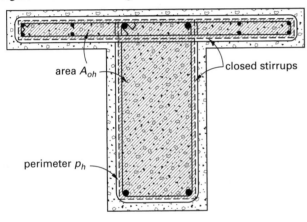

The minimum diameter, specified in ACI Sec. 11.6.6.2, is

$$d_b = 0.042s$$

$$\geq \text{no. 3 bar}$$

The minimum combined area of stirrups for combined shear and torsion is given by ACI Sec. 11.6.5.2 as

$$\frac{A_v + 2A_t}{s} = \frac{0.75\sqrt{f_c'}\,b_w}{f_{yt}} \qquad \textit{[ACI 11-23]}$$

$$\geq 50b_w/f_{yt}$$

The maximum spacing of closed stirrups is given by ACI Sec. 11.6.6.1 as

$$s = \frac{p_h}{8}$$

$$\leq 12 \text{ in}$$

In accordance with ACI Sec. 11.6.2.2, when redistribution of internal forces occurs in an indeterminate structure upon cracking, a member may be designed for the factored torsion causing cracking, which is given by

$$T_u = 4\phi\sqrt{f_c'}\left(\frac{A_{cp}^2}{p_{cp}}\right)$$

Example 1.13

A simply supported reinforced concrete beam with an overall depth of 19 in, an effective depth of 16 in, and a width of 12 in is reinforced with grade 60 bars and has a concrete compressive strength of 3000 lbf/in². Determine the combined shear and torsion reinforcement required when

(a) the factored shear force is 5 kips and the factored torsion is 2 ft-kips

(b) the factored shear force is 15 kips and the factored torsion is 4 ft-kips

Solution

(a) The area enclosed by the outside perimeter of the beam is

$$A_{cp} = (19 \text{ in})(12 \text{ in})$$

$$= 228 \text{ in}^2$$

The length of the outside perimeter of the beam is

$$p_{cp} = (2)(19 \text{ in} + 12 \text{ in})$$

$$= 62 \text{ in}$$

Torsional reinforcement is not required in accordance with ACI Sec. 11.6.1 when the factored torque does not exceed

$$T_u = \phi\sqrt{f_c'}\left(\frac{A_{cp}^2}{p_{cp}}\right)$$

$$= \frac{0.75\sqrt{3000\ \dfrac{\text{lbf}}{\text{in}^2}}\ (228\ \text{in}^2)^2}{(62\ \text{in})\left(12\ \dfrac{\text{in}}{\text{ft}}\right)\left(1000\ \dfrac{\text{lbf}}{\text{kip}}\right)}$$

$$= 2.87\ \text{ft-kips}$$

$$> 2.0 \quad \text{[Closed stirrups are not required.]}$$

The shear strength provided by the concrete was determined in Ex. 1.9 as

$$\phi V_c = 15.8\ \text{kips}$$

$$> 2V_u \quad \text{[Shear stirrups are not required.]}$$

(b) Because $\phi V_c/2 < V_u < \phi V_c$, minimum shear reinforcement is required, and because $T_u > 2.87$ ft-kips, closed stirrups are necessary. Using no. 3 stirrups with 1.5 in cover, the area enclosed by the centerline of the stirrups is

$$A_{oh} = (19\ \text{in} - 3\ \text{in} - 0.375\ \text{in})$$
$$\times (12\ \text{in} - 3\ \text{in} - 0.375\ \text{in})$$
$$= 134.77\ \text{in}^2$$

From ACI Eq. (11-21), the required area of one arm of a closed stirrup is given by

$$\frac{A_t}{s} = \frac{T_u}{1.7\phi A_{oh} f_{yt}}$$

$$= \frac{(4\ \text{ft-kips})\left(12\ \dfrac{\text{in}}{\text{ft}}\right)}{(1.7)(0.75)\left(134.77\ \text{in}^2\right)\left(60\ \dfrac{\text{kips}}{\text{in}^2}\right)}$$

$$= 0.00466\ \text{in}^2/\text{in/arm}$$

$$= 0.056\ \text{in}^2/\text{ft/arm}$$

From ACI Eq. (11-23), the governing minimum combined shear and torsion reinforcement area is given by

$$\frac{A_v + 2A_t}{s} = \frac{50 b_w}{f_{yt}} = \frac{(50)(12\ \text{in})\left(12\ \dfrac{\text{in}}{\text{ft}}\right)}{60,000\ \dfrac{\text{lbf}}{\text{in}^2}}$$

$$= 0.12\ \text{in}^2/\text{ft} \quad \text{[governs]}$$

The perimeter of the centerline of the closed stirrups is

$$p_h = (2)\big(19\ \text{in} + 12\ \text{in} - (2)(3.375\ \text{in})\big)$$

$$= 48.50\ \text{in}$$

The governing maximum permissible spacing of the closed stirrups is specified in ACI Sec. 11.6.6.1 as

$$s_{\max} = \frac{p_h}{8} = \frac{48.50\ \text{in}}{8}$$

$$= 6\ \text{in}$$

Closed stirrups consisting of two arms of no. 3 bars at 6 in spacing provides an area of

$$\frac{A}{s} = \frac{0.44\ \text{in}^2}{\text{ft}}$$

$$> 0.12 \quad \text{[satisfactory]}$$

The required area of the longitudinal reinforcement is given by ACI Eq. (11-22) as

$$A_l = \left(\frac{A_t}{s}\right) p_h \left(\frac{f_{yt}}{f_y}\right)$$

$$= \frac{\left(0.056\ \text{in}^2/\text{ft/arm}\right)(48.50\ \text{in})\left(\dfrac{60\ \dfrac{\text{kips}}{\text{in}^2}}{60\ \dfrac{\text{kips}}{\text{in}^2}}\right)}{12\ \dfrac{\text{in}}{\text{ft}}}$$

$$= 0.23\ \text{in}^2$$

Because the required value of $A_t/s = 0.00466$ in^2/in/arm is less than $25 b_w/f_{yt} = 0.0050$, the minimum permissible area of longitudinal reinforcement is given by ACI Eq. (11-24) as

$$A_{l(\min)} = \frac{5 A_{cp}\sqrt{f_c'}}{f_y} - \left(\frac{25 b_w}{f_{yt}}\right) p_h \left(\frac{f_{yt}}{f_y}\right)$$

$$= \frac{(5)\left(228\ \text{in}^2\right)\sqrt{3000\ \dfrac{\text{lbf}}{\text{in}^2}}}{60,000\ \dfrac{\text{lbf}}{\text{in}^2}}$$

$$- \left(0.0050\ \dfrac{\text{in}^2}{\text{in}}\right)(48.50\ \text{in})\left(\dfrac{60\ \dfrac{\text{kips}}{\text{in}^2}}{60\ \dfrac{\text{kips}}{\text{in}^2}}\right)$$

$$= 1.041\ \text{in}^2 - 0.242\ \text{in}^2$$

$$= 0.799\ \text{in}^2 \quad \text{[governs]}$$

Using eight no. 3 bars around the perimeter of the closed stirrups gives a longitudinal steel area of

$$A_l = 0.88\ \text{in}^2$$

$$> 0.799\ \text{in}^2 \quad \text{[satisfactory]}$$

6. CONCRETE COLUMNS

Nomenclature

A_{ch} area of core of spirally reinforced compression member measured to outside diameter of spiral   in²

A_g gross area of concrete section   in²

A_{st} total area of longitudinal reinforcement   in²

C_m a factor relating actual moment diagram to an equivalent uniform moment diagram   –

E_c modulus of elasticity of concrete

h overall thickness of member   in

I_g moment of inertia of the gross concrete section   –

k effective length factor for compression members   –

l_c length of a compression member in a frame, measured from center-to-center of the joints in the frame   ft or in

l_u unsupported length of compression member   ft or in

M_c factored moment to be used for design of compression member   ft-kips

M_1 smaller factored end moment on a compression member, positive if member is bent in single curvature, negative if bent in double curvature   ft-kips

M_{1ns} factored end moment on a compression member at the end at which M_1 acts, due to loads that cause no appreciable sidesway, calculated using a first-order elastic frame analysis   ft-kips

M_{1s} factored end moment on compression members at the end at which M_1 acts, due to loads that cause appreciable sidesway, calculated using a first-order elastic frame analysis   ft-kips

M_2 larger factored end moment on compression member, always positive   ft-kips

M_{2ns} factored end moment on compression member at the end at which M_2 acts, due to loads that cause no appreciable sidesway, calculated using a first-order elastic frame analysis   ft-kips

M_{2s} factored end moment on compression member at the end at which M_2 acts, due to loads that cause appreciable sidesway, calculated using a first-order elastic frame analysis   ft-kips

P_c critical load, $\pi^2 EI/(kl_u)^2$ kips

P_n nominal axial load strength at given eccentricity   kips

P_o nominal axial load strength at zero eccentricity   kips

P_u factored axial load at given eccentricity $\leq \phi P_n$ kips

Q stability index for a story, $\sum P_u \Delta_o / V_u l_c$ –

r radius of gyration of cross section of a compression member   in

V_{us} factored horizontal shear in a story   kips

Symbols

γ the ratio of the distance between centroids of the longitudinal reinforcement to the overall diameter of the column   –

δ_{ns} moment magnification factor for frames braced against sidesway to reflect effects of member curvature between ends of compression members   –

δ_s moment magnification factor for frames not braced against sidesway to reflect lateral drift resulting from lateral and gravity loads   –

Δ_o relative lateral deflection between the top and bottom of a story due to V_{us}, computed using a first-order elastic frame analysis   in

ρ ratio of A_{st} to A_g –

ρ_s ratio of volume of spiral reinforcement to total volume of core (out-to-out of spirals) of a spirally reinforced compression member   –

Ψ stiffness ratio at the end of a column   –

Reinforcement Requirements

ACI Sec. 10.9 limits the area of longitudinal reinforcement to not more than 8% and not less than 1% of the gross area of the section. For columns with rectangular or circular ties, a minimum of four bars is required. For columns with spirals, a minimum of six longitudinal bars is required. The minimum ratio of volume of spiral reinforcement to volume of core is given by ACI Sec. 10.9.3 as

$$\rho_s = 0.45 f_c' \left(\frac{\frac{A_g}{A_{ch}} - 1}{f_{yt}} \right) \quad \text{[ACI 10-5]}$$

In accordance with ACI Sec. 7.10, the clear spacing between spirals shall not exceed 3 in nor be less than 1 in, and the minimum diameter of the spiral is ³/₈ in.

For rectangular columns, the minimum tie size specified by ACI Sec. 7.10.5 is no. 3 for longitudinal bars of no. 10 or smaller and no. 4 for longitudinal bars larger than no. 10. The maximum vertical spacing of ties is given by

$$s_{\max} \leq 16 \times \text{longitudinal bar diameters}$$
$$\leq 48 \times \text{tie bar diameters}$$
$$\leq \text{least dimension of the column}$$

Ties shall be provided, as shown in Fig. 1.16, to support every corner and alternate bar, and no bar shall be more than 6 in clear from a supported bar.

Figure 1.16 Column Ties

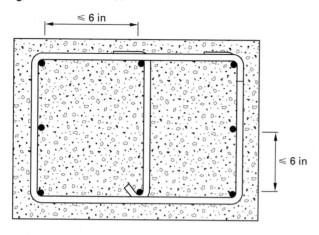

Example 1.14

A 24 in diameter spirally reinforced column with a 1.5 in cover to the spiral is reinforced with grade 60 bars and has a concrete compressive strength of 4500 lbf/in². Determine the required diameter and pitch of the spiral.

Solution

From ACI Sec. 7.10.4.2, the minimum permissible diameter of spiral reinforcement is

$$d_b = 0.375 \text{ in}$$

$$A_b = \text{area of spiral bar}$$

$$= 0.11 \text{ in}^2$$

$$A_g = \text{gross area of column}$$

$$= \frac{\pi (24 \text{ in})^2}{4}$$

$$= 452 \text{ in}^2$$

$$A_{ch} = \text{area of core}$$

$$= \frac{\pi (21 \text{ in})^2}{4}$$

$$= 346 \text{ in}^2$$

$$d_s = \text{mean diameter of spiral}$$

$$= 21 \text{ in} - 0.375 \text{ in}$$

$$= 20.625 \text{ in}$$

From ACI Eq. (10-5), the minimum allowable spiral reinforcement ratio is

$$\rho_s = \frac{0.45 f'_c \left(\dfrac{A_g}{A_{ch}} - 1 \right)}{f_{yt}}$$

$$= \frac{(0.45) \left(4.5 \ \dfrac{\text{kips}}{\text{in}^2} \right) \left(\dfrac{452 \text{ in}^2}{346 \text{ in}^2} - 1 \right)}{60 \ \dfrac{\text{kips}}{\text{in}^2}}$$

$$= 0.0103$$

$$= \frac{\text{volume of spiral bar per turn}}{\text{volume of concrete core per turn}}$$

$$= \frac{A_b \pi d_s}{A_{ch} s}$$

$$= \frac{(0.11 \text{ in}^2) \pi (20.625 \text{ in})}{(346 \text{ in}^2)(s)}$$

$$s = 2 \text{ in} \quad [\text{satisfactory}]$$

The calculated pitch lies between the maximum of 3 in and the minimum of 1 in specified in ACI Sec. 7.10.4.3.

Effective Length and Slenderness Ratio

The effective column length may be determined from the alignment charts given in ACI Sec. R10.12 and shown in Fig. 1.17.

To use the alignment charts, the stiffness ratios at each end of the column must be calculated, and this is given by ACI Fig. R10.12.1 as

$$\Psi = \frac{\sum \dfrac{E_c I_c}{l_c}}{\sum \dfrac{E_b I_b}{l_b}}$$

The subscript c refers to the columns meeting at a joint, and the subscript b refers to the beams meeting at a joint.

For a nonsway frame, the effective length factor k may be conservatively taken as unity, as indicated in ACI Sec. 10.12.1. The slenderness ratio is defined in ACI Sec. 10.11.5 as kl_u/r, and the radius of gyration is given by ACI Sec. 10.11.2 as

$$r = 0.25 \times \text{diameter of circular column}$$

$$= 0.30 \times \text{dimension of a rectangular column in the direction stability is being considered}$$

A nonsway column is defined in ACI Sec. 10.11.4.1 as one in which the secondary moments due to P-delta effects do not exceed 5% of the primary moments due to

Figure 1.17 *Alignment Charts for k[1]*

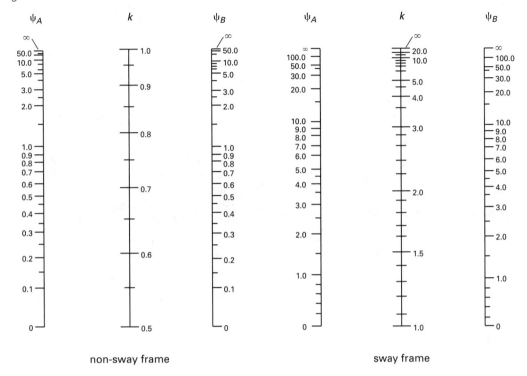

non-sway frame

sway frame

Adapted from American Concrete Institute, *Building Code Requirements for Structural Concrete (ACI Publication 318-05)*, Fig. R10.12.1.

lateral loads. ACI Sec 10.11.4.2 specifies a story within a structure as nonsway, provided that the stability index Q does not exceed 0.05 where the stability index is given by

$$Q = \frac{\sum P_u \Delta_o}{V_{us} l_c} \qquad \text{[ACI 10-6]}$$

Here, V_{us} is the story shear and $\sum P_u$ is the total vertical load on a story.

Example 1.15

Determine the slenderness ratio of columns 12 and 34 of the sway frame shown in the following illustration. The columns are 18 in square and have an unsupported height of 9 ft. All members of the frame have identical EI values.

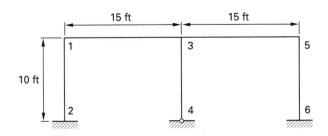

Solution

For column 12, the stiffness ratio of the fixed base is given by AISC[7] Commentary Sec. C2 as

$$\Psi_2 = 1.0$$

At joint 1, the relative stiffness value of the beam is

$$\sum \frac{E_b I_b}{l_b} = \frac{1}{15}$$

At joint 1, the relative stiffness value of the column is

$$\sum \frac{E_c I_c}{l_c} = \frac{1}{10}$$

The stiffness ratio at joint 1 is

$$\Psi_1 = \frac{\sum \dfrac{E_c I_c}{l_c}}{\sum \dfrac{E_b I_b}{l_b}} = \frac{15}{10}$$

$$= 1.5$$

From the alignment chart, for a sway frame, the effective length factor for column 12 is

$$k_{12} = 1.38$$

The radius of gyration of the column, in accordance with ACI Sec. 10.11.2, is

$$r = 0.30h$$
$$= (0.30)(18 \text{ in})$$
$$= 5.4 \text{ in}$$

The slenderness ratio of column 12 is

$$\frac{kl_u}{r} = \frac{(1.38)(9 \text{ ft})\left(12 \, \frac{\text{in}}{\text{ft}}\right)}{(5.4 \text{ in})}$$
$$= 27.6$$

For column 34, the stiffness ratio of the pinned base is given by AISC[7] Commentary Sec. C2 as

$$\Psi_4 = 10$$

At joint 3, the sum of the relative stiffness values of the two beams is

$$\sum \frac{E_b I_b}{l_b} = \frac{1}{15} + \frac{1}{15}$$
$$= \frac{2}{15}$$

At joint 3, the relative stiffness value of the column is

$$\sum \frac{E_c I_c}{l_c} = \frac{1}{10}$$

The stiffness ratio at joint 3 is

$$\Psi_3 = \frac{\sum \dfrac{E_c I_c}{l_c}}{\sum \dfrac{E_b I_b}{l_b}}$$
$$= \frac{15}{20}$$
$$= 0.75$$

From the alignment chart, for a sway frame, the effective length factor for column 34 is

$$k_{34} = 1.85$$

The slenderness ratio of column 34 is

$$\frac{kl_u}{r} = \frac{(1.85)(9 \text{ ft})\left(12 \, \frac{\text{in}}{\text{ft}}\right)}{5.4 \text{ in}}$$
$$= 37.0$$

Short Column with Axial Load

In accordance with ACI Sec. 10.13.2, a column in a sway frame is classified as a short column, and slenderness

effects may be ignored, when the slenderness ratio is less than

$$\frac{kl_u}{r} = 22$$

For a nonsway frame, a column is classified as a short column, in accordance with ACI Sec. 10.12.2, when the slenderness ratio is not more than

$$\frac{kl_u}{r} = 34 - \frac{12M_1}{M_2} \qquad \text{[ACI 10-7]}$$

The term $(34 - 12M_1/M_2) \leq 40$ and M_1/M_2 is positive if the column is bent in single curvature. When the slenderness ratio exceeds these values, a column is classified as a long column and secondary moments resulting from P-delta effects must be considered.

For a short column with spiral reinforcement, ACI Sec. 10.3.6.1 gives the design axial load capacity as

$$\phi P_n = 0.85\phi \left(\begin{array}{c} 0.85 f'_c (A_g - A_{st}) \\ + A_{st} f_y \end{array} \right) \quad [\phi = 0.70]$$

$$\text{[ACI 10-1]}$$

For a short column with lateral tie reinforcement, ACI Sec. 10.3.6.2 gives the design axial load capacity as

$$\phi P_n = 0.80\phi \left(\begin{array}{c} 0.85 f'_c (A_g - A_{st}) \\ + A_{st} f_y \end{array} \right) \quad [\phi = 0.65]$$

$$\text{[ACI 10-2]}$$

Example 1.16

An 18 in square column is reinforced with 12 no. 9 grade 60 bars and has a concrete compressive strength of 4000 lbf/in². The column, which is braced against sidesway, has an unsupported height of 9 ft and supports axial load only without end moments. Determine the lateral ties required and the design axial load capacity.

Solution

The minimum tie size specified by ACI Sec. 7.10.5.1 is no. 3 for longitudinal bars of size no. 9. ACI Sec. 7.10.5.2 specifies a tie spacing not greater than

$$h = 18 \text{ in}$$
$$48d_t = (48)(0.375 \text{ in})$$
$$= 18 \text{ in}$$
$$16d_b = (16)(1.128 \text{ in})$$
$$= 18 \text{ in}$$

From ACI Sec. 10.12.1, the effective length factor for a column braced against sidesway is

$$k = 1.0$$

The radius of gyration, in accordance with ACI Sec. 10.11.2, is

$$r = 0.3h$$
$$= (0.3)(18 \text{ in})$$
$$= 5.4 \text{ in}$$

The slenderness ratio is specified in ACI Sec. 10.11.5 as

$$\frac{kl_u}{r} = \frac{(1.0)(9 \text{ ft})\left(12 \frac{\text{in}}{\text{ft}}\right)}{5.4 \text{ in}}$$
$$= 20.0$$

In accordance with ACI Eq. (10-7), the column may be classified as a short column provided that

$$\frac{kl_u}{r} \leq 34 - \frac{12M_1}{M_2}$$
$$20 < 34$$

Where,

$$M_1 = 0$$
$$M_2 = M_{\min}$$

Hence, the column is a short column and the design axial load capacity is given by ACI Eq. (10-2) as

$$\phi P_n = 0.80\phi(0.85f_c'(A_g - A_{st}) + A_{st}f_y)$$

$$= (0.80)(0.65) \begin{pmatrix} (0.85)\left(4 \frac{\text{kips}}{\text{in}^2}\right) \\ \times \left(324 \text{ in}^2 - 12 \text{ in}^2\right) \\ + \left(12 \text{ in}^2\right)\left(60 \frac{\text{kips}}{\text{in}^2}\right) \end{pmatrix}$$

$$= 926 \text{ kips}$$

Short Column with End Moments

The axial load carrying capacity of a column decreases as end moments are applied to the column. Design of the column may then be obtained by means of a computer program based on the 2002 ACI code.[8] The 2002 code is identical to the 2005 code for column design. Alternatively, approximate design values may be obtained from the interaction diagrams of Figs. A.1 through A.6 in the appendix.

Example 1.17

A 24 in diameter tied column with a 1.5 in cover to the $^3/_8$ in diameter ties is reinforced with 14 no. 9 grade 60 bars and has a concrete compressive strength of 4000 lbf/in^2. The column, which is braced against sidesway, has an unsupported height of 9 ft and is bent in single curvature with factored end moments of $M_1 = M_2 = 400$ ft-kips. Determine the maximum axial load that the column can carry.

Solution

It is clear that the column may be classified as a short column and slenderness effects do not have to be considered. The ratio of the distance between centroids of the longitudinal reinforcement to the overall diameter of the column is

$$\gamma = \frac{24 \text{ in} - (2)(1.5 \text{ in}) - (2)(0.375 \text{ in}) - 1.125 \text{ in}}{24 \text{ in}}$$
$$= 0.80$$

The reinforcement ratio is

$$\rho = \frac{A_{st}}{A_g}$$
$$= \frac{14 \text{ in}^2}{\pi(12 \text{ in})^2}$$
$$= 0.031$$

$$\frac{M_u}{A_g h} = \frac{(400 \text{ ft-kips})\left(12 \frac{\text{in}}{\text{ft}}\right)}{\pi(12 \text{ in})^2(24 \text{ in})}$$
$$= 0.44 \text{ kip/in}^2$$

From Fig. A.2, with $\gamma = 0.75$, the design axial stress is

$$\frac{P_u}{A_g} = 1.2 \text{ kips/in}^2$$

From Fig. A.3, with $\gamma = 0.90$, the design axial stress is

$$\frac{P_u}{A_g} = 1.7 \text{ kips/in}^2$$

By interpolation, for $\gamma = 0.80$, the design axial stress is

$$\frac{P_u}{A_g} = 1.2 \frac{\text{kips}}{\text{in}^2} + \left(0.5 \frac{\text{kip}}{\text{in}^2}\right)\left(\frac{5}{15}\right)$$
$$= 1.37 \text{ kips/in}^2$$

The maximum factored axial load that the column can carry is

$$P_u = 1.37A_g$$
$$= 1.37\pi(12 \text{ in})^2$$
$$= 620 \text{ kips}$$

Long Column Without Sway

Provided the slenderness ratio of a nonsway column does not exceed 100, ACI Sec. 10.11.5 permits secondary effects in the column to be compensated for by multiplying the end moments, determined in a linear-elastic analysis, by a magnification factor given by ACI Sec. 10.12.3 as

$$\delta_{ns} = \frac{C_m}{1 - \dfrac{P_u}{0.75 P_c}}$$

$$\geq 1.0 \qquad \text{[ACI 10-9]}$$

P_u is the factored axial load, and the Euler critical load is given by

$$P_c = \frac{\pi^2 EI}{(kl_u)^2} \qquad \text{[ACI 10-10]}$$

The flexural rigidity, in accordance with ACI Sec. R10.12.3, may be taken as

$$EI = 0.25 E_c I_g \qquad \text{[ACI F]}$$

The factor C_m corrects for a non-uniform bending moment on a column and is defined by ACI Sec. 10.12.3.1 as

$$C_m = 0.6 + \frac{0.4 M_1}{M_2} \qquad \text{[ACI 10-13]}$$

$$\geq 0.4$$

$$= 1.0 \text{ for columns with transverse}$$
$$\text{loads between supports}$$

The column is now designed for the magnified moment given by ACI Sec. 10.12.3 as

$$M_c = \delta_{ns} M_2 \qquad \text{[ACI 10-8]}$$

Example 1.18

A 24 in diameter tied column with a 1.5 in cover to the $^3/_8$ in diameter ties is reinforced with 14 no. 9 grade 60 bars and has a concrete compressive strength of 4000 lbf/in^2. The column, which is braced against sidesway, has an unsupported height of 12 ft and is bent in single curvature with factored end moments of $M_1 = M_2 = 400$ ft-kips. Determine whether the column can carry a factored axial load of 700 kips.

Solution

$k = 1.0$ and $r = 6.0$ in. The slenderness ratio is

$$\frac{kl_u}{r} = \frac{(1.0)(12 \text{ ft})\left(12 \dfrac{\text{in}}{\text{ft}}\right)}{6.0 \text{ in}}$$

$$= 24$$

In accordance with ACI Eq. (10-7), the column is classified as a long column if

$$\frac{kl_u}{r} > 34 - \frac{12 M_1}{M_2}$$

$$24 > 34 - \frac{(12)(400 \text{ ft-kips})}{400 \text{ ft-kips}}$$

$$> 22$$

Hence, the column is a long column and secondary effects must be considered.

The modulus of elasticity is given by ACI Sec. 8.5.1 as

$$E_c = 57{,}000 \sqrt{f'_c}$$

$$= 57{,}000 \sqrt{4000 \ \frac{\text{lbf}}{\text{in}^2}}$$

$$= 3.61 \times 10^6 \text{ lbf/in}^2$$

The gross moment of inertia is given by

$$I_g = \frac{\pi (12 \text{ in})^4}{4}$$

$$= 16{,}288 \text{ in}^4$$

The effective flexural rigidity is given by ACI Sec. R10.12.3 as

$$EI = 0.25 E_c I_g$$

$$= (0.25)\left(3.61 \times 10^3 \ \frac{\text{kips}}{\text{in}^2}\right)(16{,}288 \text{ in}^4)$$

$$= 14.68 \times 10^6 \text{ in}^2\text{-kips}$$

The critical load is given by ACI Eq. (10-10) as

$$P_c = \frac{\pi^2 EI}{(kl_u)^2}$$

$$= \frac{\pi^2 \left(14.68 \times 10^6 \text{ in}^2\text{-kips}\right)}{\left((1.0)(12 \text{ ft})\left(12 \dfrac{\text{in}}{\text{ft}}\right)\right)^2}$$

$$= 6987 \text{ kips}$$

The moment correction factor is defined by ACI Eq. (10-13) as

$$C_m = 0.6 + \frac{0.4 M_1}{M_2}$$

$$= \frac{0.6 + (0.4)(400 \text{ ft-kips})}{400 \text{ ft-kips}}$$

$$= 1.0$$

The moment magnification factor is given by ACI Eq. (10-9) as

$$\delta_{ns} = \frac{C_m}{1 - \dfrac{P_u}{0.75 P_c}}$$

$$= \frac{1}{1 - \dfrac{700}{(0.75)(6987)}}$$

$$= 1.15$$

The magnified end moment is given by ACI Eq. (10-8) as

$$M_c = \delta_{ns} M_2$$

$$= (1.15)(400 \text{ ft-kips})$$

$$= 461 \text{ ft-kips}$$

From Ex. 1.17, $\gamma = 0.80$, $\rho = 0.031$, and

$$\frac{M_u}{A_g h} = \frac{(461 \text{ ft-kips}) \left(12 \, \dfrac{\text{in}}{\text{ft}} \right)}{\pi (12 \text{ in})^2 (24 \text{ in})}$$

$$= 0.51 \text{ kip/in}^2$$

From Fig. A.2, with $\gamma = 0.75$, the design axial stress is

$$\frac{P_u}{A_g} = 0.4 \text{ kips/in}^2$$

From Fig. A.3, with $\gamma = 0.90$, the design axial stress is

$$\frac{P_u}{A_g} = 1.2 \text{ kips/in}^2$$

By interpolation, for $\gamma = 0.80$, the allowable design axial stress is

$$\frac{P_u}{A_g} = 0.4 \, \frac{\text{kips}}{\text{in}^2} + \left(0.8 \, \frac{\text{kip}}{\text{in}^2} \right) \left(\frac{5}{15} \right)$$

$$= 0.67 \text{ kips/in}^2$$

The maximum factored axial load that the column can carry is

$$P_u = 0.67 A_g$$

$$= (0.67) \pi (12 \text{ in})^2$$

$$= 303 \text{ kips}$$

$$< 700 \text{ kips}$$

Hence, the column cannot carry the axial load of 700 kips.

Long Column with Sway

For slenderness ratios not exceeding 100, the magnification factor for end moments produced by the loads that cause sway is given by ACI Sec. 10.13.4.3 as

$$\delta_s = \frac{1}{1 - \dfrac{\sum P_u}{0.75 \sum P_c}} \qquad \textit{[ACI 10-18]}$$

$$\geq 1.0$$

$$\leq 2.5 \quad \text{[from ACI Sec. 10.13.6]}$$

Here, the summations extend over all the columns in a story. The sway moments are multiplied by the magnification factor, and the nonsway moments are added, in accordance with ACI Sec. 10.13.3, to give the final design end moments in the column of

$$M_1 = M_{1ns} + \delta_s M_{1s} \qquad \textit{[ACI 10-15]}$$

$$M_2 = M_{2ns} + \delta_s M_{2s} \qquad \textit{[ACI 10-16]}$$

In accordance with ACI Sec. 10.13.5, when $l_u/r > 35/\sqrt{P_u/f_c' A_g}$, the maximum moments do not occur at the ends of the column and the column will be designed for the magnified moment

$$M_c = \delta_{ns}(M_{2ns} + \delta_s M_{2s})$$

Example 1.19

A 24 in diameter, tied column with a 1.5 in cover to the $^3/_8$ in diameter ties is reinforced with 14 no. 9 grade 60 bars and has a concrete compressive strength of 4000 lbf/in². The column is not braced and has an unsupported height of 12 ft, an effective length factor of 1.3, and factored end moments due to sway and nonsway moments of $M_{2s} = 300$ ft-kips and $M_{2ns} = 50$ ft-kips. In the story where the column is located, the sum of the column critical loads is $\sum P_c = 29{,}600$ kips and the sum of the factored column loads is $\sum P_u = 2700$ kips. Determine whether the column can carry a factored axial load of 900 kips.

Solution

$r = 6$ in and the slenderness ratio is

$$\frac{k l_u}{r} = \frac{(1.3)(12 \text{ ft}) \left(12 \, \dfrac{\text{in}}{\text{ft}} \right)}{6 \text{ in}}$$

$$= 31.2$$

In accordance with ACI Sec. 10.13.2, the column is classified as a long column if

$$\frac{k l_u}{r} \geq 22$$

$$31.2 > 22 \quad \text{[The column is long.]}$$

Applying ACI Sec. 10.13.5,

$$\frac{35}{\sqrt{\dfrac{P_u}{f_c' A_g}}} = \frac{35}{\sqrt{\dfrac{900 \text{ kips}}{\left(4 \dfrac{\text{kips}}{\text{in}^2}\right)(452 \text{ in}^2)}}}$$

$$= 49.6$$

$$\frac{l_u}{r} = \frac{(12 \text{ ft})\left(12 \dfrac{\text{in}}{\text{ft}}\right)}{6 \text{ in}}$$

$$= 24$$

$$< 49.6 \quad [\text{Eq. (10-16) applies.}]$$

The moment magnification factor for the sway moments is given by ACI Eq. (10-18) as

$$\delta_s = \frac{1}{1 - \dfrac{\sum P_u}{0.75 \sum P_c}}$$

$$= \frac{1}{1 - \dfrac{2700 \text{ kips}}{(0.75)(29{,}600 \text{ kips})}}$$

$$= 1.14$$

The magnified end moment is given by ACI Eq. (10-16) as

$$M_2 = M_{2ns} + \delta_s M_{2s}$$

$$= (50 \text{ ft-kips}) + (1.14)(300 \text{ ft-kips})$$

$$= 392 \text{ ft-kips}$$

From Ex. 1.17, $\gamma = 0.80$, $\rho_g = 0.031$, and

$$\frac{M_2}{A_g h} = \frac{(392 \text{ ft-kips})\left(12 \dfrac{\text{in}}{\text{ft}}\right)}{(452 \text{ in}^2)(24 \text{ in})}$$

$$= 0.43 \text{ kip/in}^2$$

From Fig. A.2, with $\gamma = 0.75$, the design axial stress is

$$\frac{P_u}{A_g} = 1.3 \text{ kips/in}^2$$

From Fig. A.3, with $\gamma = 0.90$, the design axial stress is

$$\frac{P_u}{A_g} = 1.7 \text{ kips/in}^2$$

By interpolation, for $\gamma = 0.80$, the design axial stress is

$$\frac{P_u}{A_g} = 1.3 \frac{\text{kips}}{\text{in}^2} + \left(0.4 \frac{\text{kip}}{\text{in}^2}\right)\left(\frac{5}{15}\right)$$

$$= 1.43 \text{ kips/in}^2$$

The maximum factored axial load that the column can carry is

$$P_u = 1.43 A_g$$

$$= (1.43)(452 \text{ in}^2)$$

$$= 648 \text{ kips}$$

$$< 900 \text{ kips} \quad [\text{The column is unsatisfactory.}]$$

7. DEVELOPMENT AND SPLICE LENGTH OF REINFORCEMENT

Nomenclature

A_b	area of an individual bar	in^2
A_{tr}	total cross-sectional area of all transverse reinforcement that is within the spacing s and that crosses the potential plane of splitting through the reinforcement being developed	in^2
c	spacing or cover dimension	in
c_b	distance from center of bar to nearest concrete surface	in
c_b	$\leq$ one-half the center-to-center spacing of bars	in
d_b	nominal diameter of bar	in
f_{ct}	average splitting tensile compressive strength of lightweight aggregate concrete	kips/in^2
h	overall thickness of member	
K_{tr}	transverse reinforcement index, $A_{tr} f_{yt}/1500 sn$	in
l_a	additional embedment length at support or at point of inflection	in
l_d	development length, $l_{db} \times$ applicable modification factors	in
l_{db}	basic development length	in
l_{dh}	development length of standard hook in tension, measured from critical section to outside end of hook [straight embedment length between critical section and start of hook (point of tangency) plus radius of bend and one bar diameter], $l_{hb} \times$ applicable modification factors	in
l_{hb}	basic development length of standard hook in tension	in
l_s	lap splice length	in
n	number of bars or wires being spliced or developed along the plane of splitting	—
s	maximum center-to-center spacing of transverse reinforcement within l_d	in

Symbols

β_b	ratio of area of reinforcement cutoff to total area of tension reinforcement at section	

Concrete

λ lightweight aggregate concrete factor
     = 1.3 for lightweight aggregate concrete
     = $(6.7) \left(f'_c \right)^{0.5} / f_{ct} \geq 1.0$ when f_{ct} is specified
     = 1.0 for normal weight concrete

Ψ_e coating factor
     = 1.5 for epoxy-coated bars with cover $< 3d_b$ or clear spacing$< 6d_b$
     = 1.2 for all other epoxy-coated bars
     = 1.0 for uncoated bars

Ψ_s reinforcement size factor
     = 0.8 for no. 6 and smaller bars
     = 1.0 for no. 7 and larger bars

Ψ_t reinforcement location factor
     = 1.3 for horizontal bar with more than 12 in of concrete below
     = 1.0 for all other bars

Development Length of Straight Bars in Tension

The development length for tension reinforcement is given by ACI Sec. 12.2.3 as

$$\frac{l_d}{d_b} = \frac{0.075 f_y \Psi_t \Psi_e \Psi_s \lambda}{\left(\dfrac{\sqrt{f'_c}\,(c_b + K_{tr})}{d_b} \right)} \quad \text{[ACI 12-1]}$$

$$l_d \geq 12 \text{ in}$$

$$\frac{c_b + K_{tr}}{d_b} \leq 2.5$$

From ACI Sec. 12.2.4,

$$\Psi_t \Psi_e \leq 1.7$$

From ACI Sec. 12.1.2,

$$\sqrt{f'_c} \leq 100 \text{ lbf/in}^2$$

The derivation of the transverse reinforcement index is illustrated in Fig. 1.18.

ACI Sec. 12.2.2 also provides the following simplified, conservative values.

- Using a minimum clear spacing of $2d_b$, a minimum clear cover to flexural reinforcement of d_b, and in the absence of stirrups,

$$\frac{l_d}{d_b} = \frac{0.04 f_y \Psi_t \Psi_e \lambda}{\sqrt{f'_c}} \quad \text{[for } d_b \text{ no. 6 or smaller]}$$

$$\frac{l_d}{d_b} = \frac{0.05 f_y \Psi_t \Psi_e \lambda}{\sqrt{f'_c}} \quad \text{[for } d_b \text{ no. 7 or larger]}$$

- Using a minimum clear spacing of d_b, a minimum clear cover to flexural reinforcement of d_b, and minimum stirrups specified in ACI Eq. (11-13),

$$\frac{l_d}{d_b} = \frac{0.04 f_y \Psi_t \Psi_e \lambda}{\sqrt{f'_c}} \quad \text{[for } d_b \text{ no. 6 or smaller]}$$

$$\frac{l_d}{d_b} = \frac{0.05 f_y \Psi_t \Psi_e \lambda}{\sqrt{f'_c}} \quad \text{[for } d_b \text{ no. 7 or larger]}$$

- For all other cases,

$$\frac{l_d}{d_b} = \frac{0.06 f_y \Psi_t \Psi_e \lambda}{\sqrt{f'_c}} \quad \text{[for } d_b \text{ no. 6 or smaller]}$$

$$\frac{l_d}{d_b} = \frac{0.075 f_y \Psi_t \Psi_e \lambda}{\sqrt{f'_c}} \quad \text{[for } d_b \text{ no. 7 or larger]}$$

For bundled bars, ACI Sec. 12.4.1 specifies that the development length shall be that for an individual bar increased by 20% for a three-bar bundle and 33% for a four-bar bundle. No increase is required for a two-bar

Figure 1.18 *Derivation of K_{tr}*

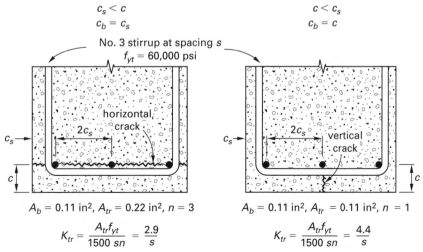

$c_s < c$
$c_b = c_s$

$c < c_s$
$c_b = c$

No. 3 stirrup at spacing s
$f_{yt} = 60{,}000$ psi

horizontal crack

$2c_s$

vertical crack

$2c_s$

$A_b = 0.11$ in^2, $A_{tr} = 0.22$ in^2, $n = 3$

$K_{tr} = \dfrac{A_{tr} f_{yt}}{1500\, sn} = \dfrac{2.9}{s}$

$A_b = 0.11$ in^2, $A_{tr} = 0.11$ in^2, $n = 1$

$K_{tr} = \dfrac{A_{tr} f_{yt}}{1500\, sn} = \dfrac{4.4}{s}$

bundle. The equivalent diameter d_b of a bundle is specified in ACI Sec. 12.4.2 as that of a bar with an area equal to that of the bundle.

ACI Sec. 12.2.5 specifies that when excess reinforcement is provided in a member, the development length may be reduced by multiplying by the factor $A_{s(\text{required})}/A_{s(\text{provided})}$. This reduction factor may not be applied when development is required for the yield strength of the reinforcement, as is the case for shrinkage and temperature reinforcement specified in ACI Sec. 7.12, integrity reinforcement specified in ACI Secs. 7.13 and 13.3.8, positive moment reinforcement specified in ACI Sec. 12.11, and tension lap splices specified in ACI Sec. 12.15.

Example 1.20

The simply supported reinforced concrete beam of normal weight concrete shown in the illustration is reinforced with grade 60 bars and has a concrete compressive strength of 3000 lbf/in². The maximum moment in the beam occurs at a point 2 ft from the end of the bars, and 10% more flexural reinforcement is provided than is required. Determine whether the development length available is satisfactory.

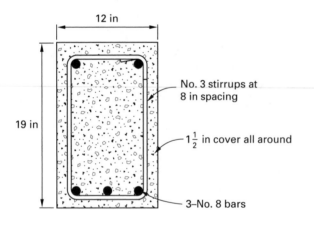

Solution

From ACI Sec. 12.2.4,

$$\Psi_t = \Psi_e = \Psi_s = \lambda = 1.0$$

The excess reinforcement factor is

$$E_{xr} = \frac{A_{s(\text{required})}}{A_{s(\text{provided})}} = \frac{100 \text{ in}^2}{110 \text{ in}^2}$$
$$= 0.91$$

The cover dimension to the flexural reinforcement is

$$c = 1.5 \text{ in} + 0.375 \text{ in} + \frac{1.0 \text{ in}}{2}$$
$$= 2.375 \text{ in}$$

The spacing dimension to the flexural reinforcement is

$$c_s = \frac{12.0 \text{ in} - (2)(1.5 \text{ in}) - (2)(0.375 \text{ in}) - 1.0 \text{ in}}{4}$$
$$= 1.81 \text{ in} \quad [\text{horizontal cracking governs}]$$
$$c_b = 1.81 \text{ in}$$

The area of transverse reinforcement crossing the horizontal crack is

$$A_{tr} = (2)(0.11 \text{ in}^2)$$
$$= 0.22 \text{ in}^2$$

The number of bars being developed along the cracking plane is

$$n = 3$$

The transverse reinforcement index is

$$K_{tr} = \frac{A_{tr} f_{yt}}{1500 s n}$$
$$= \frac{(0.22 \text{ in}^2)\left(60{,}000 \dfrac{\text{lbf}}{\text{in}^2}\right)}{(1500)(8 \text{ in})(3)}$$
$$= 0.367 \text{ in}$$
$$\frac{c_b + K_{tr}}{d_b} = \frac{1.81 \text{ in} + 0.367 \text{ in}}{1.0 \text{ in}}$$
$$= 2.18$$
$$< 2.5 \quad [\text{satisfactory}]$$

From ACI Eq. (12-1),

$$\frac{l_d}{d_b} = \frac{0.075 f_y \Psi_t \Psi_e \Psi_s \lambda E_{xr}}{\sqrt{f_c'}\,\dfrac{(c_b + K_{tr})}{d_b}}$$
$$= \frac{(0.075)\left(60{,}000 \dfrac{\text{lbf}}{\text{in}^2}\right)(1)(1)(1)(1)(0.91)}{\sqrt{3000 \dfrac{\text{lbf}}{\text{in}^2}}\,(2.18)}$$
$$= 34.3$$

The development length required is

$$l_d = 34.3 d_b$$
$$= (34.3)(1.0 \text{ in})$$
$$= 34.3 \text{ in}$$
$$> 24 \text{ in}$$

The development length available is unsatisfactory.

Development Length of Straight Bars in Compression

The basic development length for bars in compression is specified in ACI Sec. 12.3.2 as

$$l_{dc} = \frac{0.02 d_b f_y}{\sqrt{f_c'}}$$
$$\geq 0.0003 d_b f_y \quad [\text{governs when } f_c' \geq 4444 \text{ lbf/in}^2]$$

The actual development length l_d is obtained by multiplying l_{dc} by

- the excess reinforcement factor specified in ACI Sec. 12.3.3(a)

$$E_{xr} = \frac{A_{s(\text{required})}}{A_{s(\text{provided})}}$$

- the confinement factor specified in ACI Sec. 12.3.3(b)

$$C_f = 0.75$$

The confinement factor is applicable when spiral reinforcement is provided with a minimum diameter of 0.25 in and a maximum pitch of 4 in. Alternatively, no. 4 ties or larger may be provided at a maximum spacing of 4 in. The minimum allowable value for l_d is 8 in.

The requirements for bundled bars in compression are identical with those applicable to bundled bars in tension.

Example 1.21

A spirally reinforced column with a 0.25 in diameter spiral at a pitch of 4 in is reinforced with no. 9 grade 60 bars in bundles of three and has a concrete compressive strength of 4500 lbf/in². The compression reinforcement provided is 15% in excess of that required. Determine the required development length of an individual bar.

Solution

From ACI Sec. 12.4, the development length of a bar in a three-bar bundle is increased 20%.

From ACI Sec. 12.3.3, the excess reinforcement factor is

$$E_{xr} = \frac{100}{115}$$
$$= 0.87$$

And the confinement factor is

$$C_f = 0.75$$

For a concrete strength in excess of 4444 lbf/in², ACI Secs. 12.3.2 and 12.3.3 give the required development length as

$$l_d = 1.2 E_{xr} C_f l_{dc}$$
$$= (1.2)(0.87)(0.75)(0.0003)d_b f_y$$
$$= (1.2)(0.87)(0.75)(0.0003)(1.13 \text{ in}) \left(60{,}000 \frac{\text{lbf}}{\text{in}^2} \right)$$
$$= 15.9 \text{ in}$$

Development of Hooked Bars in Tension

The basic development length for hooked bars in tension is specified in ACI Sec. 12.5.2 as

$$l_{hb} = \frac{0.02 d_b f_y}{\sqrt{f'_c}}$$

The actual development length l_{dh} is obtained by multiplying l_{hb} by the factors given in ACI Sec. 12.5.3.

- the cover factor for bars not exceeding size no. 11, with side covers not less than 2.5 in and end covers not less than 2.0 in

$$C_b = 0.7 \quad \text{[for conditions specified above]}$$
$$= 1.0 \quad \text{[for all other conditions]}$$

- the tie factor for bars not exceeding size no. 11, with ties provided perpendicular to the bar being developed over the full development length at a spacing not exceeding $3d_b$; or for a 90° hook with ties provided, parallel to the bar being developed, along the tail extension of the hook plus bend at a spacing not exceeding $3d_b$.

$$T_f = 0.8$$

- the excess reinforcement factor

$$E_{xr} = \frac{A_{s(\text{required})}}{A_{s(\text{provided})}}$$

- the lightweight aggregate concrete factor specified in ACI Sec. 12.5.2

$$\lambda = 1.3$$

- the epoxy-coated reinforcement factor specified in ACI Sec. 12.5.2

$$\psi_e = 1.2$$

For hooked bars at the discontinuous end of a member with cover of less than 2.5 in, ties shall be provided, over the full development length, at a spacing not exceeding $3d_b$. The modification factor for this condition is 1.0.

In accordance with ACI Sec. 12.5.1, the minimum allowable value for the actual development length l_{dh} is

$$l_{dh} \geq 8d_b$$
$$\geq 6 \text{ in}$$

Example 1.22

Assume that the reinforced concrete beam for Ex. 1.20 is reinforced with hooked bars. Determine whether the development length provided for the hooked bars is satisfactory.

Solution

Because the bars terminate at the discontinuous end of the beam and the cover provided is less than 2.5 in, the modification factor is 1.0 and ties must be provided over the full development length at a maximum spacing of

$$s = 3d_b$$
$$= (3)(1 \text{ in})$$
$$= 3 \text{ in}$$

The excess reinforcement factor derived in Ex. 1.20 is

$$E_{xr} = 0.91$$

All other modification factors are equal to unity.

The required development length is given by ACI Secs. 12.5.1 and 12.5.2 as

$$l_{dh} = E_{xr}l_{hb}$$
$$= \frac{(0.91)(1200)d_b}{\sqrt{f_c'}}$$
$$= \frac{(0.91)(1200)(1 \text{ in})}{\sqrt{3000}}$$
$$= 20 \text{ in}$$
$$< 24 \text{ in}$$

The development length provided is satisfactory.

Curtailment of Reinforcement

ACI Secs. 12.10.3 and 12.10.4 specify that reinforcement shall extend a distance beyond the theoretical cutoff point not less than the effective depth of the member or 12 times the bar diameter and shall extend beyond the point at which it is fully stressed not less than the development length. This is illustrated in Fig. 1.19, where the four no. 9 bars are assumed to be fully stressed at the center of the simply supported beam. These requirements are not necessary at supports of simple spans or at the free ends of cantilevers.

Figure 1.19 *Curtailment of Reinforcement*

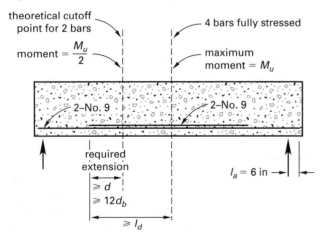

In addition to the above requirements, ACI Sec. 12.10.5 requires that one of the following conditions be satisfied at the physical cutoff point.

- The factored shear force at the cutoff point does not exceed two-thirds of the shear capacity ϕV_n.

- Additional stirrups with a minimum area of $60 b_w s / f_{yt}$ are provided along the terminated bar for a distance of $0.75d$ at a maximum spacing of $d/8\beta_b$.

- For no. 11 bars and smaller, the continuing reinforcement provides twice the flexural capacity required, and the factored shear force at the cutoff point does not exceed three-fourths of the shear capacity ϕV_n.

Example 1.23

The simply supported reinforced concrete beam, of normal weight concrete, shown in Fig. 1.19 is reinforced with four no. 9 grade 60 bars and has a concrete strength of 3000 lbf/in². The maximum moment in the beam occurs at midspan, and the flexural reinforcement is fully stressed. The bending moment reduces to 50% of its maximum value at a point 3 ft from midspan. The effective depth is 16 in, the beam width is 12 in, and the development length of the no. 9 bars may be taken as $55d_b$. The factored shear force is less than two-thirds the shear capacity of the section everywhere along the span. Determine the distance from midspan at which two of the no. 9 bars may be terminated.

Solution

From ACI Secs. 12.10.3 and 12.10.4, the physical cutoff point may be located a minimum distance from midspan given by the largest of

- $12d_b + 36 \text{ in} = (12)(1.128 \text{ in}) + 36 \text{ in}$
  $$= 49.5 \text{ in}$$

- $d + 36 \text{ in} = 16 \text{ in} + 36 \text{ in}$
  $$= 52 \text{ in}$$

- $l_d = 55d_b$
  $$= (55)(1.128 \text{ in})$$
  $$= 62 \text{ in} \quad [\text{governs}]$$

Development of Positive Moment Reinforcement

To allow for variations in the applied loads, ACI Sec. 12.11.1 requires a minimum of one-third of the positive reinforcement in a simply supported beam or one-fourth of the positive reinforcement in a continuous beam to extend not less than 6 in into the support. This is shown in Fig. 1.20.

Figure 1.20 *Positive Moment Reinforcement*

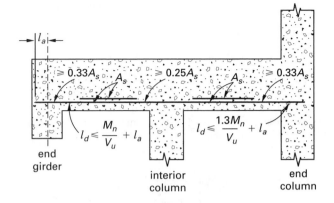

The nominal flexural strength at the support is given by ACI Sec. 10.2 as

$$M_n = A_s f_y d \left(1 - \frac{0.59 A_s f_y}{b_w d f'_c} \right)$$

$$= \left(2 \text{ in}^2 \right) \left(60 \ \frac{\text{kips}}{\text{in}^2} \right) (16 \text{ in})$$

$$\times \left(1 - \frac{(0.59) \left(2 \text{ in}^2 \right) \left(60 \ \frac{\text{kips}}{\text{in}^2} \right)}{(12 \text{ in}) (16 \text{ in}) \left(3 \ \frac{\text{kips}}{\text{in}^2} \right)} \right)$$

$$= 1684 \text{ in-kips}$$

To ensure that allowable bond stresses are not exceeded, ACI Sec. 12.11.3 requires a bar diameter to be chosen such that its development length, in the case of a beam framing into a girder, is given by

$$l_d \le \frac{M_n}{V_u} + l_a \qquad \text{[ACI 12-5]}$$

M_n is the nominal strength assuming all reinforcement is stressed to the specified yield stress f_y, V_u is the factored shear force at the section, and l_a is the embedment length beyond the center of the support.

For the case of a beam framing into a column, ACI Sec. R12.11.3 requires a bar diameter to be chosen such that its development length is given by

$$l_d \le 1.3 \frac{M_n}{V_u} + l_a$$

At a point of inflection, PI, ACI Sec. 12.11.3 requires a bar diameter to be chosen such that its development length is given by

$$l_d \le \frac{M_n}{V_u} + (\text{maximum of } d \text{ or } 12 d_b)$$

Alternatively, at a simple support, ACI Sec. 12.11.3 specifies that the reinforcement may terminate in a standard hook beyond the support.

Example 1.24

Assume that the reinforced concrete beam for Ex. 1.23, which is shown in Fig. 1.19, has a factored end reaction of 30 kips and that the beam frames into concrete girders at each end. Determine whether the two no. 9 bars at the support satisfy the requirements for local bond.

Solution

The development length for the no. 9 bars was given in Ex. 1.23 as

$$l_d = 55 d_b$$

$$= 62 \text{ in}$$

For a beam framing into a concrete girder, the appropriate factors given by ACI Sec. 12.11.3 are

$$\frac{M_n}{V_u} + l_a = \frac{1684 \text{ in-kips}}{30 \text{ kips}} + 6 \text{ in}$$

$$= 62.1 \text{ in}$$

$$> l_d$$

Local bond requirements are satisfied.

Development of Negative Moment Reinforcement

To allow for variation in the applied loads, ACI Sec. 12.12 requires a minimum of one-third of the negative reinforcement at a support to extend past the point of inflection not less than the effective depth of the beam, 12 times the bar diameter, or one-sixteenth of the clear span. This is shown in Fig. 1.21.

Figure 1.21 *Negative Moment Reinforcement*

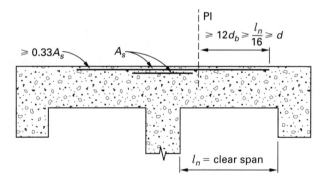

Example 1.25

The reinforced concrete continuous beam shown in the following illustration, with an effective depth of 16 in, is reinforced with grade 60 bars and has a concrete compressive strength of 3000 lbf/in². Determine the minimum length x at which the bars indicated may terminate.

Illustration for Ex. 1.25

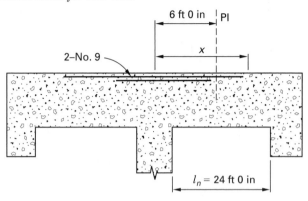

Solution

The cutoff point of the bars may be located a minimum distance beyond the point of inflection given by ACI Sec. 12.12 as the greater of

$$12d_b = (12)(1.128 \text{ in})$$
$$= 13.5 \text{ in}$$
$$d = 16 \text{ in}$$
$$\frac{l}{16} = \frac{(24 \text{ ft})(12 \text{ in})}{16}$$
$$= 18 \text{ in} \quad [\text{governs}]$$

The minimum allowable length for x is given by

$$x = 6 \text{ ft} + 1.5 \text{ ft}$$
$$= 7.5 \text{ ft}$$

Splices of Bars in Tension

Lap splices for bars in tension, in accordance with ACI Sec. 12.14, may not be used either for bars larger than no. 11 or for bundled bars. In flexural members, the transverse spacing between lap splices shall not exceed one-fifth of the lap length, or 6 in. Within a bundle, individual bar splices shall not overlap and the lap length for each bar must be increased by 20% for a three-bar bundle and 33% for a four-bar bundle.

The length of a lap splice may not be less than 12 in nor less than the values given by ACI Sec. 12.15, which are

$$\text{class A splice length} = 1.0l_d$$
$$\text{class B splice length} = 1.3l_d$$

A class A splice may be used only when the reinforcement area provided is at least twice that required and when, in addition, not more than one-half of the total reinforcement is spliced within the lap length. Otherwise, as shown in Table 1.3, a class B tension lap splice is required.

Table 1.3 *Tension Lap Splices*

$\dfrac{A_s \text{ provided}}{A_s \text{ required}}$	maximum percentage of A_s spliced within lap length	
	50	100
2 or more	class A	class B
less than 2	class B	class B

In determining the relevant development length, all applicable modifiers are used with the exception of that for excess reinforcement. The development length, l_d, is determined by using the values for clear spacing, c_s, indicated in Figs. 1.22 and 1.23.

Figure 1.22 *Value of c_s for Lap Splices*

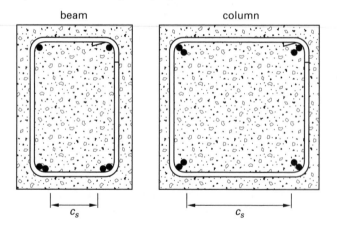

Figure 1.23 *Value of c_s in Slabs and Walls*

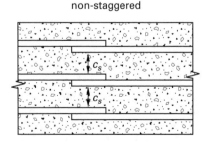
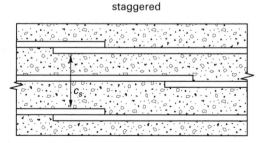

Example 1.26

A reinforced concrete beam is reinforced with two no. 8 grade 60 bars and has a concrete compressive strength of 3000 lbf/in². Both bars are lap spliced at the same location, and the development length of the no. 8 bars may be taken as $50d_b$. Determine the required splice length.

Solution

The development length is given as

$$l_d = 50d_b = (50)(1 \text{ in})$$

$$= 50 \text{ in}$$

Because both bars are spliced at the same location, a class B splice is required and the splice length is given by

$$l_s = 1.3l_d = (1.3)(50 \text{ in})$$

$$= 65 \text{ in}$$

Splices of Bars in Compression

In accordance with ACI Sec. 12.16.1, the length of a lap splice for bars in compression shall not be less than 12 in and is given by

$$l_s = 0.0005f_yd_b \quad [f_y \leq 60{,}000 \text{ lbf/in}^2]$$

$$l_s = (0.0009f_y - 24)d_b \quad [f_y > 60{,}000 \text{ lbf/in}^2]$$

An increase in the lap length of 33% is required when the concrete strength is less than 3000 lbf/in². When bars of different sizes are lap spliced, ACI Sec. 12.16.2 specifies that the lap length shall be the larger of the splice length of the smaller bar or the development length of the larger bar.

In accordance with ACI Sec. 12.17.2, lap lengths for columns may be reduced by 17% when ties are provided with an effective area of $0.0015hs$ and may be reduced by 25% in a spirally reinforced column. When tensile stress exceeding $0.5f_y$ occurs in a column, a class B tension lap splice shall be used. A class A tension lap splice is adequate provided that the tensile stress does not exceed $0.5f_y$, not more than one-half of the bars are spliced at the same location, and alternate splices are staggered by l_d; otherwise, a class B tension lap splice is required.

Example 1.27

The reinforced concrete column shown in the illustration is reinforced with grade 60 bars that are fully stressed and has a concrete compressive strength of 4000 lbf/in². The column is subjected to compressive stress only. Determine the required lap splice for the no. 8 and no. 9 bars.

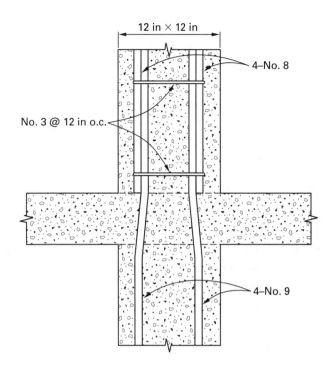

Solution

The development length of a no. 9 bar is given by ACI Secs. 12.3.2 and 12.3.3 as

$$l_d = \frac{0.02d_bf_y}{\sqrt{f_c'}}$$

$$= \frac{(0.02)(1.128 \text{ in})\left(60{,}000 \, \dfrac{\text{lbf}}{\text{in}^2}\right)}{\sqrt{4000 \, \dfrac{\text{lbf}}{\text{in}^2}}}$$

$$= 21.4 \text{ in}$$

To qualify for the 17% reduction in lap splice length of the no. 8 bars, ACI Sec. 12.17.2.4 requires the two arms of the no. 3 ties to have a minimum effective area of

$$\frac{A_t}{s} = 0.0015h$$

$$= (0.0015)(12 \text{ in})\left(12 \, \frac{\text{in}}{\text{ft}}\right)$$

$$= 0.22 \text{ in}^2/\text{ft}$$

Two arms of the no. 3 ties provide an area of

$$\frac{A_t}{s} = 0.22 \text{ in}^2/\text{ft}$$

Hence, the 17% reduction applies.

The compression lap splice length of the no. 8 bars is given by ACI Secs. 12.16.1 and 12.17.2.4 as

$$l_s = 0.0005 f_y d_b (0.83)$$
$$= (0.0005)\left(60{,}000 \; \frac{\text{lbf}}{\text{in}^2}\right)(1.0 \text{ in})(0.83)$$
$$= 25 \text{ in} \quad [\text{governs}]$$
$$> 21.4 \text{ in}$$

The required lap splice length is 25 in.

8. TWO-WAY SLAB SYSTEMS

Nomenclature

b_o	perimeter of critical section	in
c_1	size of rectangular or equivalent rectangular column, capital, or bracket measured in the direction of the span for which moments are being determined	in
c_2	size of rectangular or equivalent rectangular column, capital, or bracket measured transverse to the direction of the span for which moments are being determined	in
C	cross-sectional constant to define torsional properties $\left(\sum(1 - 0.63 x/y)x^3 y\right)/3$ The constant C for T- or L-sections shall be permitted to be evaluated by dividing the section into separate rectangular parts and summing the values for C for each part.	–
E_{cb}	modulus of elasticity of beam concrete	lbf/in^2
E_{cs}	modulus of elasticity of slab concrete	lbf/in^2
I_b	moment of inertia about centroidal axis of gross section of beam	in^4
I_s	moment of inertia about centroidal axis of gross section of slab $= h^3/12$ times width of slab defined in symbols α and β_t	in^4
l_n	length of clear span in direction that moments are being determined, measured face-to-face of supports $= l_1 - c_1 > 0.65 l_1$	ft or in
l_1	length of span in direction that moments are being determined, measured center-to-center of supports	ft or in
l_2	length of span transverse to l_1, measured center-to-center of supports	ft or in
M_o	total factored static moment	ft-kips
q_u	factored load per unit area	kips/ft^2
x	shorter overall dimension of rectangular part of cross section	in
y	longer overall dimension of rectangular part of cross section	in

Symbols

α_f	ratio of flexural stiffness of beam section to flexural stiffness of a slab width bounded laterally by center lines of adjacent panels (if any) on each side of the beam $= E_{cb}I_b/E_{cs}I_s$	
α_s	constant used to compute V_c	
α_{f1}	α_f in direction of l_1	
α_{f2}	α_f in direction of l_2	
β_c	ratio of long side to short side of column	
β_t	ratio of torsional stiffness of edge beam section to flexural stiffness of a slab width equal to span length of beam, center-to-center of supports $= E_{cb}C/2E_{cs}I_s$	

Direct Design Method

ACI Sec. 13.6.1 permits the direct design method to be used provided the following limitations apply.

- A minimum of three continuous spans is required in each direction.

- Panels are rectangular with an aspect ratio not exceeding 2.

- Successive span lengths do not differ by more than one-third of the longer span.

- Columns are not offset by more than 10% of the span.

- Loading consists of uniformly distributed gravity loads with the service live load not exceeding twice the service dead load.

- For beam supported slabs, the ratio of the beam stiffnesses in two perpendicular directions $\alpha_{f1}l_2^2/\alpha_{f2}l_1^2$ is between 0.2 and 5.0 where the moments of inertia of the equivalent beam and slab are based on the sections shown in Fig. 1.24.

Figure 1.24 *Equivalent Beam and Slab Dimensions*

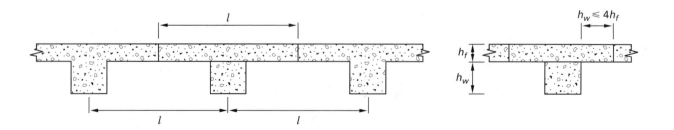

Concrete

The slab is divided into design strips, as shown in Fig. 1.25, with a column strip extending the lesser of $0.25l_1$ or $0.25l_2$ on each side of a column centerline. A middle strip consists of the remainder of the slab between column strips.

Figure 1.25 *Details of Design Strips*

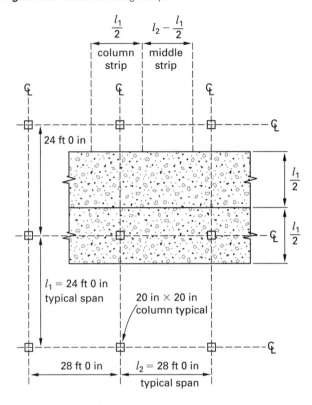

Example 1.28

A 9 in thick flat plate floor has plan dimensions between column centers of 24 ft and 28 ft as shown in Fig. 1.25. Calculate the widths of the column strip and the middle strip for the direction parallel to the 28 ft side of the panel.

Solution

From ACI Sec. 13.2, the width of the column strip is

$$w_c = 0.5l_{min} = (0.5)(24 \text{ ft})$$
$$= 12 \text{ ft}$$

The width of the middle strip in the direction of the longer span is

$$w_m = l_{min} - 0.5l_{min} = 24 \text{ ft} - 12 \text{ ft}$$
$$= 12 \text{ ft}$$

Design for Flexure

The total factored static moment on a panel is calculated by ACI Sec. 13.6.2 as

$$M_o = \frac{q_u l_2 l_n^2}{8} \qquad \text{[ACI 13-4]}$$

The total factored moment is now distributed in accordance with ACI Sec. 13.6.3 into positive and negative moments across the full width of the panel, depending on the support conditions, as shown in Fig. 1.26.

These distributed moments are now further subdivided between the column and middle strips as specified in ACI Secs. 13.6.4 through 13.6.7. That portion of the moment not attributed to a column strip is assigned to the corresponding half middle strips. The negative moment at an interior support is distributed to a column strip as shown in Table 1.4.

Table 1.4 *Percentage Distribution of Interior Negative Moment to Column Strip*

l_2/l_1	0.5	1	2
$\alpha_{f1}l_2/l_1 = 0$	75	75	75
$\alpha_{f1}l_2/l_1 \geq 1.0$	90	75	45

The negative moment at an exterior support is distributed to a column strip as shown in Table 1.5.

Table 1.5 *Percentage Distribution of Exterior Negative Moment to Column Strip*

l_2/l_1		0.5	1	2
$(\alpha_{f1}l_2/l_1) = 0$	$\beta_t = 0$	100	100	100
	$\beta_t \geq 2.5$	75	75	75
$(\alpha_{f1}l_2/l_1) \geq 1.0$	$\beta_t = 0$	100	100	100
	$\beta_t \geq 2.5$	90	75	45

The positive moment at midspan is distributed to a column strip as shown in Table 1.6.

Table 1.6 *Percentage Distribution of Positive Moment to Column Strip*

l_2/l_1	0.5	1	2
$(\alpha_{f1}l_2/l_1) = 0$	60	60	60
$(\alpha_{f1}l_2/l_1) \geq 1.0$	90	75	45

Beams at panel edges shall be assigned the percentage of column strip moment as shown in Table 1.7.

Table 1.7 *Percentage Distribution of Column Strip Moments to Edge Beam*

$\alpha_{f1}l_2/l_1$	0	≥ 1.0
% assigned to beam	0	85

Figure 1.26 *M_o Distribution Factors*

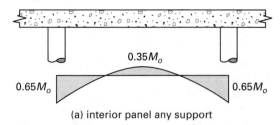

(a) interior panel any support

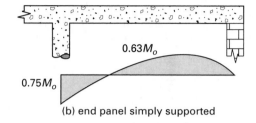

(b) end panel simply supported

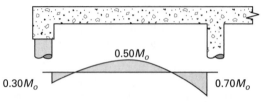

(c) flat plate, end panel with edge beam

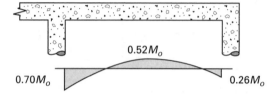

(d) flat plate, end panel without edge beam

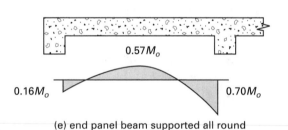

(e) end panel beam supported all round

(f) end panel fully restrained end

Example 1.29

Assume that the flat plate floor for Ex. 1.28, which is shown in Fig. 1.25, supports a total factored distributed load of 200 lbf/ft². The column size is 20 in by 20 in. Determine the factored moments in the column strip and middle strip in the direction of the longer span.

Solution

From ACI Sec. 13.6.2.5, the clear span in the direction of the longer span is

$$l_n = l_1 - c_1$$

$$= 28 \text{ ft} - \frac{20 \text{ in}}{12 \dfrac{\text{in}}{\text{ft}}}$$

$$= 26.33 \text{ ft}$$

$$> 0.65 l_1 \quad \text{[satisfactory]}$$

From ACI Sec. 13.6.2, the total factored static moment is

$$M_o = \frac{q_u l_2 l_n^2}{8}$$

$$= \frac{\left(0.20 \dfrac{\text{kip}}{\text{ft}^2}\right)(24 \text{ ft})(26.33 \text{ ft})^2}{8}$$

$$= 416 \text{ ft-kips}$$

From ACI Sec. 13.6.3 and Fig. 1.26(a), the total positive moment across the panel is

$$M_m = 0.35 M_o$$

$$= (0.35)(416 \text{ ft-kips})$$

$$= 146 \text{ ft-kips}$$

From ACI Sec. 13.6.4 and Table 1.6, $\alpha_{f1} l_2 / l_1 = 0.0$ and the column strip positive moment at midspan is

$$M_{cm} = 0.60 M_m$$

$$= (0.60)(146 \text{ ft-kips})$$

$$= 87 \text{ ft-kips}$$

From ACI Sec. 13.6.6, the middle strip positive moment at midspan is

$$M_{mm} = M_m - M_{cm}$$
$$= 146 \text{ ft-kips} - 87 \text{ ft-kips}$$
$$= 59 \text{ ft-kips}$$

From ACI Sec. 13.6.3 and Fig. 1.26(a), the total negative moment across the panel is

$$M_c = 0.65 M_o$$
$$= (0.65)(416 \text{ ft-kips})$$
$$= 270 \text{ ft-kips}$$

From ACI Sec. 13.6.4 and Table 1.4, $\alpha_{f1} l_2/l_1 = 0.0$ and the column strip negative moment at an interior support is

$$M_{cc} = 0.75 M_c$$
$$= (0.75)(270 \text{ ft-kips})$$
$$= 203 \text{ ft-kips}$$

From ACI Sec. 13.6.6, the middle strip negative moment at an interior support is

$$M_{mc} = M_c - M_{cc}$$
$$= 270 \text{ ft-kips} - 203 \text{ ft-kips}$$
$$= 67 \text{ ft-kips}$$

Design for Shear

The design for shear at column supports must consider both flexural or one-way shear and punching or two-way shear, as shown in Fig. 1.27. The flexural shear capacity of the panel, in a direction parallel to the side l_2, is given by ACI Sec. 11.3.1.1 as

$$\phi V_c = 2\phi d l_1 \sqrt{f_c'} \qquad \text{[ACI 11-3]}$$

The critical perimeter for punching shear is specified in ACI Sec. 11.12.1.2 as being a distance from the face of the column equal to one-half of the effective depth. The length of the critical perimeter is given by

$$b_o = (2)(c_1 + c_2) + 4d$$

For a corner column located less than $d/2$ from the edge of a panel, the critical perimeter is two sided. For a similarly situated edge column, it is three sided, as shown in Fig. 1.28.

Openings in a panel within ten times the thickness of the panel from the edge of a column reduce the critical perimeter, as shown in Fig. 1.29.

In accordance with ACI Sec. 13.4, openings of any size are permitted in the area common to two intersecting middle strips. In the area common to two intersecting column strips, the maximum width of opening is limited to one-eighth of the column strip width. In the area common to one column strip and one middle strip, not more than one-fourth of the reinforcement in either strip shall be interrupted by openings. In all cases, the area of reinforcement interrupted by openings shall be replaced by an equivalent amount added on the sides of the opening.

Figure 1.27 *Critical Sections for Shear*

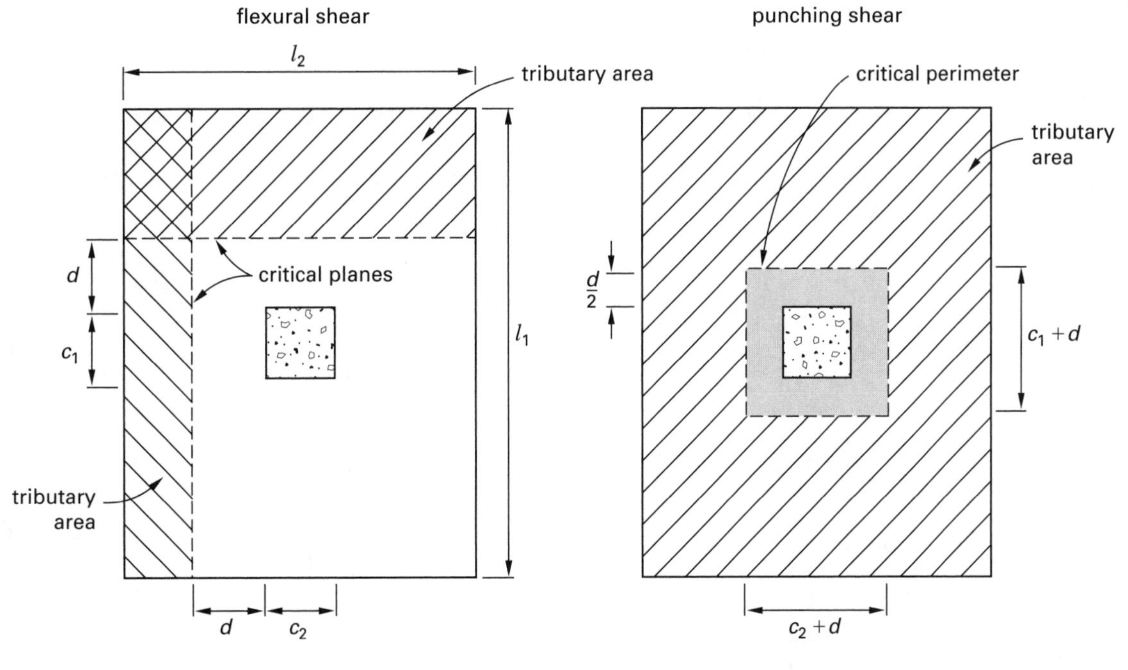

Figure 1.28 *Corner and Edge Columns*

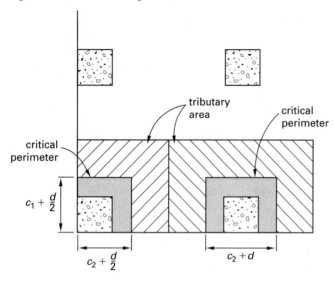

Figure 1.29 *Reduction in Critical Perimeter*

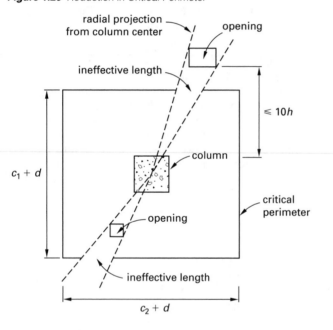

The punching shear capacity of the panel is the smallest of the three values given by ACI Sec. 11.12.2.1, which are

$$\phi V_c = 4\phi d b_o \sqrt{f'_c} \qquad \text{[ACI 11-35]}$$

$$\leq \phi d b_o \left(2 + \frac{4}{\beta}\right) \sqrt{f'_c} \qquad \text{[ACI 11-33]}$$

$$\leq \phi d b_o \left(2 + \frac{\alpha_s d}{b_o}\right) \sqrt{f'_c} \qquad \text{[ACI 11-34]}$$

$$\alpha_s = 40 \text{ for an interior column}$$

$$= 30 \text{ for an edge column}$$

$$= 20 \text{ for a corner column}$$

Example 1.30

Assume that the flat plate floor for Exs. 1.28 and 1.29 has an effective depth of 7.5 in and a concrete compressive strength of 4000 lbf/in^2. Determine whether the shear capacity of the plate is adequate for an interior column.

Solution

The critical section for flexural shear is located a distance from the center of the panel given by

$$x = \frac{l_1}{2} - \left(d + \frac{c_1}{2}\right)$$

$$= \frac{28 \text{ ft}}{2} - \frac{7.5 \text{ in}}{12 \, \frac{\text{in}}{\text{ft}}} - \frac{10 \text{ in}}{12 \, \frac{\text{in}}{\text{ft}}}$$

$$= 12.54 \text{ ft}$$

The factored applied shear at the critical section for flexural shear is

$$V_u = q_u l_2 x$$

$$= \left(0.2 \, \frac{\text{kip}}{\text{ft}^2}\right) (24 \text{ ft}) (12.54 \text{ ft})$$

$$= 60 \text{ kips}$$

The flexural shear capacity at the critical section is given by ACI Eq. (11-3) as

$$\phi V_c = 2\phi d l_2 \sqrt{f'_c}$$

$$= \frac{(2)(0.75)(7.5 \text{ in})(24 \text{ ft}) \times \left(12 \, \frac{\text{in}}{\text{ft}}\right) \sqrt{4000 \, \frac{\text{lbf}}{\text{in}^2}}}{1000 \, \frac{\text{lbf}}{\text{kip}}}$$

$$= 205 \text{ kips}$$

$$> V_u \quad \text{[satisfactory]}$$

The length of one side of the critical perimeter for punching shear is

$$b = c + d$$

$$= 20 \text{ in} + 7.5 \text{ in}$$

$$= 27.5 \text{ in}$$

The factored applied shear at the critical perimeter for punching shear is

$$V_u = q_u \left(l_1 l_2 - b^2\right)$$

$$= \left(0.2 \, \frac{\text{kip}}{\text{ft}^2}\right) \left((24 \text{ ft})(28 \text{ ft}) - \frac{(27.5 \text{ in})^2}{144 \, \frac{\text{in}^2}{\text{ft}^2}}\right)$$

$$= 133 \text{ kips}$$

The length of the critical perimeter for punching shear is

$$b_o = 4b$$
$$= (4)(27.5 \text{ in})$$
$$= 110 \text{ in}$$

The punching shear capacity of the plate is governed by ACI Eq. (11-35) as

$$\phi V_c = 4\phi db_o \sqrt{f_c'}$$
$$= \frac{(4)(0.75)(7.5 \text{ in})(110 \text{ in})\sqrt{4000 \frac{\text{lbf}}{\text{in}^2}}}{1000 \frac{\text{lbf}}{\text{kip}}}$$
$$= 157 \text{ kips}$$
$$> V_u \quad \text{[satisfactory]}$$

References

1. American Concrete Institute. *Building Code Requirements and Commentary for Structural Concrete, (ACI 318-05)*. 2005.

2. American Concrete Institute. *Commentary on Building Code Requirements for Reinforced Concrete, (ACI 318-83)*. 1985.

3. Williams, A. *Design of Reinforced Concrete Structures.*, 3rd ed. 2008.

4. American Concrete Institute. *Design Handbook: Beams, 1-Way Slabs, Brackets, Footings, Pile Caps, 2-Way Slabs and Seismic Design*. 2009.

5. Ghosh, S.K. and Domel, A.W. *Design of Concrete Buildings for Earthquake and Wind Forces*. Portland Cement Association. 1995.

6. Portland Cement Association. *Notes on ACI 318-05: Building Code Requirements for Structural Concrete*. 2005.

7. American Institute of Steel Construction. *Steel Construction Manual*, 13th ed. 2005.

8. Portland Cement Association. *PCA Notes on ACI 318-05: Building Code Requirements for Structural Concrete*. 2005.

PRACTICE PROBLEMS

1. The reinforced concrete beam shown in the illustration is continuous over four spans and integral with columns at the ends. The clear distance between supports is 15 ft, and the beam supports a factored load of 10 kips/ft. What are the design bending moments and shear forces?

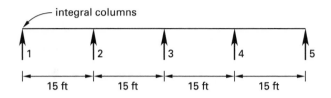

2. A reinforced concrete beam with an effective depth of 20 in and a width of 12 in is reinforced with 3 in^2 of grade 60 reinforcement and has a concrete compressive strength of 3000 lbf/in^2. What is the maximum applied ultimate moment that the beam can support?

3. The reinforced concrete beam for Prob. 2 supports a factored shear force of 9 kips at the critical section. Is shear reinforcement required?

4. A short reinforced concrete column, 20 in square, is reinforced with ten no. 9 grade 60 bars and has a concrete strength of 4000 lbf/in^2. What are the lateral ties required and the design axial load capacity?

5. A simply supported reinforced concrete beam is reinforced with no. 9 grade 60 bars in bundles of three and has a concrete compressive strength of 4000 lbf/in^2. The reinforcement provided is 10% in excess of that required and has a clear cover equal to the equivalent diameter of the bundled bars and a clear spacing of twice the equivalent diameter. What is the required development length of an individual bar?

6. A reinforced concrete flat plate floor without beams has 18 in square columns at 20 ft centers in one direction and 24 ft centers in the other direction, and supports a factored distributed load of 200 lbf/ft^2. What are the column strip and middle strip moments in the span and at the interior support of an end bay, and in the span of an interior bay? Determine the moments in the direction of the shorter span only.

SOLUTIONS

1. The shear at the face of the first interior support is given by ACI Sec. 8.3.3 as

$$V_u = \frac{1.15 w_u l_n}{2}$$

$$= \frac{(1.15)\left(10 \; \frac{\text{kips}}{\text{ft}}\right)(15 \; \text{ft})}{2}$$

$$= 86.3 \; \text{kips}$$

The shear at the face of all other supports is

$$V_u = \frac{w_u l_n}{2} = \frac{\left(10 \; \frac{\text{kips}}{\text{ft}}\right)(15 \; \text{ft})}{2}$$

$$= 75.0 \; \text{kips}$$

The bending moment is given by

$$M_u = \tau w_u l_n^2$$

Values of the bending moment coefficients τ and of the bending moments, in ft-kips units, are given in the table[1].

$\dfrac{\tau}{M_u}$	support 1	span 12	support 2	span 23	support 3
τ	$-\dfrac{1}{16}$	$\dfrac{1}{14}$	$-\dfrac{1}{10}$	$\dfrac{1}{16}$	$-\dfrac{1}{11}$
M_u	-141	161	-225	141	-205

2. The nominal moment of resistance is

$$M_n = A_s f_y d \left(1 - 0.59 \frac{A_s f_y}{b_w d f_c'}\right)$$

$$= \frac{(3 \; \text{in}^2)\left(60 \; \frac{\text{kips}}{\text{in}^2}\right)(20 \; \text{in})}{12 \; \frac{\text{in}}{\text{ft}}} \times \left(1 - (0.59)\frac{(3 \; \text{in}^2)\left(60 \; \frac{\text{kips}}{\text{in}^2}\right)}{(12 \; \text{in})(20 \; \text{in})\left(3 \; \frac{\text{kips}}{\text{in}^2}\right)}\right)$$

$$= 255.8 \; \text{ft-kips}$$

The reinforcement ratio of the beam is

$$\rho = \frac{A_s}{b_w d}$$

$$= \frac{3 \; \text{in}^2}{(12 \; \text{in})(20 \; \text{in})}$$

$$= 0.0125$$

The limiting reinforcement ratio for a tension-controlled section is

$$\rho_t = \frac{0.319 \beta_1 f_c'}{f_y}$$

$$= \frac{(0.319)(0.85)\left(\dfrac{3 \; \text{kips}}{\text{in}^2}\right)}{60 \; \dfrac{\text{kips}}{\text{in}^2}}$$

$$= 0.0136$$

$$> \rho$$

The section is tension controlled and the strength reduction factor is

$$\phi = 0.9$$

The maximum allowable ultimate moment is, then,

$$M_u = 0.9 M_n$$

$$= 230.2 \; \text{ft-kips}$$

3. The design shear capacity of the concrete section is, then,

$$\phi V_c = 2 \phi b_w d \sqrt{f_c'}$$

$$= \frac{(2)(0.75)(12 \; \text{in})(20 \; \text{in})\sqrt{3000 \; \dfrac{\text{lbf}}{\text{in}^2}}}{1000 \; \dfrac{\text{lbf}}{\text{kip}}}$$

$$= 19.7 \; \text{kips}$$

$$> 2 V_u$$

Shear reinforcement is not required.

4. The design axial load capacity is

$$\phi P_n = 0.80 \phi \left(0.85 f_c' (A_g - A_{st}) + A_{st} f_y\right)$$

$$= (0.80)(0.65)$$
$$\times \left(\begin{array}{l} (0.85)\left(4 \; \dfrac{\text{kips}}{\text{in}^2}\right)\left((400 \; \text{in}^2) - (10 \; \text{in}^2)\right) \\ + (10 \; \text{in}^2)\left(60 \; \dfrac{\text{kips}}{\text{in}^2}\right) \end{array}\right)$$

$$= 1002 \; \text{kips}$$

The minimum allowable tie size is

$$d_t = \text{no. 3 bar}$$

The maximum tie spacing shall not be greater than

$$h = 20 \; \text{in}$$

$$48 d_t = (48)(0.375 \; \text{in})$$

$$= 18 \; \text{in}$$

$$16 d_b = (16)(1.128 \; \text{in})$$

$$= 18 \; \text{in} \quad [\text{governs}]$$

5. The development length of a bar in a three-bar bundle is increased 20%.

The excess reinforcement factor is

$$E_{xr} = \frac{100}{110}$$

$$= 0.91$$

The required development length is given by

$$l_d = 1.2E_{xr}d_b\left(\frac{0.05f_y\Psi_t\Psi_e\lambda}{\sqrt{f_c'}}\right)$$

$$= \frac{(1.2)(0.91)(1.13 \text{ in})(0.05)\left(60{,}000 \ \frac{\text{lbf}}{\text{in}^2}\right)}{\sqrt{4000 \ \frac{\text{lbf}}{\text{in}^2}}}$$

$$= 59 \text{ in}$$

6. The clear span is

$$l_n = l_1 - c_1 = 20 \text{ ft} - 1.5 \text{ ft}$$

$$= 18.5 \text{ ft}$$

$$> 0.65l_1 \quad \text{[satisfactory]}$$

The total factored static moment is

$$M_o = \frac{q_u l_2 l_n^2}{8}$$

$$= \frac{\left(0.2 \ \frac{\text{kip}}{\text{ft}^2}\right)(24 \text{ ft})(18.5 \text{ ft})^2}{8}$$

$$= 205 \text{ ft-kips}$$

The relevant coefficients are

$$\alpha_{f1} = \alpha_{f2} = \beta_t = 0$$

$$\frac{\alpha_{f1}l_2}{l_1} = 0$$

The total static moment is distributed as shown in the table.

strip	coefficient/ moment	end span	interior support	interior span
full width	distribution coeff.	0.52	0.70	0.35
	moment	107	144	72
column strip	distribution coeff.	0.60	0.75	0.60
	moment	64	108	43
middle strip	distribution coeff.	0.40	0.25	0.40
	moment	43	36	29

2 Foundations and Retaining Structures

1. Strip Footing 2-1
2. Isolated Column with Square Footing 2-5
3. Isolated Column with Rectangular Footing . 2-10
4. Combined Footing 2-12
5. Strap Footing 2-16
6. Cantilever Retaining Wall 2-20
7. Counterfort Retaining Wall 2-25
   Practice Problems 2-27
   Solutions 2-28

1. STRIP FOOTING

Nomenclature

A_1	loaded area at base of column	ft^2
A_2	area of the base of the pyramid, with side slopes of 1:2, formed within the footing by the loaded area	ft^2
b_o	perimeter of critical section for punching shear	in
B	length of strip footing parallel to wall, length of short side of a rectangular footing	ft
c	length of side of column	in
D	dead load	kips
e	eccentricity with respect to center of footing	in
e'	eccentricity with respect to edge of footing, $(L/2 - e)$	in
h	depth of footing	in
H	lateral force due to earth pressure	kips
l	distance between column centers	ft
L	length of strip footing perpendicular to wall, length of long side of a rectangular footing, length of side of a square footing	ft
L	live load	kips
P	column axial service load	kips
P_{bn}	nominal bearing strength	kips
P_D	column axial service dead load	kips
P_L	column axial service live load	kips
P_u	column axial factored load	kips
q	soil bearing pressure due to service loads	lbf/ft^2
q_s	equivalent bearing pressure due to service loads	lbf/ft^2
q_u	net factored pressure acting on footing	lbf/ft^2
w_c	weight of concrete	lbf/ft^3
x	distance from edge of footing or center of column to critical section	ft
x_o	distance from edge of property line to centroid of service loads	ft

Symbols

β	ratio of long side to short side of loaded area	

Pressure Distribution

To determine soil pressure under a footing, the self-weight of the footing is added to the applied load from the wall. The footing dimensions are adjusted to ensure that the soil pressure as calculated from the unfactored loads does not exceed the allowable pressure. To determine the forces acting on a footing, the net pressure is required as determined from the applied wall load only. For a strip footing of length B parallel to the wall, the net pressure acting on the footing, as shown in Fig. 2.1, is given by

$$q = \frac{P}{BL} \quad [e = 0]$$

$$q = \frac{P\left(1 \pm \dfrac{6e}{L}\right)}{BL} \quad [e \le \tfrac{L}{6}]$$

$$q_{\max} = \frac{2P}{3Be'} \quad [e > \tfrac{L}{6}]$$

Design of a footing is based on the net factored applied loads, determined in accordance with ACI Sec. 9.2.

Example 2.1

The 18 in deep strip footing shown in the following illustration supports a 12 in concrete wall that is offset 1 ft from the center of the footing. The applied service loads are indicated in the figure, and the allowable soil pressure is 5000 lbf/ft^2. Determine the required footing dimensions, the soil pressure under the footing, and the net factored pressure acting on the footing.

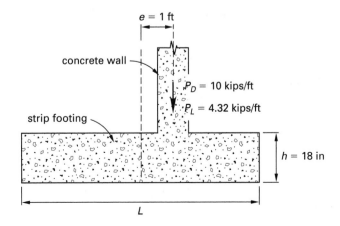

Figure 2.1 *Net Pressure Distribution on a Footing*

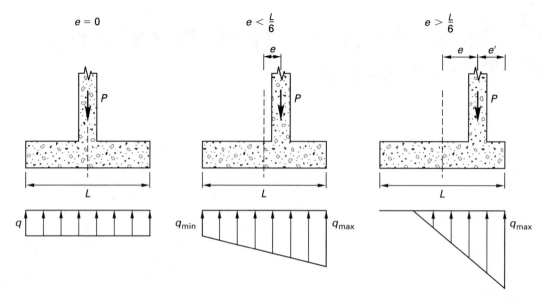

Solution

The total applied service load per foot run is

$$P_1 = P_D + P_L$$
$$= 10{,}000 \text{ lbf} + 4320 \text{ lbf}$$
$$= 14{,}320 \text{ lbf}$$

The self-weight of the footing per foot run is

$$P_2 = w_c h L$$
$$= \left(150 \; \frac{\text{lbf}}{\text{ft}^3}\right)(1.5 \text{ ft})(L)$$
$$= 225L \text{ lbf}$$

Assuming that $e < L/6$, the maximum soil pressure is given by

$$q_{\max} = \frac{P_1\left(1 + \dfrac{6e}{L}\right)}{L} + \frac{P_2}{L}$$
$$= \frac{(14{,}320 \text{ lbf})\left(1 + \dfrac{6}{L}\right)}{L} + 225 \; \frac{\text{lbf}}{\text{ft}}$$
$$= 5000 \text{ lbf/ft}^2 \quad \text{[as given]}$$

Solving for the footing length gives

$$L = 6 \text{ ft}$$

The maximum soil pressure under the footing is

$$q_{\max} = \frac{P_1\left(1 + \dfrac{6e}{L}\right)}{BL} + \frac{P_2}{BL}$$
$$= \frac{(14{,}320 \text{ lbf})\left(1 + \dfrac{(6)(1 \text{ ft})}{6 \text{ ft}}\right)}{(1 \text{ ft})(6 \text{ ft})} + \frac{1350 \text{ lbf}}{(1 \text{ ft})(6 \text{ ft})}$$
$$= 5000 \text{ lbf/ft}^2$$

The minimum soil pressure under the footing is

$$q_{\min} = \frac{P_1\left(1 - \dfrac{6e}{L}\right)}{BL} + \frac{P_2}{BL}$$
$$= 0 + \frac{1350 \text{ lbf}}{(1 \text{ ft})(6 \text{ ft})}$$
$$= 225 \text{ lbf/ft}^2$$

The net factored load on the footing, in accordance with ACI Sec. 9.2, is

$$P_u = 1.2P_D + 1.6P_L$$
$$= (1.2)(10 \text{ kips}) + (1.6)(4.32 \text{ kips})$$
$$= 18.91 \text{ kips}$$

The maximum net factored pressure acting on the footing is

$$q_{u(\max)} = \frac{P_u\left(1 + \dfrac{6e}{L}\right)}{BL}$$
$$= \frac{(18.91 \text{ kips})(2)}{(1 \text{ ft})(6 \text{ ft})}$$
$$= 6.30 \text{ kips/ft}^2$$

The minimum net factored pressure acting on the footing is

$$q_{u(\min)} = \frac{P_u\left(1 - \dfrac{6e}{L}\right)}{BL}$$
$$= 0$$

Figure 2.2 *Critical Sections for Flexure and Shear*

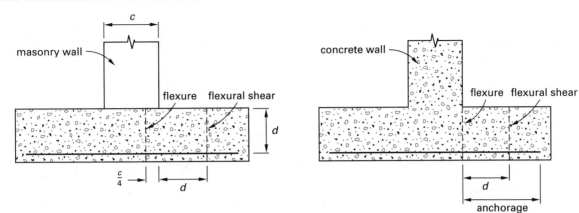

Design for Flexural Shear

The critical section for flexural shear is defined in ACI Secs. 15.5.2 and 11.1.3.1 as being located a distance d from the face of the concrete or masonry wall, as shown in Fig. 2.2. The shear strength of the footing is determined in accordance with ACI Sec. 11.3.

Example 2.2

The strip footing for Ex. 2.1 has an effective depth of 14 in and a concrete strength of 3000 lbf/in². Using the following illustration, determine whether the shear capacity is adequate.

Illustration for Ex. 2.2

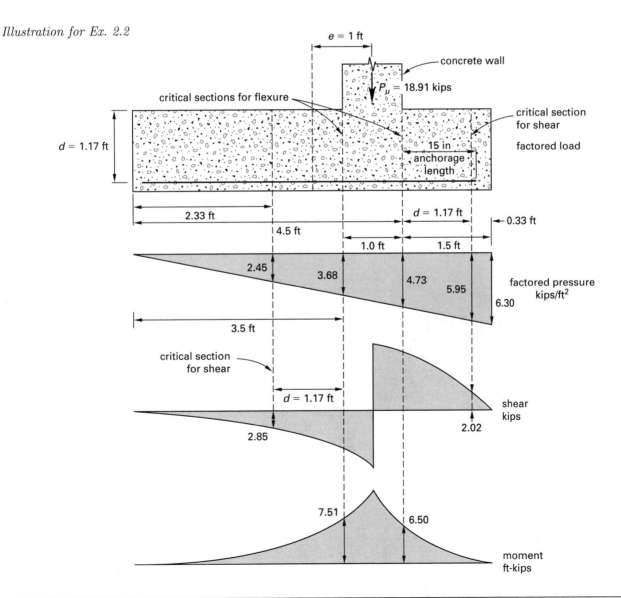

Solution

The net factored pressure acting on the footing is shown in the illustration, and the pressure at the critical section for shear on the right side of the footing is

$$q = \frac{q_u \, (5.67 \text{ ft})}{6 \text{ ft}}$$

$$= \frac{\left(6.30 \, \dfrac{\text{kips}}{\text{ft}^2}\right) (5.67 \text{ ft})}{6 \text{ ft}}$$

$$= 5.95 \text{ kips/ft}^2$$

For a 1 ft strip, the factored shear force at the critical section is

$$V_u = \frac{(0.33 \text{ ft}) \, (1 \text{ ft}) \, (q + q_u)}{2}$$

$$= \frac{(0.33 \text{ ft}) \, (1 \text{ ft}) \left(5.95 \, \dfrac{\text{kips}}{\text{ft}^2} + 6.30 \, \dfrac{\text{kips}}{\text{ft}^2}\right)}{2}$$

$$= 2.02 \text{ kips}$$

The pressure at the critical section for shear on the left side of the footing is

$$q = \frac{q_u \, (2.33 \text{ ft})}{6 \text{ ft}}$$

$$= \frac{\left(6.30 \, \dfrac{\text{kips}}{\text{ft}^2}\right) (2.33 \text{ ft})}{6 \text{ ft}}$$

$$= 2.45 \text{ kips/ft}^2$$

For a 1 ft strip, the factored shear force at the critical section is

$$V_u = \frac{(2.33 \text{ ft}) \, (1 \text{ ft}) \, q}{2}$$

$$= \frac{(2.33 \text{ ft}) \, (1 \text{ ft}) \left(2.45 \, \dfrac{\text{kips}}{\text{ft}^2}\right)}{2}$$

$$= 2.85 \text{ kips} \quad \text{[governs]}$$

The shear capacity of the footing is given by ACI Eq. (11-3) as

$$\phi V_c = 2\phi b d \sqrt{f_c'}$$

$$= \frac{(2)(0.75)(12 \text{ in})(14 \text{ in})\sqrt{3000 \, \dfrac{\text{lbf}}{\text{in}^2}}}{1000 \, \dfrac{\text{lbf}}{\text{kip}}}$$

$$= 13.80 \text{ kips}$$

$$> V_u \quad \text{[satisfactory]}$$

Design for Flexure

The critical section for flexure is defined in ACI Sec. 15.4.2 as being located at the face of a concrete wall and halfway between the center and the face of a masonry wall, as shown in Fig. 2.2. The required reinforcement area is determined in accordance with ACI Sec. 10.2. The minimum ratio $\rho_{\min}$ of reinforcement area to gross concrete area is specified in ACI Secs. 10.5.1 and 7.12.2, for both main reinforcement and distribution reinforcement, as 0.0018 for grade 60 bars. The maximum spacing of the main reinforcement shall not exceed 18 in or three times the footing depth. The diameter of bar provided must be such that the development length does not exceed the available anchorage length. Distribution reinforcement may be spaced at a maximum of 18 in or five times the footing depth.

Example 2.3

Determine the reinforcement required in the strip footing for Exs. 2.1 and 2.2.

Solution

As shown in the figure in Ex. 2.2, on the right side of the footing the factored pressure at the critical section for flexure, which is at the right face of the wall, is given by

$$q = \frac{q_u \, (4.5 \text{ ft})}{6 \text{ ft}} = \frac{\left(6.30 \, \dfrac{\text{kips}}{\text{ft}^2}\right) (4.5 \text{ ft})}{6 \text{ ft}}$$

$$= 4.73 \text{ kips/ft}^2$$

For a 1 ft strip, the factored moment at the critical section is

$$M_u = \frac{(1.5 \text{ ft})^2 \, (1 \text{ ft}) \, (q + 2q_u)}{6}$$

$$= \frac{(1.5 \text{ ft})^2 \, (1 \text{ ft}) \left(\left(4.73 \, \dfrac{\text{kips}}{\text{ft}^2}\right) + (2)\left(6.30 \, \dfrac{\text{kips}}{\text{ft}^2}\right)\right)}{6}$$

$$= 6.50 \text{ ft-kips}$$

As shown in the figure in Ex. 2.2, on the left side of the footing the factored pressure at the critical section for flexure, which is at the left face of the wall, is given by

$$q = \frac{q_u \, (3.5 \text{ ft})}{6 \text{ ft}} = \frac{\left(6.30 \, \dfrac{\text{kips}}{\text{ft}^2}\right) (3.5 \text{ ft})}{6 \text{ ft}}$$

$$= 3.68 \text{ kips/ft}^2$$

For a 1 ft strip, the factored moment at the critical section is

$$M_u = \frac{(3.5 \text{ ft})^2 \, (1 \text{ ft}) \, q}{6}$$

$$= \frac{(3.5 \text{ ft})^2 \, (1 \text{ ft}) \left(3.68 \, \dfrac{\text{kips}}{\text{ft}^2}\right)}{6}$$

$$= 7.51 \text{ ft-kips} \quad \text{[governs]}$$

Assuming a tension-controlled section, the required reinforcement ratio is given by

$$\rho = \frac{0.85 f_c' \left(1 - \sqrt{1 - \dfrac{M_u}{0.383 bd^2 f_c'}}\right)}{f_y}$$

$$= \frac{(0.85)\left(3\ \dfrac{\text{kips}}{\text{in}^2}\right)}{60\ \dfrac{\text{kips}}{\text{in}^2}} \times \left(1 - \sqrt{1 - \dfrac{\begin{array}{c}(7.51\ \text{ft-kips})\left(12\ \dfrac{\text{in}}{\text{ft}}\right)\\ \times \left(1000\ \dfrac{\text{lbf}}{\text{kip}}\right)\end{array}}{\begin{array}{c}(0.383)(12\ \text{in})(14\ \text{in})^2\\ \times \left(3000\ \dfrac{\text{lbf}}{\text{in}^2}\right)\end{array}}}\right)$$

$$= 0.0007$$

The minimum reinforcement area governs and is given by ACI Sec. 7.12.2 as

$$\begin{aligned} A_s &= 0.0018bh \\ &= (0.0018)(12\ \text{in})(18\ \text{in}) \\ &= 0.39\ \text{in}^2/\text{ft} \end{aligned}$$

Providing no. 4 bars at 6 in on center for both main and distribution reinforcement gives

$$A_s = 0.40\ \text{in}^2/\text{ft} \quad \text{[satisfactory]}$$

$$\Psi_t = \Psi_e = \lambda = 1 \quad \left[\begin{array}{c}\text{uncoated bottom bars}\\ \text{in normal weight concrete}\end{array}\right]$$

$$\Psi_s = 0.8 \quad \text{[for no. 4 bars]}$$

$$\frac{c_b + K_{tr}}{d_b} = 2.5 \quad \text{[from ACI Sec. 12.2.3]}$$

ACI Eq. (12-1) for development length reduces to

$$\begin{aligned} l_d &= \frac{(0.075)(0.8 d_b f_y)}{(2.5)\sqrt{f_c'}} = \frac{0.06 d_b \left(60{,}000\ \dfrac{\text{lbf}}{\text{in}^2}\right)}{(2.5)\sqrt{3000\ \dfrac{\text{lbf}}{\text{in}^2}}} \\ &= 26.3 d_b \\ &= 13.2\ \text{in} \quad \text{[for no. 4 bars]} \\ &< 15\ \text{in anchorage length provided} \quad \text{[satisfactory]} \end{aligned}$$

2. ISOLATED COLUMN WITH SQUARE FOOTING

Nomenclature

b_1	width of the critical perimeter measured in the direction of M_u, $c_1 + d$	in
b_2	width of the critical perimeter measured perpendicular to b_1, $c_2 + d$	in
b_o	length of the critical perimeter, $2(b_1 + b_2)$	in
B	length of the footing measured perpendicular to L	ft
c_1	width of the column measured in the direction of M_u	in
c_2	width of the column measured perpendicular to c_1	in
d	average effective depth	in
J_c	polar moment of inertia of critical perimeter	in^4
J_c/y	$(b_1 d(b_1 + 3b_2) + d^3)/3$ [for a footing with central column as specified by ACI Sec. R11.12.6.2]	in^3
L	length of the footing measured in the direction of M_u	ft
M_u	moment applied to the column	ft-kips
V_u	factored shear force acting on the critical perimeter, $P_u(1 - b_1 b_2/BL)$	kips
y	distance from the centroid of the critical perimeter to edge of critical perimeter, $b_1/2$ [for footing with central column]	in

Symbols

γ_v	fraction of the applied moment transferred by shear as specified by ACI Sec. 11.12.6.1 and Sec. 13.5.3.2, $1 - 1/(1 + 0.67\sqrt{b_1/b_2})$	

Design for Punching Shear

The critical perimeter for punching shear is specified in ACI Secs. 15.5.2 and 11.12.1.2 and illustrated in Fig. 2.3. For a concrete or masonry column, the critical perimeter is a distance from the face of the column equal to one-half the effective depth. For a steel column with a base plate, the critical perimeter is one-half the effective depth from a plane halfway between the face of the column and the edge of the base plate. The punching shear strength of the footing is determined by ACI Sec. 11.12.2.1 as

$$\phi V_c = 4\phi d b_o \sqrt{f_c'} \qquad \text{[ACI 11-35]}$$

When $\beta > 2$,

$$\phi V_c = \phi d b_o \left(2 + \frac{4}{\beta}\right)\sqrt{f_c'} \qquad \text{[ACI 11-33]}$$

$$\phi = 0.75$$

The depth of the footing is usually governed by the punching shear capacity.

When the column supports only an axial load, P_u, shear stress at the critical perimeter is uniformly distributed around the critical perimeter. When, in addition to the axial load, a bending moment, M_u, is applied to the

Figure 2.3 *Critical Perimeter for Punching Shear*

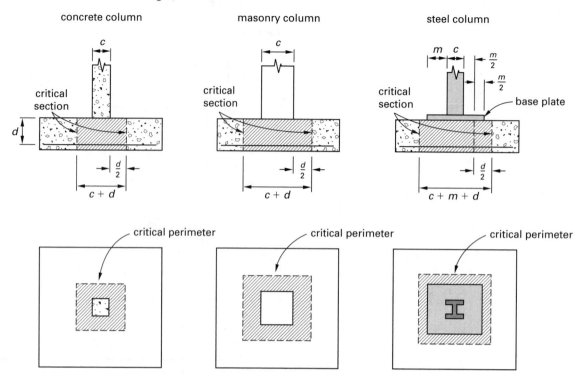

column, an eccentric shear stress is introduced into the critical section with the maximum value occurring on the face nearest the largest bearing pressure. When both axial load and bending moment occur, the shear stress due to both conditions are combined as specified in ACI Sec. R11.12.6.2 to give a maximum value of

$$v_u = \frac{V_u}{db_o} + \frac{\gamma_v M_u y}{J_c}$$

Example 2.4

A 6 ft square reinforced concrete footing with an effective depth of 12 in supports a 12 in square column with a factored axial load of 200 kips. Assuming the concrete strength is 3000 lbf/in^2, determine whether the punching shear capacity of the footing is satisfactory.

Solution

The net factored pressure on the footing is

$$q_u = \frac{P_u}{LB}$$

$$= \frac{200 \text{ kips}}{(6 \text{ ft})(6 \text{ ft})}$$

$$= 5.56 \text{ kips/ft}^2$$

The length of the critical perimeter is

$$b_o = (4)(c + d)$$

$$= (4)(12 \text{ in} + 12 \text{ in})$$

$$= 96 \text{ in}$$

$$= 8 \text{ ft}$$

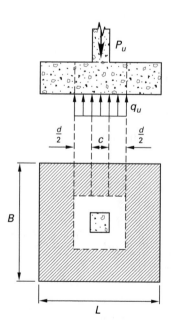

Shear at the critical perimeter is

$$V_u = P_u - q_u(c+d)^2$$

$$= 200 \text{ kips} - \left(5.56 \, \frac{\text{kips}}{\text{ft}^2}\right)(2 \text{ ft})^2$$

$$= 177.7 \text{ kips}$$

Shear capacity of the footing is given by ACI Eq. (11-35) as

$$\phi V_c = 4\phi d b_o \sqrt{f_c'}$$

$$= \frac{(4)(0.75)(12 \text{ in})(96 \text{ in})\sqrt{3000 \, \frac{\text{lbf}}{\text{in}^2}}}{1000 \, \frac{\text{lbf}}{\text{kip}}}$$

$$= 189.3 \text{ kips}$$

$$> V_u \quad [\text{satisfactory}]$$

Design for Flexural Shear

For concrete and masonry columns, the location of the critical section for flexural shear is identical with that for a strip footing and is shown in Fig. 2.2. As shown in Fig. 2.4, the critical section for a steel column with a base plate is located an effective depth from the plane halfway between the face of the column and edge of the base plate.

Figure 2.4 *Critical Sections for a Footing with Steel Base Plate*

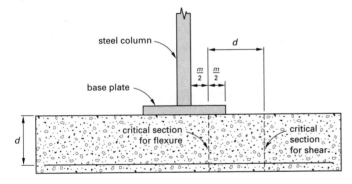

Example 2.5

For the reinforced concrete footing of Ex. 2.4, determine whether the flexural-shear capacity is adequate.

Solution

The distance from the edge of the footing of the critical section for flexural shear is

$$x = \frac{L}{2} - \frac{c}{2} - d$$

$$= \frac{6 \text{ ft}}{2} - \frac{1 \text{ ft}}{2} - 1 \text{ ft}$$

$$= 1.5 \text{ ft}$$

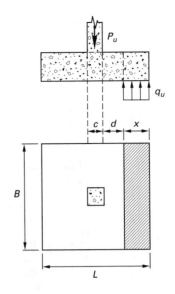

The factored shear force at this section is

$$V_u = q_u B x$$

$$= \left(5.56 \, \frac{\text{kips}}{\text{ft}^2}\right)(6 \text{ ft})(1.5 \text{ ft})$$

$$= 50.0 \text{ kips}$$

The shear capacity of the footing is given by ACI Eq. (11-3) as

$$\phi V_c = 2\phi b d \sqrt{f_c'}$$

$$= \frac{(2)(0.75)(72 \text{ in})(12 \text{ in})\sqrt{3000 \, \frac{\text{lbf}}{\text{in}^2}}}{1000 \, \frac{\text{lbf}}{\text{kip}}}$$

$$= 71.0 \text{ kips}$$

$$> V_u \quad [\text{satisfactory}]$$

Design for Flexure

For concrete and masonry columns, the location of the critical section for flexure is identical with that for a strip footing and is shown in Fig. 2.2. As shown in Fig. 2.4, the critical section for flexure for a steel column with a base plate is at a plane halfway between the face of the column and the edge of the base plate.

Example 2.6

Determine the reinforcement required in the square footing for Ex. 2.4. The depth of the footing is 15.5 in, and the reinforcement is grade 60.

Solution

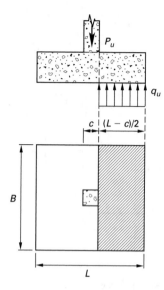

The factored moment at the critical section for flexure, which is at the face of the column, is given by

$$M_u = \frac{q_u B \left(\dfrac{L}{2} - \dfrac{c}{2}\right)^2}{2}$$

$$= \frac{\left(5.56 \ \dfrac{\text{kips}}{\text{ft}^2}\right)(6 \ \text{ft})(3 \ \text{ft} - 0.5 \ \text{ft})^2}{2}$$

$$= 104.3 \ \text{ft-kips}$$

Assuming a tension-controlled section, the required reinforcement ratio is given by

$$\rho = \frac{(0.85)\,(f_c')\left(1 - \sqrt{1 - \dfrac{M_u}{0.383 b d^2 f_c'}}\right)}{f_y}$$

$$= \frac{(0.85)\left(3 \ \dfrac{\text{kips}}{\text{in}^2}\right)}{60 \ \dfrac{\text{kips}}{\text{in}^2}}$$

$$\times \left(1 - \sqrt{\frac{1 - (104.3 \ \text{ft-kips})\left(12 \ \dfrac{\text{in}}{\text{ft}}\right) \times \left(1000 \ \dfrac{\text{lbf}}{\text{kip}}\right)}{(0.383)\,(72 \ \text{in})\,(12 \ \text{in})^2 \times \left(3000 \ \dfrac{\text{lbf}}{\text{in}^2}\right)}}\right)$$

$$= 0.0023$$

The maximum allowable reinforcement ratio for a tension-controlled section is given by

$$\rho_t = 0.319\beta_1 \frac{f_c'}{f_y}$$

$$= 0.0136$$

$$> \rho \quad \left[\begin{array}{c}\text{satisfactory, the section} \\ \text{is tension-controlled}\end{array}\right]$$

The required reinforcement area is

$$A_s = \rho b d$$

$$= (0.0023)(72 \ \text{in})(12 \ \text{in})$$

$$= 1.99 \ \text{in}^2$$

Providing ten no. 4 bars gives a reinforcement area of

$$A_s = 2.0 \ \text{in}^2 \quad \text{[satisfactory]}$$

The minimum allowable reinforcement area is given by ACI Sec. 7.12.2 as

$$A_{s(\min)} = 0.0018 b h$$

$$= (0.0018)(72 \ \text{in})(15.5 \ \text{in})$$

$$= 2.0 \ \text{in}^2 \quad \text{[satisfactory]}$$

From Ex. 2.3, the development length of a no. 4 bar is

$$l_d = 13.2 \ \text{in}$$

The anchorage length provided is

$$l_a = \frac{L}{2} - \frac{c}{2} - \text{end cover}$$

$$= 36 \ \text{in} - 6 \ \text{in} - 3 \ \text{in}$$

$$= 27 \ \text{in}$$

$$> l_d \quad \text{[satisfactory]}$$

Transfer of Force at Base of Column

In accordance with ACI Sec. 15.8, load transfer between a reinforced concrete column and the footing may be effected by bearing on concrete and by reinforcement. The bearing capacity of the column concrete at the interface is given by ACI Sec. 10.17.1 as

$$\phi P_{bn} = 0.85\phi f_c' A_1$$

$$= 0.553 f_c' A_1 \quad \text{[for } \phi = 0.65\text{]}$$

A_1 is the area of the column.

Figure 2.5 *Bearing on Footing Concrete*

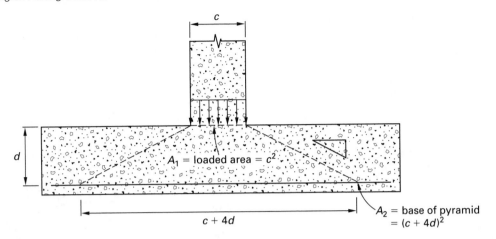

Refer to Fig. 2.5. The bearing capacity of the footing concrete at the interface is given by ACI Sec. 10.17.1 as

$$\phi P_{bn} = 0.85\phi f'_c A_1 \sqrt{\frac{A_2}{A_1}}$$

$$\le (0.85\phi f'_c A_1)(2)$$

A_2 = area of the base of the pyramid, with side slopes of 1:2, formed within the footing by the column base

In accordance with ACI Sec. R15.8.1.2, when the bearing strength at the base of the column or at the top of the footing is exceeded, reinforcement must be provided to carry the excess load. This reinforcement may be provided by dowels or extended longitudinal bars, and the capacity of this reinforcement is

$$\phi P_s = \phi A_s f_y$$

A minimum area of reinforcement is required across the interface, and this is given by ACI Sec. 15.8.2.1 as

$$A_{s(\min)} = 0.005 A_1$$

Example 2.7

Assume that the column in Ex. 2.4 has a concrete compressive strength of 3000 lbf/in^2 and carries a factored axial load of 280 kips. Design the dowels required at the interface.

Solution

The bearing capacity of the column concrete is given by ACI Sec. 10.17 as

$$\phi P_{bn} = 0.553 f'_c A_1 = (0.553)\left(3\ \frac{\text{kips}}{\text{in}^2}\right)(144\ \text{in}^2)$$

$$= 239\ \text{kips}$$

Excess column load to be carried by dowels is

$$\phi P_s = P_u - \phi P_{bn} = 280\ \text{kips} - 239\ \text{kips}$$

$$= 41\ \text{kips}$$

The required area of dowels is given by

$$A_{s(\text{reqd})} = \frac{\phi P_s}{0.65 f_y} = \frac{41\ \text{kips}}{(0.65)\left(60\ \dfrac{\text{kips}}{\text{in}^2}\right)}$$

$$= 1.05\ \text{in}^2$$

Providing four no. 5 dowels gives an area of

$$A_{s(\text{prov})} = 1.24\ \text{in}^2 \quad [\text{satisfactory}]$$

The minimum dowel area allowed is given by ACI Sec. 15.8.2.1 as

$$A_{s(\min)} = 0.005 A_1 = (0.005)(144\ \text{in}^2)$$

$$= 0.72\ \text{in}^2$$

$$< A_{s(\text{prov})} \quad [\text{satisfactory}]$$

Allowing for the excess reinforcement provided, the development length of the dowels in the column and in the footing is given by ACI Sec. 12.3.2 as

$$l_{dc} = \frac{\left(\dfrac{A_{s(\text{reqd})}}{A_{s(\text{prov})}}\right)(0.02 d_b f_y)}{\sqrt{f'_c}}$$

$$= \frac{\left(\dfrac{1.05\ \text{in}^2}{1.24\ \text{in}^2}\right)(0.02)(0.63\ \text{in})\left(60{,}000\ \dfrac{\text{lbf}}{\text{in}^2}\right)}{\sqrt{3000\ \dfrac{\text{lbf}}{\text{in}^2}}}$$

$$= 12\ \text{in}$$

This length exceeds the minimum length of 8 in specified in ACI Sec. 12.3.1 and is satisfactory. In the footing, the length of the base of the pyramid, with side slopes of 1:2, formed within the footing by the loaded area is

$$L_p = c + 4d = 12\ \text{in} + (4)(12\ \text{in})$$

$$= 60\ \text{in}$$

$$< L \quad [\text{satisfactory}]$$

The area of the base of the pyramid is

$$A_2 = L_p^2 = (60 \text{ in})^2 = 3600 \text{ in}^2$$

$$\sqrt{\frac{A_2}{A_1}} = \sqrt{\frac{3600}{144}} = 5$$

Use a maximum value of 2.

Then, bearing capacity of the footing concrete is given by ACI Sec. 10.17 as

$$\phi P_{bn} = 2(0.553 f_c' A_1)$$

$$= 2(0.553)\left(3 \frac{\text{kips}}{\text{in}^2}\right)(144 \text{ in}^2)$$

$$= 478 \text{ kips}$$

$$> P_u \quad [\text{satisfactory}]$$

3. ISOLATED COLUMN WITH RECTANGULAR FOOTING

Nomenclature

A_b	area of reinforcement in central band	in^2
A_s	total required reinforcement area	in^2
B	length of short side of a rectangular footing	ft
c_1	length of short side of a rectangular column	in
c_2	length of long side of a rectangular column	in
L	length of long side of a rectangular footing	ft

Symbols

β	ratio of the long side to the short side of the footing L/B

Design for Flexure

Bending moments are calculated at the critical sections in both the longitudinal and transverse directions. The reinforcement required in the longitudinal direction is distributed uniformly across the width of the footing. Part of the reinforcement required in the transverse direction is concentrated in a central band width equal

to the length of the short side of the footing, as shown in Fig. 2.6.

The area of reinforcement required in the central band is given by ACI Sec. 15.4.4.2 as

$$A_b = \frac{2A_s}{\beta + 1}$$

The remainder of the reinforcement required in the transverse direction is

$$A_r = \frac{A_s(\beta - 1)}{\beta + 1}$$

This is distributed uniformly on each side of the center band.

Example 2.8

The reinforced concrete footing shown in the following illustration is 10 ft long and 7 ft wide, has an effective depth of 12 in and an overall depth of 16 in, and has a concrete compressive strength of 5000 lbf/in^2. The footing supports a column with dimensions 12 in by 18 in and is reinforced with grade 60 bars. Determine the transverse reinforcement required when the net factored pressure acting on the footing is 4.8 kips/ft^2.

Solution

The factored moment in the transverse direction at the critical section, which is at the face of the column, is

$$M_u = \frac{q_u L\left(\dfrac{B}{2} - \dfrac{c_1}{2}\right)^2}{2}$$

$$= \frac{\left(4.8 \dfrac{\text{kips}}{\text{ft}^2}\right)(10 \text{ ft})(3.5 \text{ ft} - 0.5 \text{ ft})^2}{2}$$

$$= 216 \text{ ft-kips}$$

Figure 2.6 *Rectangular Footing: Reinforcement Areas*

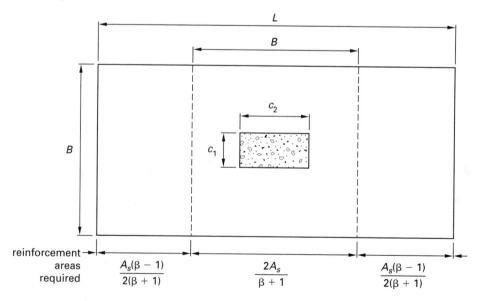

Illustration for Ex. 2.8

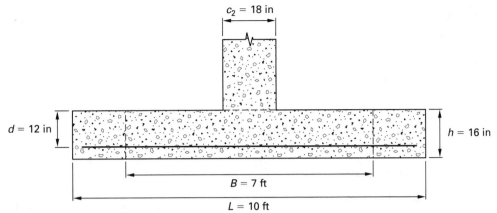

Assuming a tension-controlled section, the required reinforcement ratio is given by

$$\rho = \frac{0.85 f'_c \left(1 - \sqrt{1 - \dfrac{M_u}{0.383 b d^2 f'_c}}\right)}{f_y}$$

$$= \frac{(0.85)\left(5\ \dfrac{\text{kips}}{\text{in}^2}\right)}{60\ \dfrac{\text{kips}}{\text{in}^2}}$$

$$\times \left(1 - \sqrt{1 - \cfrac{(216\ \text{ft-kips})\left(12\ \dfrac{\text{in}}{\text{ft}}\right)\times \left(1000\ \dfrac{\text{lbf}}{\text{kip}}\right)}{(0.383)(120\ \text{in})(12\ \text{in})^2 \times \left(5000\ \dfrac{\text{lbf}}{\text{in}^2}\right)}}\right)$$

$$= 0.0028$$

The maximum allowable reinforcement ratio for a tension-controlled section is given by

$$\rho_t = 0.319 \beta_1 \frac{f'_c}{f_y}$$

$$= 0.0213$$

$$> \rho \quad \begin{bmatrix}\text{satisfactory, the section}\\ \text{is tension-controlled}\end{bmatrix}$$

The minimum allowable reinforcement area is given by ACI Sec. 7.12.2 as

$$A_{s(\text{min})} = 0.0018bh = (0.0018)(120\ \text{in})(16\ \text{in})$$

$$= 3.46\ \text{in}^2$$

The required reinforcement area is

$$A_s = \rho b d = (0.0028)(120\ \text{in})(12\ \text{in})$$

$$= 4.08\ \text{in}^2$$

$$> A_{s(\text{min})} \quad \text{[satisfactory]}$$

The reinforcement required in the central 7 ft band width is

$$A_b = \frac{2A_s}{\beta + 1} = \frac{(2)\left(4.08\ \text{in}^2\right)}{\dfrac{10\ \text{ft}}{7\ \text{ft}} + 1}$$

$$= 3.36\ \text{in}^2$$

Providing 11 no. 5 bars gives a reinforcement area of

$$A_{b(\text{prov})} = 3.41\ \text{in}^2 \quad \text{[satisfactory]}$$

The remaining reinforcement is

$$A_r = A_s - A_{b(\text{prov})}$$

$$= 4.08\ \text{in}^2 - 3.41\ \text{in}^2$$

$$= 0.67\ \text{in}^2$$

Providing two no. 4 bars on each side of the central band gives a reinforcement area of

$$A_{r(\text{prov})} = (4)(0.20\ \text{in}^2)$$

$$= 0.80\ \text{in}^2 \quad \text{[satisfactory]}$$

By inspection, the anchorage length provided is adequate.

4. COMBINED FOOTING

Pressure Distribution

The combined footing shown in Fig. 2.7 has one column adjacent to the property line. The length of the footing is adjusted until the footing centroid coincides with the centroid of the service loads on the two columns to provide a uniform soil pressure. The footing width is adjusted to ensure that the soil bearing pressure does not exceed the allowable pressure.

Figure 2.7 *Combined Footing with Applied Service Loads*

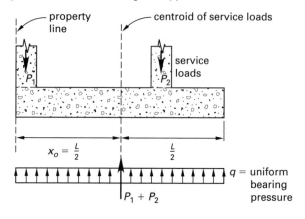

Example 2.9

Determine the plan dimensions required for the combined footing shown in the following illustration to provide a uniform soil bearing pressure of 4000 lbf/ft^2 under the service loads indicated.

Solution

Allowing for the self-weight of the footing, the maximum allowable equivalent soil bearing pressure is

$$q_e = q - w_c h = 4000 \ \frac{\text{lbf}}{\text{ft}^2} - \frac{\left(150 \ \frac{\text{lbf}}{\text{ft}^3}\right) (27 \text{ in})}{12 \ \frac{\text{in}}{\text{ft}}}$$

$$= 3663 \text{ lbf/ft}^2$$

$$= 3.663 \text{ kips/ft}^2$$

The centroid of the column service loads is located a distance x_o from the property line, which is obtained by taking moments about the property line and is given by

$$x_o = \frac{0.5 P_1 + 15.5 P_2}{P_1 + P_2}$$

$$= \frac{(0.5 \text{ ft}) (200 \text{ kips}) + (15.5 \text{ ft}) (300 \text{ kips})}{500 \text{ kips}}$$

$$= 9.5 \text{ ft}$$

The length of footing required to produce a uniform bearing pressure on the soil is

$$L = 2x_o = (2)(9.5 \text{ ft}) = 19 \text{ ft}$$

The width of footing to produce a uniform pressure on the soil of 4000 lbf/ft^2 is

$$B = \frac{P_1 + P_2}{q_e L} = \frac{500 \text{ kips}}{\left(3.663 \ \frac{\text{kips}}{\text{ft}^2}\right) (19 \text{ ft})} = 7.2 \text{ ft}$$

Design for Punching Shear

The critical perimeter for punching shear in a combined footing is identical with that in an isolated column footing and is located a distance from the face of the column equal to one-half the effective depth. The net factored pressure on the footing must be determined from the factored applied column loads, as shown in Fig. 2.8. It will not necessarily be uniform unless the ratios of the factored loads to service loads on both columns are identical.

Example 2.10

The combined footing for Ex. 2.9 has a concrete strength of 5000 lbf/in^2 and a factored load on each column that is 1.5 times the service load. Determine whether the punching shear capacity is adequate.

Illustration for Ex. 2.9

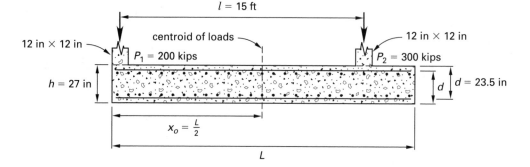

Figure 2.8 *Combined Footing with Applied Factored Loads*

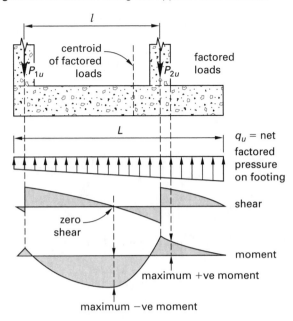

Solution

Because the ratios of the factored loads to service loads on both columns are identical, the net factored pressure on the footing is uniform and has a value of

$$q_u = 1.5q_e$$

$$= (1.5)\left(3.663 \ \frac{\text{kips}}{\text{ft}^2}\right)$$

$$= 5.5 \ \text{kips/ft}^2$$

For column no. 1, the factored load is

$$P_{1u} = 1.5P_1$$

$$= (1.5)(200 \ \text{kips})$$

$$= 300 \ \text{kips}$$

The length of the critical perimeter, as shown in the following illustration, is

$$b_o = (c + d) + (2)\left(c + \frac{d}{2}\right)$$

$$= 12 \ \text{in} + 23.5 \ \text{in} + (2)\left(12 \ \text{in} + \frac{23.5 \ \text{in}}{2}\right)$$

$$= 35.5 \ \text{in} + (2)(23.75 \ \text{in})$$

$$= 83 \ \text{in}$$

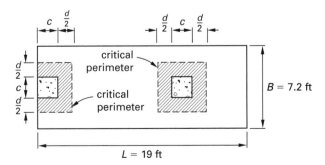

The punching shear force at the critical perimeter is

$$V_u = P_{1u} - q_u\,(c + d)\left(c + \frac{d}{2}\right)$$

$$= 300 \ \text{kips} - \frac{\left(5.5 \ \dfrac{\text{kips}}{\text{ft}^2}\right)(35.5 \ \text{in})\,(23.75 \ \text{in})}{144 \ \dfrac{\text{in}^2}{\text{ft}^2}}$$

$$= 268 \ \text{kips}$$

The punching shear capacity is given by ACI Eq. (11-35) as

$$\phi V_c = 4\phi d b_o \sqrt{f_c'}$$

$$= \frac{(4)\,(0.75)\,(23.5 \ \text{in})\,(83 \ \text{in})\sqrt{5000 \ \dfrac{\text{lbf}}{\text{in}^2}}}{1000 \ \dfrac{\text{lbf}}{\text{kip}}}$$

$$= 414 \ \text{kips}$$

$$> V_u \quad \text{[satisfactory]}$$

For column no. 2, the factored load is

$$P_{2u} = 1.5P_2 = (1.5)(300 \ \text{kips}) = 450 \ \text{kips}$$

The length of the critical perimeter is

$$b_o = (4)(c + d) = (4)(35.5 \ \text{in}) = 142 \ \text{in}$$

The punching shear force at the critical perimeter is

$$V_u = P_{2u} - q_u(c + d)^2$$

$$= 450 \ \text{kips} - \frac{\left(5.5 \ \dfrac{\text{kips}}{\text{ft}^2}\right)(35.5 \ \text{in})^2}{144 \ \dfrac{\text{in}^2}{\text{ft}^2}}$$

$$= 402 \ \text{kips}$$

The punching shear capacity is given by ACI Eq. (11-35) as

$$\phi V_c = 4\phi d b_o \sqrt{f_c'} = \frac{(414 \ \text{kips})\,(142 \ \text{in})}{83 \ \text{in}}$$

$$= 708 \ \text{kips}$$

$$> V_u \quad \text{[satisfactory]}$$

Design for Flexural Shear

The critical section for flexural shear in a combined footing is identical with that in an isolated column footing and is located a distance d from the face of the column. The shear force at the critical section is determined from the shear force diagram, as shown in Fig. 2.8. The depth of the footing is usually governed by flexural shear.

Example 2.11

Determine whether the flexural shear capacity is adequate for the combined footing of Ex. 2.10.

Solution

At the center of column no. 1, the shear force is

$$V_1 = P_{1u} - \frac{q_u B c}{2}$$

$$= 300 \text{ kips} - \frac{\left(5.5 \ \frac{\text{kips}}{\text{ft}^2}\right)(7.2 \text{ ft})(1 \text{ ft})}{2}$$

$$= 280 \text{ kips}$$

At the center of column no. 2, the shear force is

$$V_2 = V_1 - q_u B l$$

$$= 280 \text{ kips} - \left(5.5 \ \frac{\text{kips}}{\text{ft}^2}\right)(7.2 \text{ ft})(15 \text{ ft})$$

$$= -314 \text{ kips}$$

The shear force diagram is shown in the following illustration, and the critical flexural shear is a distance $(d + c/2)$ from the center of column no. 2.

The critical flexural shear at this section is

$$V_u = V_2 - q_u B \left(d + \frac{c}{2}\right)$$

$$= 314 \text{ kips} - \frac{\left(5.5 \ \frac{\text{kips}}{\text{ft}^2}\right)(7.2 \text{ ft})(29.5 \text{ in})}{12 \ \frac{\text{in}}{\text{ft}}}$$

$$= 217 \text{ kips}$$

The flexural-shear capacity of the footing is given by ACI Eq. (11-3) as

$$\phi V_c = 2\phi b d \sqrt{f_c'}$$

$$= \frac{\begin{array}{c}(2)(0.75)(7.2 \text{ ft})\left(12 \ \frac{\text{in}}{\text{ft}}\right) \\ \times (23.5 \text{ in}) \sqrt{5000 \ \frac{\text{lbf}}{\text{in}^2}}\end{array}}{1000 \ \frac{\text{lbf}}{\text{kip}}}$$

$$= 215 \text{ kips}$$

$$\approx V_u \quad \text{[satisfactory]}$$

Illustration for Ex. 2.11

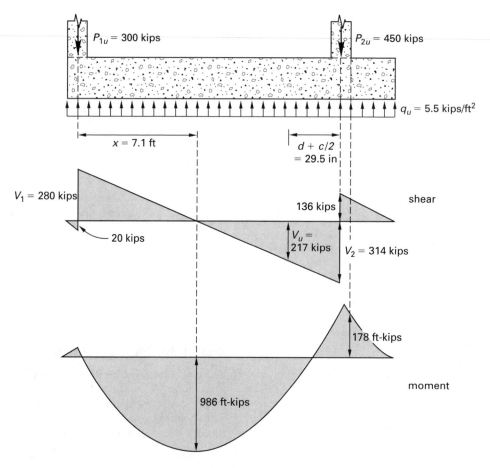

Design for Flexure

The footing is designed in the longitudinal direction as a beam continuous over two supports. As shown in Fig. 2.8, the maximum negative moment occurs at the section of zero shear. The maximum positive moment occurs at the outside face of column no. 2. In the transverse direction, it is assumed that the footing cantilevers about the face of both columns. The reinforcement required is concentrated under each column in a band width equal to the length of the shorter side. The area of reinforcement required in the band width under column no. 1 is given by ACI Sec. 15.4.4.2 as

$$A_{1b} = \frac{2A_s P_{1u}}{(\beta + 1)(P_{1u} + P_{2u})}$$

Example 2.12

Determine the required longitudinal and transverse grade 60 reinforcement for the combined footing of Ex. 2.11.

Solution

From Ex. 2.11, the point of zero shear is a distance from the center of column no. 1 given by

$$x = \frac{V_1}{q_u B}$$

$$= \frac{280 \text{ kips}}{\left(5.5 \, \frac{\text{kips}}{\text{ft}^2}\right)(7.2 \text{ ft})}$$

$$= 7.1 \text{ ft}$$

The maximum negative moment at this point is

$$M_u = P_{1u}x - \frac{q_u B \left(x + \frac{c}{2}\right)^2}{2}$$

$$= (300 \text{ kips})(7.1 \text{ ft})$$

$$\quad - \frac{\left(5.5 \, \frac{\text{kips}}{\text{ft}^2}\right)(7.2 \text{ ft})(7.6 \text{ ft})^2}{2}$$

$$= 986 \text{ ft-kips}$$

For the reinforcement in the top of the footing, assuming a tension-controlled section, the reinforcement ratio required is

$$\rho = \frac{0.85 f'_c \left(1 - \sqrt{1 - \dfrac{M_u}{0.383 b d^2 f'_c}}\right)}{f_y}$$

$$= \frac{(0.85)\left(5 \, \dfrac{\text{kips}}{\text{in}^2}\right)}{60 \, \dfrac{\text{kips}}{\text{in}^2}}$$

$$\times \left(1 - \sqrt{1 - \frac{(986 \text{ ft-kips})\left(12 \, \dfrac{\text{in}}{\text{ft}}\right) \times \left(1000 \, \dfrac{\text{lbf}}{\text{kip}}\right)}{(0.383)(86.4 \text{ in})(23.5 \text{ in})^2 \times \left(5000 \, \dfrac{\text{lbf}}{\text{in}^2}\right)}}\,\right)$$

$$= 0.0048$$

The maximum allowable reinforcement ratio for a tension-controlled section is given by

$$\rho_t = 0.319 \beta_1 \frac{f'_c}{f_y}$$

$$= 0.0213$$

$$> \rho \quad \begin{bmatrix} \text{satisfactory, the section} \\ \text{is tension-controlled} \end{bmatrix}$$

The required reinforcement area in the top of the footing is

$$A_s = \rho b d$$

$$= (0.0048)(86.4 \text{ in})(23.5 \text{ in})$$

$$= 9.75 \text{ in}^2$$

Providing ten no. 9 bars gives an area of 10 in^2 (satisfactory).

The minimum permissible reinforcement area is given by ACI Sec. 7.12.2 as

$$A_{s(\min)} = 0.0018 b h$$

$$= (0.0018)(86.4 \text{ in})(27 \text{ in})$$

$$= 4.20 \text{ in}^2$$

$$< A_s \quad [\text{satisfactory}]$$

The maximum positive moment at the outside face of column no. 2 is

$$M_u = \frac{q_u B (L - l - c)^2}{2}$$

$$= \frac{\left(5.5 \, \dfrac{\text{kips}}{\text{ft}^2}\right)(7.2 \text{ ft})(3 \text{ ft})^2}{2}$$

$$= 178 \text{ ft-kips}$$

For the reinforcement in the bottom of the footing, assuming a tension-controlled section, the reinforcement ratio required is

$$\rho = \frac{0.85 f_c' \left(1 - \sqrt{1 - \dfrac{M_u}{0.383 bd^2 f_c'}}\right)}{f_y}$$

$$= \frac{(0.85)\left(5 \, \dfrac{\text{kips}}{\text{in}^2}\right)}{60 \, \dfrac{\text{kips}}{\text{in}^2}} \times \left(1 - \sqrt{1 - \dfrac{(178 \, \text{ft-kips})\left(12 \, \dfrac{\text{in}}{\text{ft}}\right) \times \left(1000 \, \dfrac{\text{lbf}}{\text{kip}}\right)}{(0.383)(86.4 \, \text{in})(23.5 \, \text{in})^2 \times \left(5000 \, \dfrac{\text{lbf}}{\text{in}^2}\right)}}\right)$$

$$= 0.00083$$

The combined area of reinforcement in the top and bottom of the footing exceeds the minimum required value of $0.0018bh$, and in accordance with ACI Sec. 10.5.3, the required reinforcement area in the bottom of the footing is

$$A_s = 1.33 \rho bd$$
$$= (1.33)(0.00083)(86.4 \, \text{in})(23.5 \, \text{in})$$
$$= 2.24 \, \text{in}^2$$

Providing ten no. 5 bars gives an area of 3.10 in^2 (satisfactory).

The factored moment in the transverse direction at the face of the columns is

$$M_u = \frac{q_u L \left(\dfrac{B}{2} - \dfrac{c}{2}\right)^2}{2}$$
$$= \frac{\left(5.5 \, \dfrac{\text{kips}}{\text{ft}^2}\right)(19 \, \text{ft})(3.1 \, \text{ft})^2}{2}$$
$$= 502 \, \text{ft-kips}$$

For the transverse reinforcement in the bottom of the footing, assuming a tension-controlled section, the reinforcement ratio required is

$$\rho = \frac{0.85 f_c' \left(1 - \sqrt{1 - \dfrac{M_u}{0.383 bd^2 f_c'}}\right)}{f_y}$$

$$= \frac{(0.85)\left(5 \, \dfrac{\text{kips}}{\text{in}^2}\right)}{60 \, \dfrac{\text{kips}}{\text{in}^2}} \times \left(1 - \sqrt{1 - \dfrac{(502 \, \text{ft-kips})\left(12 \, \dfrac{\text{in}}{\text{ft}}\right) \times \left(1000 \, \dfrac{\text{lbf}}{\text{kip}}\right)}{(0.383)(228 \, \text{in})(23.5 \, \text{in})^2 \times \left(5000 \, \dfrac{\text{lbf}}{\text{in}^2}\right)}}\right)$$

$$= 0.00089$$

The minimum permissible reinforcement area governs, and the reinforcement area required in both the top and bottom of the footing transversely is

$$A_s = \frac{0.0018bh}{2}$$
$$= \frac{(0.0018)\left(12 \, \dfrac{\text{in}}{\text{ft}}\right)(27 \, \text{in})}{2}$$
$$= 0.29 \, \text{in}^2/\text{ft}$$

Providing no. 4 bars at 8 in centers gives an area of 0.30 in^2/ft (satisfactory).

5. STRAP FOOTING

Nomenclature

A_1	base area of pad footing no. 1	in^2
A_2	base area of pad footing no. 2	in^2
B_S	length of short side of strap	ft
B_1	length of short side of pad footing no. 1	ft
B_2	length of short side of pad footing no. 2	ft
h_S	depth of strap	in
h_1	depth of pad footing no. 1	in
h_2	depth of pad footing no. 2	in
l	distance between column centers	ft
l_R	distance between soil reactions	ft
L_S	length of long side of strap	ft
L_1	length of long side of pad footing no. 1	ft
L_2	length of long side of pad footing no. 2	ft
R_1	soil reaction under pad footing no. 1	kips
R_2	soil reaction under pad footing no. 2	kips
w_c	unit weight of concrete	lbf/ft^3
W_S	weight of strap beam	kips
W_1	weight of pad footing no. 1	kips
W_2	weight of pad footing no. 2	kips

Pressure Distribution

The strap footing shown in Fig. 2.9 has the strap beam, which connects the two pad footings, underlaid by a layer of Styrofoam™ so that the soil pressure under the strap may be considered negligible. Because of the stiffness of the strap beam, the strap and pad footings act as a rigid body producing uniform soil pressure under the pad footings. The base areas of the two pad footings may be adjusted to produce equal soil pressure q under both footings.

The total service load acting is

$$\sum P = P_1 + P_2 + W_1 + W_2 + W_S$$

$$q = \frac{\sum P}{A_1 + A_2}$$

The soil reactions act at the center of the pad footings and are given by

$$R_1 = qA_1$$
$$R_2 = qA_2$$

Pad footing no. 2 is located symmetrically with respect to column no. 2 so that the lines of action of P_2 and R_2 are coincident.

$$l_R = l + \frac{c_1}{2} - \frac{B_1}{2}$$

$$L_S = l_R - \frac{B_1 + B_2}{2}$$

Equating vertical forces gives

$$R_2 = \sum P - R_1 \quad \text{[equilibrium equation no. 1]}$$

Taking moments about the center of pad footing no. 2 gives

$$R_1 = \frac{P_1 l + W_1 l_R + \dfrac{W_S (L_S + B_2)}{2}}{l_R}$$
[equilibrium equation no. 2]

To determine suitable dimensions that will give a soil bearing pressure equal to the allowable pressure q, suitable values are selected for h_1, h_2, h_S, B_1, B_2, and B_S. Hence, l_R and L_S are determined and

$$W_S = w_c L_S B_S h_S$$

An initial estimate is made of R_1, and hence

$$A_1 = \frac{R_1}{q}$$

$$W_1 = w_c A_1 h_1$$

An initial estimate is made of R_2, and hence

$$A_2 = \frac{R_2}{q}$$

$$W_2 = w_c A_2 h_2$$

$$\sum P = P_1 + P_2 + W_1 + W_2 + W_S$$

Substituting in the two equilibrium equations provides revised estimates of R_1 and R_2, and the process is repeated until convergence is reached.

Figure 2.9 *Strap Footing with Applied Service Loads*

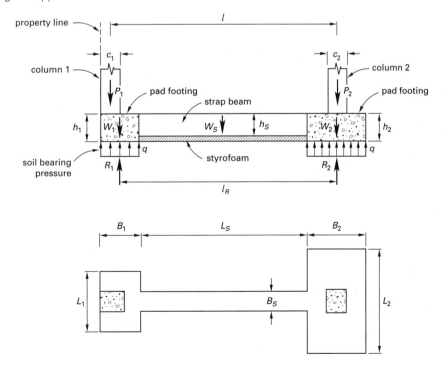

Example 2.13

Determine the plan dimensions required for the strap footing shown in the following illustration to provide a uniform bearing pressure of 3000 lbf/ft^2 under both pad footings for the service loads indicated.

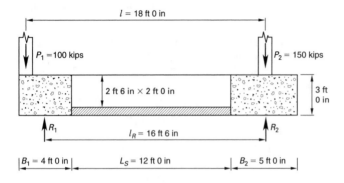

Solution

From the dimensions indicated in the illustration,

$$W_S = w_c L_S B_s h_S$$
$$= \left(0.15 \; \frac{\text{kip}}{\text{ft}^3}\right)(12 \text{ ft})(2 \text{ ft})(2.5 \text{ ft})$$
$$= 9 \text{ kips}$$

Assuming that $R_1 = 134$ kips, then

$$A_1 = \frac{R_1}{q} = \frac{134 \text{ kips}}{3 \; \dfrac{\text{kips}}{\text{ft}^2}} = 44.67 \text{ ft}^2$$

$$W_1 = w_c A_1 h_1$$
$$= \left(0.15 \; \frac{\text{kip}}{\text{ft}^3}\right)(44.67 \text{ ft}^2)(3 \text{ ft})$$
$$= 20.1 \text{ kips}$$

Assuming that $R_2 = 171$ kips, then

$$A_2 = \frac{R_2}{q} = \frac{171 \text{ kips}}{3 \; \dfrac{\text{kips}}{\text{ft}^2}} = 57 \text{ ft}^2$$

$$W_2 = w_c A_2 h_2$$
$$= \left(0.15 \; \frac{\text{kip}}{\text{ft}^3}\right)(57 \text{ ft}^2)(3 \text{ ft})$$
$$= 25.7 \text{ kips}$$

$$\sum P = P_1 + P_2 + W_1 + W_2 + W_S$$
$$= 100 \text{ kips} + 150 \text{ kips} + 20.1 \text{ kips}$$
$$\quad + 25.7 \text{ kips} + 9 \text{ kips}$$
$$= 304.8 \text{ kips}$$

Equating vertical forces gives

$$R_2 = \sum P - R_1$$
$$= 304.8 \text{ kips} - 134 \text{ kips}$$
$$= 170.8 \text{ kips}$$
$$\approx 171 \text{ kips} \quad [\text{satisfactory}]$$

Taking moments about the center of pad footing no. 2 gives

$$R_1 = \frac{P_1 l + W_1 l_R + \dfrac{W_S (L_S + B_2)}{2}}{l_R}$$

$$= \frac{\begin{array}{c}(100 \text{ kips})(18 \text{ ft}) \\ + (20.1 \text{ kips})(16.5 \text{ ft}) + \dfrac{(9 \text{ kips})(17 \text{ ft})}{2}\end{array}}{16.5 \text{ ft}}$$

$$= 133.8 \text{ kips}$$
$$\approx 134 \text{ kips} \quad [\text{satisfactory}]$$

The initial estimates were sufficiently accurate, and the required pad footing areas are

$$A_1 = 44.67 \text{ ft}^2$$
$$A_2 = 57 \text{ ft}^2$$

Design of Strap Beam for Shear

The factored forces acting on the footing are shown in Fig. 2.10. The total factored load on the footing is

$$\sum P_u = P_{1u} + P_{2u} + W_{1u} + W_{2u} + W_{Su}$$

Taking moments about the center of pad footing no. 2 gives

$$R_{1u} = \frac{P_{1u} l + W_{1u} l_R + \dfrac{W_{Su} (L_S + B_2)}{2}}{l_R}$$

Equating vertical forces gives

$$R_{2u} = \sum P_u - R_{1u}$$

The shear at the left end of the strap is

$$V_{Su} = R_{1u} - P_{1u} - W_{1u}$$

The shear at the right end of the strap is

$$V'_{Su} = P_{2u} + W_{2u} - R_{2u}$$

Figure 2.10 *Factored Forces on Strap Footing*

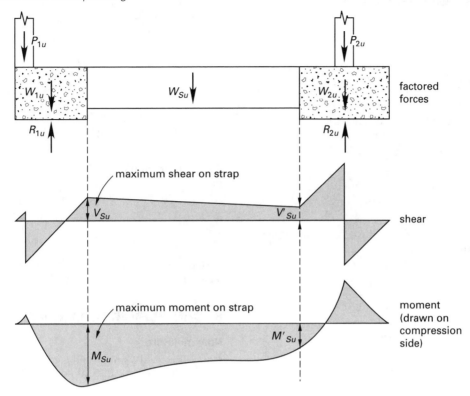

Example 2.14

The strap footing for Ex. 2.13 has a concrete strength of 3000 lbf/in^2 and a factored load on each column that is 1.5 times the service load. The strap beam has an effective depth of 27.5 in. Determine whether the shear capacity is adequate.

Solution

The factored forces are

$$P_{1u} = 1.5P_1 = 150 \text{ kips}$$
$$P_{2u} = 1.5P_2 = 225 \text{ kips}$$
$$W_{1u} = 1.2W_1 = 24 \text{ kips}$$
$$W_{2u} = 1.2W_2 = 31 \text{ kips}$$
$$W_{Su} = 1.2W_S = 11 \text{ kips}$$
$$\sum P_u = 441 \text{ kips}$$

$$R_{1u} = \frac{P_{1u}l + W_{1u}l_R + \dfrac{W_{Su}(L_S + B_2)}{2}}{l_R}$$

$$= \frac{(150 \text{ kips})(18 \text{ ft}) + (24 \text{ kips})(16.5 \text{ ft})}{16.5 \text{ ft}} + \dfrac{(11 \text{ kips})(17 \text{ ft})}{2}$$

$$= 193 \text{ kips}$$

$$R_{2u} = \sum P_u - R_{1u}$$
$$= 248 \text{ kips}$$

The shear at the right end of the strap is

$$V'_{Su} = P_{2u} + W_{2u} - R_{2u}$$
$$= 225 \text{ kips} + 31 \text{ kips} - 248 \text{ kips}$$
$$= 8 \text{ kips}$$

The shear at the left end of the strap is

$$V_{Su} = R_{1u} - P_{1u} - W_{1u}$$
$$= 193 \text{ kips} - 150 \text{ kips} - 24 \text{ kips}$$
$$= 19 \text{ kips} \quad [\text{governs}]$$

The design shear capacity of the strap beam is given by ACI Eq. (11-3) as

$$\phi V_c = 2\phi bd\sqrt{f'_c}$$

$$= \frac{(2)(0.75)(24 \text{ in})(27.5 \text{ in})\sqrt{3000 \dfrac{\text{lbf}}{\text{in}^2}}}{1000 \dfrac{\text{lbf}}{\text{kip}}}$$

$$= 54 \text{ kips}$$
$$> 2V_{Su} \quad [\text{No shear reinforcement is required.}]$$

Design of Strap Beam for Flexure

From Fig. 2.10, the factored moment at the left end of the strap is

$$M_{Su} = P_{1u}\left(B_1 - \frac{c_1}{2}\right) - \frac{(R_{1u} - W_{1u})(B_1)}{2}$$

The factored moment at the right end of the strap is

$$M'_{Su} = \frac{(R_{2u} - W_{2u} - P_{2u})(B_2)}{2}$$

Example 2.15

Determine the required grade 60 flexural reinforcement for the strap beam of Ex. 2.14.

Solution

The factored moment at the right end of the strap is

$$\begin{aligned}
M'_{Su} &= \frac{(R_{2u} - W_{2u} - P_{2u})(B_2)}{2}\\
&= (8 \text{ kips})(2.5 \text{ ft})\\
&= 20 \text{ ft-kips}
\end{aligned}$$

The factored moment at the left end of the strap is

$$\begin{aligned}
M_{Su} &= P_{1u}\left(B_1 - \frac{c_1}{2}\right) - \frac{(R_{1u} - W_{1u})(B_1)}{2}\\
&= (150 \text{ kips})(3.5 \text{ ft}) - (169 \text{ kips})(2 \text{ ft})\\
&= 187 \text{ ft-kips} \quad \text{[governs]}
\end{aligned}$$

Assuming a tension-controlled section, the required reinforcement ratio is

$$\rho = \frac{0.85 f'_c \left(1 - \sqrt{1 - \dfrac{M_{Su}}{0.383bd^2 f'_c}}\right)}{f_y}$$

$$= \frac{(0.85)\left(3\dfrac{\text{kips}}{\text{in}^2}\right)}{60\dfrac{\text{kips}}{\text{in}^2}}$$

$$\times\left(1 - \sqrt{1 - \dfrac{(187 \text{ ft-kips})\left(12\dfrac{\text{in}}{\text{ft}}\right)}{(0.383)(24 \text{ in})(27.5 \text{ in})^2 \times \left(1000\dfrac{\text{lbf}}{\text{kip}}\right)}{\times\left(3000\dfrac{\text{lbf}}{\text{in}^2}\right)}}\right)$$

$$= 0.0023$$

The controlling minimum reinforcement ratio is given by ACI Secs. 10.5.1 and 10.5.3 as the lesser of the following results.

$$\begin{aligned}
\rho_{\min} &= \frac{200}{f_y} = \frac{200}{60,000}\\
&= 0.0033
\end{aligned}$$

$$\rho_{\min} = \left(\frac{4}{3}\right)(0.0023) = 0.0031 \quad \text{[governs]}$$

The reinforcement required in the top of the strap beam is

$$\begin{aligned}
A_s &= bd\rho_{\min}\\
&= (24 \text{ in})(27.5 \text{ in})(0.0031)\\
&= 2.05 \text{ in}^2
\end{aligned}$$

Providing four no. 7 bars gives an area of 2.4 in^2 (satisfactory).

6. CANTILEVER RETAINING WALL

Nomenclature

F	frictional force at underside of base	kips
$\bar{h}$	equivalent additional height of fill, w/γ_S	ft
h_B	depth of base	in
h_K	height of shear key	in
h_T	total height of retaining wall, $h_B + L_W$	in
h_W	stem thickness	in
H_A	total active earth pressure behind wall	kips
H_L	total pressure behind wall due to live load surcharge	kips
H_P	total passive earth pressure in front of wall	kips
K_A	Rankine coefficient of active earth pressure $1 - \sin\phi/1 + \sin\phi$	–
K_P	Rankine coefficient of passive earth pressure $1 + \sin\phi/1 - \sin\phi$	–
L_B	length of base	ft
L_H	length of heel	ft
L_T	length of toe	ft
L_W	height of stem	ft
p_A	lateral pressure due to a fluid of density p_A, $K_A\gamma_S$	lbf/ft^2
p_L	lateral pressure due to live load surcharge, wK_A	lbf/ft^2
p_P	lateral pressure due to a fluid of density p_P, $K_P\gamma_S$	lbf/ft^2
q	earth pressure under the base	lbf/ft^2
w	live load surcharge	lbf/ft^2
W_B	weight of base	kips
W_K	weight of key	kips
W_L	weight of surcharge	kips
W_S	weight of backfill	kips
W_W	weight of stem	kips

Symbols

γ_S	density of backfill	lbf/ft^3
μ	coefficient of friction	–
ϕ	angle of internal friction	degrees

Pressure Distribution

Figure 2.11 shows the forces acting on a cantilever retaining wall. The total active earth pressure behind the wall is given by Rankine's theory as

$$H_A = \frac{p_A h_T^2}{2} = \frac{K_A \gamma_S h_T^2}{2} = \frac{\left(\dfrac{1 - \sin\phi}{1 + \sin\phi}\right)\gamma_S h_T^2}{2}$$

$$= \frac{30 h_T^2}{2} \quad [\text{for } \gamma_S = 110 \text{ lbf/ft}^3 \text{ and } \phi = 35°]$$

= pressure exerted by a fluid of density 30 lbf/ft^3

The total active earth pressure acts at a height of $h_T/3$ above the base.

The total surcharge pressure behind the wall due to a live load surcharge of w lbf/ft^2 is

$$H_L = p_L h_T = w K_A h_T = \frac{w p_A h_T}{\gamma_S}$$

The surcharge may be represented by an equivalent height of fill given by

$$\bar{h} = \frac{w}{\gamma_S}$$
$$H_L = p_A \bar{h} h_T$$

The total surcharge pressure acts at a height of $h_T/2$ above the base.

The total passive earth pressure in front of the wall is

$$H_P = \frac{p_P h_K^2}{2} = \frac{K_P \gamma_S h_K^2}{2} = \frac{\left(\dfrac{1 + \sin\phi}{1 - \sin\phi}\right)\gamma_S h_K^2}{2}$$

$$= \frac{400 h_K^2}{2} \quad [\text{for } \gamma_S = 110 \text{ lbf/ft}^3 \text{ and } \phi = 35°]$$

= pressure exerted by a fluid of density 400 lbf/ft^3

The total passive earth pressure acts at a height of $h_K/3$ above the bottom of the key. The frictional force acting on the underside of the base is given by

$$F = \mu \sum W$$

$\sum W$ is the total weight of the retaining wall plus backfill plus live load surcharge.

A factor of safety of 1.5 against sliding is required, which gives

$$F + H_P \geq 1.5 \text{ times the active pressure acting}$$
$$\text{from top of wall to bottom of key}$$

A factor of safety of 1.5 is required for overturning about the toe.

Figure 2.11 Cantilever Retaining Wall with Applied Service Loads

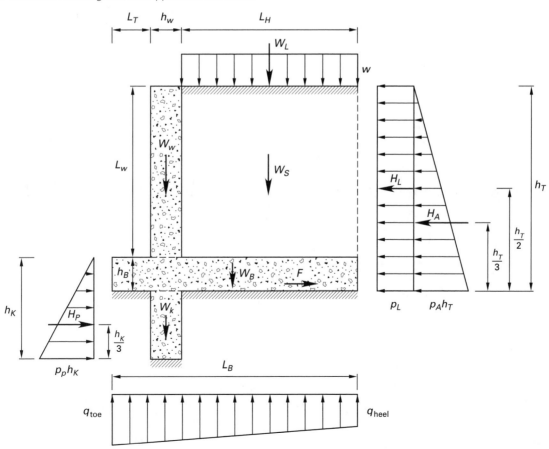

Example 2.16

The retaining wall shown in the following illustration retains soil with a unit weight of 110 lbf/ft³ and an equivalent fluid pressure of 30 lbf/ft² per foot. The live load surcharge behind the wall is equivalent to an additional height of 2 ft of fill. Passive earth pressure may be assumed equivalent to a fluid pressure of 450 lbf/ft², and the coefficient of friction at the underside of the base is 0.4. Determine the factors of safety against sliding and overturning and the bearing pressure distribution under the base.

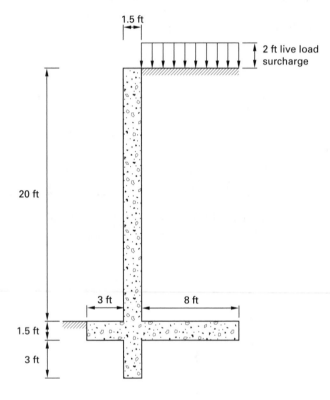

Solution

The lateral pressures from the backfill and the surcharge are

$$H_A = \frac{p_A h_T^2}{2} = \frac{\left(30 \ \dfrac{\text{lbf}}{\text{ft}^2}\right)(21.5 \text{ ft})^2}{2} = 6934 \text{ lbf}$$

$$H_L = p_A h_T \bar{h} = \left(30 \ \frac{\text{lbf}}{\text{ft}^2}\right)(21.5 \text{ ft})(2 \text{ ft}) = 1290 \text{ lbf}$$

Taking moments about the toe gives

$$\begin{aligned} M_o &= \frac{H_A h_T}{3} + \frac{H_L h_T}{2} \\ &= \frac{(6934 \text{ lbf})(21.5 \text{ ft})}{3} + \frac{(1290 \text{ lbf})(21.5 \text{ ft})}{2} \\ &= 63{,}561 \text{ ft-lbf} \end{aligned}$$

The gravity loads acting are

$$\begin{aligned} W_W + W_K &= w_c h_W L_W + w_c (h_K - h_B)(h_W) \\ &= \left(150 \ \frac{\text{lbf}}{\text{ft}^3}\right)(1.5 \text{ ft})(20 \text{ ft}) \\ &\quad + \left(150 \ \frac{\text{lbf}}{\text{ft}^3}\right)(3 \text{ ft})(1.5 \text{ ft}) \\ &= 5175 \text{ lbf} \end{aligned}$$

$$\begin{aligned} W_B &= w_c h_B L_B \\ &= \left(150 \ \frac{\text{lbf}}{\text{ft}^3}\right)(1.5 \text{ ft})(12.5 \text{ ft}) \\ &= 2813 \text{ lbf} \end{aligned}$$

$$\begin{aligned} W_L + W_S &= \gamma_S (h_T - h_B + \bar{h})(L_H) \\ &= \left(110 \ \frac{\text{lbf}}{\text{ft}^3}\right)(20 + 2)(8 \text{ ft}) \\ &= 19{,}360 \text{ lbf} \end{aligned}$$

$$\begin{aligned} \sum W &= 5175 \text{ lbf} + 2813 \text{ lbf} + 19{,}360 \text{ lbf} \\ &= 27{,}348 \text{ lbf} \end{aligned}$$

The distance of the resultant vertical load from the toe is

$$x_o = \frac{\begin{aligned} &(W_W + W_K)\left(L_T + \frac{h_W}{2}\right) + \frac{W_B L_B}{2} \\ &+ (W_L + W_S)\left(L_T + h_W + \frac{L_H}{2}\right) \end{aligned}}{\sum W}$$

$$= \frac{\begin{aligned} &(5175 \text{ lbf})(3.75 \text{ ft}) + (2813 \text{ lbf})(6.25 \text{ ft}) \\ &+ (19{,}360 \text{ lbf})(8.5 \text{ ft}) \end{aligned}}{27{,}348 \text{ lbf}}$$

$$= 7.37 \text{ ft}$$

The factor of safety against overturning is

$$\frac{x_o \sum W}{M_o} = \frac{(7.37 \text{ ft})(27{,}348 \text{ lbf})}{63{,}561 \text{ ft-lbf}}$$

$$= 3.2$$

$$> 1.5 \quad [\text{satisfactory}]$$

The eccentricity of all applied loads about the toe is

$$\begin{aligned} e' &= \frac{x_o \sum W - M_o}{\sum W} = \frac{(7.37)(27{,}348) - 63{,}561}{27{,}348} \\ &= 5.05 \text{ ft} \end{aligned}$$

The eccentricity of all applied loads about the midpoint of the base is

$$\begin{aligned} e &= \frac{L_B}{2} - e' = \frac{12.5 \text{ ft}}{2} - 5.05 \text{ ft} \\ &= 1.20 \text{ ft} \quad [\text{within middle third}] \end{aligned}$$

The pressure under the base is given by

$$q = \frac{\sum W \left(1 \pm \frac{6e}{L_B}\right)}{BL_B}$$

$$= \frac{(27{,}348 \text{ lbf}) \left(1 \pm \frac{(6)(1.2 \text{ ft})}{12.5 \text{ ft}}\right)}{(1 \text{ ft})(12.5 \text{ ft})}$$

$$q_{\text{toe}} = 3448 \text{ lbf/ft}^2$$

$$q_{\text{heel}} = 928 \text{ lbf/ft}^2$$

The frictional resistance under the base is

$$F = \mu \sum W = (0.4)(27{,}348 \text{ lbf})$$
$$= 10{,}939 \text{ lbf}$$

The passive pressure in front of the wall is

$$H_P = \frac{p_P h_K^2}{2} = \frac{\left(450 \frac{\text{lbf}}{\text{ft}^2}\right)(4.5 \text{ ft})^2}{2}$$
$$= 4556 \text{ lbf}$$

The lateral pressures behind the wall from the backfill and surcharge are

$$H_A = \frac{\left(30 \frac{\text{lbf}}{\text{ft}^2}\right)(24.5 \text{ ft})^2}{2} = 9004 \text{ lbf}$$

$$H_L = \left(30 \frac{\text{lbf}}{\text{ft}^2}\right)(24.5 \text{ ft}) = 1470 \text{ lbf}$$

The factor of safety against sliding is

$$\frac{F + H_P}{H_A + H_L} = \frac{10{,}939 \text{ lbf} + 4556 \text{ lbf}}{9004 \text{ lbf} + 1470 \text{ lbf}}$$
$$= 1.48$$
$$\approx 1.5 \quad [\text{satisfactory}]$$

Design for Shear and Flexure

To determine the shear and flexure at the critical sections in the wall, the soil pressure under the footing is recalculated by using the factored forces given by ACI Sec. 9.2 as

$$U = 1.2D + 1.6L + 1.6H \qquad [ACI\ 9\text{-}2]$$

Shear is generally not critical. The location of the critical section for flexure in the stem is at the base of the stem; for flexure in the toe, at the front face of the stem; and for flexure in the heel, at the rear face of the stem.

Example 2.17

Determine the reinforcement areas required in the stem, toe, and heel of the retaining wall for Ex. 2.16. The concrete strength is 3000 lbf/in², and grade 60 reinforcement is provided.

Solution

The factored overturning moment about the toe is

$$M_{ou} = 1.6M_o = (1.6)(63{,}561 \text{ ft-lbf})$$
$$= 101{,}698 \text{ ft-lbf}$$

The factored total vertical load is

$$\sum W_u = (1.2)(W_W + W_B + W_S + W_K) + 1.6W_L$$
$$= (1.2)(5175 \text{ lbf} + 2813 \text{ lbf} + 17{,}600 \text{ lbf})$$
$$+ (1.6)(1760 \text{ lbf})$$
$$= 6210 \text{ lbf} + 3376 \text{ lbf} + 21{,}120 \text{ lbf} + 2816 \text{ lbf}$$
$$= 33{,}522 \text{ lbf}$$

The factored restoring moment is

$$M_{Ru} = (6210 \text{ lbf})(3.75 \text{ ft}) + (3376 \text{ lbf})(6.25 \text{ ft})$$
$$+ (21{,}120 \text{ lbf})(8.5 \text{ ft}) + (2816 \text{ lbf})(8.5 \text{ ft})$$
$$= 247{,}844 \text{ ft-lbf}$$

The eccentricity of the factored loads about the toe is

$$e'_u = \frac{M_{Ru} - M_{ou}}{\sum W_u} = \frac{247{,}844 \text{ ft-lbf} - 101{,}698 \text{ ft-lbf}}{33{,}522 \text{ lbf}}$$
$$= 4.36 \text{ ft}$$

The eccentricity of the factored loads about the midpoint of the base is

$$e_u = \frac{L_B}{2} - e' = \frac{12.5 \text{ ft}}{2} - 4.36 \text{ ft}$$
$$= 1.89 \text{ ft} \quad [\text{within middle third}]$$

The factored pressure under the base is given by

$$q_u = \frac{\sum W_u \left(1 \pm \frac{6e_u}{L_B}\right)}{BL_B}$$

$$= \frac{(33{,}522 \text{ lbf}) \left(1 \pm \frac{(6)(1.89 \text{ ft})}{12.5 \text{ ft}}\right)}{(1 \text{ ft})(12.5 \text{ ft})}$$

$$q_{u(\text{toe})} = 5115 \text{ lbf/ft}^2$$

$$q_{u(\text{heel})} = 249 \text{ lbf/ft}^2$$

The factored pressure distribution under the base is shown in the following illustration, and the maximum factored bending moment in the toe is

$$M_u = \frac{(L_T)^2 B \left(q_{uF} + 2q_{u(\text{toe})}\right)}{6} - \frac{1.2W_B (L_T)^2}{2L_B}$$

$$= \frac{(3 \text{ ft})^2 (1 \text{ ft}) \left(3947 \frac{\text{lbf}}{\text{ft}^2} + (2)\left(5115 \frac{\text{lbf}}{\text{ft}^2}\right)\right)}{6}$$

$$- \frac{(3376 \text{ lbf})(3 \text{ ft})^2}{(2)(12.5 \text{ ft})}$$

$$= 20{,}050 \text{ ft-lbf}$$

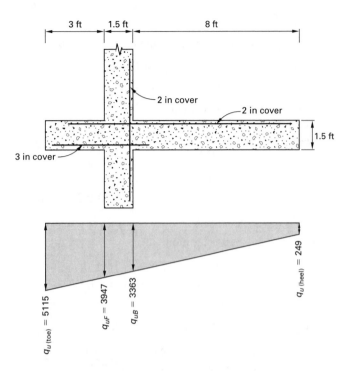

Assuming a tension-controlled section, the required reinforcement ratio is

$$\rho = \frac{0.85 f_c' \left(1 - \sqrt{1 - \dfrac{M_u}{0.383 bd^2 f_c'}}\right)}{f_y}$$

$$= \frac{(0.85)\left(3\,\dfrac{\text{kips}}{\text{in}^2}\right)}{60\,\dfrac{\text{kips}}{\text{in}^2}} \times \left(1 - \sqrt{1 - \dfrac{(20{,}050 \text{ ft-lbf})\left(12\,\dfrac{\text{in}}{\text{ft}}\right)}{(0.383)(12 \text{ in})(14.5 \text{ in})^2 \times \left(3000\,\dfrac{\text{lbf}}{\text{ft}^2}\right)}}\right)$$

$$= 0.0018$$

The maximum allowable reinforcement ratio for a tension-controlled section is given by

$$\rho_t = 0.319 \beta_1 \frac{f_c'}{f_y}$$

$$= 0.0136$$

$$> \rho \quad \text{[satisfactory, the section is tension-controlled]}$$

$$A_s = \rho b d$$

$$= (0.0018)(12 \text{ in})(14.5 \text{ in})$$

$$= 0.314 \text{ in}^2$$

$$A_{s(\text{min})} = 0.0018 bh$$

$$= (0.0018)(12 \text{ in})(18 \text{ in})$$

$$= 0.389 \text{ in}^2 \quad \text{[governs]}$$

The maximum factored bending moment in the heel is

$$M_u = \frac{L_H \left(1.2 W_S + 1.6 W_L + \dfrac{1.2 W_B L_H}{L_B}\right)}{2}$$

$$- \frac{L_H^2 B \left(q_{uB} + 2 q_{u(\text{heel})}\right)}{6}$$

$$= \frac{(8 \text{ ft})\left(\begin{array}{c} 21{,}120 \text{ lbf} + 2816 \text{ lbf} \\ + \dfrac{(3376 \text{ lbf})(8 \text{ ft})}{12.5 \text{ ft}} \end{array}\right)}{2}$$

$$- \frac{(8 \text{ ft})^2 (1 \text{ ft})\left(3363\,\dfrac{\text{lbf}}{\text{ft}^2} + (2)\left(249\,\dfrac{\text{lbf}}{\text{ft}^2}\right)\right)}{6}$$

$$= 63{,}203 \text{ ft-lbf}$$

Assuming a tension-controlled section, the required reinforcement ratio is

$$\rho = \frac{0.85 f_c' \left(1 - \sqrt{1 - \dfrac{M_u}{0.383 bd^2 f_c'}}\right)}{f_y}$$

$$= \frac{(0.85)\left(3\,\dfrac{\text{kips}}{\text{in}^2}\right)}{60\,\dfrac{\text{kips}}{\text{in}^2}} \times \left(1 - \sqrt{1 - \dfrac{(63{,}203 \text{ ft-lbf})\left(12\,\dfrac{\text{in}}{\text{ft}}\right)}{(0.383)(12 \text{ in})(15.5 \text{ in})^2 \times \left(3000\,\dfrac{\text{lbf}}{\text{in}^2}\right)}}\right)$$

$$= 0.0052$$

$$< \rho_t \quad \begin{bmatrix} \text{satisfactory, the section} \\ \text{is tension-controlled} \end{bmatrix}$$

$$A_s = \rho b d$$

$$= (0.0052)(12 \text{ in})(15.5 \text{ in})$$

$$= 0.97 \text{ in}^2$$

The maximum factored bending moment in the stem is

$$M_u = (1.6)\left(\frac{p_A (L_W)^3}{6} + \frac{p_A \bar{h} (L_W)^2}{2}\right)$$

$$= (1.6)\left(\frac{\left(30\,\dfrac{\text{lbf}}{\text{ft}^2}\right)(20 \text{ ft})^3}{6} \right.$$

$$\left. + \frac{\left(30\,\dfrac{\text{lbf}}{\text{ft}^2}\right)(2 \text{ ft})(20 \text{ ft})^2}{2}\right)$$

$$= 83{,}200 \text{ ft-lbf}$$

The required reinforcement ratio is

$$\rho = 0.0070 < \rho_t \quad \text{[satisfactory]}$$

$$A_s = \rho bd = (0.0075)(12 \text{ in})(15.5 \text{ in}) = 1.30 \text{ in}^2$$

7. COUNTERFORT RETAINING WALL

Nomenclature

a	lever-arm of resisting couple in counterfort	ft
b_c	width of counterfort	in
C	compression force in resisting couple	kips
l_c	clear height of counterfort	ft
l_n	clear span between counterforts	ft
q	earth pressure at a depth of $0.6l_c = 0.6p_A l_c$	lbf/ft
s_c	spacing, center to center, of counterforts, $l_n + b_c$	ft
T	tension force in resisting couple	kips

Design of Stem and Base

The stem spans horizontally between counterforts and cantilevers from the base. For ratios of l_n/l_c between 0.5 and 1.0, the stem may be designed for a value of the earth pressure at a depth of $0.6l_c$. The horizontal span moments are given by ACI Sec. 8.3.3 as $ql_n^2/11$ at counterfort supports and $ql_n^2/16$ between counterforts. The cantilever moment at the base is $0.035p_A l_c^3$. The distribution of bending moment in the stem is shown in Fig. 2.12, and more precise values of moment may be obtained from tabulated coefficients.[1,2]

The base slab is similarly designed for the net factored pressure as a slab spanning longitudinally between counterforts.

Example 2.18

For the counterfort retaining wall shown in the following illustration, determine the design moments in the stem. The fill behind the wall has an equivalent fluid pressure of 40 lbf/ft²/ft.

Figure 2.12 *Details of Counterfort Retaining Wall*

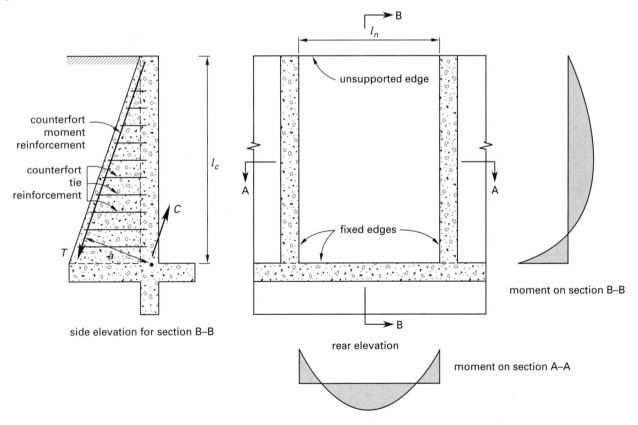

Illustration for Ex. 2.18

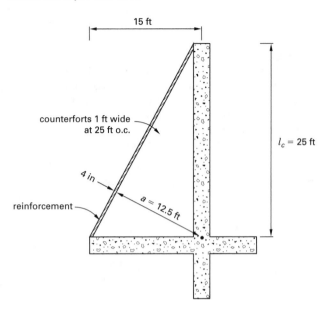

Solution

The lateral earth pressure acting on a 1 ft horizontal strip at a depth of 0.6 times the stem height is

$$q = 0.6p_A l_c$$
$$= (0.6)\left(40\ \frac{\text{lbf}}{\text{ft}^2}\right)(25\ \text{ft})$$
$$= 600\ \text{lbf/ft}$$

At the counterfort supports, the factored design moment is given by ACI Secs. 8.3.3 and 9.2.1 as

$$M_u = \frac{1.6ql_n^2}{11} = \frac{(1.6)\left(0.6\ \frac{\text{kip}}{\text{ft}}\right)(24\ \text{ft})^2}{11}$$
$$= 50.27\ \text{ft-kips}$$

Between counterforts, the factored design moment is

$$M_u = \frac{1.6ql_n^2}{16} = 34.56\ \text{ft-kips}$$

The factored design cantilever moment at the base is

$$M_u = 1.6(0.035p_A l_c^3)$$
$$= (1.6)(0.035)$$
$$\times \left(0.04\ \frac{\text{kip}}{\text{ft}^2}\right)(25\ \text{ft})^3$$
$$= 35.00\ \text{ft-kips}$$

Design of Counterforts

The bending moment produced by the earth pressure at the base of the stem is resisted by the couple produced by the tension in the reinforcement at the rear of the counterfort and the compression in the stem concrete. As shown in Fig. 2.12, the lever-arm of the couple acts at right angles to the reinforcement. The thrust produced by the earth pressure acting on the rear face of the stem is resisted by the horizontal ties in the counterfort.

Example 2.19

For the counterfort retaining wall of Ex. 2.18, determine the reinforcement area required in the rear of the counterfort and the tie area required at the base of the counterfort. Grade 60 reinforcement is provided, and the concrete compressive strength is 3000 kips/in^2.

Solution

The factored moment produced by the earth pressure at the base of the stem over one bay is

$$M_u = \frac{1.6p_A s_c l_c^3}{6}$$
$$= \frac{(1.6)\left(40\ \dfrac{\text{lbf}}{\text{ft}^2}\right)(25\ \text{ft})(25\ \text{ft})^3}{(6)\left(1000\ \dfrac{\text{lbf}}{\text{kip}}\right)}$$
$$= 4167\ \text{ft-kips}$$

The reinforcement area required in the rear of the counterfort to resist this moment is

$$A_s = \frac{M_u}{\phi a f_y}$$
$$= \frac{4167\ \text{ft-kips}}{(0.9)(12.5\ \text{ft})\left(60\ \dfrac{\text{kips}}{\text{in}^2}\right)}$$
$$= 6.17\ \text{in}^2$$

The factored lateral pressure from the backfill on a 1 ft horizontal strip at the base of the stem over one bay is

$$Q_u = 1.6p_A l_c l_n$$
$$= \frac{(1.6)\left(40\ \dfrac{\text{lbf}}{\text{ft}^2}\right)(25\ \text{ft})(24\ \text{ft})}{1000\ \dfrac{\text{lbf}}{\text{kip}}}$$
$$= 38.40\ \text{kips}$$

The tie reinforcement area required at the base of the counterfort to resist this lateral pressure is

$$A_s = \frac{Q_u}{\phi f_y} = \frac{38.40\ \text{kips}}{(0.9)\left(60\ \dfrac{\text{kips}}{\text{in}^2}\right)} = 0.72\ \text{in}^2$$

References

1. Portland Cement Association. *Rectangular Concrete Tanks*. 1998.

2. Reynolds, C.E., Steedman, J.C., and Threlfall, J.C. *Reinforced Concrete Designer's Handbook*. Spon, London. 2007.

PRACTICE PROBLEMS

1. What is the factored net pressure on the footing shown in the following illustration?

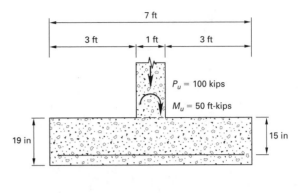

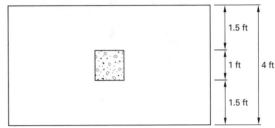

2. The rectangular footing of Prob. 1 has a concrete strength of 4000 lbf/in^2. Is the punching shear capacity adequate?

3. Is the flexural-shear capacity adequate for the rectangular footing of Prob. 1?

4. What is the area of grade 60 reinforcement required in the direction of the applied moment for the rectangular footing of Prob. 1?

5. For the retaining wall of Ex. 2.17, what is the minimum area of horizontal reinforcement required in the stem? Use no. 3 grade 60 bars.

SOLUTIONS

1. The equivalent eccentricity is

$$e = \frac{M_u}{P_u} = \frac{50 \text{ ft-kips}}{100 \text{ kips}} = 0.5 \text{ ft}$$

$$< \frac{L}{6} \quad \text{[within middle third]}$$

The net factored pressure on the footing is

$$q_u = \frac{P_u \left(1 \pm \dfrac{6e}{L}\right)}{BL}$$

$$= \frac{(100 \text{ kips}) \left(1 \pm \dfrac{(6)(0.5 \text{ ft})}{7 \text{ ft}}\right)}{(4 \text{ ft})(7 \text{ ft})}$$

$$q_{u(\max)} = 5.10 \text{ kips/ft}^2$$

$$q_{u(\min)} = 2.04 \text{ kips/ft}^2$$

2. The length of the critical perimeter is

$$b_o = (4)(c + d) = (4)(12 \text{ in} + 15 \text{ in})$$

$$= 108 \text{ in} = 9 \text{ ft}$$

Shear at the critical perimeter is

$$V_u = P_u - (0.5)\left(q_{u(\max)} + q_{u(\min)}\right)\left(\frac{b_o}{4}\right)^2$$

$$= 100 \text{ kips} - (0.5)\left(5.10 \frac{\text{kips}}{\text{ft}^2} + 2.04 \frac{\text{kips}}{\text{ft}^2}\right)$$

$$\times (2.25 \text{ ft})^2$$

$$= 82 \text{ kips}$$

The polar moment of inertia of the critical perimeter is

$$\frac{J_c}{y} = \frac{b_1 d(b_1 + 3b_2) + d^3}{3} \quad \left[\begin{array}{c}\text{for a footing with a}\\\text{central column}\end{array}\right]$$

$$= \frac{(27 \text{ in})(15 \text{ in})((27 \text{ in}) + (3)(27 \text{ in}) + (15 \text{ in})^3}{3}$$

$$= 15{,}705 \text{ in}^3$$

The fraction of the column moment transferred by shear is

$$\gamma_v = 1 - \frac{1}{1 + 0.67\sqrt{\dfrac{b_1}{b_2}}}$$

$$= 1 - \frac{1}{1.67}$$

$$= 0.40$$

The combined shear stress due to the applied axial load and the column moment is

$$v_u = \frac{V_u}{db_o} + \frac{\gamma_v M_u y}{J_c}$$

$$= \frac{(82 \text{ kips})\left(1000 \dfrac{\text{lbf}}{\text{kip}}\right)}{(15 \text{ in})(108 \text{ in})}$$

$$+ \frac{(0.4)(600 \text{ in-kips})\left(1000 \dfrac{\text{lbf}}{\text{kip}}\right)}{15{,}705 \text{ in}^3}$$

$$= 51 + 15$$

$$= 66 \text{ lbf/in}^2$$

The ratio of the long side to the short side of the column is

$$\beta_c = \frac{c_2}{c_1}$$

$$= \frac{12 \text{ in}}{12 \text{ in}}$$

$$= 1.00$$

$$< 2$$

The allowable shear stress for two-way action is given by ACI Eq. (11-35) as

$$\phi v_c = 4\phi\sqrt{f_c'}$$

$$\phi = \text{strength reduction factor}$$

$$= 0.75 \text{ from ACI Sec. 9.3}$$

$$\phi v_c = (4)(0.75)\sqrt{4000 \frac{\text{lbf}}{\text{in}^2}}$$

$$= 190 \text{ lbf/in}^2$$

$$> v_u \quad \text{[satisfactory]}$$

3. The distance of the critical section for flexural shear from the edge of the footing is

$$x = \frac{L}{2} - \frac{c}{2} - d = \frac{7 \text{ ft}}{2} - \frac{1 \text{ ft}}{2} - 1.25 \text{ ft}$$

$$= 1.75 \text{ ft}$$

The net factored pressure on the footing at this section is

$$q_{ux} = q_{u(\max)} - \frac{x\left(q_{u(\max)} - q_{u(\min)}\right)}{L}$$

$$= 5.10 \frac{\text{kips}}{\text{ft}^2} - \frac{(1.75 \text{ ft})\left(5.10 \dfrac{\text{kips}}{\text{ft}^2} - 2.04 \dfrac{\text{kips}}{\text{ft}^2}\right)}{7 \text{ ft}}$$

$$= 5.10 \frac{\text{kips}}{\text{ft}^2} - (1.75 \text{ ft})\left(0.437 \frac{\text{kips}}{\text{ft}^3}\right)$$

$$= 4.34 \text{ kips/ft}^2$$

Foundations

The factored shear force at the critical section is

$$V_u = \frac{Bx\left(q_{u(\text{max})} + q_{ux}\right)}{2}$$

$$= \frac{(4\text{ ft})(1.75\text{ ft})\left(5.10\ \dfrac{\text{kips}}{\text{ft}^2} + 4.34\ \dfrac{\text{kips}}{\text{ft}^2}\right)}{2}$$

$$= 33.04\text{ kips}$$

The flexural-shear capacity of the footing is given by ACI Eq. (11-3) as

$$\phi V_c = 2\phi Bd\sqrt{f_c'}$$

$$= \frac{(2)(0.75)(48\text{ in})(15\text{ in})\sqrt{4000\ \dfrac{\text{lbf}}{\text{in}^2}}}{1000\ \dfrac{\text{lbf}}{\text{kip}}}$$

$$= 68\text{ kips}$$

$$> V_u \quad [\text{satisfactory}]$$

4. The net factored pressure on the footing at the face of the column is

$$q_{uc} = q_{u(\text{max})} - \frac{\left(\dfrac{L}{2} - \dfrac{c}{2}\right)\left(q_{u(\text{max})} - q_{u(\text{min})}\right)}{L}$$

$$= 5.10\ \frac{\text{kips}}{\text{ft}^2} - (3.5\text{ ft} - 0.5\text{ ft})\left(0.437\ \frac{\text{kips}}{\text{ft}^3}\right)$$

$$= 3.79\text{ kips/ft}^2$$

The factored moment at the face of the column is

$$M_u = \frac{B\left(\dfrac{L}{2} - \dfrac{c}{2}\right)^2\left(2q_{u(\text{max})} + q_{uc}\right)}{6}$$

$$= \frac{(4\text{ ft})(3\text{ ft})^2\left((2)\left(5.10\ \dfrac{\text{kips}}{\text{ft}^2}\right) + 3.79\ \dfrac{\text{kips}}{\text{ft}^2}\right)}{6}$$

$$= 83.94\text{ ft-kips}$$

Assuming a tension-controlled section, the required reinforcement ratio is

$$\rho = \frac{0.85f_c'\left(1 - \sqrt{1 - \dfrac{M_u}{0.383bd^2f_c'}}\right)}{f_y}$$

$$= \frac{(0.85)\left(4\ \dfrac{\text{kips}}{\text{in}^2}\right)}{60\ \dfrac{\text{kips}}{\text{in}^2}}$$
$$\times\left(1 - \sqrt{1 - \dfrac{(83.94\text{ ft-kips})\left(12\ \dfrac{\text{in}}{\text{ft}}\right)\times\left(1000\ \dfrac{\text{lbf}}{\text{kip}}\right)}{(0.383)(48\text{ in})(15\text{ in})^2\times\left(4000\ \dfrac{\text{lbf}}{\text{in}^2}\right)}}\right)$$

$$= 0.0018$$

$$A_s = \rho Bd$$

$$= (0.0018)(48\text{ in})(15\text{ in})$$

$$= 1.30\text{ in}^2$$

The minimum reinforcement area is given by ACI Sec. 7.12.2 as

$$A_{s(\text{min})} = 0.0018Bh$$

$$= (0.0018)(48\text{ in})(19\text{ in})$$

$$= 1.64\text{ in}^2 \quad [\text{governs}]$$

5. From ACI Sec. 14.3.3, the required ratio of horizontal reinforcement in the stem is

$$\rho_{\text{hor}} = 0.0020$$

$$A_{sh} = \rho_{\text{hor}}bh$$

$$= (0.0020)(12\text{ in})(18\text{ in})$$

$$= 0.432\text{ in}^2/\text{ft}$$

3 Prestressed Concrete Design

1. Design Stages 3-1
2. Design for Shear and Torsion 3-12
3. Prestress Losses 3-17
4. Composite Construction 3-21
5. Load Balancing Procedure 3-26
6. Statically Indeterminate Structures 3-27
   Practice Problems 3-30
   Solutions 3-31

1. DESIGN STAGES

Nomenclature

a	depth of equivalent rectangular stress block	in
A_{ct}	area of concrete section between the centroid and extreme tension fiber	in^2
A_g	area of concrete section	in^2
A_{ps}	area of prestressed reinforcement in tension zone	in^2
A_s	area of nonprestressed tension reinforcement	in^2
A'_s	area of compression reinforcement	in^2
b	width of compression face of member	in
c	distance from extreme compression fiber to neutral axis	in
C_u	total compression force in equivalent rectangular stress block	lbf
d	distance from extreme compression fiber to centroid of nonprestressed reinforcement	in
d'	distance from extreme compression fiber to centroid of compression reinforcement	in
d_p	distance from extreme compression fiber to centroid of prestressed reinforcement as defined in Fig. 3.8	–
e	eccentricity of prestressing force	in
E_c	modulus of elasticity of concrete	kips/in^2
E_p	modulus of elasticity of prestressing tendon	kips/in^2
f_{be}	bottom fiber stress at service load after allowance for all prestress losses	lbf/in^2
f_{bi}	bottom fiber stress immediately after prestress transfer and before time-dependent prestress losses	lbf/in^2
f'_c	specified compressive strength of concrete	lbf/in^2
f'_{ci}	compressive strength of concrete at time of prestress transfer	lbf/in^2
f_{ps}	stress in prestressed reinforcement at nominal strength	kips/in^2
f_{pu}	specified tensile strength of prestressing tendons	kips/in^2

f_{py}	specified yield strength of prestressing tendons	kips/in^2
f_r	modulus of rupture of concrete	lbf/in^2
f_s	permissible stress in prestressed reinforcement at the jacking end	kips/in^2
f_{se}	effective stress in prestressed reinforcement after allowance for all prestress losses	kips/in^2
f_{si}	permissible stress in prestressed reinforcement immediately after prestress transfer	kips/in^2
f_{te}	top fiber stress at service loads after allowance for all prestress losses	lbf/in^2
f_{ti}	top fiber stress immediately after prestress transfer and before time-dependent prestress losses	lbf/in^2
f_y	specified yield strength of nonprestressed reinforcement	kips/in^2
h	height of section	in
I_g	moment of inertia of gross concrete section	in^4
l	span length	ft
M_{cr}	cracking moment strength	ft-kips
M_D	bending moment due to superimposed dead load	ft-kips
M_G	bending moment due to self-weight of member	ft-kips
M_L	bending moment due to superimposed live load	ft-kips
M_n	nominal flexural strength	ft-kips
M_S	bending moment due to sustained load	ft-kips
M_T	bending moment due to total load	ft-kips
M_u	factored moment	ft-kips
P_e	prestressing force after all losses	kips
P_i	initial prestressing force	kips
r	distance of tendon from the neutral axis	in
S_b	section modulus of the concrete section referred to the bottom fiber	in^3
S_t	section modulus of the concrete section referred to the top fiber	in^3
w_c	unit weight of concrete	lbf/ft^3
$\bar{y}$	height of centroid of the concrete section	in

Symbols

β_1	compression zone factor	
γ_p	factor for type of prestressing tendon	
ε_c	strain at extreme compression fiber at nominal strength, 0.003	
ε_p	prestrain in prestressed reinforcement due to the final prestress	
ε_s	strain produced in prestressed reinforcement by the ultimate loading	
ϕ	strength reduction factor	

General Requirements

In accordance with ACI Sec. R18.2, prestressed concrete members shall be designed for the three design stages: transfer, service, and ultimate. At the *transfer design stage*, a prestressing force is applied to the member. Immediate prestress losses occur as a result of elastic deformation of the concrete and, in the case of post-tensioned concrete, anchor set and friction losses. At the *serviceability design stage*, all time-dependent prestress losses have occurred as a result of creep and shrinkage of the concrete and relaxation of the tendon stress. At the *strength design stage*, a rectangular stress block is assumed with a maximum strain in the concrete of 0.003.

Transfer Design Stage

The permissible stresses at transfer are specified in ACI Sec. 18.4.1. As shown in Fig. 3.1, the initial prestressing force mobilizes the self-weight of the member producing the stresses.

$$f_{ti} = P_i \left(\frac{1}{A_g} - \frac{e}{S_t} \right) + \frac{M_G}{S_t} = P_i R_t + \frac{M_G}{S_t}$$

$$\geq -6\sqrt{f'_{ci}} \quad \begin{bmatrix} \text{at ends of simply supported beams} \\ \text{without auxiliary reinforcement} \end{bmatrix}$$

$$\geq -3\sqrt{f'_{ci}} \quad \begin{bmatrix} \text{at all other locations without} \\ \text{auxiliary reinforcement} \end{bmatrix}$$

P_i = force in prestressing tendon immediately after prestress transfer

$$= A_{ps} f_{si}$$

$$R_t = \frac{1}{A_g} - \frac{e}{S_t}$$

$$f_{bi} = P_i \left(\frac{1}{A_g} + \frac{e}{S_b} \right) - \frac{M_G}{S_b} = P_i R_b - \frac{M_G}{S_b}$$

$$\leq 0.60 f'_{ci}$$

$$R_b = \frac{1}{A_g} + \frac{e}{S_b}$$

The permissible stresses are shown in Fig. 3.2.

Figure 3.2 *Specified Concrete Stress at Transfer*

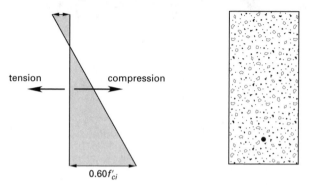

$-6\sqrt{f'_{ci}}$ ...at member ends without auxiliary reinforcement

$-3\sqrt{f'_{ci}}$ ...at other locations without auxiliary reinforcement

tension   compression

$0.60 f'_{ci}$

In accordance with ACI Sec. 18.5.1, the permissible stress in the prestressing tendon immediately after transfer is

$$f_{si} = 0.82 f_{py}$$

$$\leq 0.74 f_{pu}$$

At post-tensioning anchorages and couplers, the permissible stress is

$$f_{si} = 0.70 f_{pu}$$

The maximum permissible stress due to the tendon jacking force is

$$f_s = 0.94 f_{py}$$

$$\leq 0.80 f_{pu}$$

The permissible tendon stresses are shown in Fig. 3.3.

Example 3.1

The pre-tensioned beam shown in the following illustration is simply supported over a span of 20 ft and has a concrete strength at transfer of 4500 lbf/in². Determine the magnitude and location of the initial prestressing

Figure 3.1 *Transfer Design Stage*

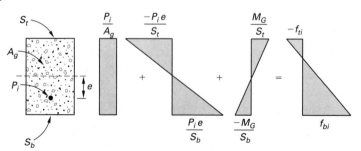

Figure 3.3 *Specified Stress in Prestressing Tendons*

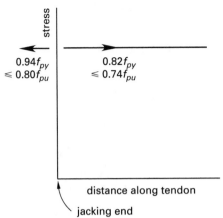 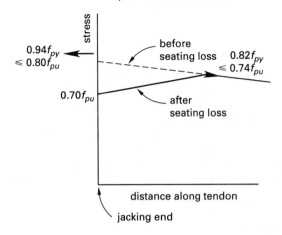

force required to produce satisfactory stresses at mid-span, immediately after transfer, without using auxiliary reinforcement.

Solution

The properties of the concrete section are

$$A_g = 72 \text{ in}^2 = 0.5 \text{ ft}^2$$

$$I_g = 863 \text{ in}^4$$

$$\bar{y} = 4.67 \text{ in}$$

$$S_t = 118 \text{ in}^3$$

$$S_b = 185 \text{ in}^3$$

At midspan, the self-weight moment is

$$M_G = \frac{w_c A_g l^2}{8}$$

$$= \frac{\left(150 \dfrac{\text{lbf}}{\text{ft}^3}\right)(0.5 \text{ ft}^2)(20 \text{ ft})^2 \left(12 \dfrac{\text{in}}{\text{ft}}\right)}{8}$$

$$= 45{,}000 \text{ in-lbf}$$

At midspan, the permissible tensile stress in the top fiber without auxiliary reinforcement is given by ACI Sec. 18.4.1 as

$$f_{ti} = -3\sqrt{f'_{ci}} = -3\sqrt{4500 \ \dfrac{\text{lbf}}{\text{in}^2}}$$

$$= -201 \text{ lbf/in}^2 = \frac{P_i}{A_g} - \frac{P_i e}{S_t} + \frac{M_G}{S_t}$$

$$= \frac{P_i}{72 \text{ in}^2} - \frac{P_i e}{118 \text{ in}^3} + \frac{45{,}000 \text{ in-lbf}}{118 \text{ in}^3}$$

$$-582 \ \frac{\text{lbf}}{\text{in}^2} = \frac{P_i}{72 \text{ in}^2} - \frac{P_i e}{118 \text{ in}^3} \quad \text{[Eq. 1]}$$

At midspan, the permissible compressive stress in the bottom fiber is given by ACI Sec. 18.4.1 as

$$f_{bi} = 0.6 f'_{ci} = 2700 \text{ lbf/in}^2$$

$$= \frac{P_i}{A_g} + \frac{P_i e}{S_b} - \frac{M_G}{S_b}$$

$$= \frac{P_i}{72 \text{ in}^2} + \frac{P_i e}{185 \text{ in}^3} - \frac{45{,}000 \text{ in-lbf}}{185 \text{ in}^3}$$

$$2943 \ \frac{\text{lbf}}{\text{in}^2} = \frac{P_i}{72 \text{ in}^2} + \frac{P_i e}{185 \text{ in}^3} \quad \text{[Eq. 2]}$$

Illustration for Ex. 3.1

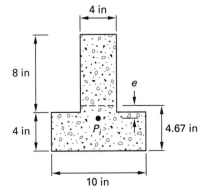

Solving Eqs. [1] and [2] gives

$$P_i = 113{,}056 \text{ lbf}$$
$$e = 2.25 \text{ in}$$

Auxiliary Reinforcement

ACI Sec. 18.4.1 specifies that when the computed tensile stress exceeds the permissible stress, bonded auxiliary reinforcement shall be provided to resist the total tensile force in the concrete. The tensile force is computed by using the properties of the uncracked concrete section, and in accordance with ACI Sec. R18.4.1, the permissible stress in the auxiliary reinforcement is $0.6f_y$ or 30 kips/in^2 maximum.

Figure 3.4 *Determination of Tensile Force*

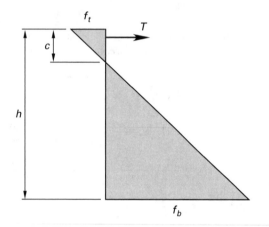

From Fig. 3.4, the depth to the location of zero stress is given by

$$c = \frac{hf_t}{f_t + f_b}$$

The tensile force in the concrete is

$$T = \frac{cf_t b}{2}$$

The area of auxiliary reinforcement required is given by ACI Sec. R18.4.1 as

$$A_s = \frac{T}{0.6f_y} \geq \frac{T}{30 \frac{\text{kips}}{\text{in}^2}}$$

Example 3.2

The pre-tensioned beam of Ex. 3.1 is prestressed with tendons providing an initial prestressing force of 110,100 lbf at an eccentricity of 2.37 in at midspan. Determine the area of grade 60 auxiliary reinforcement required.

Solution

$$R_b = \frac{1}{A_g} + \frac{e}{S_b} = \frac{1}{72 \text{ in}^2} + \frac{2.37 \text{ in}}{185 \text{ in}^3}$$
$$= 0.0267 \text{ 1/in}^2$$

$$R_t = \frac{1}{A_g} - \frac{e}{S_t} = \frac{1}{72 \text{ in}^2} - \frac{2.37 \text{ in}}{118 \text{ in}^3}$$
$$= -0.0062 \text{ 1/in}^2$$

The top and bottom fiber stresses are

$$f_t = P_i R_t + \frac{M_G}{S_t}$$
$$= (110{,}100 \text{ lbf}) \left(-0.0062 \frac{1}{\text{in}^2} \right)$$
$$\quad + \frac{45{,}000 \text{ in-lbf}}{118 \text{ in}^3}$$
$$= -301 \text{ lbf/in}^2$$

This is less than the minimum permissible value of $-3\sqrt{f_c'} = -201 \text{ lbf/in}^2$ determined in Ex. 3.1, and auxiliary reinforcement is required.

$$f_b = P_i R_b - \frac{M_G}{S_b}$$
$$= (110{,}100 \text{ lbf}) \left(0.0267 \frac{1}{\text{in}^2} \right)$$
$$\quad - \frac{45{,}000 \text{ in-lbf}}{185 \text{ in}^3}$$
$$= 2696 \text{ lbf/in}^2$$

This is less than the maximum permissible value of $0.6f_{ci}' = 2700 \text{ lbf/in}^2$ determined in Ex. 3.1 and is satisfactory.

Depth to the neutral axis is obtained from Fig. 3.4 as

$$c = \frac{hf_t}{f_t + f_b} = \frac{(12 \text{ in}) \left(301 \frac{\text{lbf}}{\text{in}^2} \right)}{301 \frac{\text{lbf}}{\text{in}^2} + 2696 \frac{\text{lbf}}{\text{in}^2}}$$
$$= 1.21 \text{ in}$$

The tensile force in the concrete is

$$T = \frac{cf_t b}{2} = \frac{(1.21 \text{ in}) \left(301 \frac{\text{lbf}}{\text{in}^2} \right)(4 \text{ in})}{2}$$
$$= 728 \text{ lbf}$$

The area of auxiliary reinforcement required is obtained from ACI Sec. R18.4.1 as

$$A_s = \frac{T}{30{,}000 \frac{\text{lbf}}{\text{in}^2}} = 0.024 \text{ in}^2$$

Figure 3.5 *Serviceability Design Stage After all Losses*

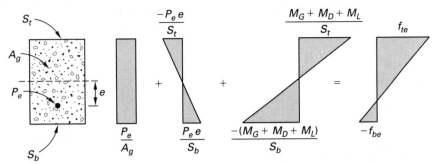

Serviceability Design Stage

The permissible stresses under service loads after all prestressing losses have occurred are specified in ACI Secs. 18.3.3 and 18.4.2. The stress conditions are shown in Fig. 3.5, and the stresses are given by

$$f_{te} = P_e R_t + \frac{M_G + M_D + M_L}{S_t}$$

$$\leq 0.45 f_c' \quad \text{[for sustained loads]}$$

$$\leq 0.60 f_c' \quad \text{[for total loads]}$$

$$f_{be} = P_e R_b - \frac{M_G + M_D + M_L}{S_b}$$

$$= -7.5\sqrt{f_c'} \quad \text{[for class U member]}$$

$$= -12\sqrt{f_c'} \quad \text{[for class T member]}$$

$$< -12\sqrt{f_c'} \quad \text{[for class C member]}$$

In accordance with ACI Sec. R18.3.3, stresses at the serviceability design stage in class U and class T members may be computed using uncracked section properties, and no crack control measures are necessary. Stresses in class C members are computed using cracked section properties, and crack control measures are necessary as specified in ACI Secs. 10.6.4, 10.6.7, and 18.4.4.1. Deflections for class U members are based on uncracked section properties, and for class T and class C members are based on the cracked transformed section properties, as specified in ACI Sec. 9.5.4.2.

$$P_e = \text{force in prestressing tendon at service loads after allowance for all losses}$$

$$= A_{ps} f_{se}$$

The permissible stresses are shown in Fig. 3.6.

Figure 3.6 *Permissible Concrete Stress at Service load*

0.45f_c'...for sustained loads

0.60f_c'...for total loads

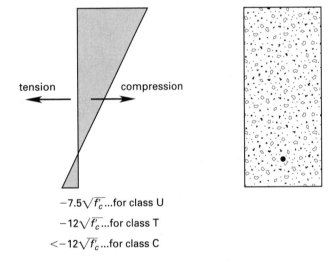

$-7.5\sqrt{f_c'}$...for class U

$-12\sqrt{f_c'}$...for class T

$< -12\sqrt{f_c'}$...for class C

Example 3.3

The class U pre-tensioned beam of Ex. 3.1 has a long-term loss in prestress of 25% and a 28 day compressive strength of 6000 lbf/in². The initial prestressing force is $P_i = 112{,}850$ lbf with an eccentricity of $e = 2.25$ in. Determine the maximum bending moment the beam can carry if the sustained load is 75% of the total superimposed load.

Solution

The relevant parameters are $e = 2.25$ in, $R_b = 0.0261$ in^{-2}, $P_i = 112{,}850$ lbf, $P_i R_b = 2940$ lbf/in², and $R_t = -0.00518$ in^{-2}.

The permissible tensile stress at midspan, in the bottom fiber, due to the total load is given by ACI Sec. 18.4.2 as

$$f_{be} = -7.5\sqrt{f_c'}$$

$$= -7.5\sqrt{6000\ \frac{\text{lbf}}{\text{in}^2}}$$

$$= -581\ \text{lbf/in}^2$$

$$= P_e R_b - \frac{M_G}{S_b} - \frac{M_T}{S_b}$$

$$= (0.75)\left(2940\ \frac{\text{lbf}}{\text{in}^2}\right) - 243\ \frac{\text{lbf}}{\text{in}^2} - \frac{M_T}{185\ \text{in}^3}$$

$$M_T = 470{,}455\ \text{in-lbf}$$

The permissible compressive stress at midspan, in the top fiber, due to the sustained load is given by ACI Sec. 18.4.2 as

$$f_{te} = 0.45 f_c'$$

$$= (0.45)\left(6000\ \frac{\text{lbf}}{\text{in}^2}\right)$$

$$= 2700\ \text{lbf/in}^2$$

$$= P_e R_t + \frac{M_G}{S_t} + \frac{M_S}{S_t}$$

$$= (0.75)\left(-584\ \frac{\text{lbf}}{\text{in}^2}\right) + 381\ \frac{\text{lbf}}{\text{in}^2} + \frac{0.75 M_T}{118\ \text{in}^3}$$

$$M_T = 433{,}770\ \text{in-lbf}$$

The permissible compressive stress at midspan, in the top fiber, due to the total load is given by ACI Sec. 18.4.2 as

$$f_{te} = 0.60 f_c' = (0.60)\left(6000\ \frac{\text{lbf}}{\text{in}^2}\right)$$

$$= 3600\ \frac{\text{lbf}}{\text{in}^2} = P_e R_t + \frac{M_G}{S_t} + \frac{M_T}{S_t}$$

$$= (0.75)\left(-584\ \frac{\text{lbf}}{\text{in}^2}\right) + 381\ \frac{\text{lbf}}{\text{in}^2} + \frac{M_T}{118\ \text{in}^3}$$

$$M_T = 431{,}530\ \text{in-lbf}\quad\text{[governs]}$$

Cracking Moment

The *cracking moment* is the moment that, when applied to the member after all losses have occurred, will cause cracking in the bottom fiber. From Fig. 3.7, equating the bottom fiber stresses gives a value for the modulus of rupture of

$$f_r = \frac{M_{cr}}{S_b} - P_e R_b = 7.5\sqrt{f_c'}\quad\text{[ACI 9-10]}$$

$$M_{cr} = S_b (P_e R_b + f_r)$$

In accordance with ACI Sec. 18.8.2,

$$\phi M_n \geq 1.2 M_{cr}$$

Example 3.4

For the pretensioned beam of Ex. 3.3, determine the cracking moment strength.

Solution

The modulus of rupture is given by ACI Sec. 9.5.2.3 as

$$f_r = 7.5\sqrt{f_c'}$$

$$= 7.5\sqrt{6000\ \frac{\text{lbf}}{\text{in}^2}}$$

$$= 581\ \text{lbf/in}^2$$

The cracking moment strength is

$$M_{cr} = S_b (P_e R_b + f_r)$$

$$= (185\ \text{in}^3)\left(\begin{array}{c}(0.75)\left(2940\ \dfrac{\text{lbf}}{\text{in}^2}\right)\\[1mm] +\ 581\ \dfrac{\text{lbf}}{\text{in}^2}\end{array}\right)$$

$$= 515{,}410\ \text{in-lbf}$$

Strength Design Stage

In accordance with ACI Sec. 18.7.3, nonprestressed reinforcement is assumed to contribute to the ultimate-moment of resistance of the section at its yield strength. Equating the longitudinal forces shown in Fig. 3.8 gives

$$0.85 f_c' ab = A_{ps} f_{ps} + A_s f_y - A_s' f_y$$

Figure 3.7 Cracking Moment

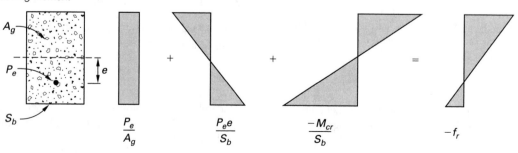

Figure 3.8 *Strain Distribution and Internal Forces at Flexural Failure*

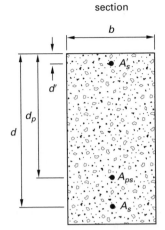
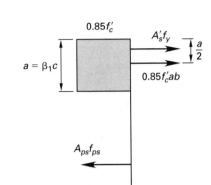

The nominal flexural strength of the member is

$$M_n = A_{ps}f_{ps}\left(d_p - \frac{a}{2}\right) + A_sf_y\left(d - \frac{a}{2}\right) + A'_sf_y\left(\frac{a}{2} - d'\right)$$

When the section does not contain auxiliary reinforcement, this strength reduces to

$$M_n = A_{ps}f_{ps}\left(d_p - \frac{0.59A_{ps}f_{ps}}{bf'_c}\right)$$

Approximate values of f_{ps} in terms of the reinforcement index may be determined in accordance with ACI Sec. 18.7.2, provided that $f_{se} \geq 0.5f_{pu}$ and that all the prestressing tendons are located in the tensile zone.

The reinforcement indices are given by ACI Sec. 2.1 as

ω = reinforcement index of nonprestressed tension reinforcement

$$= \frac{\rho f_y}{f'_c}$$

ω' = reinforcement index of compression reinforcement

$$= \frac{\rho' f_y}{f'_c}$$

The reinforcement ratios are given by ACI Sec. 2.1 as

ρ = ratio of nonprestressed tension reinforcement

$$= \frac{A_s}{bd}$$

ρ' = ratio of compression reinforcement

$$= \frac{A'_s}{bd}$$

ρ_p = ratio of prestressed reinforcement

$$= \frac{A_{ps}}{bd_p}$$

Example 3.5

For the pre-tensioned beam of Ex. 3.1, which has a 28 day compressive strength of 6000 lbf/in², determine the maximum possible value of the nominal flexural strength for a tension-controlled section.

Solution

The height of the centroid of the section is given in Ex. 3.1 as

$$\bar{y} = 4.67 \text{ in}$$

The eccentricity of the prestressing force is given in Ex. 3.1 as

$$e = 2.25 \text{ in}$$

The height of the section is given in Ex. 3.1 as

$$h = 12 \text{ in}$$

Then, the distance from the extreme compression fiber to the centroid of prestressed reinforcement is given by

$$\begin{aligned} d_p &= h - \bar{y} + e \\ &= 12 \text{ in} - 4.67 \text{ in} + 2.25 \text{ in} \\ &= 9.58 \text{ in} \end{aligned}$$

The maximum depth of the rectangular stress block for a tension-controlled section is given by ACI Sec. 10.3.4 as

$$\begin{aligned} a &= 0.375\beta_1 d_p \\ &= (0.375)(0.75)(9.58 \text{ in}) \\ &= 2.69 \text{ in} \end{aligned}$$

The maximum nominal flexural strength is

$$\begin{aligned} M_n &= (0.85f'_c ab)\left(d_p - \frac{a}{2}\right) \\ &= (0.85)\left(6000 \ \frac{\text{lbf}}{\text{in}^2}\right)(2.69 \text{ in})(4 \text{ in})(8.23 \text{ in}) \\ &= 451{,}630 \text{ in-lbf} \end{aligned}$$

$$\phi M_n < 1.2M_{cr} \quad \text{[unsatisfactory]}$$

Flexural Strength of Members with Bonded Tendons

For bonded tendons, ACI Sec. 18.7.2 gives the value of the stress in the prestressed reinforcement at nominal strength as

$$f_{ps} = f_{pu}\left(1 - \left(\frac{\gamma_p}{\beta_1}\right)\left(\rho_p\left(\frac{f_{pu}}{f_c'}\right) + \frac{d(\omega - \omega')}{d_p}\right)\right)$$

[ACI 18-3]

The factor for type of prestressing tendon is given by ACI Sec. 18.7.2 as

$\gamma_p = 0.55$ for deformed bars with $f_{py}/f_{pu} \geq 0.80$

$\quad = 0.40$ for stress-relieved wire and strands, and plain bars with $f_{py}/f_{pu} \geq 0.85$

$\quad = 0.28$ for low-relaxation wire and strands with $f_{py}/f_{pu} \geq 0.90$

When compression reinforcement is taken into account while calculating f_{ps} by ACI Eq. (18-3),

$$0.17 \leq \left(\rho_p\left(\frac{f_{pu}}{f_c'}\right) + \frac{d(\omega - \omega')}{d_p}\right)$$

$$d' \leq 0.15d_p$$

When the section contains no auxiliary reinforcement, the value for f_{ps} reduces to

$$f_{ps} = f_{pu}\left(1 - \frac{\gamma_p \rho_p f_{pu}}{\beta_1 f_c'}\right)$$

Example 3.6

The pre-tensioned beam shown in the following illustration is simply supported over a span of 30 ft and has a 28 day concrete strength of 6000 lbf/in². The area of the low-relaxation prestressing tendons provided is 0.765 in² with a specified tensile strength of 270 kips/in², a yield strength of 243 kips/in², and an effective stress of 150 kips/in² after all losses. Determine the nominal flexural strength of the beam.

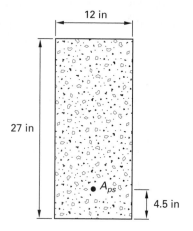

Solution

The relevant properties of the beam are

$$A_g = 324 \text{ in}^2$$

$$S_b = 1458 \text{ in}^3$$

$$e = \frac{h}{2} - 4.5 \text{ in} = \frac{27 \text{ in}}{2} - 4.5 \text{ in}$$

$$= 9 \text{ in}$$

$$R_b = \frac{1}{A_g} + \frac{e}{S_b} = \frac{1}{324 \text{ in}^2} + \frac{9 \text{ in}}{1458 \text{ in}^3}$$

$$= 0.00926 \text{ 1/in}^2$$

The factor for this type of prestressing tendon is given by ACI Sec. 18.7.2 as

$$\gamma_p = 0.28 \quad [\text{for } f_{py}/f_{pu} \geq 0.9]$$

$$\rho_p = \frac{A_{ps}}{bd_p} = \frac{0.765 \text{ in}^2}{(12 \text{ in})(22.5 \text{ in})}$$

$$= 0.00283$$

From ACI Sec. 10.2.7.3, the compression zone factor is given by

$$\beta_1 = 0.75$$

$$M_{cr} = S_b(P_e R_b + f_r)$$

$$= (1458 \text{ in}^3)\left(\begin{array}{c} \left(0.765 \text{ in}^2\right)\left(150 \frac{\text{kips}}{\text{in}^2}\right) \\ \times (0.00926 \text{ in}^{-2}) + 0.581 \frac{\text{kips}}{\text{in}^2} \end{array}\right)$$

$$= 2400 \text{ in-kips}$$

From ACI Eq. (18-3),

$$f_{ps} = f_{pu}\left(1 - \frac{\gamma_p \rho_p f_{pu}}{\beta_1 f_c'}\right)$$

$$= \left(270 \frac{\text{kips}}{\text{in}^2}\right)\left(1 - \frac{(0.28)(0.00283) \times \left(270 \frac{\text{kips}}{\text{in}^2}\right)}{(0.75)\left(6 \frac{\text{kips}}{\text{in}^2}\right)}\right)$$

$$= 257 \text{ kips/in}^2$$

The depth of the stress block is given by

$$a = \frac{A_{ps}f_{ps}}{0.85f_c'b}$$

$$= \frac{\left(0.765 \text{ in}^2\right)\left(257 \frac{\text{kips}}{\text{in}^2}\right)}{(0.85)\left(6 \frac{\text{kips}}{\text{in}^2}\right)(12 \text{ in})}$$

$$= 3.21 \text{ in}$$

The maximum depth of the stress block for a tension-controlled section is given by ACI Sec. 10.3.4 as

$$a_t = 0.375\beta_1 d_p$$
$$= (0.375)(0.75)(22.5)$$
$$= 6.33 \text{ in}$$
$$> a$$

Hence the section is tension-controlled and $\phi = 0.9$.

The nominal moment of resistance of the section is given by

$$M_n = A_{ps}f_{ps}\left(d_p - \frac{0.59A_{ps}f_{ps}}{bf'_c}\right)$$

$$= \left(0.765 \text{ in}^2\right)\left(257 \frac{\text{kips}}{\text{in}^2}\right)$$

$$\times \left(22.5 \text{ in} - \frac{\begin{array}{c}(0.59)\left(0.765 \text{ in}^2\right)\\ \times \left(257 \frac{\text{kips}}{\text{in}^2}\right)\end{array}}{(12 \text{ in})\left(6 \frac{\text{kips}}{\text{in}^2}\right)} \right)$$

$$= 4106 \text{ in-kips}$$

$$\phi M_n > 1.2M_{cr} \quad \text{[satisfactory]}$$

Flexural Strength of Members with Unbonded Tendons

For unbonded tendons and a span-to-depth ratio ≤ 35, ACI Sec. 18.7.2 gives

$$f_{ps} = f_{se} + 10{,}000 + \frac{f'_c}{100\rho_p} \quad \text{[ACI 18-4]}$$

$$\leq f_{py}$$
$$\leq f_{se} + 60{,}000$$

For unbonded tendons and a span-to-depth ratio > 35, ACI Sec. 18.7.2 gives

$$f_{ps} = f_{se} + 10{,}000 + \frac{f'_c}{300\rho_p} \quad \text{[ACI 18-5]}$$

$$\leq f_{py}$$
$$\leq f_{se} + 30{,}000$$

In accordance with ACI Sec. 18.9, auxiliary bonded reinforcement is required near the extreme tension fiber in all beams with unbonded tendons. The minimum area required is independent of the grade of steel and is given by

$$A_s = 0.004A_{ct} \quad \text{[ACI 18-6]}$$

A_{ct} is the area of the concrete section between the centroid of the section and the extreme tension fiber, as shown in Fig. 3.9.

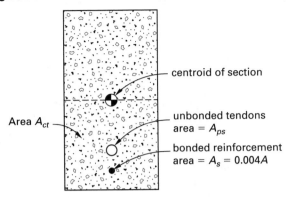

Figure 3.9 *Bonded Reinforcement Area*

In flat slabs, when the tensile stress due to dead load plus live load is less than $-2\sqrt{f'_c}$, auxiliary reinforcement at a stress of $0.5f_y$ must be provided to resist the total tensile force in the concrete.

Example 3.7

The post-tensioned beam shown in the following illustration is simply supported over a span of 30 ft and has a 28 day concrete strength of 6000 lbf/in². The area of the low-relaxation unbonded tendons provided is 0.765 in² with a specified tensile strength of 270 kips/in², a yield strength of 243 kips/in², and an effective stress of 150 kips/in² after all losses. The area of the grade 60 auxiliary reinforcement provided is 0.8 in². Determine the nominal flexural strength of the beam.

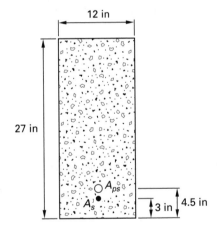

Solution

Because $f_{se}/f_{pu} > 0.5$, the method of ACI Sec. 18.7.2 may be used. The ratio of prestressed reinforcement is

$$\rho_p = \frac{A_{ps}}{bd_p}$$

$$= \frac{0.765 \text{ in}^2}{(12 \text{ in})(22.5 \text{ in})}$$

$$= 0.00283$$

From ACI Sec. 18.7.2, the stress in the unbonded tendons at nominal strength is

$$f_{ps} = f_{se} + 10 + \frac{f'_c}{100\rho_p}$$

$$= 150 \ \frac{\text{kips}}{\text{in}^2} + 10 \ \frac{\text{kips}}{\text{in}^2} + \frac{6 \ \frac{\text{kips}}{\text{in}^2}}{(100)\,(0.00283)}$$

$$= 181 \text{ kips/in}^2$$

$$< f_{py} \quad \text{[satisfactory]}$$

$$< f_{se} + 60 \quad \text{[satisfactory]}$$

The minimum area of auxiliary reinforcement required is specified by ACI Sec. 18.9 as

$$A_s = 0.004 A_{ct}$$

$$= (0.004)(12 \text{ in})(13.5 \text{ in})$$

$$= 0.648 \text{ in}^2$$

$$< 0.80 \text{ in}^2 \quad \text{[satisfactory]}$$

Assuming full use of the auxiliary reinforcement, the depth of the stress block is

$$a = \frac{A_{ps}f_{ps} + A_s f_y}{0.85 f'_c b}$$

$$= \frac{(0.765 \text{ in}^2)\left(181 \ \dfrac{\text{kips}}{\text{in}^2}\right) + (0.8 \text{ in}^2)\left(60 \ \dfrac{\text{kips}}{\text{in}^2}\right)}{(0.85)\left(6 \ \dfrac{\text{kips}}{\text{in}^2}\right)(12 \text{ in})}$$

$$= 3.04 \text{ in}$$

The maximum depth of the stress block for a tension-controlled section is given by ACI Sec. 10.3.4 as

$$a_t = 0.375\beta_1 d_p$$

$$= (0.375)(0.75)(22.5 \text{ in})$$

$$= 6.33 \text{ in}$$

$$> a \quad \text{[section is tension-controlled]}$$

The nominal flexural strength is

$$M_n = A_{ps}f_{ps}\left(d_p - \frac{a}{2}\right) + A_s f_y\left(d - \frac{a}{2}\right)$$

$$= (0.765 \text{ in}^2)\left(181 \ \frac{\text{kips}}{\text{in}^2}\right)(22.5 \text{ in} - 1.52 \text{ in})$$

$$\quad + (0.80 \text{ in}^2)\left(60 \ \frac{\text{kips}}{\text{in}^2}\right)(24 \text{ in} - 1.52 \text{ in})$$

$$= 3980 \text{ in-kips}$$

$$\phi M_n > 1.2 M_{cr} \quad \text{[satisfactory]}$$

Flexural Strength of Members Using Strain Compatibility

When the approximate methods of determining flexural strength are inapplicable, the strain analysis technique shown in Fig. 3.10 is used. An initial estimate is made of the location of the neutral axis, and assuming the maximum strain is 0.003, the strains in the tendons are determined to be $\varepsilon_s = 0.003r/c$. To these strains is added the prestrain in the tendons due to the final prestress in each tendon, which is $\varepsilon_p = f_{se}/E_p$. The force in each tendon is determined by using a stress-strain curve for the tendons. The total tensile force is compared with the compressive force in the concrete, and the location of the neutral axis is adjusted until equilibrium is obtained. Summing moments of forces about the neutral axis provides the flexural strength.

Figure 3.10 *Flexural Strength by Strain Compatibility*

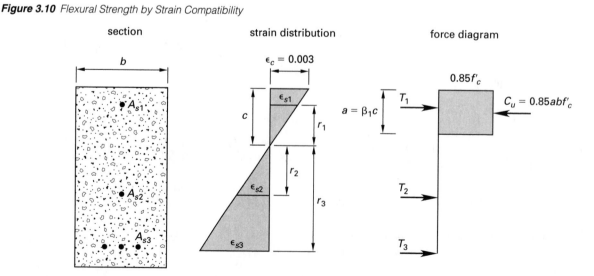

Example 3.8

The pre-tensioned beam shown in the following illustration has a 28 day concrete strength of 6000 lbf/in^2 and is pre-tensioned with five $1/2$ in diameter strands. The area of each strand is 0.153 in^2 with a specified tensile strength of 270 $kips/in^2$ and an effective stress of 150 $kips/in^2$ after all losses. Using the idealized stress-strain curve shown, determine the nominal flexural strength of the beam.

Solution

Assume that the depth to the neutral axis is $c = 4$ in. Then, the depth of the equivalent stress block is

$$a = \beta_1 c = (0.75)(4 \text{ in}) = 3 \text{ in}$$

The total compressive force in the concrete stress block is obtained from Fig. 3.10 as

$$C_u = 0.85 f'_c (ab - A_{s1})$$
$$= (0.85) \left(6 \, \frac{\text{kips}}{\text{in}^2} \right) ((3 \text{ in})(12 \text{ in}) - 0.153 \text{ in}^2)$$
$$= 183 \text{ kips}$$

For an effective final prestress in each tendon of 150 $kips/in^2$, the prestrain in each tendon is

$$\varepsilon_p = \frac{f_{se}}{E_p} = \frac{150 \, \dfrac{\text{kips}}{\text{in}^2}}{28{,}000 \, \dfrac{\text{kips}}{\text{in}^2}} = 5.36 \times 10^{-3}$$

The total strain in each tendon is given by

$$\varepsilon = \varepsilon_s + \varepsilon_p$$
$$= (0.003) \left(\frac{r}{c} \right) + \varepsilon_p$$

The tendons reach their specified tensile strength at a strain of

$$\varepsilon_{pu} = 14 \times 10^{-3} \quad \text{[from stress-strain curve]}$$

$$\varepsilon_{s1} = (0.003) \left(\frac{-2.5 \text{ in}}{4 \text{ in}} \right) + 5.36 \times 10^{-3}$$
$$= 3.49 \times 10^{-3}$$

$$\varepsilon_{s2} = (0.003) \left(\frac{18.5 \text{ in}}{4 \text{ in}} \right) + 5.36 \times 10^{-3}$$
$$= 19.24 \times 10^{-3} \quad \text{[exceeds } \varepsilon_{pu}\text{]}$$

$$\varepsilon_{s3} = (0.003) \left(\frac{21.5 \text{ in}}{4 \text{ in}} \right) + 5.36 \times 10^{-3} \quad \text{[exceeds } \varepsilon_{pu}\text{]}$$
$$= 21.49 \times 10^{-3}$$

The force in each tendon is given by

$$T = A_s f_s$$
$$T_1 = A_{s1} \varepsilon_{s1} E_p$$
$$= \left(0.153 \text{ in}^2 \right) \left(3.49 \times 10^{-3} \right) \left(28 \times 10^3 \, \frac{\text{kips}}{\text{in}^2} \right)$$
$$= 15 \text{ kips}$$

$$T_2 = \left(0.153 \text{ in}^2 \right) \left(270 \, \frac{\text{kips}}{\text{in}^2} \right) = 41 \text{ kips}$$

$$T_3 = (3) \left(0.153 \text{ in}^2 \right) \left(270 \, \frac{\text{kips}}{\text{in}^2} \right) = 124 \text{ kips}$$

$$\sum T = 15 + 41 + 124 = 180 \text{ kips}$$
$$\approx C_u \quad \text{[satisfactory]}$$

Taking moments about the neutral axis gives

$$M_n = T_1 (-r_1) + T_2 r_2 + T_3 r_3 + C_u \left(c - \frac{a}{2} \right)$$
$$= (15 \text{ kips})(-2.5 \text{ in}) + (41 \text{ kips})(18.5 \text{ in})$$
$$\quad + (124 \text{ kips})(21.5 \text{ in})$$
$$\quad + (183 \text{ kips})(4 \text{ in} - 1.5 \text{ in})$$
$$= 3845 \text{ in-kips}$$

Illustration for Ex. 3.8

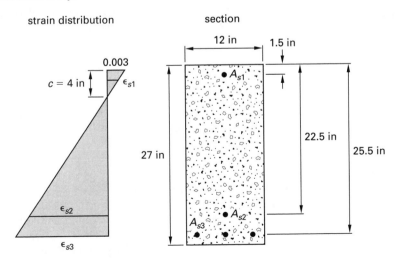

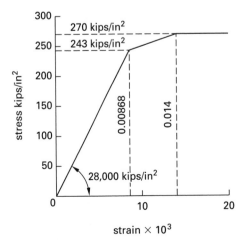

2. DESIGN FOR SHEAR AND TORSION

Nomenclature

A_{cp}	area enclosed by outside perimeter of concrete cross section	in^2
A_l	total area of longitudinal reinforcement to resist torsion	in^2
A_o	gross area enclosed by shear flow	in^2
A_{oh}	gross area enclosed by centerline of the outermost closed transverse torsional reinforcement	in^2
A_{ps}	area of prestressed reinforcement in tension zone	in^2
A_t	area of one leg of a closed stirrup resisting torsion within a spacing s	in^2
A_v	area of shear reinforcement within a spacing s	in^2
b_w	web width	in
d	distance from extreme compression fiber to centroid of tension reinforcement $\geq 0.8h$	in
d_{bl}	diameter of longitudinal torsional reinforcement	in
d_p	actual distance from extreme compression fiber to centroid of prestressing tendons as defined in Fig 3.8	in
f_d	tensile stress at bottom fiber of section due to unfactored dead load	kips/in^2
f_{pc}	compressive stress in concrete, due to final prestressing force, at the centroid of the section	kips/in^2
f_{pe}	compressive stress in concrete, due to final prestressing force, at the bottom fiber of the section	kips/in^2
f_{pu}	specified strength of prestressing tendons	kips/in^2
f_{se}	effective stress in prestressing reinforcement after allowance for all prestressing losses	kips/in^2
f_y	yield strength of longitudinal torsional reinforcement	kips/in^2
f_{yt}	yield strength of transverse torsional reinforcement	kips/in^2
h	overall thickness of member	in
M_{cre}	moment causing flexural cracking at section	in-kips
M_{max}	maximum factored moment at section due to externally applied loads	in-kips
M_u	factored moment at section	in-kips
p_{cp}	outside perimeter of concrete cross section	in
p_h	perimeter of centerline of outermost closed transverse torsional reinforcement	in

s	spacing of shear or torsion reinforcement in direction parallel to longitudinal reinforcement	in
S_b	section modulus of the section referred to the bottom fiber	in^3
T_n	nominal torsional moment strength	in-kips
T_u	factored torsional moment at section	in-kips
V_c	nominal shear strength provided by concrete	kips
V_{ci}	nominal shear strength provided by concrete when diagonal cracking results from combined shear and moment	kips
V_{cw}	nominal shear strength provided by concrete when diagonal cracking results from excessive principal tensile stress in the web	kips
V_d	shear force at section due to unfactored dead load	kips
V_i	factored shear force at section due to externally applied loads applied simultaneously with M_{max}	kips
V_p	vertical component of effective prestress force at section	kips
V_s	nominal shear strength provided by shear reinforcement	kips
V_u	factored shear force at section	kips

Symbols

λ	correction factor related to unit weight of concrete	
μ	coefficient of friction	
ρ_w	$A_s/b_w d$	
ϕ	strength reduction factor, 0.75 for shear and torsion	

Design for Shear

The nominal shear capacity of shear reinforcement perpendicular to the member is given by ACI Sec. 11.5.7.2 as

$$V_s = \frac{A_v f_y d}{s} \qquad \text{[ACI 11-15]}$$

The nominal shear strength of the shear reinforcement is limited by ACI Sec. 11.5.7.9 to a value of

$$V_s = 8 b_w d \sqrt{f_c'}$$

If additional shear capacity is required, the size of the concrete section must be increased. ACI Sec. 11.5.5 limits the spacing of the stirrups to a maximum value of $0.75h$ or 24 in, and when the value of V_s exceeds $4 b_w d \sqrt{f_c'}$, the spacing is reduced to a maximum value of $0.375h$ or 12 in.

When $f_{se} \geq 0.4f_{pu}$, a conservative and usually sufficiently accurate value of the nominal shear capacity of the concrete section is given by ACI Sec. 11.4.2 as

$$V_c = \left(0.60\sqrt{f_c'} + \frac{700V_u d_p}{M_u}\right)b_w d \quad \text{[ACI 11-9]}$$

$$\leq 5b_w d\sqrt{f_c'}$$

$$\geq 2b_w d\sqrt{f_c'}$$

$$\frac{V_u d_p}{M_u} \leq 1.0$$

M_u is the factored moment occurring simultaneously with V_u at the section being analyzed, and d_p is the distance from the extreme compression fiber to the centroid of the prestressed reinforcement.

The depth d is the distance from the extreme compression fiber to centroid of prestressed and nonprestressed tension reinforcement, but need not be taken less than $0.8h$.

In accordance with ACI Sec. 11.1.1, the combined shear capacity of the concrete section and the shear reinforcement is

$$\phi V_n = \phi V_c + \phi V_s$$

When the applied factored shear force V_u is less than $\phi V_c/2$, the concrete section is adequate to carry the shear without any shear reinforcement. Within the range $\phi V_c/2 \leq V_u \leq \phi V_c$, a minimum area of shear reinforcement is specified by ACI Sec. 11.5.6; this area is given by the smaller of the results of the following equations.

$$A_{v(\min)} = 0.75\sqrt{f_c'}\frac{b_w s}{f_{yt}} \quad \text{[ACI 11-13]}$$

$$= \frac{50 b_w s}{f_{yt}}$$

$$A_{v(\min)} = \frac{A_{ps}f_{pu}s\sqrt{\dfrac{d}{b_w}}}{80 f_{yt}d} \quad \text{[for } f_{se} > 0.4f_{pu}\text{]} \quad \text{[ACI 11-14]}$$

ACI Sec. 11.1.3 specifies that when the support reaction produces a compressive stress in the member, the critical section for shear is located at a distance from the support equal to one-half the overall depth. This location of the critical section is applicable provided that loads are applied near or at the top of the beam and no concentrated load is applied within a distance from the support equal to the effective depth.

Example 3.9

The post-tensioned beam shown in the following illustration has a 28 day concrete strength of 6000 lbf/in² and is tensioned with five ¹/₂ in diameter strands. The area of each strand is 0.153 in² with a specified tensile strength of 270 kips/in² and an effective stress of 150 kips/in² after all losses. The cable centroid, as shown, is parabolic in shape, and the value of $V_u d_p/M_u$ = 1.0 at section A-A. Determine the nominal shear capacity at section A-A.

Solution

The equation of the parabolic cable profile is

$$y = \frac{gx^2}{a^2} = \frac{(10.5 \text{ in})\,x^2}{\left((15 \text{ ft})\left(12\,\dfrac{\text{in}}{\text{ft}}\right)\right)^2}$$

At section A-A, the rise of the cable is given by

$$y_A = \frac{(10.5 \text{ in})\left((15 \text{ ft})\left(12\,\dfrac{\text{in}}{\text{ft}}\right) - 23 \text{ in}\right)^2}{\left((15 \text{ ft})\left(12\,\dfrac{\text{in}}{\text{ft}}\right)\right)^2}$$

$$= 8 \text{ in}$$

The actual depth of the cable is

$$d_p = h - y_A - y_o$$
$$= 27 \text{ in} - 8 \text{ in} - 4.5 \text{ in} = 14.5 \text{ in}$$

The effective depth of the section is

$$d = 0.8h$$
$$= (0.8)(27 \text{ in})$$
$$= 21.6 \text{ in} \quad \text{[governs]}$$

Illustration for Ex. 3.9

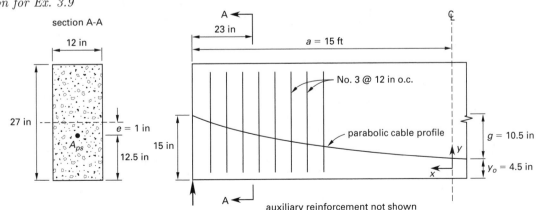

The nominal shear capacity is given by ACI Eq. (11-9) as

$$V_c = \left(0.6\sqrt{f_c'} + \frac{700 V_u d_p}{M_u}\right) b_w d$$

$$= \left(0.6\sqrt{6000 \; \frac{\text{lbf}}{\text{in}^2}} + (700)(1.0)\right)$$

$$\times \left(\frac{(12 \text{ in})(21.6 \text{ in})}{1000 \; \frac{\text{lbf}}{\text{kip}}}\right)$$

$$= 193 \text{ kips}$$

$$\leq 5 b_w d \sqrt{f_c'}$$

$$= \frac{(5)(12 \text{ in}) \; 21.6 \text{ in} \sqrt{6000 \; \frac{\text{lbf}}{\text{in}^2}}}{1000 \; \frac{\text{lbf}}{\text{kip}}}$$

$$= 100 \text{ kips} \quad [\text{governs}]$$

The nominal shear capacity of the stirrups provided is given by ACI Eq. (11-15) as

$$V_s = \frac{A_v f_{yt} d}{s}$$

$$= \frac{(0.22 \text{ in}^2)\left(60 \; \frac{\text{kips}}{\text{in}^2}\right)(21.6 \text{ in})}{12 \text{ in}}$$

$$= 24 \text{ kips}$$

The total nominal shear capacity is given by ACI Eq. (11-2) as

$$V_n = V_c + V_s$$

$$= 100 \text{ kips} + 24 \text{ kips}$$

$$= 124 \text{ kips}$$

Flexure-Shear and Web-Shear Cracking

A more precise value of the nominal shear capacity is provided by the lesser value of V_{ci} or V_{cw} given by ACI Secs. 11.4.3.1 and 11.4.3.2. For flexure-shear cracking, the nominal shear capacity is given by

$$V_{ci} = 0.6 b_w d_p \sqrt{f_c'} + V_d + \frac{V_i M_{cre}}{M_{\max}} \quad \textit{[ACI 11-10]}$$

$$\geq 1.7 b_w d \sqrt{f_c'}$$

V_d is the shear force at the section due to the unfactored dead load, $M_{\max}$ is the maximum factored moment at the section due to the externally applied loads, V_i is the factored shear force at the section associated with $M_{\max}$, and d is the distance from the top fiber of the section to the centroid of the prestressing tendons, but not less than $0.8h$. The cracking moment due to the unfactored external applied loads is given by

$$M_{cre} = S_b \left(6 \sqrt{f_c'} + f_{pe} - f_d\right) \quad \textit{[ACI 11-11]}$$

f_d is the tensile stress at the bottom fiber of the section, due to the unfactored dead load, and f_{pe} is the compressive stress in the concrete, due to the final prestressing force, at the bottom fiber of the section.

For a uniformly loaded member, ACI Sec. R11.4.3 gives the variant of ACI Eq. (11-10).

$$V_{ci} = 0.6 b_w d \sqrt{f_c'} + \frac{V_u M_{ct}}{M_u}$$

M_u is the total factored moment at the section, V_u is the factored shear force associated with M_u, and the total unfactored moment, including dead load, required to cause cracking is given by

$$M_{ct} = S_b \left(6 \sqrt{f_c'} + f_{pe}\right)$$

For composite members, in accordance with ACI Sec. R11.4.3, the applicable expressions are ACI Eqs. (11-10) and (11-11), with the stress f_d determined from the unfactored dead load resisted by the precast unit and the unfactored superimposed dead load resisted by the composite member. Similarly, M_d is the bending moment at the section due to the unfactored dead load acting on the precast unit plus the moment due to the unfactored superimposed dead load acting on the composite member; V_d is the unfactored shear force associated with M_d. Then,

$$V_i = V_u - V_d$$

$$M_{\max} = M_u - M_d$$

For web-shear cracking, the nominal shear capacity is given by ACI Sec. 11.4.3.2 as

$$V_{cw} = b_w d_p \left(3.5 \sqrt{f_c'} + 0.3 f_{pc}\right) + V_p$$

$$\textit{[ACI 11-12]}$$

f_{pc} is the compressive stress in the concrete due to the final prestressing force at the centroid of the section, and V_p is the vertical component of the effective prestress force at the section, in kips.

Example 3.10

For the post-tensioned beam of Ex. 3.9, determine the nominal shear capacity at section A-A by using ACI Eqs. (11-10) and (11-12). The unfactored bending moment at section A-A due to dead load and live load is 500 in-kips.

Solution

At section A-A, the slope of the cable is given by

$$\frac{dy}{dx} = 2\left(\frac{gx}{a^2}\right)$$

$$= 2\left(\frac{(10.5 \text{ in})\left((15 \text{ ft})\left(12 \; \frac{\text{in}}{\text{ft}}\right) - 23 \text{ in}\right)}{(180 \text{ in})^2}\right)$$

$$= 0.102$$

The vertical component of the final effective prestressing force at section A-A is

$$V_p = A_{ps} f_{se} \frac{dy}{dx}$$

$$= (0.765 \text{ in}^2) \left(150 \frac{\text{kips}}{\text{in}^2}\right) (0.102)$$

$$= 11.7 \text{ kips}$$

The compressive stress in the concrete, due to the final prestressing force, at the centroid of the section is

$$f_{pc} = \frac{P_e}{A_g}$$

$$= \frac{(0.765 \text{ in}^2) \left(150 \frac{\text{kips}}{\text{in}^2}\right)}{324 \text{ in}^2}$$

$$= 0.354 \text{ kips/in}^2$$

The nominal web-shear capacity is given by ACI Eq. (11-12) as

$$V_{cw} = b_w d_p \left(3.5\sqrt{f_c'} + 0.3 f_{pc}\right) + V_p$$

$$= (12 \text{ in})(21.6 \text{ in}) \left(\begin{array}{c} (3.5)\sqrt{6000 \frac{\text{lbf}}{\text{in}^2}} \\ + (0.3)\left(354 \frac{\text{lbf}}{\text{in}^2}\right) \end{array} \right)$$

$$+ 11{,}700 \text{ lbf}$$

$$= 110{,}000 \text{ lbf}$$

$$= 110 \text{ kips}$$

At section A-A, the cable eccentricity is

$$e = \frac{h}{2} - y_A - y_o$$

$$= 13.5 \text{ in} - 8 \text{ in} - 4.5 \text{ in}$$

$$= 1.0 \text{ in}$$

$$R_b = \frac{1}{A_g} + \frac{e}{S_b}$$

$$= \frac{1}{324 \text{ in}^2} + \frac{1 \text{ in}}{1458 \text{ in}^3}$$

$$= 0.00377 \text{ 1/in}^2$$

The compressive stress in the bottom fiber, at section A-A, due to the final prestressing force is

$$f_{pe} = P_e R_b$$

$$= (0.765 \text{ in}^2) \left(150 \frac{\text{kips}}{\text{in}^2}\right) \left(0.00377 \frac{1}{\text{in}^2}\right)$$

$$= 0.433 \text{ kips/in}^2$$

The applied moment required to produce cracking at section A-A is given by modified ACI Eq. (11-11) as

$$M_{ct} = S_b \left(6\sqrt{f_c'} + f_{pe}\right)$$

$$= (1458 \text{ in}^3) \left(\frac{6\sqrt{6000 \frac{\text{lbf}}{\text{in}^2}}}{1000 \frac{\text{lbf}}{\text{kip}}} + 0.433 \frac{\text{kips}}{\text{in}^2} \right)$$

$$= 1309 \text{ in-kips}$$

$$> 500 \text{ in-kips}$$

This moment exceeds the given unfactored applied moment at section A-A, and hence flexural cracking does not occur at section A-A; ACI Eq. (11-10) is not applicable, and ACI Eq. (11-12) governs.

Design for Torsion

The design provisions for torsion in prestressed concrete are similar to those for reinforced concrete. In accordance with ACI Sec. 11.6.1, torsional effects may be neglected, and closed stirrups and longitudinal torsional reinforcement are not required when the factored torque does not exceed

$$T_u = \phi\sqrt{f_c'} \left(\frac{A_{cp}^2}{p_{cp}}\right) \sqrt{1 + \frac{f_{pc}}{4\sqrt{f_c'}}}$$

When this value is exceeded, reinforcement shall be provided to resist the full torsion. When both shear and torsion reinforcements are required, the sum of the individual areas must be provided.

ACI Sec. 11.6.3.6 specifies the required area of one leg of a closed stirrup as

$$\frac{A_t}{s} = \frac{T_u}{2\phi A_o f_{yt} \cot(37.5°)}$$

$$= \frac{T_u}{1.7\phi A_{oh} f_{yt} \cot(37.5°)}$$

The corresponding area of longitudinal reinforcement required is specified in ACI Secs. 11.6.3.7 and R11.6.3.10 as

$$A_l = \left(\frac{A_t p_h f_{yt}}{f_y s}\right) \cot^2(37.5°) \qquad \textit{[ACI 11-22]}$$

$$\geq \frac{5 A_{cp} \sqrt{f_c'}}{f_y} - \frac{A_t p_h f_{yt}}{f_y s} \qquad \textit{[ACI 11-24]}$$

$$\frac{A_t}{s} \geq \frac{25 b_w}{f_{yt}}$$

The minimum diameter is

$$d_{bl} = \frac{s}{24} \text{ in}$$

$$\geq \text{ no. 3 bar}$$

The minimum combined area of stirrups for combined shear and torsion is given by ACI Sec. 11.6.5.2 as

$$\frac{A_v + 2A_t}{s} = \frac{0.75\sqrt{f'_c}b_w}{f_{yt}}$$

$$\geq \frac{50b_w}{f_{yt}} \qquad \text{[ACI 11-23]}$$

The maximum spacing of closed stirrups is given by ACI Sec. 11.6.6.1 as

$$s = \frac{p_h}{8} \text{ in}$$

$$\leq 12 \text{ in}$$

When redistribution is possible in an indeterminate structure, the nominal torsional capacity of the member, in accordance with ACI Sec. 11.6.2.2, need not exceed

$$T_n = 4\sqrt{f'_c}\left(\frac{A^2_{cp}}{p_{cp}}\right)\sqrt{1 + \frac{f_{pc}}{4\sqrt{f'_c}}}$$

Example 3.11

The post-tensioned beam for Ex. 3.9 is subjected to a factored shear force of 88 kips and a factored torsion of 100 in-kips at section A-A. Determine the combined shear and torsion reinforcement required.

Solution

The area enclosed by the outside perimeter of the beam is

$$A_{cp} = (27 \text{ in})(12 \text{ in}) = 324 \text{ in}^2$$

The length of the outside perimeter of the beam is

$$p_{cp} = (2)(27 \text{ in} + 12 \text{ in}) = 78 \text{ in}$$

The compressive stress at the centroid, due to the final prestressing force, was determined in Ex. 3.10 as

$$f_{pc} = 354 \text{ lbf/ in}^2$$

Torsional reinforcement is not required in accordance with ACI Sec. 11.6.1 when the factored torque does not exceed

$$T_u = \phi\sqrt{f'_c}\left(\frac{A^2_{cp}}{p_{cp}}\right)\sqrt{1 + \frac{f_{pc}}{4\sqrt{f'_c}}}$$

$$= \left(0.75\sqrt{6000\ \frac{\text{lbf}}{\text{in}^2}}\left(\frac{(324 \text{ in}^2)^2}{(78 \text{ in})\left(1000\ \frac{\text{lbf}}{\text{kip}}\right)}\right)\right)$$

$$\times \sqrt{1 + \frac{354\ \frac{\text{lbf}}{\text{in}^2}}{4\sqrt{6000\ \frac{\text{lbf}}{\text{in}^2}}}}$$

$$= 115 \text{ in-kips}$$

$$> 100 \text{ in-kips} \qquad \begin{bmatrix} \text{Closed stirrups are} \\ \text{not required.} \end{bmatrix}$$

The shear strength provided by the concrete was determined in Ex. 3.9 as

$$V_c = 100 \text{ kips}$$

From ACI Eqs. (11-1) and (11-2), the required nominal capacity of the shear reinforcement is

$$V_s = \frac{V_u}{\phi} - V_c = \frac{88 \text{ kips}}{0.75} - 100 \text{ kips}$$

$$= 17.33 \text{ kips}$$

The minimum permissible area of shear reinforcement is the smaller value given by ACI Eqs. (11-13) and (11-14).

$$\frac{A_{v(\min)}}{s} = \frac{A_{ps}f_{pu}\sqrt{\dfrac{d}{b_w}}}{80f_{yt}d} \qquad \text{[ACI 11-14]}$$

$$= \frac{(0.765 \text{ in}^2)\left(270\ \frac{\text{kips}}{\text{in}^2}\right)\sqrt{\dfrac{21.6 \text{ in}}{12 \text{ in}}}}{(80)\left(60\ \frac{\text{kips}}{\text{in}^2}\right)(21.6 \text{ in})}$$

$$= 0.0027 \text{ in}^2/\text{in} \quad \text{[governs]}$$

or

$$\frac{A_{v(\min)}}{s} = \frac{50b_w}{f_{yt}} = \frac{(50)(12 \text{ in})}{60{,}000\ \dfrac{\text{lbf}}{\text{in}^2}}$$

$$= 0.010 \text{ in}^2/\text{in}$$

or

$$\frac{A_{v(\min)}}{s} = \frac{0.75b_w\sqrt{f'_c}}{f_{yt}} \qquad \text{[ACI 11-13]}$$

$$= \frac{(0.75)(12 \text{ in})\sqrt{6000\ \dfrac{\text{lbf}}{\text{in}^2}}}{60{,}000\ \dfrac{\text{lbf}}{\text{in}^2}}$$

$$= 0.012$$

From ACI Eq. (11-15), the shear reinforcement required is

$$\frac{A_v}{s} = \frac{V_s}{f_{yt}d} = \frac{17.33 \text{ kips}}{\left(60\ \dfrac{\text{kips}}{\text{in}^2}\right)(21.6 \text{ in})}$$

$$= 0.014 \text{ in}^2/\text{in}$$

$$> \frac{A_{v(\min)}}{s} \quad \text{[satisfactory]}$$

Provide no. 3 stirrups at 15 in spacing, which gives

$$\frac{A_v}{s} = \frac{0.22 \text{ in}^2}{15 \text{ in}} = 0.015 \text{ in}^2/\text{in}$$

$$> 0.014 \quad \text{[satisfactory]}$$

3. PRESTRESS LOSSES

Nomenclature

A_{ps}	area of prestressing tendon	in^2
c	anchor set	in
C	factor for relaxation losses	–
E_{ci}	modulus of elasticity of concrete at time of initial prestress	kips/in^2
E_p	modulus of elasticity of prestressing tendon	kips/in^2
f_{pd}	compressive stress at level of tendon centroid after elastic losses and including sustained dead load	lbf/in^2
f_{pi}	compressive stress at level of tendon centroid after elastic losses	lbf/in^2
f_{pp}	compressive stress at level of tendon centroid before elastic losses	lbf/in^2
g	sag of prestressing tendon	in
H	ambient relative humidity	%
J	factor for relaxation losses	–
K	wobble friction coefficient per foot of prestressing tendon	–
K_{re}	factor for relaxation losses	kips/in^2
K_{sh}	factor for shrinkage losses accounting for elapsed time between completion of casting and transfer of prestressing force	–
l_c	length of prestressing tendon affected by anchor seating loss	ft
l_{px}	length of prestressing tendon from jacking end to any point x measured along the curve	ft
m	loss of force per foot of cable due to friction	kips/ft
n_i	E_p/E_{ci}–	
p_{cp}	outside perimeter of the concrete cross section	in
P_c	prestressing tendon force at a distance of l_c from the jacking end	kips
P_i	prestressing tendon force after elastic losses	kips
P_p	prestressing tendon force before elastic losses	kips
P_{pj}	prestressing tendon force at jacking end	kips
P_{px}	prestressing tendon force at a distance of l_{px} from the jacking end	kips

$P_{\Delta c}$	loss of tendon force due to anchor set	kips
$P_{\Delta cr}$	loss of tendon force due to creep	kips
$P_{\Delta el}$	loss of tendon force due to elastic shortening	kips
$P_{\Delta re}$	loss of tendon force due to relaxation	kips
$P_{\Delta sh}$	loss of tendon force due to shrinkage	kips
R	radius of curvature of tendon profile	ft

Symbols

α	angular change in radians of tendon profile from jacking end to any point x	radians
ε_{sh}	basic shrinkage strain	–
μ	curvature friction coefficient	–

Friction Losses

Friction losses occur in post-tensioned members due to friction and unintentional out-of-straightness of the ducts and are determined by ACI Sec. 18.6.2.1 as

$$P_{pj} = P_{px}\exp(Kl_{px} + \mu\alpha) \qquad \textit{[ACI 18-1]}$$

For a value of $(Kl_{px} + \mu\alpha)$ not greater than 0.3, this expression reduces to

$$P_{pj} = P_{px}(1 + Kl_{px} + \mu\alpha) \qquad \textit{[ACI 18-2]}$$

Rearranging the terms gives

$$P_{px} = P_{pj}(1 - Kl_{px} - \mu\alpha)$$

Example 3.12

The post-tensioned beam shown in the following illustration has a prestressing cable consisting of five $^1/_2$ in diameter low-relaxation strands. Each strand has an area of 0.153 in^2, a yield strength of 243 kips/in^2, and a tensile strength of 270 kips/in^2. The cable centroid, as shown, is parabolic in shape and is stressed simultaneously from both ends with a jacking force of 159 kips. The value of the wobble friction coefficient is 0.0015/ft, and the curvature friction coefficient is 0.25. Determine the cable force at midspan of the member before elastic losses.

Illustration for Ex. 3.12

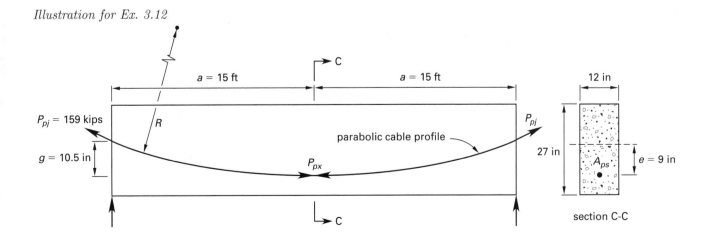

Solution

The nominal radius of the cable profile is

$$R = \frac{a^2}{2g} = \frac{(15 \text{ ft})^2 \left(12 \, \frac{\text{in}}{\text{ft}} \right)}{(2)(10.5 \text{ in})}$$
$$= 129 \text{ ft}$$

The cable length along the curve, from the jacking end to midspan, is

$$l_{px} = a + \frac{g^2}{3a} = 15 \text{ ft} + \frac{\left(\dfrac{10.5 \text{ in}}{12 \, \frac{\text{in}}{\text{ft}}} \right)^2}{(3)(15 \text{ ft})}$$
$$= 15.02 \text{ ft}$$

The angular change of the cable profile over this length is

$$\alpha = \frac{l_{px}}{R} = \frac{15.02 \text{ ft}}{129 \text{ ft}}$$
$$= 0.117 \text{ radians}$$
$$(Kl_{px} + \mu\alpha) = (0.0015)(15.02 \text{ ft})$$
$$+ (0.25)(0.117 \text{ radians})$$
$$= 0.052$$
$$< 0.3 \quad [\text{ACI Eq. (18-2) is applicable.}]$$

The cable force at midspan is given by

$$P_{px} = P_{pj}(1 - Kl_{px} - \mu\alpha) = (159 \text{ kips})(1 - 0.052)$$
$$= 151 \text{ kips}$$

Anchor Seating Loss

Anchor seating loss results from the slip or set that occurs in the anchorage when the prestressing force is transferred to the anchor device. In a pre-tensioned tendon with an anchor set of c, the loss in prestressing is constant along the cable, as shown in Fig. 3.11. From Fig. 3.11, the anchor set is obtained as

$$c = \frac{\text{shaded area}}{A_{ps}E_p} = \frac{P_{\Delta c}l}{A_{ps}E_p}$$

and $P_{\Delta c}$ may be determined.

In a post-tensioned tendon, friction in the duct resists the inward movement of the tendon[1] and limits the affected zone to the length l_c shown in Fig. 3.12. The anchor set is obtained from Fig. 3.12 as

$$c = \frac{\text{shaded area}}{A_{ps}E_p} = \frac{ml_c^2}{A_{ps}E_p}$$
$$l_c^2 = \frac{cA_{ps}E_p}{m}$$
$$P_c = P_{pj} - ml_c$$

Example 3.13

The post-tensioned beam of Ex. 3.12 has tendon anchorages with a cable set of 0.05 in. Determine the residual tendon force, after anchorage and before elastic and long-term losses occur, at the jacking end and at a distance of l_c from the jacking end.

Solution

From Ex. 3.12, the stress loss per foot due to friction is

$$m = \frac{P_{pj} - P_{px}}{l_{px}} = \frac{159 \text{ kips} - 151 \text{ kips}}{15.02 \text{ ft}}$$
$$= 0.532 \text{ kips/ft}$$
$$l_c^2 = \frac{cA_{ps}E_p}{m}$$
$$= \frac{(0.05 \text{ in})(0.765 \text{ in}^2) \left(28 \times 10^3 \, \frac{\text{kips}}{\text{in}^2} \right)}{\left(0.532 \, \frac{\text{kips}}{\text{ft}} \right) \left(12 \, \frac{\text{in}}{\text{ft}} \right)}$$
$$l_c = 13 \text{ ft}$$

Figure 3.11 *Seating Loss in a Pre-Tensioned Tendon*

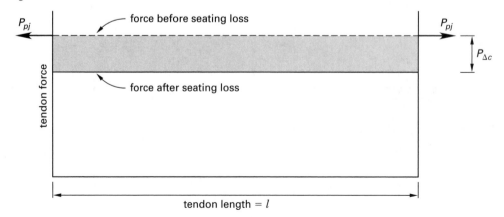

Figure 3.12 *Seating Loss in a Post-Tensioned Tendon*

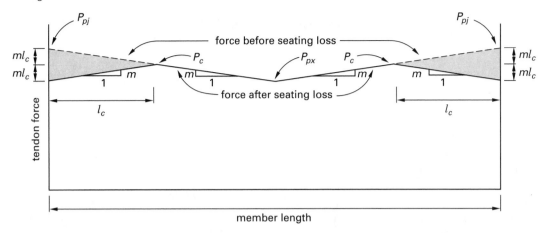

The cable force at a distance of l_c from the jacking end is

$$P_c = P_{pj} - ml_c$$

$$= (159 \text{ kips}) - \left(0.532 \frac{\text{kips}}{\text{ft}}\right)(13 \text{ ft})$$

$$= 152 \text{ kips}$$

The cable force at the jacking end after anchoring is

$$P_{pj(\text{anc})} = P_{pj} - 2ml_c = 145 \text{ kips}$$

Elastic Shortening Losses

Losses occur in a prestressed concrete beam at transfer due to the elastic shortening of the concrete at the level of the centroid of the prestressing tendons. The concrete stress at the level of the centroid of the prestressing tendons after elastic shortening is

$$f_{pi} = P_i\left(\frac{1}{A_g} + \frac{e^2}{I_g}\right) - \frac{eM_G}{I_g}$$

Conservatively, the concrete stress at the level of the centroid of the prestressing tendons before elastic shortening is

$$f_{pi} = f_{pp} = P_p\left(\frac{1}{A_g} + \frac{e^2}{I_g}\right) - \frac{eM_G}{I_g}$$

In a pre-tensioned member with transfer occuring simultaneously in all tendons, the loss of prestressing force is

$$P_{\Delta el} = n_i A_{ps} f_{pi}$$

In a post-tensioned member with only one tendon, no loss from elastic shortening occurs.

In a post-tensioned member with several tendons stressed sequentially, the maximum loss occurs in the first tendon stressed, and no loss occurs in the last tendon stressed. The total loss is then one-half the value for a pre-tensioned member or

$$P_{\Delta el} = \frac{n_i A_{ps} f_{pi}}{2}$$

Example 3.14

The pre-tensioned beam shown in the following illustration is simply supported over a span of 30 ft and has a concrete strength at transfer of 4500 lbf/in². Five $\frac{1}{2}$ in diameter low-relaxation strands are provided, each with an area of 0.153 in². The initial force in each tendon after anchor seating loss is 32 kips. Determine the loss of prestressing force due to elastic shortening.

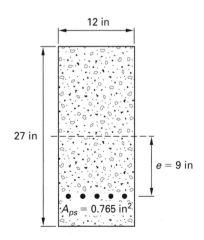

Solution

From ACI Sec. 8.5, the modulus of elasticity of the concrete at transfer is

$$E_{ci} = 57\sqrt{f'_{ci}} = 57\sqrt{4500 \; \frac{\text{lbf}}{\text{in}^2}} = 3824 \; \text{kips/in}^2$$

$$n_i = \frac{E_p}{E_{ci}} = \frac{28 \times 10^3 \; \frac{\text{kips}}{\text{in}^2}}{3824 \; \frac{\text{kips}}{\text{in}^2}} = 7.32$$

$$f_{pp} = P_p \left(\frac{1}{A_g} + \frac{e^2}{I_g} \right) - \frac{eM_G}{I_g}$$

$$= (5)(32 \; \text{kips}) \left(\frac{1}{324 \; \text{in}^2} + \frac{(9 \; \text{in})^2}{19{,}683 \; \text{in}^4} \right)$$

$$- \frac{(9 \; \text{in})(455 \; \text{in-kip})}{19{,}683 \; \text{in}^4}$$

$$= 1.152 \; \frac{\text{kips}}{\text{in}^2} - 0.208 \; \frac{\text{kip}}{\text{in}^2}$$

$$= 0.944 \; \text{kip/in}^2$$

The total loss of prestressing force is

$$P_{\Delta el} \approx n_i A_{ps} f_{pp}$$

$$= (7.32)(5) \left(0.153 \; \text{in}^2 \right) \left(0.944 \; \frac{\text{kip}}{\text{in}^2} \right)$$

$$= 5.3 \; \text{kips}$$

Creep Losses

Creep occurs in a prestressed concrete member as a result of the sustained compressive stress. The concrete stress at the level of the centroid of the prestressing tendons after elastic shortening and allowing for sustained deadload is

$$f_{pd} = P_i \left(\frac{1}{A_g} + \frac{e^2}{I_g} \right) - \frac{eM_G}{I_g} - \frac{eM_D}{I_g}$$

For post-tensioned members with transfer at 28 days, the creep loss is given by[2,3]

$$P_{\Delta cr} = 1.6 n A_{ps} f_{pd}$$

For pre-tensioned members with transfer at 3 days, the creep loss is given by

$$P_{\Delta cr} = 2.0 n A_{ps} f_{pd}$$

Example 3.15

For the post-tensioned beam of Ex. 3.12, the cable force at midspan after elastic losses is 151 kips and the 28 day concrete strength is 6000 lbf/in². The superimposed dead load moment is 800 in-kips. Determine the loss of prestressing force due to creep.

Solution

From ACI Sec. 8.5, the modulus of elasticity of the concrete at 28 days is

$$E_c = 57\sqrt{f'_c} = 57\sqrt{6000 \; \frac{\text{lbf}}{\text{in}^2}}$$

$$= 4415 \; \text{kips/in}^2$$

$$n = \frac{E_p}{E_c} = \frac{28 \times 10^3 \; \frac{\text{kips}}{\text{in}^2}}{4415 \; \frac{\text{kips}}{\text{in}^2}}$$

$$= 6.34$$

$$f_{pd} = P_i \left(\frac{1}{A_g} + \frac{e^2}{I_g} \right) - \frac{eM_G}{I_g} - \frac{eM_D}{I_g}$$

$$= (151 \; \text{kips}) \left(\frac{1}{324 \; \text{in}^2} + \frac{(9 \; \text{in})^2}{19{,}683 \; \text{in}^4} \right)$$

$$- \frac{(9 \; \text{in})(455 \; \text{in-kips} + 800 \; \text{in-kips})}{19{,}683 \; \text{in}^4}$$

$$= 1.087 \; \frac{\text{kips}}{\text{in}^2} - 0.574 \; \frac{\text{kip}}{\text{in}^2}$$

$$= 0.514 \; \text{kip/in}^2$$

The loss of prestressing force due to creep is

$$P_{\Delta cr} = 1.6 n A_{ps} f_{pd}$$

$$= (1.6)(6.34) \left(0.765 \; \text{in}^2 \right) \left(0.514 \; \frac{\text{kip}}{\text{in}^2} \right)$$

$$= 4.0 \; \text{kips}$$

Shrinkage Loss

The shrinkage of a concrete member with time produces a corresponding loss of prestress. The basic shrinkage strain is given by[2,3]

$$\varepsilon_{sh} = 8.2 \times 10^{-6} \; \text{in/in}$$

Allowing for the ambient relative humidity H and the ratio of the member's volume to surface area A_g/p_{cp}, the shrinkage loss for a pre-tensioned member is

$$P_{\Delta sh} = A_{ps} \varepsilon_{sh} E_p \left(1 - \frac{0.06 A_g}{p_{cp}} \right) (100 - H)$$

For a post-tensioned member with transfer after some shrinkage has already occurred, the shrinkage loss is

$$P_{\Delta sh} = K_{sh} A_{ps} \varepsilon_{sh} E_p \left(1 - \frac{0.06 A_g}{p_{cp}} \right) (100 - H)$$

Example 3.16

The post-tensioned beam of Ex. 3.15 is located in an area with an ambient relative humidity of 55% and transfer is effected 7 days after the completion of curing, giving a value[3] of 0.77 for K_{sh}. Determine the loss of prestressing force due to shrinkage.

Solution

The shrinkage loss is given by

$$P_{\Delta sh} = K_{sh} A_{ps} \varepsilon_{sh} E_p \left(1 - \frac{0.06 A_g}{p_{cp}} \right) (100 - H)$$

$$= (0.77) \left(0.765 \text{ in}^2 \right) \left(8.2 \times 10^{-6} \frac{\text{in}}{\text{in}} \right)$$

$$\times \left(28 \times 10^3 \frac{\text{kips}}{\text{in}^2} \right)$$

$$\times \left(1 - \frac{(0.06 \text{ in}^{-1}) (324 \text{ in}^2)}{78 \text{ in}} \right) (100 - 55)$$

$$= 4.6 \text{ kips}$$

Relaxation Losses

A prestressing tendon is subjected to relaxation over time. The loss in prestress depends on the tendon properties and the initial force in the tendon and on the losses due to creep, shrinkage, and elastic shortening. The relaxation loss is given by[2,3]

$$P_{\Delta re} = \left(A_{ps} K_{re} - J(P_{\Delta cr} + P_{\Delta sh} + P_{\Delta el}) \right)(C)$$

Example 3.17

For the post-tensioned beam of Ex. 3.15, the values of the relevant parameters are[3] $K_{re} = 5$ kips/in^2, $J = 0.04$, and $C = 0.90$. Determine the loss of prestressing force due to relaxation.

Solution

The loss due to relaxation is

$$P_{\Delta re} = \left(A_{ps} K_{re} - J(P_{\Delta cr} + P_{\Delta sh} + P_{\Delta el}) \right)(C)$$

$$= \left(\begin{array}{c} (0.765 \text{ in}^2) \left(5 \dfrac{\text{kips}}{\text{in}^2} \right) \\ -(0.04)(4.0 + 4.6 + 0) \end{array} \right)(0.90)$$

$$= 3.1 \text{ kips}$$

4. COMPOSITE CONSTRUCTION

Nomenclature

A_c	area of precast surface or area of contact surface for horizontal shear	in^2
A_v	area of ties within a distance s	in^2
A_{vf}	area of friction reinforcement	in^2
b_f	actual flange width	in
$b_{f(\text{eff})}$	effective flange width	in
$b_{f(\text{tran})}$	transformed flange width	in
b_v	width of girder at contact surface	in
b_w	width of girder web	in
E_f	modulus of elasticity of flange concrete	kips/in^2
E_w	modulus of elasticity of precast girder concrete	kips/in^2
f_{cb}	stress in the bottom fiber of composite section	kips/in^2
$f_{ci(\text{flan})}$	stress in the flange at interface of composite section	kips/in^2
$f_{ci(\text{web})}$	stress in the girder at interface of composite section	kips/in^2
f_{ct}	stress in the top fiber of composite section	kips/in^2
h	depth of composite section	in
h_f	depth of flange of composite section	in
h_w	depth of precast girder	in
I_{cc}	moment of inertia of composite section	in^4
l	span length	ft
M_F	moment due to flange concrete	ft-kips
M_G	moment due to precast girder self-weight	ft-kips
M_{Prop}	moment due to removal of props	ft-kips
M_{Sht}	moment due to formwork	ft-kips
M_W	moment due to superimposed dead plus live load	ft-kips
n	modular ratio E_w/E_f	
P_{Δ}	total loss of prestress	kips
s	spacing of ties	in
S	spacing of precast girder	ft
S_{cb}	section modulus at bottom of composite section	in^3
S_{ci}	section modulus at interface of composite section	in^3
S_{ct}	section modulus at top of composite section	in^3
V_{nh}	nominal horizontal shear strength	lbf
V_u	factored shear force at section	lbf

Symbols

ρ_v	ratio of tie reinforcement area to area of contact surface $A_v/b_v s$	

Section Properties

The effective width of the flange of a composite section, as shown in Fig. 3.13, is limited by ACI Sec. 8.10 to the least of

- $\dfrac{l}{4}$

- $b_w + 16 h_f$

- S

When the 28 day compressive strengths of the precast section and the flange are different, the transformed section properties are obtained, as shown in Fig. 3.14, by dividing by the modular ratio $n = E_w/E_f$. The stresses calculated in the flange by using the transformed section properties are converted to actual stresses by dividing by n.

Prestressed Concrete

Figure 3.13 *Effective Flange Width*

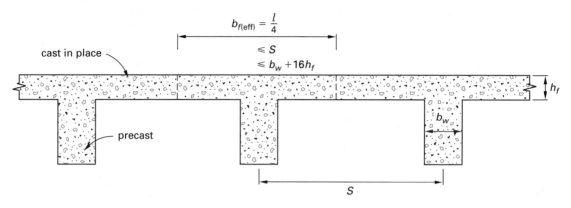

Figure 3.14 *Transformed Flange Width*

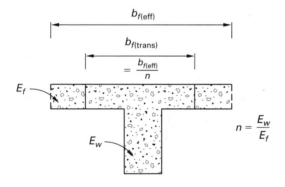

$$n = \frac{E_w}{E_f}$$

Example 3.18

The following illustration shows a composite beam with an effective span of 25 ft. The beam is an interior beam in the floor of a commercial building. The precast, pretensioned girder has a 28 day concrete strength of 6000 lbf/in², and the flange has a 28 day concrete strength of 3000 lbf/in². Determine the transformed section properties of the composite section.

Solution

From ACI Sec. 8.5,

$$E_w = 57\sqrt{f_c'} = 57\sqrt{6000 \ \frac{\text{lbf}}{\text{in}^2}}$$
$$= 4415 \ \text{kips/in}^2$$

$$E_f = 57\sqrt{f_c'} = 57\sqrt{3000 \ \frac{\text{lbf}}{\text{in}^2}}$$
$$= 3122 \ \text{kips/in}^2$$

$$n = \frac{E_w}{E_f} = \frac{4415 \ \dfrac{\text{kips}}{\text{in}^2}}{3122 \ \dfrac{\text{kips}}{\text{in}^2}} = 1.41$$

The effective flange width is limited to the least of

- $b_{f(\text{eff})} = \dfrac{l}{4} = \dfrac{(25 \ \text{ft})\left(12 \ \dfrac{\text{in}}{\text{ft}}\right)}{4}$
  $= 75 \ \text{in}$
- $b_{f(\text{eff})} = b_w + 16h_f = 4 \ \text{in} + (16)(2 \ \text{in})$
  $= 36 \ \text{in}$
- $b_{f(\text{eff})} = b_f = 30 \ \text{in}$ [governs]

Illustration for Ex. 3.18

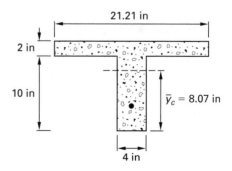

The transformed flange width is

$$b_{f(\text{tran})} = \frac{b_{f(\text{eff})}}{n} = \frac{30 \text{ in}}{1.41} = 21 \text{ in}$$

The relevant properties of the precast girder are

$$A_g = 40 \text{ in}^2$$
$$I_g = 333 \text{ in}^4$$
$$S_b = S_t = 66.67 \text{ in}^3$$

The properties of the transformed section are obtained as shown in the following table.

part	A (in^2)	y (in)	I (in^4)	Ay (in^3)	Ay (in^4)
girder	40	5	333	200	1000
flange	42	11	14	462	5082
total	82	–	347	662	6082

$$\bar{y}_c = \frac{\sum Ay}{\sum A} = \frac{662 \text{ in}^3}{82 \text{ in}^2} = 8.07 \text{ in}$$

$$I_{cc} = \sum I + \sum Ay^2 - \bar{y}_c^2 \sum A$$
$$= 347 \text{ in}^4 + 6082 \text{ in}^4 - 5344 \text{ in}^4$$
$$= 1085 \text{ in}^4$$

$$S_{ct} = \frac{I_{cc}}{h - \bar{y}_c} = \frac{1085 \text{ in}^4}{3.93 \text{ in}} = 276 \text{ in}^3$$

$$S_{ci} = \frac{I_{cc}}{h_w - \bar{y}_c} = \frac{1085 \text{ in}^4}{1.93 \text{ in}} = 562 \text{ in}^3$$

$$S_{cb} = \frac{I_{cc}}{\bar{y}_c} = \frac{1085 \text{ in}^4}{8.07 \text{ in}} = 134 \text{ in}^3$$

Horizontal Shear Requirements

To ensure composite action, full transfer of horizontal shear at the interface is necessary, and ACI Sec. 17.5.3 specifies that the factored shear force at a section shall not exceed

$$V_u = \phi V_{nh} \qquad \textit{[ACI 17-1]}$$

ACI Sec. 17.5.3.1 specifies that when the interface is intentionally roughened, the nominal horizontal shear strength is given by

$$V_{nh} = 80 b_v d$$

ACI Sec. 17.5.3.2 specifies that when the interface is smooth with minimum ties provided across the interface to give $A_v/s = 50 b_w / f_y$,

$$V_{nh} = 80 b_v d$$

ACI Sec. 17.5.3.3 specifies that when the interface is roughened to $^1/_4$ in amplitude with minimum ties provided across the interface to give $A_v/s = 50 b_w / f_y$,

$$V_{nh} = (260 + 0.6 \rho_v f_y) \lambda b_v d$$
$$\leq 500 b_v d$$

The correction factor related to unit weight of concrete is given by ACI Sec. 11.7.4.3 as

$$\lambda = 1.0 \quad \text{[for normal weight concrete]}$$
$$= 0.85 \quad \text{[for sand-lightweight concrete]}$$
$$= 0.75 \quad \text{[for all lightweight concrete]}$$

ACI Sec. 17.5.3.4 requires that when the factored shear force exceeds $\phi(500 b_v d)$, the design shall be based on the shear-friction method given in ACI Sec. 11.7.4, with the nominal horizontal shear strength given by

$$V_{nh} = A_{vf} f_y \mu \qquad \textit{[ACI 11-25]}$$
$$\leq 0.2 f'_c A_c$$
$$\leq 800 A_c$$

The coefficient of friction is given by ACI Sec. 11.7.4.3 as

$$\mu = 1.0\lambda \quad \begin{bmatrix} \text{interface roughened to} \\ \text{an amplitude of } ^1/_4 \text{ in} \end{bmatrix}$$
$$= 0.6\lambda \quad \text{[interface not roughened]}$$

In accordance with ACI Sec. 17.6.1, the tie spacing shall not exceed four times the least dimension of the supported element, nor exceed 24 in.

Example 3.19

The composite section of Ex. 3.18 has a factored shear force of $V_u = 14$ kips at the critical section. Determine the area of grade 60 ties at a spacing of 12 in required at the interface that is intentionally roughened to an amplitude of $^1/_4$ in.

Solution

From ACI Sec. 17.5.2 and 17.5.3.3,

$$d = 0.8h = (0.8)(12 \text{ in}) = 9.6 \text{ in}$$
$$500\phi b_v d = \left(500 \ \frac{\text{lbf}}{\text{in}^2}\right)(0.75)(4 \text{ in})(9.6 \text{ in})$$
$$= 14{,}400 \text{ lbf}$$
$$> V_u \quad \text{[ACI Sec. 17.5.3.3 applies.]}$$
$$V_u = 14{,}000 \text{ lbf}$$
$$= \phi(260 + 0.6 \rho_v f_y) \lambda b_v d$$
$$= (0.75)\left(260 \ \frac{\text{lbf}}{\text{in}^2} + 0.6 \rho_v \left(60{,}000 \ \frac{\text{lbf}}{\text{in}^2}\right)\right)$$
$$\times (1.0)(4 \text{ in})(9.6 \text{ in})$$
$$\rho_v = 0.0063$$

The required area of vertical ties is

$$A_v = \rho_v b_v s = (0.0063)(4 \text{ in})(12 \text{ in}) = 0.30 \text{ in}^2$$

Non-Propped Construction

In non-propped construction, the precast section supports its own self-weight, the formwork required to support the cast-in-place flange, and the weight of the flange. It may be conservatively assumed that all prestress losses occur before the flange is cast. As shown in Fig. 3.15, the composite section is subjected to the forces produced by removal of the formwork and by the superimposed applied load.

Figure 3.15 *Non-Propped Construction*

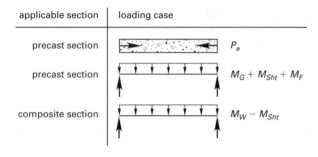

applicable section	loading case	
precast section		P_e
precast section		$M_G + M_{Sht} + M_F$
composite section		$M_W - M_{Sht}$

Example 3.20

The precast, pre-tensioned girder of the composite section of Ex. 3.18 is prestressed with an initial prestressing force of 65 kips. The total loss of prestress is 20% and may be assumed to occur before the flange is cast. The weight of the formwork to support the flange is 25 lbf/ft, the superimposed applied load is 250 lbf/ft, and the precast section is not propped. Determine the stresses at midspan in the composite section.

Solution

$$M_G = \frac{wl^2}{8} = \frac{\left(150 \; \frac{\text{lbf}}{\text{ft}^3}\right)(40 \text{ in}^2)(25 \text{ ft})^2}{(8)\left(12 \; \frac{\text{in}}{\text{ft}}\right)}$$
$$= 39{,}060 \text{ in-lbf}$$

$$M_{Sht} = \frac{wl^2}{8} = \frac{\left(25 \; \frac{\text{lbf}}{\text{ft}}\right)(25 \text{ ft})^2\left(12 \; \frac{\text{in}}{\text{ft}}\right)}{8}$$
$$= 23{,}440 \text{ in-lbf}$$

$$M_F = \frac{wl^2}{8} = \frac{\left(150 \; \frac{\text{lbf}}{\text{ft}^3}\right)(60 \text{ in}^2)(25 \text{ ft})^2}{(8)\left(12 \; \frac{\text{in}}{\text{ft}}\right)}$$
$$= 58{,}600 \text{ in-lbf}$$

$$M_W = \frac{wl^2}{8} = \frac{\left(250 \; \frac{\text{lbf}}{\text{ft}}\right)(25 \text{ ft})^2\left(12 \; \frac{\text{in}}{\text{ft}}\right)}{8}$$
$$= 234{,}400 \text{ in-lbf}$$

$$P_e = 0.8P_i = (0.8)(65 \text{ kips})$$
$$= 52 \text{ kips}$$

The prestressing force is applied at a height of $h_w/3$, and the stresses in the precast section after casting the flange are

$$f_t = \frac{M_G + M_{Sht} + M_F}{S_t}$$
$$= \frac{121{,}100 \text{ in-lbf}}{66.67 \text{ in}^3}$$
$$= 1816 \text{ lbf/in}^2$$

$$f_b = \frac{2P_e}{A_g} - \frac{M_G + M_{Sht} + M_F}{S_b}$$
$$= \frac{(2)(52{,}000 \text{ lbf})}{40 \text{ in}^2} - 1816 \; \frac{\text{lbf}}{\text{in}^2}$$
$$= 784 \text{ lbf/in}^2$$

The stresses in the composite section due to all loads are

$$f_{ct} = \frac{M_W - M_{Sht}}{nS_{ct}} = \frac{210{,}960 \text{ in-lbf}}{(1.41)\left(276 \text{ in}^3\right)}$$
$$= 542 \text{ lbf/in}^2$$

$$f_{cb} = f_b - \frac{M_W - M_{Sht}}{S_{cb}} = 784 - \frac{210{,}960 \text{ in-lbf}}{134 \text{ in}^3}$$
$$= -790 \text{ lbf/in}^2$$

$$f_{ci(\text{flan})} = \frac{M_W - M_{Sht}}{nS_{ci}} = \frac{210{,}960 \text{ in-lbf}}{(1.41)\left(562 \text{ in}^3\right)}$$
$$= 266 \text{ lbf/in}^2$$

$$f_{ci(\text{web})} = f_t + \frac{M_W - M_{Sht}}{S_{ci}} = 1816 + \frac{210{,}960 \text{ in-lbf}}{562 \text{ in}^3}$$
$$= 2191 \text{ lbf/in}^2$$

Propped Construction

In propped construction, the weight of the formwork and the flange act on the propped precast girder, producing moments in the girder and reactions in the props. As shown in Fig. 3.16, removal of the props is equivalent to applying forces, equal and opposite to the reactions in the props, to the composite section. The superimposed load is carried by the composite section.

Figure 3.16 *Propped Construction*

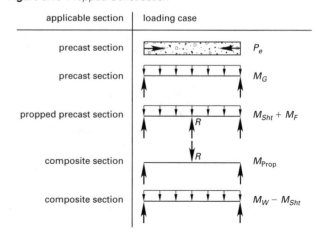

applicable section	loading case	
precast section		P_e
precast section		M_G
propped precast section		$M_{Sht} + M_F$
composite section		M_{Prop}
composite section		$M_W - M_{Sht}$

When four or more props are used, the precast section may be considered continuously supported, and no stresses are produced in the precast girder by the weight of the formwork and the flange. Similarly, no stresses are produced by the removal of the formwork. On the removal of the props, the weight of the flange is carried by the composite section, as shown in Fig. 3.17.

Figure 3.17 *Continuously Supported Section*

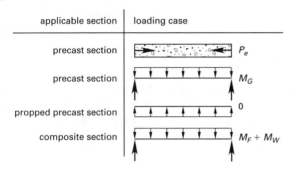

Example 3.21

Before placing the formwork to support the flange and casting the flange, the precast, pre-tensioned girder of the composite section of Ex. 3.20 is propped at midspan. Determine the stresses at midspan in the composite section.

Solution

The prestressing force is applied at a height of $h/3$, and the stresses in the precast girder before propping are

$$f_t = \frac{M_G}{S_t} = \frac{39{,}060 \text{ in-lbf}}{66.67 \text{ in}^3}$$
$$= 586 \text{ lbf/in}^2$$
$$f_b = \frac{2P_e}{A_c} - \frac{M_G}{S_b} = \frac{(2)(52{,}000 \text{ lbf})}{40 \text{ in}^2} - 586 \frac{\text{lbf}}{\text{in}^2}$$
$$= 2014 \text{ lbf/in}^2$$

The central prop creates a continuous beam with two spans of 12.5 ft each. The reaction on the prop due to the formwork and the flange concrete is

$$R = 1.25wl$$
$$= (1.25)\left(25 \frac{\text{lbf}}{\text{ft}} + 62.5 \frac{\text{lbf}}{\text{ft}}\right)(12.5 \text{ ft})$$
$$= 1367 \text{ lbf}$$

The moment in the precast girder at midspan, due to the formwork and flange concrete, is

$$M_{Sht} + M_F = \frac{wl^2}{8}$$
$$= \frac{\left(87.5 \frac{\text{lbf}}{\text{ft}}\right)(12.5 \text{ ft})^2 \left(12 \frac{\text{in}}{\text{ft}}\right)}{8}$$
$$= 20{,}510 \text{ in-lbf}$$

The stresses in the precast girder after casting the flange are

$$f_t' = 586 \frac{\text{lbf}}{\text{in}^2} - \frac{20{,}510 \text{ in-lbf}}{66.67 \text{ in}^3}$$
$$= 278 \text{ lbf/in}^2$$
$$f_b' = 2014 \frac{\text{lbf}}{\text{in}^2} + \frac{20{,}510 \text{ in-lbf}}{66.67 \text{ in}^3}$$
$$= 2321 \text{ lbf/in}^2$$

Removing the prop produces a moment at midspan of

$$M_{\text{Prop}} = \frac{Rl}{4}$$
$$= \frac{(1367 \text{ lbf})(25 \text{ ft})\left(12 \frac{\text{in}}{\text{ft}}\right)}{4}$$
$$= 102{,}530 \text{ in-lbf}$$

The stresses in the composite section due to all loads are

$$f_{ct} = \frac{M_{\text{Prop}} + M_W - M_{Sht}}{nS_{ct}}$$
$$= \frac{313{,}485 \text{ in-lbf}}{(1.41)\left(276 \text{ in}^3\right)}$$
$$= 805 \text{ lbf/in}^2$$
$$f_{cb} = f_b' - \frac{M_{\text{Prop}} + M_W - M_{Sht}}{S_{cb}}$$
$$= 2321 \frac{\text{lbf}}{\text{in}^2} - \frac{313{,}485 \text{ in-lbf}}{134 \text{ in}^3}$$
$$= -18 \text{ lbf/in}^2$$
$$f_{ci(\text{flan})} = \frac{M_{\text{Prop}} + M_W - M_{Sht}}{nS_{ci}}$$
$$= \frac{313{,}485 \text{ in-lbf}}{(1.41)\left(562 \text{ in}^3\right)}$$
$$= 396 \text{ lbf/in}^2$$
$$f_{ci(\text{web})} = f_t' + \frac{M_{\text{Prop}} + M_W - M_{Sht}}{S_{ci}}$$
$$= 278 \frac{\text{lbf}}{\text{in}^2} + \frac{313{,}485 \text{ in-lbf}}{562 \text{ in}^3}$$
$$= 836 \text{ lbf/in}^2$$

5. LOAD BALANCING PROCEDURE

Nomenclature

f_c	concrete stress	lbf/in^2
g	sag of prestressing tendon	in
M_B	balancing load moment due to w_B	in-lbf
M_O	out-of-balance moment due to w_O, $M_W - M_B$	in-lbf
M_W	applied load moment due to w_W	in-lbf
P	prestressing force	kips
w_B	balancing load produced by prestressing tendon	kips/ft
w_O	out-of-balance load, $w_W - w_B$	kips/ft
w_W	superimposed applied load	kips/ft

Design Technique

The prestressing tendon of the beam shown in Fig. 3.18 has a parabolic profile and produces a uniform upward pressure of

$$w_B = \frac{8Pg}{l^2}$$

If the total downward load on the beam is equal to w_B, the net load is zero and a uniform compressive stress of $f_c = P/A_g$ is produced in the beam. If the downward load is not fully balanced by the upward force, the out-of-balance moment is

$$M_O = M_w - M_B$$

The stress in the concrete is then given by

$$f_c = \frac{P}{A_g} \pm \frac{M_O}{S}$$

Balancing loads produced by alternative tendon profiles are available[4,5,6] and are shown in Fig. 3.19. This technique also facilitates the calculation of deflections.

Figure 3.18 *Load Balancing Method*

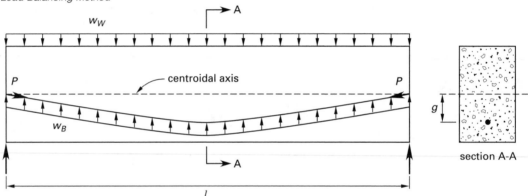

Figure 3.19 *Alternative Tendon Profiles*

cable profile	balancing load	deflection
	$w_B = \dfrac{8Pg}{l^2}$	$\dfrac{5w_B l^4}{384EI}$
	$M_B = Pg$	$\dfrac{M_B l^2}{8EI}$
	$W_B = \dfrac{4Pg}{l}$	$\dfrac{W_B l^3}{48EI}$
	$W_B = \dfrac{Pg}{a}$	$\dfrac{W_B a\,(3l^2 - 4a^2)}{24EI}$

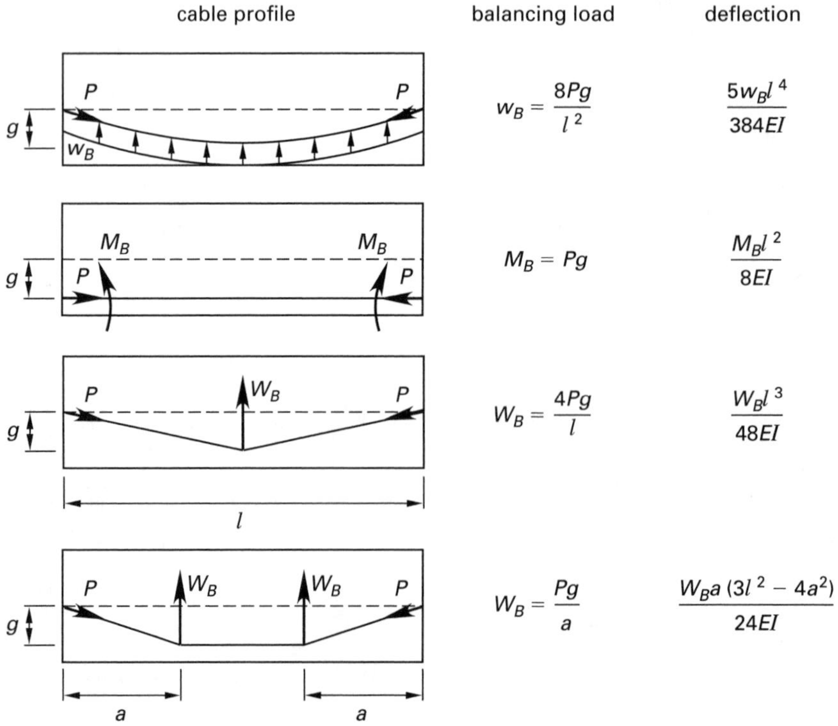

Example 3.22

The post-tensioned beam shown in the following illustration supports a uniformly distributed load, including the weight of the beam, of 0.75 kip/ft. The tendon has a parabolic profile.

(a) Determine the prestressing force required in the tendon to balance the applied load exactly, and determine the resulting stress in the beam.

(b) Determine the stresses in the beam at midspan when an additional distributed load of 0.75 kip/ft is applied to the beam.

Solution

(a) The sag of the tendon is

$$g = 13.5 \text{ in} - 4.5 \text{ in} = 9 \text{ in}$$

The prestressing force required to balance the applied load exactly is

$$P = \frac{w_W l^2}{8g}$$
$$= \frac{\left(0.75 \dfrac{\text{kip}}{\text{ft}}\right)(30 \text{ ft})^2 \left(12 \dfrac{\text{in}}{\text{ft}}\right)}{(8)(9 \text{ in})}$$
$$= 112.5 \text{ kips}$$

The uniform compressive stress throughout the beam is

$$f_c = \frac{P}{A_g} = \frac{112{,}500 \text{ lbf}}{324 \text{ in}^2}$$
$$= 347 \text{ lbf/in}^2$$

(b) The out-of-balance moment produced at midspan by an additional load of 0.75 kip/ft is

$$M_O = \frac{w_O l^2}{8} = \frac{\left(0.75 \dfrac{\text{kip}}{\text{ft}}\right)(30 \text{ ft})^2 \left(12 \dfrac{\text{in}}{\text{ft}}\right)}{8}$$
$$= 1013 \text{ in-kips}$$

The resultant stresses at midspan are

$$f_{be} = f_c - \frac{M_O}{S}$$
$$= 347 \frac{\text{lbf}}{\text{in}^2} - \frac{1{,}013{,}000 \text{ in-lbf}}{1458 \text{ in}^3}$$
$$= 347 \frac{\text{lbf}}{\text{in}^2} - 694 \frac{\text{lbf}}{\text{in}^2}$$
$$= -347 \text{ lbf/in}^2$$
$$f_{te} = f_c + \frac{M_O}{S} = 347 \frac{\text{lbf}}{\text{in}^2} + 694 \frac{\text{lbf}}{\text{in}^2}$$
$$= 1041 \text{ lbf/in}^2$$

6. STATICALLY INDETERMINATE STRUCTURES

Nomenclature

e'	resultant cable eccentricity	in
m	moment produced by unit value of the redundant	in-kips
M_R	resultant moment due to prestressing force and secondary effects, $Pe+M_S$	in-kips
M_S	moment produced by secondary effects	in-kips
R	reaction, support restraint	kips

Design Principles

Prestressing an indeterminate structure may result in secondary moments, due to the support restraints, that produce the resultant moment

$$M_R = Pe + M_S$$

In the two-span beam shown in Fig. 3.20, the support restraint R_2 is taken as the redundant and a release introduced at 2 to produce the cut-back structure. Applying the prestressing force to the cut-back structure produces the primary moment $M_P = Pe$. Applying the unit value of R_2 to the cut-back structure produces

Illustration for Ex. 3.22

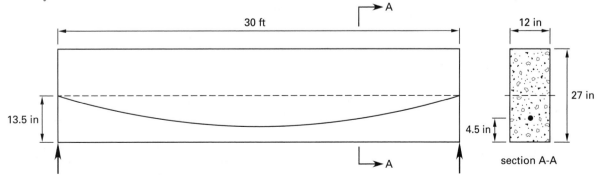

Figure 3.20 *Continuous Beam*

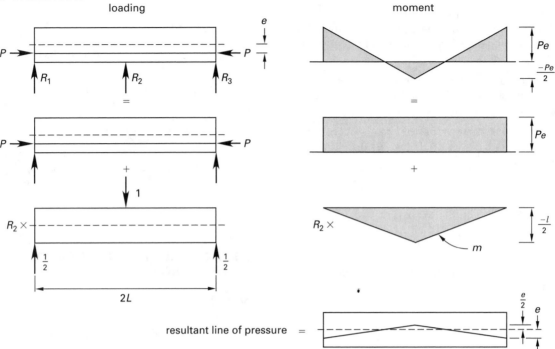

the moment diagram m, and the secondary moment is $M_S = R_2 m$. From the compatibility of displacements,

$$\left(\frac{Pe}{EI}\right) \int m dx = -\left(\frac{R_2}{EI}\right) \int m^2 dx$$

$$m = \frac{x}{2}$$

$$\frac{Pel^2}{2} = -\frac{R_2 l^3}{6}$$

$$R_2 = -\frac{3Pe}{l}$$

The secondary moment is given by

$$M_S = R_2 m = -\frac{3Pex}{2l}$$

The primary moment is given by

$$M_P = Pe$$

The resultant moment is given by

$$M_R = M_P + M_S = Pe - \frac{3Pex}{2l}$$

The resultant line of pressure, as shown in Fig. 3.20, is given by

$$e' = \frac{M_R}{P} = e - \frac{3ex}{2l}$$

At midspan,

$$M_{S(\text{midspan})} = -\frac{3Pe}{2}$$

$$M_{R(\text{midspan})} = Pe - \frac{3Pe}{2} = -\frac{Pe}{2}$$

$$e'_{(\text{midspan})} = e - \frac{3e}{2} = -\frac{e}{2}$$

A tendon with an initial eccentricity of e' produces no secondary effects in the member and is termed the *concordant cable*. Similarly, as shown in Fig. 3.21, the bending moment diagram for the external loads on a continuous beam is also a concordant profile because no support restraints will be produced. In addition, a concordant profile may be modified by means of a linear transformation by varying the location of the tendon at interior supports, as shown in Fig. 3.21, without changing the resultant moment.

Figure 3.21 *Concordant Tendon Profile*

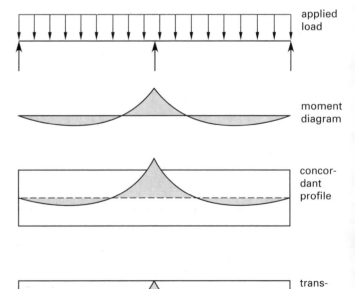

Example 3.23

The post-tensioned two-span beam shown in the following illustration supports a uniformly distributed load, including the weight of the beam, of 0.75 kip/ft. The tendon profile is parabolic in each span and is located in span 12 as indicated.

(a) Determine the prestressing force required in the tendon and the required sag of the tendon in span 23 to balance the applied load exactly. Then, determine the resulting stress in the beam.

(b) Determine the stresses in the beam at the central support when an additional distributed load of 0.75 kip/ft is applied to the beam. Then, determine the location of the resultant line of pressure at the central support.

Solution

(a) The sag of the tendon in span 12 is given by

$$g_{12} = e_4 + \frac{e_2}{2}$$

$$= 3.65 \text{ in} + \frac{10.7 \text{ in}}{2}$$

$$= 9 \text{ in}$$

The prestressing force required to balance exactly the applied load in span 12 is

$$P = \frac{w_w l_{12}^2}{8g_{12}}$$

$$= \frac{\left(0.75 \dfrac{\text{kip}}{\text{ft}}\right)(30 \text{ ft})^2 \left(12 \dfrac{\text{in}}{\text{ft}}\right)}{(8)(9 \text{ in})}$$

$$= 112.5 \text{ kips}$$

The required sag of the tendon in span 23 is given by

$$g_{23} = \frac{w_w l_{23}^2}{8P} = \frac{\left(0.75 \dfrac{\text{kip}}{\text{ft}}\right)(40 \text{ ft})^2 \left(12 \dfrac{\text{in}}{\text{ft}}\right)}{(8)(112.5 \text{ kips})}$$

$$= 16 \text{ in}$$

$$e_5 = g_{23} - \frac{e_2}{2} = 16 \text{ in} - \frac{10.7 \text{ in}}{2}$$

$$= 10.65 \text{ in}$$

The uniform compressive stress throughout the beam is

$$f_c = \frac{P}{A_g} = \frac{112{,}500 \text{ lbf}}{324 \text{ in}^2} = 347 \text{ lbf/in}^2$$

(b) Allowing for the hinges at supports 1 and 3, the fixed-end moments produced by the additional load of 0.75 kip/ft are

$$M_{F21} = \frac{w_w l_{12}^2}{8} \quad \text{[clockwise]}$$

$$= Pg_{12}$$

$$= (112.5 \text{ kips})(9 \text{ in})$$

$$= 1013 \text{ in-kips}$$

$$M_{F23} = -Pg_{23} \quad \text{[counterclockwise]}$$

$$= -(112.5 \text{ kips})(16 \text{ in})$$

$$= -1800 \text{ in-kips}$$

The fixed-end moments are distributed as shown in the table, allowing for the hinges at the supports to eliminate carryover to ends 1 and 3.

joint	2	
member	21	23
relative $\frac{EI}{l}$	$\frac{3}{30}$	$\frac{3}{40}$
distribution factors	$\frac{4}{7}$	$\frac{3}{7}$
FEM	1013	−1800
distribution	449	+337
final moments	1463	−1463

The final moment at support 2 due to the distributed load is

$$M_{O2} = 1463 \text{ in-kips}$$

Illustration for Ex. 3.23

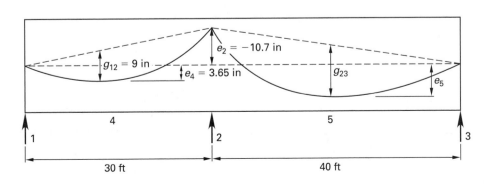

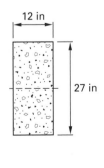

The resultant stresses at the central support are

$$f_{te} = f_c - \frac{M_{O2}}{S}$$
$$= 347 \ \frac{\text{lbf}}{\text{in}^2} - \frac{1,463,000 \ \text{in-lbf}}{1458 \ \text{in}^3}$$
$$= -656 \ \text{lbf/in}^2$$

$$f_{be} = f_c + \frac{M_{O2}}{S}$$
$$= 347 \ \frac{\text{lbf}}{\text{in}^2} + \frac{1,463,000 \ \text{in-lbf}}{1458 \ \text{in}^3}$$
$$= 1350 \ \text{lbf/in}^2$$

The location of the resultant line of pressure at the central support is

$$e_2' = -\frac{M_{O2}}{P} = -\frac{1463 \ \frac{\text{in}}{\text{kips}}}{112.5 \ \text{kips}} = -13 \ \text{in}$$

References

1. The Concrete Society. *Post-Tensioned Flat Slab Design Handbook.* 1984.

2. Zia, P. et al. "Estimating Prestress Losses." *Concrete International: Design and Construction* (1)6: 32–38. 1979, June.

3. Portland Cement Association. *Notes on ACI 318–05: Building Code Requirements for Structural Concrete.* 2005.

4. Lin, T.Y. "Load Balancing Method for Design and Analysis of Prestressed Concrete Structures." *Proceedings American Concrete Institute* 60: 719–742. 1963.

5. Precast/Prestressed Concrete Institute. *PCI Design Handbook.* 2004.

6. Freyermuth, C.L. and Schoolbred, R.A. *Post-Tensioned Prestressed Concrete.* Portland Cement Association. 1967.

PRACTICE PROBLEMS

1. The pre-tensioned beam shown in the illustration is simply supported over a span of 20 ft and has a concrete strength at transfer of 4500 lbf/in². What are the magnitude and location of the initial prestressing force required to produce satisfactory stresses at midspan, immediately after transfer, without using auxiliary reinforcement?

2. The class U pre-tensioned beam of Prob. 1 has a long-term loss in prestress of 25% and a 28 day compressive strength of 6000 lbf/in²; normal cover is provided to the tendons. What is the maximum bending moment the beam can carry if all the superimposed load is sustained?

3. For the pre-tensioned beam of Prob. 1, what is the cracking moment strength?

4. The pre-tensioned beam of Prob. 1 is prestressed with low-relaxation tendons. The area of the low-relaxation prestressing tendons provided is 0.306 in² with a specified tensile strength of 270 kips/in² and a yield strength of 243 kips/in². What is the nominal flexural strength of the beam?

5. The pre-tensioned beam of Prob. 1 supports two concentrated loads each of 2.5 kips, as shown. What are (a) the prestressing force required in the tendons to balance the applied loads exactly and (b) the resulting stress in the beam?

6. What are the stresses in the beam of Prob. 5 at midspan due to the loads W and self-weight?

Illustration for Prob. 1

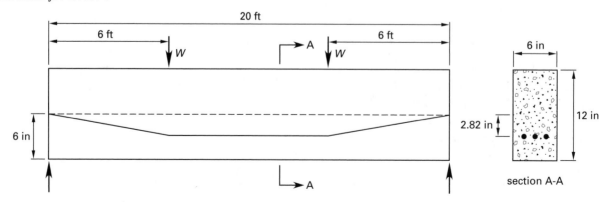

SOLUTIONS

1. The properties of the concrete section are

$$A_g = 72 \text{ in}^2$$
$$S_t = S_b = 144 \text{ in}^3$$

At midspan, the self-weight moment is

$$
\begin{aligned}
M_G &= \frac{w_c A_g l^2}{8} \\
&= \frac{\left(150 \frac{\text{lbf}}{\text{ft}^3}\right)\left(\frac{72}{144 \text{ ft}^2}\right)(20 \text{ ft})^2 \left(12 \frac{\text{in}}{\text{ft}}\right)}{8} \\
&= 45{,}000 \text{ in-lbf}
\end{aligned}
$$

At midspan, the permissible tensile stress in the top fiber without auxiliary reinforcement is given by ACI Sec. 18.4.1 as

$$
\begin{aligned}
f_{ti} = -3\sqrt{f'_{ci}} &= -3\sqrt{4500 \frac{\text{lbf}}{\text{in}^2}} \\
= -201 \text{ lbf/in}^2 &= \frac{P_i}{A_g} - \frac{P_i e}{S_t} + \frac{M_G}{S_t} \\
&= \frac{P_i}{72 \text{ in}^2} - \frac{P_i e}{144 \text{ in}^3} + \frac{45{,}000 \text{ in-lbf}}{144 \text{ in}^3} \\
-514 \frac{\text{lbf}}{\text{in}^2} &= \frac{P_i}{72 \text{ in}^2} - \frac{P_i e}{144 \text{ in}^3} \quad \text{[Eq. 1]}
\end{aligned}
$$

At midspan, the permissible compressive stress in the bottom fiber is given by ACI Sec. 18.4.1 as

$$
\begin{aligned}
f_{bi} = 0.6 f'_{ci} &= 2700 \text{ lbf/in}^2 \\
&= \frac{P_i}{A_g} + \frac{P_i e}{S_b} - \frac{M_G}{S_b} \\
&= \frac{P_i}{72 \text{ in}^2} + \frac{P_i e}{144 \text{ in}^3} - \frac{45{,}000 \text{ in-lbf}}{144 \text{ in}^3} \\
3013 \frac{\text{lbf}}{\text{in}^2} &= \frac{P_i}{72 \text{ in}^2} + \frac{P_i e}{144 \text{ in}^3} \quad \text{[Eq. 2]}
\end{aligned}
$$

Solving Eqs. [1] and [2] gives

$$P_i = 90 \text{ kips}$$
$$e = 2.82 \text{ in}$$

2. The permissible compressive stress at midspan, in the top fiber, due to the sustained load is

$$
\begin{aligned}
f_{te} &= 0.45 f'_c \\
&= (0.45)\left(6000 \frac{\text{lbf}}{\text{in}^2}\right) \\
&= 2700 \text{ lbf/in}^2 \\
&= 0.75 P_i \left(\frac{1}{A_g} - \frac{e}{S_t}\right) + \frac{M_G}{S_t} + \frac{M_T}{S_t} \\
&= (0.75)\left(-514 \frac{\text{lbf}}{\text{in}^2}\right) + \left(313 \frac{\text{lbf}}{\text{in}^2}\right) + \frac{M_T}{144 \text{ in}^3} \\
M_T &= 399{,}240 \text{ in-lbf}
\end{aligned}
$$

The permissible tensile stress at midspan, in the bottom fiber, due to the total load is

$$
\begin{aligned}
f_{be} &= -7.5\sqrt{f'_c} \\
&= -7.5\sqrt{6000 \frac{\text{lbf}}{\text{in}^2}} \\
&= -581 \text{ lbf/in}^2 \\
&= 0.75 P_i \left(\frac{1}{A_g} + \frac{e}{S_b}\right) - \frac{M_G}{S_b} - \frac{M_T}{S_b} \\
&= (0.75)\left(3013 \frac{\text{lbf}}{\text{in}^2}\right) - 313 \frac{\text{lbf}}{\text{in}^2} - \frac{M_T}{144 \text{ in}^3} \\
M_T &= 363{,}996 \text{ in-lbf} \quad \text{[governs]}
\end{aligned}
$$

3. The modulus of rupture is

$$
\begin{aligned}
f_r &= 7.5\sqrt{f'_c} \\
&= 7.5\sqrt{6000 \frac{\text{lbf}}{\text{in}^2}} \\
&= 581 \text{ lbf/in}^2
\end{aligned}
$$

The cracking moment strength is

$$
\begin{aligned}
M_{cr} &= S_b \left(P_e R_b + f_r\right) \\
&= \frac{(144 \text{ in}^3)\left((0.75)\left(3013 \frac{\text{lbf}}{\text{in}^2}\right) + 581 \frac{\text{lbf}}{\text{in}^2}\right)}{1000 \frac{\text{lbf}}{\text{kip}}} \\
&= 409 \text{ in-kips}
\end{aligned}
$$

4. The relevant properties of the beam are

$$
\begin{aligned}
\gamma_p &= 0.28 \quad \text{[for } f_{py}/f_{pu} \geq 0.9] \\
\rho_p &= \frac{A_{ps}}{b d_p} \\
&= \frac{0.306 \text{ in}^2}{(6 \text{ in})(8.82 \text{ in})} \\
&= 0.00578 \\
\beta_1 &= 0.75 \quad \text{[from ACI Sec. 10.2.7.3]}
\end{aligned}
$$

From ACI Eq. (18-3),

$$
\begin{aligned}
f_{ps} &= f_{pu}\left(1 - \frac{\gamma_p \rho_p f_{pu}}{\beta_1 f'_c}\right) \\
&= \left(270 \frac{\text{kips}}{\text{in}^2}\right)\left(1 - \frac{(0.28)(0.00578) \times \left(270 \frac{\text{kips}}{\text{in}^2}\right)}{(0.75)\left(6 \frac{\text{kips}}{\text{in}^2}\right)}\right) \\
&= 244 \text{ kips/in}^2
\end{aligned}
$$

The depth of the stress block is given by

$$a = \frac{A_{ps}f_{ps}}{0.85f'_c b}$$

$$= \frac{(0.306 \text{ in}^2)\left(244 \ \dfrac{\text{kips}}{\text{in}^2}\right)}{(0.85)\left(6 \ \dfrac{\text{kips}}{\text{in}^2}\right)(6 \text{ in})}$$

$$= 2.44 \text{ in}$$

The maximum depth of the stress block for a tension-controlled section is given by ACI Sec. 10.3.4 as

$$a_t = 0.375\beta_1 d_p$$
$$= (0.375)(0.75)(8.82)$$
$$= 2.48 \text{ in}$$
$$> a$$

Hence the section is tension-controlled and $\phi = 0.9$.

The maximum nominal flexural strength is

$$M_n = (0.85f'_c ab)\left(d_p - \frac{a}{2}\right)$$

$$= (0.85)\left(6 \ \frac{\text{kips}}{\text{in}^2}\right)(2.44 \text{ in})(6 \text{ in})$$

$$\times \left(8.82 \text{ in} - \frac{2.44 \text{ in}}{2}\right)$$

$$= 567 \text{ in-kips}$$

5. The sag of the tendon is

$$g = 2.82 \text{ in}$$

(a) The prestressing force required in the tendons to balance the applied load exactly is

$$P = \frac{Wa}{g}$$

$$= \frac{(2.5 \text{ kips})(72 \text{ in})}{2.82 \text{ in}}$$

$$= 63.8 \text{ kips}$$

(b) The uniform compressive stress throughout the beam is

$$f_c = \frac{P}{A_g} = \frac{63{,}800 \text{ lbf}}{72 \text{ in}^2}$$

$$= 886 \text{ lbf/in}^2$$

6. The moment produced at midspan by the beam self-weight is

$$M_G = \frac{w_G l^2}{8}$$

$$= \frac{\left(75 \ \dfrac{\text{lbf}}{\text{ft}}\right)(20 \text{ ft})^2 \left(12 \ \dfrac{\text{in}}{\text{ft}}\right)}{8}$$

$$= 45{,}000 \text{ in-lbf}$$

The resultant stresses in the beam at midspan are

$$f_{be} = f_c - \frac{M_G}{S}$$

$$= 886 \ \frac{\text{lbf}}{\text{in}^2} - \frac{45{,}000 \text{ in-lbf}}{144 \text{ in}^3}$$

$$= 886 \ \frac{\text{lbf}}{\text{in}^2} - 313 \ \frac{\text{lbf}}{\text{in}^2}$$

$$= 573 \text{ lbf/in}^2$$

$$f_{te} = f_c + \frac{M_G}{S}$$

$$= 886 \ \frac{\text{lbf}}{\text{in}^2} + 313 \ \frac{\text{lbf}}{\text{in}^2}$$

$$= 1199 \text{ lbf/in}^2$$

4 Structural Steel Design

1. Load and Resistance Factor Design 4-1
2. Design for Flexure 4-2
3. Design for Shear 4-8
4. Design of Compression Members 4-10
5. Plastic Design 4-24
6. Design of Tension Members 4-29
7. Design of Bolted Connections 4-33
8. Design of Welded Connections 4-38
9. Plate Girders 4-42
10. Composite Beams 4-49
    Practice Problems 4-54
    Solutions 4-55

1. LOAD AND RESISTANCE FACTOR DESIGN

Nomenclature

D	dead loads	kips or kips/ft
E	earthquake load	kips or kips/ft
H	load due to lateral pressure	kips/ft^2
L	live loads due to occupancy	kips or kips/ft
L_r	roof live load	kips or kips/ft
Q	load effect produced by service load	kips
R	load due to rainwater or ice	kips or kips/ft
R_n	nominal strength	–
S	snow load	kips or kips/ft
U	required strength to resist factored loads	–
W	wind load	kips or kips/ft

Symbols

γ	load factor
ϕ	resistance factor

Required Strength

The required ultimate strength of a member consists of the most critical combination of factored loads applied to the member. Factored loads consist of working, or service, loads multiplied by the appropriate load factors. In accordance with LRFD[1] Sec. B2, load combinations shall be as stipulated by the applicable building code. The required strength $\sum \gamma Q$ is defined by seven combinations in IBC[2] Sec. 1605.2.1 as

$$\sum \gamma Q = 1.4D \qquad \text{[IBC 16-1]}$$

$$\sum \gamma Q = 1.2D + 1.6L \\ + (0.5)(L_r \text{ or } S \text{ or } R) \qquad \text{[IBC 16-2]}$$

$$\sum \gamma Q = 1.2D + (1.6)(L_r \text{ or } S \text{ or } R) \\ + (0.5L^* \text{ or } 0.8W) \qquad \text{[IBC 16-3]}$$

$$\sum \gamma Q = 1.2D + 1.6W + 0.5L^* \\ + (0.5)(L_r \text{ or } S \text{ or } R) \qquad \text{[IBC 16-4]}$$

$$\sum \gamma Q = 1.2D \pm 1.0E + 0.5L^* \\ + 0.2S^{**} \qquad \text{[IBC 16-5]}$$

$$\sum \gamma Q = 0.9D + 1.6W + 1.6H \qquad \text{[IBC 16-6]}$$

$$\sum \gamma Q = 0.9D + 1.0E + 1.6H \qquad \text{[IBC 16-7]}$$

*Replace $0.5L$ with $1.0L$ for garages, places of public assembly, and areas where $L > 100$ lbf/ft^2.
**Replace $0.2S$ with $0.7S$ for roof configurations that do not shed snow.

Example 4.1

The following illustration shows a typical frame of a six-story office building. The loading on the frame is as follows.

roof dead load, including cladding and columns, w_{Dr}	$= 1.2$ kips/ft
roof live load, w_{Lr}	$= 0.4$ kip/ft
floor dead load, including cladding and columns, w_D	$= 1.6$ kips/ft
floor live load, w_L	$= 1.25$ kips/ft
horizontal wind pressure, p_h	$= 1.0$ kip/ft
vertical wind pressure, p_v	$= 0.5$ kip/ft

Determine the maximum and minimum design loads on the columns.

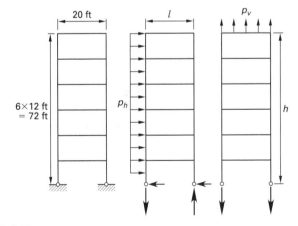

Solution

The axial load on one column due to the dead load is

$$D = \frac{l(w_{Dr} + 5w_D)}{2}$$

$$= \frac{(20 \text{ ft})\left(1.2\,\dfrac{\text{kips}}{\text{ft}} + (5 \text{ stories})\left(1.6\,\dfrac{\text{kips}}{\text{ft}}\right)\right)}{2}$$

$$= 92 \text{ kips}$$

Steel

The axial load on one column due to the roof live load is

$$L_r = \frac{lw_{Lr}}{2} = \frac{(20 \text{ ft}) \left(0.4 \frac{\text{kips}}{\text{ft}} \right)}{2}$$
$$= 4 \text{ kips}$$

The axial load on one column due to the floor live load is

$$L = \frac{5lw_L}{2} = \frac{(5) (20 \text{ ft}) \left(1.25 \frac{\text{kips}}{\text{ft}} \right)}{2}$$
$$= 62.5 \text{ kips}$$

The axial load on one column due to horizontal wind pressure is

$$W_h = \pm \frac{p_h h^2}{2l} = \pm \frac{\left(1 \frac{\text{kip}}{\text{ft}} \right) (72 \text{ ft})^2}{(2) (20 \text{ ft})}$$
$$= \pm 130 \text{ kips}$$

The axial load on one column due to the vertical wind pressure is

$$W_v = \frac{-p_v l}{2} = -\frac{\left(0.5 \frac{\text{kip}}{\text{ft}} \right) (20 \text{ ft})}{2}$$
$$= -5 \text{ kips}$$

From IBC Eq. (16-2), the maximum design load on a column is

$$\sum \gamma Q = 1.2D + 1.6L + 0.5L_r$$
$$= (1.2)(92 \text{ kips}) + (1.6)(62.5 \text{ kips})$$
$$+ (0.5)(4 \text{ kips})$$
$$= 212 \text{ kips} \quad \text{[compression]}$$

Alternatively, from IBC Eq. (16-4),

$$\sum \gamma Q = 1.2D + 1.6W + 0.5L + 0.5L_r$$
$$= (1.2)(92 \text{ kips}) + (1.6)(130 \text{ kips})$$
$$+ (1.6)(-5 \text{ kips}) + (0.5)(62.5 \text{ kips})$$
$$+ (0.5)(4 \text{ kips})$$
$$= 344 \text{ kips} \quad \text{[governs]}$$

From IBC Eq. (16-6), the minimum design load on a column is

$$\sum \gamma Q = 0.9D - 1.6W_h + 1.6W_v$$
$$= (0.9)(92 \text{ kips}) - (1.6)(130 \text{ kips})$$
$$+ (1.6)(-5 \text{ kips})$$
$$= -133 \text{ kips} \quad \text{[tension]}$$

Design Strength

The design strength of a member consists of the nominal, or theoretical ultimate, strength of the member R_n multiplied by the appropriate resistance factor ϕ. The resistance factor is defined in LRFD specifications as

$$\phi_b = 0.90 \quad \text{[for flexure]}$$
$$\phi_v = 1.0 \quad \text{[for shear in webs of I-shaped members]}$$
$$\phi_v = 0.90 \quad \text{[for shear in all other conditions]}$$
$$\phi_c = 0.90 \quad \text{[for compression]}$$
$$\phi_t = 0.90 \quad \text{[for tensile yielding]}$$
$$\phi_t = 0.75 \quad \text{[for tensile fracture]}$$

To ensure structural safety, LRFD Sec. B3 specifies that

$$\phi R_n \geq \sum \gamma Q$$

The LRFD load tables incorporate the appropriate values of ϕ and provide a direct value of the design strength.

Example 4.2

A pin-ended column of grade A50 steel 14 ft long is subjected to a factored axial load of $\sum \gamma Q = 440$ kips. Determine the lightest adequate W shape.

Solution

From LRFD Table 4-1, for an effective height of 14 ft, a W10 × 49 column provides the design axial strength.

$$\phi R_n = \phi_c P_n$$
$$= 471 \text{ kips}$$
$$> \sum \gamma Q \quad \text{[satisfactory]}$$

2. DESIGN FOR FLEXURE

Nomenclature

BF	tabulated factor used to calculate the design flexural strength for unbraced lengths between L_p and L_r	–
C_b	bending coefficient dependent on the moment gradient	–
F_r	compressive residual stress in the flange (10 kips/in² for rolled sections; 16.5 kips/in² for welded sections)	–
F_y	specified minimum yield stress	kips/in²
I_y	moment of inertia about the y-axis	in⁴
L_b	length between braces	ft or in
L_m	limiting laterally unbraced length for full plastic bending capacity ($C_b > 1.0$)	ft or in
L_p	limiting laterally unbraced length for full plastic bending capacity ($C_b = 1.0$)	ft or in
L_r	limiting laterally unbraced length for inelastic lateral torsional buckling	ft or in

M_A absolute value of moment at quarter
point of the unbraced beam
segment   ft-kips

M_B absolute value of moment at centerline of the unbraced beam segment   ft-kips

M_C absolute value of moment at three-
quarter point of the unbraced
beam segment   ft-kips

M_{max} absolute value of maximum moment
in the unbraced beam segment   ft-kips

M_n nominal flexural strength   ft-kips

M_p plastic bending moment   ft-kips

M_r limiting buckling moment   ft-kips

M_r required bending moment   ft-kips

M_u required flexural strength   ft-kips

R_m cross-section monosymmetry
parameter   –

S elastic section modulus   in^3

Z plastic section modulus   in^3

Symbols

λ_p limiting slenderness parameter
for compact element

λ_r limiting slenderness parameter
for noncompact element

Plastic Moment of Resistance

When a compact, laterally braced steel beam is loaded to the stage when the extreme fibers reach yield, as shown in Fig. 4.1(a), the applied moment, ignoring residual stresses, is given by

$$M_y = SF_y$$

Taking into account the residual stress in the beam, the applied moment at first yielding is given by

$$M_r = 0.7F_yS$$

Figure 4.1 *Stress Distribution in W Shape*

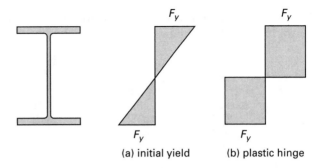

(a) initial yield   (b) plastic hinge

Continued loading eventually results in the stress distribution shown in Fig. 4.1(b), a plastic hinge is formed, and the beam cannot sustain any further increase in loading. The nominal strength of the member is given by

$$M_n = M_p = ZF_y$$

The shape factor is defined as

$$\frac{M_p}{M_y} = \frac{Z}{S} \approx 1.12 \quad \text{[for a W shape]}$$

A shape that is compact ensures that full plasticity will be achieved prior to flange or web local buckling. Compactness criteria are given in LRFD Table B4.1. Most W shapes are compact, and tabulated values of $\phi_b M_p$, in Part 3 of LRFD, allow for any reduction due to noncompactness. Adequate lateral bracing of a member ensures that full plasticity will be achieved prior to lateral torsional buckling occurring.

Example 4.3

Determine the plastic section modulus and the shape factor for the steel section shown in the following illustration. Assume that the section is compact and adequately braced.

Illustration for Ex. 4.3

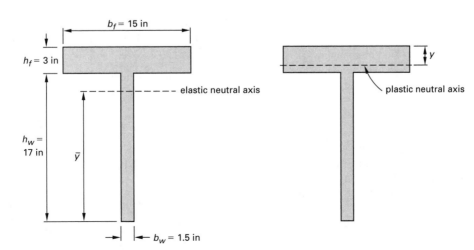

Solution

The properties of the elastic section are obtained as shown in the following table.

part	A (in^2)	y (in)	I (in^4)	Ay (in^3)	Ay^2 (in^4)
flange	45.0	18.5	34	832	15,401
web	25.5	8.5	614	217	1842
total	70.5	–	648	1049	17,243

$$\bar{y} = \frac{\sum Ay}{\sum A} = \frac{1049 \text{ in}^3}{70.5 \text{ in}^2}$$

$$= 14.9 \text{ in}$$

$$I = \sum I + \sum Ay^2 - \bar{y}^2 \sum A$$

$$= 648 \text{ in}^4 + 17{,}243 \text{ in}^4 - 15{,}652 \text{ in}^4$$

$$= 2239 \text{ in}^4$$

$$S_b = \frac{I}{\bar{y}} = \frac{2239 \text{ in}^4}{14.9 \text{ in}}$$

$$= 150 \text{ in}^3$$

The location of the plastic neutral axis is obtained by equating areas above and below the axis. The depth of the plastic neutral axis is given by

$$yb_f = (h_f - y)\, b_f + h_w b_w$$

$$y = \frac{\sum A}{2b_f} = \frac{70.5 \text{ in}^2}{(2)(15 \text{ in})}$$

$$= 2.35 \text{ in}$$

The plastic section modulus is obtained by taking moments of areas about the plastic neutral axis

$$Z = \frac{y^2 b_f}{2} + \frac{(h_f - y)^2 b_f}{2} + A_w \left(\frac{h_w}{2} + h_f - y \right)$$

$$= \frac{(2.35 \text{ in})^2 (15 \text{ in})}{2} + \frac{(0.65 \text{ in})^2 (15 \text{ in})}{2}$$

$$\quad + (25.5 \text{ in}^2)(9.15 \text{ in})$$

$$= 278 \text{ in}^3$$

The shape factor is

$$\frac{Z}{S} = \frac{278}{150} = 1.85$$

Lateral Support (Assuming C_b = 1.0)

A compact section subjected to uniform bending moment will develop its full plastic moment capacity, provided that the laterally unsupported segment length is

$$L_b \leq L_p$$

$$M_n = M_p$$

The value of L_p is defined in LRFD Sec. F2.2 and is tabulated in LRFD Part 3. The value of $\phi_b M_p$ is also tabulated in Part 3.

Example 4.4

A simply supported beam of grade 50 steel is laterally braced at 4 ft intervals. If the beam is subjected to a uniform factored bending moment of 270 ft-kips, with $C_b = 1.0$, determine

(a) the lightest adequate W shape

(b) the W shape with the minimum allowable depth

Solution

(a) From LRFD Table 3-2, the lightest satisfactory section is a W16 × 40 that has

$$\phi M_p = 274 \text{ ft-kips}$$

$$> 270 \text{ ft-kips} \quad \text{[satisfactory]}$$

$$L_p = 5.55 \text{ ft}$$

$$> 4 \text{ ft} \quad \text{[satisfactory]}$$

(b) From LRFD Table 3-2, the W shape with the minimum depth is a W10 × 60 that has

$$\phi_b M_p = 280 \text{ ft-kips}$$

$$> 270 \text{ ft-kips} \quad \text{[satisfactory]}$$

$$L_p = 9.08 \text{ ft}$$

$$> 4 \text{ ft} \quad \text{[satisfactory]}$$

$L_p < L_b \leq L_r$

When the laterally unsupported segment length equals L_r, the nominal flexural strength is given by

$$M_n = M_r = 0.7 F_y S_x$$

The value of L_r is defined in LRFD Eq. (F2–6) and is tabulated in LRFD Part 3. The value of $\phi_b M_r$ is also tabulated in LRFD Part 3.

The nominal flexural strength for an unbraced length between L_p and L_r is given by LRFD Sec. F2.2a as

$$M_n = C_b \left(M_p - \frac{(M_p - M_r)(L_b - L_p)}{L_r - L_p} \right)$$

$$\phi_b M_n = C_b \left(\phi_b M_p - (BF)(L_b - L_p) \right)$$

$$\leq \phi_b M_p$$

The variation of nominal flexural strength with unbraced length is shown in Fig. 4.2.

Figure 4.2 *Variation of M_n with L_b for $C_b = 1.0$*

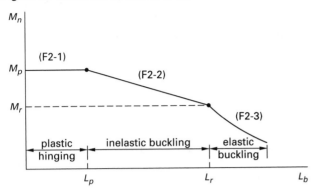

Example 4.5

A simply supported W16 × 40 beam of grade 50 steel is laterally braced at 6 ft intervals and is subjected to a uniform bending moment with $C_b = 1.0$. Determine the design flexural strength of the beam.

Solution

From LRFD Table 3-2, a W16 × 40 has

$$\phi_b M_p = 274 \text{ ft-kips}$$
$$\phi_b M_r = 170 \text{ ft-kips}$$
$$L_p = 5.55 \text{ ft}$$
$$< 6 \text{ ft}$$
$$L_r = 15.9 \text{ ft}$$
$$> 6 \text{ ft}$$
$$BF = 10.1 \text{ kips}$$
$$\phi_b M_n = C_b \left(\phi_b M_p - (BF)(L_b - L_p) \right)$$
$$= (1.0) \left((274 \text{ ft-kips}) - (10.1 \text{ kips})(0.45 \text{ ft}) \right)$$
$$= 269 \text{ ft-kips}$$

$L_b \geq L_r$

When the laterally unsupported segment length equals or exceeds L_r, the nominal flexural strength is governed by elastic lateral torsional buckling. The nominal flexural strength is equal to the critical elastic moment M_n and is defined in LRFD Eq. (F2–3). Values of ϕM_n are graphed in LRFD Table 3-10.

Example 4.6

A simply supported beam of grade 50 steel is laterally braced at 31 ft intervals. Determine the lightest adequate W shape if the beam is subjected to a uniform factored bending moment of 190 ft-kips and $C_b = 1.0$.

Solution

From LRFD Table 3-10, a W12 × 58 braced at 31 ft intervals has

$$\phi_b M_n = 196 \text{ ft-kips}$$
$$> 190 \text{ ft-kips} \quad [\text{satisfactory}]$$

Bending Coefficient

The bending coefficient C_b accounts for the influence of moment gradient on lateral torsional buckling. The value for C_b may conservatively be taken as unity, and this is applicable when a beam segment is bent in single curvature with a uniform bending moment. Other moment diagrams give larger values for C_b, which is defined in LRFD Sec. F1 as

$$C_b = \frac{12.5 R_m M_{\max}}{2.5 M_{\max} + 3 M_A + 4 M_B + 3 M_C}$$

[LRFD F1-1]

The cross-section monosymmetry parameter is given by

$$R_m = 1.0 \quad [\text{for doubly symmetric members}]$$
$$= 1.0 \quad \left[\begin{array}{c} \text{for singly symmetric members} \\ \text{in single curvature} \end{array} \right]$$
$$= 0.5 + 2 \left(\frac{I_{yc}}{I_y} \right)^2 \quad \left[\begin{array}{c} \text{for singly symmetric members} \\ \text{in reverse curvature} \end{array} \right]$$

The weak-axis moment of inertia is given by

I_{yc} = weak-axis moment of inertia of the compression flange, for members in single curvature

= weak-axis moment of inertia of the smaller flange, for members in double curvature

The terms used in determining C_b are illustrated in Fig. 4.3, and typical values are shown in Fig. 4.4.

Figure 4.3 *Determination of C_b*

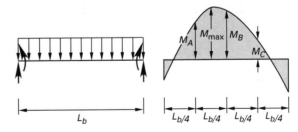

Example 4.7

The factored loads, including the beam self-weight, acting on a simply supported beam with a cantilever overhang are shown in the following illustration. The beam is laterally braced at the supports and at the end of the cantilever. Determine the relevant values of C_b.

Figure 4.4 *Typical Values of C_b*

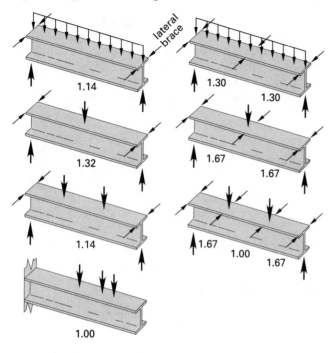

Illustration for Ex. 4.7

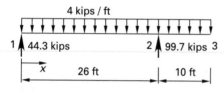

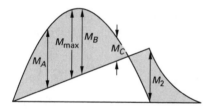

Solution

The bending moment at support 2 is

$$M_2 = \frac{wL^2}{2} = \frac{\left(4 \ \frac{\text{kips}}{\text{ft}}\right)(10 \ \text{ft})^2}{2}$$
$$= 200 \ \text{ft-kips}$$

The free bending moment in span 12 is

$$M_{\text{span}} = \frac{wL^2}{8} = \frac{\left(4 \ \frac{\text{kips}}{\text{ft}}\right)(26 \ \text{ft})^2}{8}$$
$$= 338 \ \text{ft-kips}$$

For the cantilever, LRFD Sec. F1 specifies that
$$C_b = 1.00$$
For span 12, the relevant terms are

$$M_{12} = V_1 x - \frac{wx^2}{2}$$

$$= (44.3 \ \text{kips})(x) - \frac{\left(4 \ \frac{\text{kips}}{\text{ft}}\right)(x^2)}{2}$$

$$\frac{dM_{12}}{dx} = 44.3 \ \text{kips} - \left(4 \ \frac{\text{kips}}{\text{ft}}\right)(x)$$

M_{12} is a maximum at $x = 11.1$ ft.

$$M_{\text{max}} = (44.3 \ \text{kips})(11.1 \ \text{ft}) - \frac{\left(4 \ \frac{\text{kips}}{\text{ft}}\right)(11.1 \ \text{ft})^2}{2}$$
$$= 245 \ \text{ft-kips}$$
$$M_A = 0.75 M_{\text{span}} - 0.25 M_2$$
$$= 204 \ \text{ft-kips}$$
$$M_B = M_{\text{span}} - 0.5 M_2$$
$$= 238 \ \text{ft-kips}$$
$$M_C = 0.75 M_{\text{span}} - 0.75 M_2$$
$$= 104 \ \text{ft-kips}$$

The bending coefficient is given by LRFD Eq. (F1–1) as

$$C_b = \frac{12.5 R_m M_{\text{max}}}{2.5 M_{\text{max}} + 3 M_A + 4 M_B + 3 M_C}$$

$$= \frac{(12.5)(1.0)(245 \ \text{ft-kips})}{\begin{array}{l}(2.5)(245 \ \text{ft-kips}) + (3)(204 \ \text{ft-kips}) \\ + (4)(238 \ \text{ft-kips}) + (3)(104 \ \text{ft-kips})\end{array}}$$

$$= 1.23$$

Variation of M_n with L_b

The effect of C_b on the variation of M_n with L_b is shown in Fig. 4.5. The unbraced length for which $M_n = M_p$ is extended from L_p to

$$L_m = L_p + \frac{(C_b M_p - M_p)(L_r - L_p)}{C_b (M_p - M_r)}$$

In addition, for $L_b > L_m$, the nominal flexural strength is increased from M_n for the case of $C_b = 1.0$ to $C_b M_n$.

Figure 4.5 *Variation of M_n with L_b for $C_b > 1.0$*

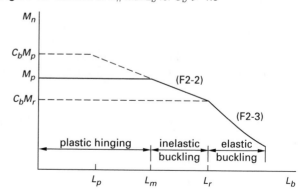

Example 4.8

Determine the lightest adequate W shape for the beam of Ex. 4.7 using grade 50 steel.

Solution

For section 23, from LRFD Table 3-2, a W18 × 40 has

$$\phi_b M_p = 294 \text{ ft-kips}$$
$$\phi_b M_r = 180 \text{ ft-kips}$$
$$L_p = 4.49 \text{ ft}$$
$$< 10 \text{ ft}$$
$$L_r = 13.1 \text{ ft}$$
$$> 10 \text{ ft}$$
$$BF = 13.3 \text{ kips}$$
$$\phi_b M_n = C_b \left(\phi_b M_p - (BF)(L_b - L_p)\right)$$
$$= (1.0)\left((294 \text{ ft-kips}) - (13.3 \text{ kips})(5.51 \text{ ft})\right)$$
$$= 221 \text{ ft-kips}$$
$$> 200 \text{ ft-kips} \quad [\text{satisfactory}]$$

For section 12, the equivalent design flexural strength required is

$$\phi_b M_n = \frac{M_{\max}}{C_b} = \frac{245 \text{ ft-kips}}{1.23}$$
$$= 199 \text{ ft-kips}$$

From LRFD Table 3-10, a W12 × 58 with an unbraced length of 26 ft has

$$\phi_b M_n = 227 \text{ ft-kips}$$
$$> 199 \text{ ft-kips} \quad [\text{satisfactory}]$$

Hence, the W12 × 58 governs the design.

Continuous Beams

LRFD App. 1 Sec. 1.3 allows advantage to be taken of the redistribution of bending moment that occurs in continuous beams. The negative moments at supports may be reduced by 10% provided that the positive moments are increased by 10% of the average adjacent support moments.

Example 4.9

The factored loading, including the beam self-weight, acting on a three-span continuous beam is shown in the following illustration. Continuous lateral support is provided to the beam. Determine the lightest adequate W shape using grade 50 steel.

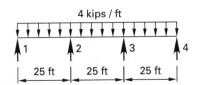

Solution

From LRFD Table 3-23,

$$M_2 = \text{moment at interior support}$$
$$= -0.10 w L^2$$
$$= (-0.10)\left(4 \frac{\text{kips}}{\text{ft}}\right)(25 \text{ ft})^2$$
$$= -250 \text{ ft-kips}$$
$$M_{12} = 0.08 w L^2$$
$$= (0.08)\left(4 \frac{\text{kips}}{\text{ft}}\right)(25 \text{ ft})^2$$
$$= 200 \text{ ft-kips}$$

Allowing for redistribution in accordance with LRFD App. 1 Sec. 1.3, the required design flexural strengths are

$$M_{u12} = 200 \text{ ft-kips} + \frac{(0.1)(0 + 250 \text{ ft-kips})}{2}$$
$$= 212.5 \text{ ft-kips}$$
$$M_{u2} = (0.9)(-250 \text{ ft-kips})$$
$$= -225 \text{ ft-kips} \quad [\text{governs}]$$

From LRFD Table 3-2, a W18 × 35 has

$$\phi_b M_p = 249 \text{ ft-kips}$$
$$> 225 \text{ ft-kips} \quad [\text{satisfactory}]$$

Biaxial Bending

A beam subjected to bending moment about both the x- and y-axes may be designed in accordance with LRFD Sec. H1.1 by using the interaction expression

$$\frac{M_{rx}}{\phi_b M_{nx}} + \frac{M_{ry}}{\phi_b M_{ny}} \leq 1.00 \quad \textit{[LRFD H1-1b]}$$

Example 4.10

A simply supported W16 × 36 beam with a span of 15 ft is subjected to a uniformly distributed factored vertical load of 4 kips/ft and a horizontal factored concentrated load of 3 kips applied at midspan. Determine whether the beam is adequate if the beam of grade 50 steel is laterally braced at the supports.

Solution

The maximum bending moments due to the factored loads are

$$M_{rx} = \frac{w L^2}{8} = \frac{\left(4 \frac{\text{kips}}{\text{ft}}\right)(15 \text{ ft})^2}{8}$$
$$= 112.5 \text{ ft-kips}$$
$$M_{ry} = \frac{W L}{4} = \frac{(3 \text{ kips})(15 \text{ ft})}{4}$$
$$= 11.3 \text{ ft-kips}$$

For bending about the x-axis with an unbraced length of 15 ft, from LRFD Table 3-10, a W16 × 36 has

$$\phi_b M_{nx} = C_b(151 \text{ ft-kips})$$
$$= (1.14)(151 \text{ ft-kips})$$
$$= 172 \text{ ft-kips}$$

For bending about the y-axis, from LRFD Table 3-4, a W16 × 36 has

$$\phi_b M_{ny} = 40.5 \text{ ft-kips}$$

The left side of the interaction equation is

$$\frac{M_{rx}}{\phi_b M_{nx}} + \frac{M_{ry}}{\phi_b M_{ny}} = \frac{112.5 \text{ ft-kips}}{172 \text{ ft-kips}} + \frac{11.3 \text{ ft-kips}}{40.5 \text{ ft-kips}}$$
$$= 0.93$$
$$< 1.0 \quad [\text{satisfactory}]$$

3. DESIGN FOR SHEAR

Nomenclature

A_{gt}	gross area subject to tension	in²
A_{gv}	gross area subject to shear	in²
A_{nt}	net area subject to tension	in²
A_{nv}	net area subject to shear	in²
A_w	web area	in²
C_v	web shear coefficient	–
d	overall depth of member	in
d_b	nominal bolt diameter	in
d_h	diameter of bolt hole	in
F_u	specified minimum tensile strength	kips/in²
h	for rolled shapes, the distance between flanges less the corner radius; for built-up sections, the clear distance between flanges	in
k	distance from outer face of flange to web toe of fillet	in
N	length of bearing	in
R	nominal reaction	kips
s	bolt spacing	in
t_f	flange thickness	in
t_w	web thickness	in
U_{bs}	reduction coefficient	–
V_n	nominal shear strength	kips
V_u	required shear strength	kips

Shear in Beam Webs

The shear in rolled W shape beams is resisted by the area of the web that is defined as

$$A_w = dt_w$$

It is assumed that the shear stress is uniformly distributed over this area, and for a slenderness ratio

$h/t_w \leq 2.24\sqrt{E/F_y}$, the nominal shear strength is governed by yielding of the web. This is the case for most W shapes, and the nominal shear strength is given by LRFD Sec. G2.1 as

$$V_n = 0.6F_y A_w C_v \qquad [\text{LRFD G2-1}]$$
$$C_v = 1.0$$

The design shear strength is

$$\phi_v V_n = 1.0 V_n$$

Example 4.11

Check the adequacy in shear of the W12 × 53 grade 50 beam of Ex. 4.7 at support 2.

Solution

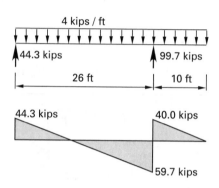

The shear force diagram is shown in the illustration, and the required shear strength is

$$V_u = 59.7 \text{ kips}$$

The design shear strength is obtained from LRFD Table 3-2 as

$$\phi_v V_n = 125 \text{ kips}$$
$$> V_u \quad [\text{satisfactory}]$$

Block Shear

When the end of a beam is coped, failure may occur by block shear, or web tear out. As shown in Fig. 4.6, this is a combination of shear along the vertical plane and tension along the horizontal plane.

The resistance to block shear is given by LRFD Eq. (J4–5) as

$$\phi R_n = \phi \left(0.6F_u A_{nv} + U_{bs} F_u A_{nt} \right)$$
$$\leq \phi \left(0.6F_y A_{gv} + U_{bs} F_u A_{nt} \right)$$

The reduction coefficient is given by

$$U_{bs} = 1.0 \text{ for uniform tension stress}$$
$$= 0.5 \text{ for non-uniform tension stress}$$

Figure 4.6 *Block Shear in a Coped Beam*

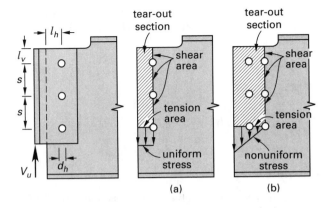

(a)   (b)

The resistance factor is

$$\phi = 0.75$$

Example 4.12

Determine the resistance to block shear of the coped W16 × 40 grade A36 beam shown in Fig. 4.6(a). The relevant dimensions are $l_h = l_v = 1.5$ in and $s = 3$ in. The bolt diameter is $3/4$ in.

Solution

The hole diameter for a $3/4$ in diameter bolt is defined in LRFD Sec. D3.2 and Table J3.3 as

$$d_h = d_b + \frac{1}{8} \text{ in} = 0.75 \text{ in} + 0.125 \text{ in}$$
$$= 0.875 \text{ in}$$
$$t_w = 0.305 \text{ in}$$
$$A_{nv} = t_w \left(l_v + 2s - 2.5d_h\right)$$
$$= t_w\big((1.5 \text{ in}) + (2)(3.0 \text{ in}) - (2.5)(0.875 \text{ in})\big)$$
$$= 5.31t_w \text{ in}^2$$
$$A_{nt} = t_w \left(l_h - 0.5d_h\right)$$
$$= t_w\big((1.5 \text{ in}) - (0.5)(0.875 \text{ in})\big)$$
$$= 1.06t_w \text{ in}^2$$

The tensile stress is uniform and the reduction coefficient is

$$U_{bs} = 1.0$$
$$U_{bs}F_u A_{nt} = \left(58 \, \frac{\text{kips}}{\text{in}^2}\right)\left(1.06t_w \text{ in}^2\right)$$
$$= 61.48t_w \text{ kips}$$
$$0.6F_u A_{nv} = (0.6)\left(58 \, \frac{\text{kips}}{\text{in}^2}\right)\left(5.31t_w \text{ in}^2\right)$$
$$= 184.79t_w \text{ kips}$$
$$A_{gv} = t_w(l_v + 2s)$$
$$= t_w\left(1.5 \text{ in} + (2)(3 \text{ in})\right)$$
$$= 7.5t_w \text{ in}^2$$
$$0.6F_y A_{gv} = (0.6)\left(36 \, \frac{\text{kips}}{\text{in}^2}\right)\left(7.5t_w \text{ in}^2\right)$$
$$= 162t_w \text{ kips} \quad \text{[governs]}$$
$$< 0.6F_u A_{nv}$$

Hence, shear yielding governs and the resistance to block shear is given by LRFD Eq. (J4-5) as

$$\phi R_n = \phi(0.6F_y A_{gv} + U_{bs}F_u A_{nt})$$
$$= (0.75)(0.305 \text{ in})\left(162 \, \frac{\text{kips}}{\text{in}} + 61.48 \, \frac{\text{kips}}{\text{in}}\right)$$
$$= 51.12 \text{ kips}$$

Local Web Yielding

As shown in Fig. 4.7, a bearing plate may be used to distribute concentrated loads applied to the flange to prevent local web yielding. The load is assumed dispersed, at a gradient of 2.5 to 1.0, to the web toe of fillet. For loads applied at a distance of not more than d from the end of the beam, the nominal strength is given by LRFD Sec. J10.2 as

$$R_n = (2.5k + N)F_y t_w \qquad \text{[LRFD J10-3]}$$

Figure 4.7 *Local Web Yielding*

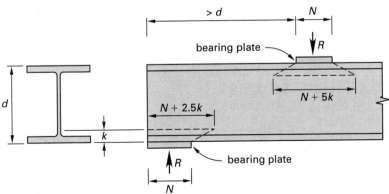

For loads applied at a distance of more than d from the end of the beam, the nominal strength is given as

$$R_n = (5k + N)F_y t_w \qquad \text{[LRFD J10-2]}$$

The design strength is given by ϕR_n with $\phi = 1.0$. LRFD Table 9-4 tabulates values of

$$\phi R_1 = \phi \left(2.5 k F_y t_w \right)$$
$$\phi R_2 = \phi \left(F_y t_w \right)$$

Example 4.13

Determine the resistance to local web yielding of a W40 × 331 grade 50 beam with a 4.0 in long bearing plate at the end of the beam.

Solution

LRFD Eq. (J10-3) is applicable, and the design strength is

$$\phi R_n = \phi(2.5k + N)F_y t_w$$
$$= \phi \left(R_1 + (4.0 \text{ in})R_2 \right)$$
$$= 504 \text{ kips} + (4.0 \text{ in}) \left(61.0 \, \frac{\text{kips}}{\text{in}} \right)$$
$$\text{[using values from LRFD Table 9-4]}$$
$$= 748 \text{ kips}$$

Web Crippling

For a concentrated load applied at a distance of not less than $d/2$ from the end of the beam, the nominal strength against web crippling is given by LRFD Sec. J10.3 as

$$R_n = 0.80 t_w^2 \left(1 + (3) \left(\frac{N}{d} \right) \left(\frac{t_w}{t_f} \right)^{1.5} \right) \sqrt{\frac{EF_y t_f}{t_w}}$$

$$\text{[LRFD J10-4]}$$

For loads applied at a distance of less than $d/2$ from the end of the beam and for $N/d \leq 0.2$, the nominal strength is given by

$$R_n = 0.40 t_w^2 \left(1 + (3) \left(\frac{N}{d} \right) \left(\frac{t_w}{t_f} \right)^{1.5} \right) \sqrt{\frac{EF_y t_f}{t_w}}$$

$$\text{[LRFD J10-5a]}$$

For loads applied at a distance of less than $d/2$ from the end of the beam and for $N/d > 0.2$, the nominal strength is given by

$$R_n = 0.40 t_w^2 \left(1 + \left((4) \left(\frac{N}{d} \right) - 0.2 \right) \left(\frac{t_w}{t_f} \right)^{1.5} \right) \sqrt{\frac{EF_y t_f}{t_w}}$$

$$\text{[LRFD J10-5b]}$$

The design strength is given by $\phi_r R_n$ with $\phi_r = 0.75$. Using values of $\phi_r R_3$, $\phi_r R_4$, $\phi_r R_5$, and $\phi_r R_6$, tabulated in LRFD Table 9-4, reduces LRFD Eq. (J10-5a) to

$$\phi_r R_n = \phi_r R_3 + N \left(\phi_r R_4 \right)$$

LRFD Eq. (J10–5b) becomes

$$\phi_r R_n = \phi_r R_5 + N \left(\phi_r R_6 \right)$$

Example 4.14

Determine the design web crippling strength of a W40 × 331 grade 50 beam with a 3.25 in long bearing plate at the end of the beam.

Solution

For $N/d < 0.2$, the applicable expression is

$$\phi_r R_n = \phi_r R_3 + N \left(\phi_r R_4 \right)$$
$$= 710 \text{ kips} + (3.25 \text{ in}) \left(22.7 \, \frac{\text{kips}}{\text{in}} \right)$$
$$\text{[using values from LRFD Table 9-4]}$$
$$= 784 \text{ kips}$$

4. DESIGN OF COMPRESSION MEMBERS

Nomenclature

A	area of member	in^2
A_g	gross area of member	in^2
B	base plate width	in
B_1	moment magnification factor applied to the primary moments to account for the curvature of the members, with lateral translation inhibited, as defined in LRFD Eq. (C2-2)	–
B_2	moment magnification factor applied to the primary moments to account for the translation of the members, with lateral translation permitted, as defined in LRFD Eq. (C2-3)	–
b_f	flange width of rolled beam	in
b_x	$8/9\phi_b M_{nx} \times 10^3$	(ft-kips)$^{-1}$
b_y	$8/9\phi_b M_{ny} \times 10^3$	(ft-kips)$^{-1}$
C_b	bending coefficient	–
C_m	reduction factor given by LRFD Eq. (C2-4)	–
d	depth of rolled beam	in
E	modulus of elasticity	kips/in^2
EA*	reduced value of EA	kips
EI*	reduced value of EI	kips-in^2

F_e	elastic critical buckling stress	kips/in^2
F_{cr}	critical stress	kips/in^2
F_y	yield stress	kips/in^2
G	ratio of total column stiffness framing into a joint to that of the total girder stiffness framing into the joint	–
H	horizontal force	kips
H_p	lateral load required to produce the design story drift	kips
I	moment of inertia	in^4
K	effective length factor	–
K_1	effective-length factor in the plane of bending	–
K_2	effective-length factor in the plane of bending for an unbraced frame	–
L	story height	ft
L	actual unbraced length in the plane of bending	ft
m	cantilever dimension for base plate along the length of the plate	in
M_a	augmented moment	ft-kips
M_{lt}	calculated first-order factored moment in a member, due to lateral translation of the frame only	ft-kips
M_{nt}	calculated first-order factored moment assuming no lateral translation of the frame	ft-kips
M_{nx}	nominal flexural strength about the strong axis in the absence of axial load	ft-kips
M_{ny}	nominal flexural strength about the weak axis in the absence of axial load	ft-kips
M_r	required second-order factored flexural strength using LRFD load combinations	ft-kips
M_{rx}	required second-order factored bending moment about the strong axis	ft-kips
M_{ry}	required second-order factored bending moment about the weak axis	ft-kips
M_1	smaller moment at end of unbraced length of member, calculated from a first-order analysis	ft-kips
M_2	larger moment at end of unbraced length of member, calculated from a first-order analysis	ft-kips
n	cantilever dimension for base plate along the width of the plate	in
N	base plate length	in
N_i	notional lateral load applied at level I	kips
p	$1/\phi_c P_n \times 10^3$	kips^{-1}
P_a	augmented axial force	kips
P_e	Euler buckling strength	kips
P_{e1}	Euler buckling strength of the member in the plane of bending as defined in LRFD Eq. (C2-5)	kips
P_{lt}	calculated first-order factored axial force in a member, due to lateral translation of the frame only	kips
P_n	nominal axial strength using LRFD load combinations	kips
P_{nt}	first-order factored axial force assuming no lateral translation of the frame	kips
P_r	required second-order axial strength using LRFD load combinations	kips
P_y	member yield strength	kips
r	radius of gyration	in
R_M	system parameter	–
t_{req}	required base plate thickness	in
Y_i	gravity load applied at level I independent of loads from above	kips

Symbols

Δ_a	interstory drift due to applied loads	in
Δ_H	first-order interstory drift due to lateral forces	in
Δ_{oh}	translational deflection of the story under consideration	in
Δ_p	permissible interstory drift	in
$\Delta_{1\text{st}}$	first-order drift	in
$\Delta_{2\text{nd}}$	second-order drift	in
λ	F_y/F_e	–
$\sum H$	story shear produced by the lateral forces used to compute Δ_H	kips
$\sum(I_c/L_c)$	the sum of the I/L values for all columns meeting at a joint	in^3
$\sum(I_g/L_g)$	the sum of the I/L values for all girders meeting at a joint	in^3
$\sum P_{e2}$	sum, for all columns in a story of a moment frame, of the Euler buckling strength	kips
$\sum P_{nt}$	total factored vertical load supported by the story, including gravity columns loads	kips
τ_b	stiffness reduction coefficient	–
ϕ_b	resistance factor for flexure	–
$\phi_b M_n$	available flexural strength	–
ϕ_c	resistance factor for compression	–
$\phi_c P_n$	available axial compression strength	–

Effective Length

In the design of compression members, the effective length factor K is used to account for the influence of restraint conditions at each end of a column. The K factor is used to equate the nominal strength of a compression member of length L to that of an equivalent pin-ended member of length KL. The nominal strength of a compression member is dependent on the slenderness ratio KL/r, which is limited to a maximum recommended value of 200.

LRFD Table C–C2.2 specifies effective length factors for well-defined conditions of restraint; these are illustrated in Fig. 4.8. These values may only be used in simple cases when the tabulated end conditions are closely approximated in practice.

Figure 4.8 *Effective Length Factors*

illus.	end conditions	theoretical	design
		k	
(a)	both ends pinned	1	1.00
(b)	both ends built in	0.5	0.65
(c)	one end pinned, one end built in	0.7	0.8
(d)	one end built in, one end free	2	2.10
(e)	one end built in, one end fixed against rotation but free to translate	1	1.20
(f)	one end pinned, one end fixed against rotation but free to translate	2	2.0

(a) (b) (c)

(d) (e) (f)

For compression members in a truss, an effective length factor of 1.0 is used. For load bearing web stiffeners on a girder, LRFD Sec. J10.8 specifies an effective length factor of 0.75. For columns in a rigid frame that is adequately braced, LRFD Sec. C1.3a specifies a conservative value for the effective length factor of 1.0. For compression members forming part of a frame with rigid joints, LRFD Commentary Sec. C2 presents alignment charts for determining the effective length for the two

conditions of sideway prevented and sideway permitted; these charts are illustrated in Fig. 4.9.

To use the alignment charts, the stiffness ratio at the two ends of the column under consideration must be determined. This ratio is defined as

$$G = \frac{\sum \left(\dfrac{I_c}{L_c} \right)}{\sum \left(\dfrac{I_g}{L_g} \right)}$$

For a braced frame with rigid joints, the girders are bent in single curvature and the alignment charts are based on a stiffness value of $2EI/L$. If one end of a girder is pinned, its stiffness is $3EI/L$; if one end is fixed, its stiffness is $4EI/L$. Hence, for these two cases, the (I_g/L_g) values are multiplied by 1.5 and 2.0, respectively. For a sway frame with rigid joints, the girders are bent in double curvature and the alignment charts are based on a stiffness value of $6EI/L$. If one end of a girder is pinned, its stiffness is $3EI/L$ and the (I_g/L_g) values are multiplied by 0.5. For a column with a pinned base, LRFD Commentary Sec. C2 specifies a stiffness ratio of $G = 10$. For a column with a fixed base, LRFD Commentary Sec. C2 specifies a stiffness ratio of $G = 1$.

Example 4.15

The sway frame shown in the following illustration consists of members having identical I/L values. Determine the effective length factors of columns 12 and 34.

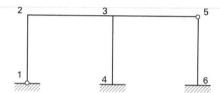

Solution

For the pinned connection at joint 1, LRFD Commentary Sec. C2 specifies a stiffness ratio of $G_1 = 10$.

At joint 2,

$$G_2 = \frac{\sum \left(\dfrac{I_c}{L_c} \right)}{\sum \left(\dfrac{I_g}{L_g} \right)} = \frac{1.0 \text{ in}^3}{1.0 \text{ in}^3} = 1.0$$

From the alignment chart for sway frames, the effective length factor is

$$K_{12} = 1.9$$

Allowing for the pinned end at joint 5, the sum of the adjusted relative stiffness values for the two girders connected to joint 3 is

$$\sum \left(\frac{I_g}{L_g} \right) = 1.0 \text{ in}^3 + 0.5 \text{ in}^3$$
$$= 1.5 \text{ in}^3$$

Figure 4.9 *Alignment Charts for Effective Length Factors*

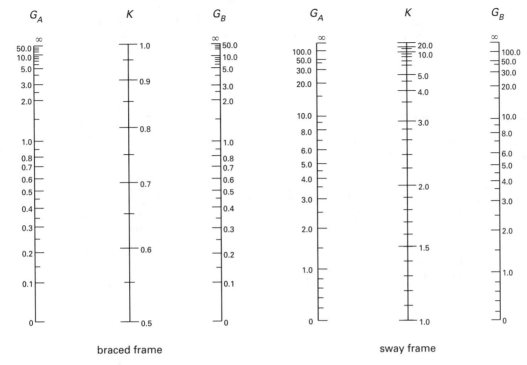

braced frame   sway frame

Adapted from American Institute of Steel Construction, *Steel Construction Manual*, Figs. C-C2.3 and C-C2.4.

Hence, the stiffness ratio at joint 3 is given by

$$G_3 = \frac{\sum\left(\dfrac{I_c}{L_c}\right)}{\sum\left(\dfrac{I_g}{L_g}\right)} = \frac{1.0 \text{ in}^3}{1.5 \text{ in}^3}$$

$$= 0.67$$

For the fixed connection at joint 4, LRFD Commentary Sec. C2 specifies a stiffness ratio of $G_4 = 1.0$. Hence, from the alignment chart for a sway frame, the effective length factor for column 34 is

$$K_{34} = 1.27$$

Axially Loaded Members

The design strength in compression is given by

$$\phi_c P_n = 0.90 A_g F_{cr}$$

For a short column with $KL/r \leq 4.71(E/F_y)^{0.5}$ and $F_e \geq 0.44F_y$, inelastic instability governs and LRFD Eq. (E3-2) defines the critical stress as

$$F_{cr} = (0.658^{\lambda})F_y$$

The parameter λ is defined by

$$\lambda = \frac{F_y}{F_e}$$

The elastic critical buckling stress is given by LRFD Eq. (E3-4) as

$$F_e = \frac{\pi^2 E}{\left(\dfrac{KL}{r}\right)^2}$$

For a long column with $KL/r > 4.71(E/F_y)^{0.5}$ and $F_e < 0.44F_y$, elastic instability governs and the critical stress is given by LRFD Eq. (E3-3) as

$$F_{cr} = 0.877F_e$$

Once the governing slenderness ratio of a column is established, the design stress $\phi_c F_{cr}$ may be obtained directly from LRFD Table 4-22 for steel members with a yield stress of 35, 36, 42, 46, or 50 kips/in^2.

Values of the design axial strength are tabulated in LRFD Table 4-1 for rolled sections W14 and smaller with respect to r_y for varying effective lengths. These tabulated values may be used directly when $(KL/r)_y$ exceeds $(KL/r)_x$.

Example 4.16

Determine the lightest W12 grade 50 column that will support a factored load of 850 kips. The column is 12 ft high, is pinned at each end, and has no intermediate bracing about either axis.

Solution

From LRFD Table 4-1, a W12 × 79 column with an effective length of 12 ft has a design axial strength of

$$\phi_c P_n = 887 \text{ kips} \quad [\text{satisfactory}]$$

Equivalent Effective Length

When the effective lengths of a column about the x- and y-axes are different, the strength of the column must be investigated with respect to both axes. Dividing the effective length about the x-axis by the ratio r_x / r_y provides an equivalent effective length about the y-axis.

Example 4.17

Determine the design axial strength of a W12 × 106 grade 50 column that is 12 ft high, pinned at each end, and braced at midheight about the y-axis.

Solution

The effective length about the y-axis is

$$KL_y = 6 \text{ ft}$$

The effective length about the x-axis is

$$KL_x = 12 \text{ ft}$$

From LRFD Table 4-1, a W12 × 106 column has a value of

$$\frac{r_x}{r_y} = 1.76$$

The equivalent effective length about the major axis with respect to the y-axis is

$$(KL_y)_{\text{equiv}} = \frac{KL_x}{\dfrac{r_x}{r_y}} = \frac{12 \text{ ft}}{1.76}$$

$$= 6.8 \text{ ft} \quad [\text{governs}]$$

$$> KL_y$$

From LRFD Table 4-1, a W12 × 106 column with an effective length $(KL_y)_{\text{equiv}}$ of 6.8 ft has a design axial strength of

$$\phi_c P_n = 1334 \text{ kips}$$

Alternative Yield Stress

For built-up sections and laced compression members, LRFD Table 4-22, tabulates $\phi_c F_{cr}$ against KL/r for steel with yield stresses of 35, 36, 42, 46, and 50 kips/in², respectively.

Example 4.18

A laced column consisting of four 5 × 5 × $^1/_2$ angles of grade A36 steel is shown in the following illustration. The column may be considered a single integral member and is 20 ft high with pinned ends. Determine the maximum design axial load.

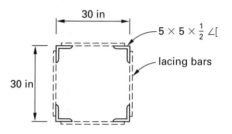

Solution

The relevant properties of a 5 × 5 × $^1/_2$ angle are

$$A = 4.75 \text{ in}^2$$
$$I = 11.3 \text{ in}^4$$
$$y = 1.42 \text{ in}$$

The relevant properties of the laced column are

$$\sum A = 4A$$
$$= (4)\left(4.75 \text{ in}^2\right)$$
$$= 19 \text{ in}^2$$

$$\sum I = 4I + \sum A \left(\frac{d}{2} - y\right)^2$$
$$= (4)\left(11.3 \text{ in}^4\right) + \left(19 \text{ in}^2\right)(15 \text{ in} - 1.42 \text{ in})^2$$
$$= 3549 \text{ in}^4$$

The radius of gyration of the laced column is

$$r = \sqrt{\frac{\sum I}{\sum A}} = \sqrt{\frac{3549 \text{ in}^4}{19 \text{ in}^2}}$$
$$= 13.67 \text{ in}$$

The slenderness ratio of the laced column is

$$\frac{KL}{r} = \frac{(1.0)(20 \text{ ft})\left(12 \dfrac{\text{in}}{\text{ft}}\right)}{13.67 \text{ in}}$$
$$= 17.56$$
$$< 200 \quad [\text{satisfactory}]$$

From LRFD Table 4-22, the design stress is

$$\phi_c F_{cr} = 31.9 \text{ kips/in}^2$$

The design axial strength is

$$\phi_c P_n = \phi_c F_{cr} \sum A$$
$$= \left(31.9 \ \frac{\text{kips}}{\text{in}^2}\right)(19 \ \text{in}^2)$$
$$= 606 \ \text{kips}$$

Composite Columns

Concrete filled hollow structural sections and concrete encased rolled steel sections reinforced with longitudinal and lateral reinforcing bars are designed by using LRFD Sec. I2. Values of the design axial strength for typical sizes of column are tabulated in LRFD Part 4.

Example 4.19

Determine the least weight rectangular composite column using an HSS section filled with 5000 lbf/in² concrete that can support a factored load of 730 kips. The column is 15 ft high and is pinned at each end.

Solution

From LRFD Table 4-14, an HSS $14 \times 10 \times {}^5/_{16}$ in with an effective length of 15 ft has a design axial strength of

$$\phi_c P_n = 732 \ \text{kips} \quad [\text{satisfactory}]$$

Second-Order Effects

In accordance with LRFD Sec. C2, the design of compression members must take into account secondary effects. The secondary moments and axial forces caused by the P-delta effects must be added to the primary moments and axial forces in a member, which were obtained by a first-order analysis. The P-delta effects are the result of the two separate effects P-δ and P-Δ and, as shown in Fig. 4.10, the final forces in a frame (including secondary effects), may be obtained as the summation of the two analyses, sway and non-sway.

Figure 4.10 *Determination of Secondary Effects*

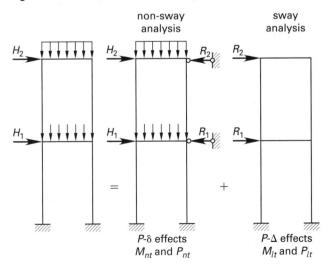

The P-δ effect produces an amplified moment due to the eccentricity of the axial force with respect to the displaced center line of the member. This is termed the *member effect*. The moment magnification factor which, when applied to the primary moments, accounts for the P-δ effect and is termed B_1.

The P-Δ effect produces an amplified moment due to the drift in a sway frame. This is termed the *frame effect*. The moment magnification factor which, when applied to the primary moments, accounts for the P-Δ effect and is termed B_2.

The required final second-order forces are then given by LRFD Eqs. (C2-1a) and (C2-1b) as

$$M_r = B_1 M_{nt} + B_2 M_{lt}$$
$$P_r = P_{nt} + B_2 P_{lt}$$

The moment magnification factor to account for the member effect, assuming no lateral translation of the frame, is defined in LRFD Eq. (C2-2) as

$$B_1 = \frac{C_m}{1 - \dfrac{P_r}{P_{e1}}}$$
$$\geq 1.0$$

When $B_1 \leq 1.05$, the required second-order moment is conservatively given by

$$M_r = B_2(M_{nt} + M_{lt})$$

The Euler buckling strength of the member in the plane of bending is defined in LRFD Eq. (C2-5) as

$$P_{e1} = \frac{\pi^2 EI}{(K_1 L)^2}$$

The effective-length factor in the plane of bending, for a member in a frame with lateral translation inhibited, is given by

$$K_1 = 1.0$$

The reduction factor is given by LRFD Eq. (C2-4) as

$$C_m = 0.6 - 0.4 \left(\frac{M_1}{M_2}\right) \quad \begin{bmatrix} \text{for a member not subjected} \\ \text{to transverse loading} \\ \text{between supports} \end{bmatrix}$$

$$= 1.0 \quad \begin{bmatrix} \text{for a member transversely loaded} \\ \text{between supports} \end{bmatrix}$$

$$= 1.0 \quad \begin{bmatrix} \text{for a member bent in single curvature} \\ \text{under uniform bending moment} \end{bmatrix}$$

$$\frac{M_1}{M_2} = +\text{ve} \quad [\text{for a member bent in reverse curvature}]$$

$$= -\text{ve} \quad [\text{for a member bent in single curvature}]$$

The moment magnification factor to account for the frame effect, with lateral translation of the frame allowed, is defined in LRFD Eq. (C2-3) as

$$B_2 = \frac{1}{1 - \dfrac{\sum P_{nt}}{\sum P_{e2}}}$$

$$= \frac{\Delta_{2nd}}{\Delta_{1st}}$$

$$\geq 1.0$$

The sum of the Euler buckling strength, for all columns in a story of a frame is given by

$$\sum P_{e2} = \sum \left(\frac{\pi^2 EI}{(K_2 L)^2} \right) \quad \begin{bmatrix} \text{from LRFD Eq. (C2-6a)} \\ \text{for a moment frame} \end{bmatrix}$$

$$= \left(\frac{R_m L}{\Delta_H} \right) \sum H \quad \begin{bmatrix} \text{from LRFD Eq. (C2-6b)} \\ \text{for all types of lateral} \\ \text{load resisting systems} \end{bmatrix}$$

The system parameter factor is given by

$$R_M = 1.0 \quad \text{[for a braced frame]}$$

$$= 0.85 \quad \begin{bmatrix} \text{for moment frame and} \\ \text{combined systems} \end{bmatrix}$$

When a limit is placed on the drift index Δ_H/L the amplification factor B_2 may be determined by using this limit in LRFD Eq. (C2-6b). LRFD Table 4-1 tabulates values of $P_e(KL)^2/10^4$ for W-shapes with a yield stress of 50 ksi.

As shown in Fig. 4.10, two first-order analyses are required in order to determine both M_{nt} and M_{lt}. In the first analysis, imaginary horizontal restraints are introduced at each floor level to prevent lateral translation. The factored loads are then applied, the primary moments, M_{nt}, are calculated, and the magnitudes of the imaginary restraints, R, are determined. In the second analysis, the frame is analyzed for the reverse of the imaginary restraints in order to determine the primary moments M_{lt}. Applying LRFD Eqs. (C2-1a) and (C2-1b) provides the final second-order forces.

Alternatively, P-delta effects may be directly determined in a rigorous second-order frame analysis and the members may then be designed directly for the calculated axial force and bending moment. The principal of superposition is not valid in a second-order analysis and separate analyses are necessary for each combination of factored loads.

Analysis Methods

Four methods are detailed in the *LRFD Manual* for determining the secondary effects in steel frames. These include the following techniques.

- the effective length method, or second-order elastic analysis, detailed in LRFD Sec. C2.2a

- direct analysis method detailed in LRFD App. 7

- first-order elastic analysis detailed in LRFD Sec. C2.2b

- simplified method detailed in LRFD Part 2

Effective Length Method

This method is restricted by LRFD Sec. C2.2 to structures with a sidesway amplification factor of

$$\frac{\Delta_{2nd}}{\Delta_{1st}} \leq 1.5$$

In accordance with LRFD Sec. C2.2a, the design forces may be determined either by a rigorous second-order computer analysis or by amplifying the results of a first-order analysis. A rigorous second-order analysis is required when

$$B_1 > 1.2$$

In applying the method, the factored loads are applied to the structure using the nominal stiffness of the members. To account for initial imperfections in the members, minimum lateral loads are applied at each story, in accordance with LRFD Sec. C2.2(a)(3). These loads are given by

$$N_i = 0.002 Y_i$$

In designing the members for the calculated second-order forces, the appropriate effective length factor K, as defined in LRFD Table C-C2.2 or calculated in accordance with LRFD Commentary Sec. C2, must be adopted.

Example 4.20

The first-order member forces produced in the outer column of a frame by the governing factored load combination are shown in the following illustration. Determine the second-order member forces in the column using the effective length method. The column consists of a W12 × 79 section with a yield stress of 50 kips/in^2. The bay length is 25 ft and the beams consist of W21 × 62 sections with a yield stress of 50 kips/in^2. No intermediate bracing is provided to the column about either axis.

Illustration for Ex. 4.20

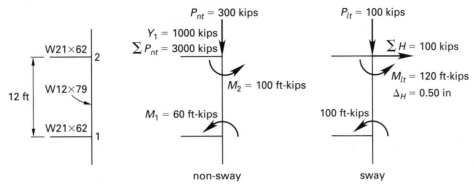

Solution

The sum of the gravity loads applied at the story, independent of loads from the upper stories is

$$Y_1 = 1000 \text{ kips}$$

The notional lateral load on the story, specified by LRFD Section C2.2(a)(3), is

$$\begin{aligned} N_1 &= 0.002Y_1 \\ &= (0.002)(1000 \text{ kips}) \\ &= 2 \text{ kips} \\ &< H \end{aligned}$$

The notional lateral load is not applicable.

Determine B_1.

The results of the first-order, non-sway analysis are shown in the illustration and the reduction factor is given by LRFD Eq. (C2-4) as

$$\begin{aligned} C_m &= 0.6 - 0.4\left(\frac{M_1}{M_2}\right) \\ &= 0.6 - (0.4)\left(\frac{60 \text{ ft-kips}}{100 \text{ ft-kips}}\right) \\ &= 0.36 \end{aligned}$$

The effective length factor in the plane of bending for a non-sway frame is given by

$$K_1 = 1.0$$

From LRFD Table 4-1, the Euler buckling strength of a W12 × 79 column in the plane of bending is given by

$$\begin{aligned} P_{e1} &= \frac{\pi^2 EI}{(K_1 L)^2} \\ &= \frac{(18{,}900)(10^4)}{((1.00)(144 \text{ in}))^2} \\ &= 9115 \text{ kips} \end{aligned}$$

Assume the required second-order axial force in the column is

$$P_r = 415 \text{ kips}$$

The moment magnification factor to account for the member effect, assuming no lateral translation of the frame, is defined in LRFD Eq. (C2-2) as

$$\begin{aligned} B_1 &= \frac{C_m}{1 - \dfrac{P_r}{P_{e1}}} \\ &= \frac{0.36}{1 - \dfrac{415 \text{ kips}}{9115 \text{ kips}}} \\ &= 1.0 \quad [\text{minimum}] \end{aligned}$$

Determine B_2.

The results of the first-order sway analysis are shown in the illustration, and the sum of the Euler buckling strength for all columns in the story, is given by LRFD Eq. (C2-6b) as

$$\begin{aligned} \sum P_{e2} &= \left(\frac{R_m L}{\Delta_H}\right) \sum H \\ &= (0.85)\left(\frac{144 \text{ in}}{0.50 \text{ in}}\right)(100 \text{ kips}) \\ &= 24{,}480 \text{ kips} \end{aligned}$$

The moment magnification factor to account for the frame effect, with lateral translation of the frame allowed, is given by LRFD Eq. (C2-3) as

$$\begin{aligned} B_2 &= \frac{1}{1 - \dfrac{\sum P_{nt}}{\sum P_{e2}}} \\ &= \frac{1}{1 - \dfrac{3000 \text{ kips}}{24{,}480 \text{ kips}}} \\ &= 1.14 \end{aligned}$$

Calculate the second-order member forces.

The second-order member forces for the frame are given by LRFD Eqs. (C2-1a) and (C2-1b) as

$$M_r = B_1 M_{nt} + B_2 M_{lt}$$
$$= (1.0)(100 \text{ ft-kips}) + (1.14)(120 \text{ ft-kips})$$
$$= 237 \text{ ft-kips}$$
$$P_r = P_{nt} + B_2 P_{lt}$$
$$= 300 \text{ kips} + (1.14)(100 \text{ kips})$$
$$= 414 \text{ kips}$$

Direct Analysis Method

The direct analysis method is applicable to all types of structures and, in accordance with LRFD Sec. C2.2, the direct analysis method must be used when

$$\frac{\Delta_{2nd}}{\Delta_{1st}} > 1.5$$

The design forces may be determined either by a rigorous second-order computer analysis, or by amplifying the results of a first-order analysis. In the analyses, the effective length factor for all members is taken as $K = 1.0$.

In applying the method, the factored loads are applied to the structure using reduced flexural and axial stiffness of members that contribute to the lateral stability of the structure. The reduced stiffnesses account for elastic instability and inelastic softening effects, and are given by LRFD Eqs. (A-7-2) and (A-7-3) as

$$EI^* = 0.8\tau_b EI$$
$$EA^* = 0.8EA$$

The stiffness reduction coefficient is given by

$$\tau_b = 1.0 \quad [\text{for } P_r \le 0.5P_y]$$
$$= 4\left(\frac{P_r}{P_y}\right)\left(1 - \frac{P_r}{P_y}\right) \quad [P_r > 0.5P_y]$$

Alternatively, when $P_r > 0.5P_y$, the stiffness reduction factor τ_b may be taken as 1.0 provided that the actual lateral loads are increased by a notional lateral load of

$$N_i = 0.001Y_i$$

To account for initial imperfections in the members, minimum lateral loads are applied at each story, in accordance with LRFD App. Sec. 7.3(2). These loads are given by

$$N_i = 0.002Y_i$$

The notional loads are additive to the applied lateral loads when the sidesway amplification ratio is

$$\frac{\Delta_{2nd}}{\Delta_{1st}} > 1.5$$

or

$$> 1.71 \quad [\text{using the reduced elastic stiffness}]$$

The notional loads are minimum lateral loads when

$$\frac{\Delta_{2nd}}{\Delta_{1st}} \le 1.5$$

In designing the members of the frame for the calculated second-order forces, the appropriate effective length factor is specified in LRFD App. Sec. 7.1, as $K = 1.0$ for all members.

Example 4.21

The first-order member forces produced in the outer column of a frame by the governing factored load combination are shown in the following illustration. Determine the second-order member forces in the column using the direct analysis method. The column consists of a $W12 \times 79$ section with a yield stress of 50 kips/in². The bay length is 25 ft and the beams consist of $W21 \times 62$ sections with a yield stress of 50 kips/in². No intermediate bracing is provided to the column about either axis.

Illustration for Ex. 4.21

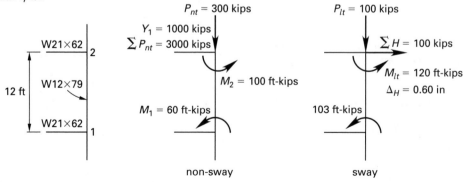

Solution

The column yield strength is

$$P_y = AF_y$$

$$= (23.2 \text{ in}^2)\left(50 \; \frac{\text{kips}}{\text{in}^2}\right)$$

$$= 1160 \text{ kips}$$

Assuming that the second-order axial load is

$$P_r = 415 \text{ kips}$$

$$\frac{P_r}{P_y} = \frac{415 \text{ kips}}{1160 \text{ kips}}$$

$$= 0.36$$

$$< 0.5$$

The stiffness reduction factor is given by LRFD App. Sec. 7.3(3) as

$$\tau_b = 1.0$$

The flexural and axial stiffness of the column is reduced as specified in LRFD Eqs. (A-7-2) and (A-7-3) to give

$$EI^* = 0.8\tau_b EI$$

$$= (0.8)(1.0)\left(29{,}000 \; \frac{\text{kips}}{\text{in}^2}\right)(662 \text{ in}^4)$$

$$= 15.36 \times 10^6 \text{ kip-in}^2$$

$$EA^* = 0.8EA$$

$$= (0.8)\left(29{,}000 \; \frac{\text{kips}}{\text{in}^2}\right)(23.2 \text{ in}^2)$$

$$= 53.82 \times 10^4 \text{ kips}$$

The sum of the gravity loads applied at the story, independent of loads from the upper stories is

$$Y_1 = 1000 \text{ kips}$$

The notional lateral load on the story is given by LRFD App. Sec. 7.3(2) as

$$N_i = 0.002Y_i$$

$$= (0.002)(1000 \text{ kips})$$

$$= 2 \text{ kips}$$

The actual applied lateral load is

$$H = 100 \text{ kips}$$

$$> N_i$$

Provided that the amplification factor B_2 is not greater than 1.5, the notional load is not added to the applied lateral load.

Applying the lateral loads to the frame with the stiffness of the columns reduced, the interstory drift is 0.6 in.

Determine B_1.

Assume the required second-order axial force in the column is

$$P_r = 415 \text{ kips}$$

The moment magnification factor to account for the member effect, assuming no lateral translation of the frame, is determined in Ex. 4.20 as

$$B_1 = \frac{C_m}{1 - \dfrac{P_r}{P_{e1}}}$$

$$= \frac{0.36}{1 - \dfrac{415 \text{ kips}}{9115 \text{ kips}}}$$

$$= 1.0 \quad [\text{minimum}]$$

Determine B_2.

The results of the first-order sway analysis are shown in the illustration, and the sum of the Euler buckling strength, for all columns in the story, is given by LRFD Eq. (C2-6b) as

$$\Sigma P_{e2} = \left(\frac{R_m L}{\Delta_H}\right)\sum H$$

$$= (0.85)\left(\frac{144 \text{ in}}{0.60 \text{ in}}\right)(100 \text{ kips})$$

$$= 20{,}400 \text{ kips}$$

The moment magnification factor to account for the frame effect, with lateral translation of the frame allowed, is given by LRFD Eq. (C2-3) as

$$B_2 = \frac{1}{1 - \dfrac{\sum P_{nt}}{\sum P_{e2}}}$$

$$= \frac{1}{1 - \dfrac{3000 \text{ kips}}{20{,}400 \text{ kips}}}$$

$$= 1.17$$

Calculate the second-order member forces.

The second-order member forces for the frame are given by LRFD Eqs. (C2-1a) and (C2-1b) as

$$M_r = B_1 M_{nt} + B_2 M_{lt}$$

$$= (1.0)(100 \text{ ft-kips}) + (1.17)(120 \text{ ft-kips})$$

$$= 240 \text{ ft-kips}$$

$$P_r = P_{nt} + B_2 P_{lt}$$

$$= 300 \text{ kips} + (1.17)(100 \text{ kips})$$

$$= 417 \text{ kips}$$

First-Order Elastic Analysis

The first-order elastic analysis method is specified in LRFD Sec. C2.2b. This method is restricted by LRFD Sec. C2.2 to structures with a sidesway amplification factor of

$$\frac{\Delta_{2nd}}{\Delta_{1st}} \leq 1.5$$

The design forces are determined by a first-order analysis with only notional loads applied to the structure.

A limit is placed on the required axial compressive strength such that

$$P_r = \text{required axial strength}$$
$$\leq 0.5 P_y$$
$$P_y = \text{member yield strength}$$
$$= A F_y$$

To account for initial imperfections in the members, notional lateral loads are applied at each story and are given by

$$N_i = \frac{2.1 \Delta_{1st} Y_i}{L}$$
$$\geq 0.0042 Y_i$$

The notional loads are additive to the applied lateral loads.

The non-sway amplification of column moments is considered by applying the B_1 amplifier to the total member moments.

In designing the members of the frame for the calculated second-order forces, the appropriate effective length factor is specified in LRFD Section C2.2b as $K = 1.0$ for all members.

Example 4.22

The factored loads acting on the outer column of a frame are shown in the following illustration. Determine the required strength of the column using the first-order elastic analysis method. The column consists of a W12 × 79 section with a yield stress of 50 kips/in².

The bay length is 25 ft and the beams consist of W21 × 62 sections with a yield stress of 50 kips/in². No intermediate bracing is provided to the column about either axis.

Solution

The column yield strength is

$$P_y = A F_y$$
$$= (23.2 \text{ in}^2) \left(50 \; \frac{\text{kips}}{\text{in}^2} \right)$$
$$= 1160 \text{ kips}$$

Assuming that the required axial load is

$$P_r = 417 \text{ kips}$$
$$\frac{P_r}{P_y} = \frac{417 \text{ kips}}{1160 \text{ kips}}$$
$$= 0.36$$
$$< 0.5$$

In accordance with LRFD Sec. C2.2b, a first-order elastic analysis is permissible.

Determine the augmented loads.

The first-order interstory drift due to the design loads is determined in Ex. 4.20 as

$$\Delta_{1st} = 0.5 \text{ in}$$

The sum of the gravity loads, applied at the story, independent of loads from the upper stories is

$$Y_1 = 1000 \text{ kips}$$

The notional lateral load on the story is given by LRFD Sec. C2.2b as

$$N_i = 0.0042 Y_i$$
$$= (0.0042)(1000 \text{ kips})$$
$$= 4.2 \text{ kips}$$

Illustration for Ex. 4.22

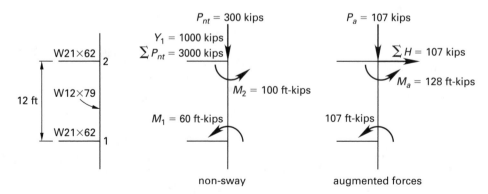

But not less than

$$N_i = \frac{2.1\Delta_{1st}Y_i}{L}$$
$$= \frac{(2.1)(0.5 \text{ in})(1000 \text{ kips})}{144 \text{ in}}$$
$$= 7.3 \text{ kips} \quad [\text{governs}]$$

This is additive to the applied lateral loads, and the augmented loads and the resulting forces on the column are indicated in the illustration.

Determine B_1.

The moment magnification factor to account for the member effect, assuming no lateral translation of the frame, is determined in Ex. 4.20 as

$$B_1 = 1.0$$

Calculate the required forces. The required forces are given by

$$M_r = B_1(M_{nt} + M_a)$$
$$= (1.0)(100 \text{ ft-kips} + 128 \text{ ft-kips})$$
$$= 228 \text{ ft-kips}$$
$$P_r = P_{nt} + P_a$$
$$= 300 \text{ kips} + 107 \text{ kips}$$
$$= 407 \text{ kips}$$

Simplified Method

The simplified method is specified in LRFD Part 2 and requires a first-order analysis only. The method is restricted to structures with a sidesway amplification factor of

$$\frac{\Delta_{2nd}}{\Delta_{1st}} \leq 1.5$$

In addition, the ratio of the sway and non-sway amplification factors is restricted to

$$\frac{B_1}{B_2} \leq 1.0$$

For members not subjected to transverse loading, it is unlikely that B_1 will be greater than B_2.

In applying the method, the factored loads are applied to the structure using the nominal stiffness of the members, and a first-order analysis performed. To account for initial imperfections in the members, minimum lateral loads are applied at each story in accordance with LRFD Sec. C2.2(a)(3). These loads are given by

$$N_i = 0.002Y_i$$

Required strengths are determined by multiplying the forces obtained from the first-order analysis by tabulated values of B_2. These tabulated values are a function of the design story drift limit and the ratio of the total story gravity load to the lateral load that produces the drift limit.

From the first-order elastic analysis, the lateral load required to produce the design story drift is determined, and the ratio of the total story gravity load to the lateral load that produces this drift limit is calculated. Using this value and the value of the design story drift, the appropriate amplification factor B_2 is obtained from the table in LRFD Part 2.

In designing the members of the frame for the calculated forces, the appropriate effective length factor is specified as $K = 1.0$ for all members, provided that the amplification factor does not exceed 1.1.

For cases where the amplification factor exceeds 1.1, the effective length factors for the members are determined by analysis.

Example 4.23

The factored loads acting on the outer column of a frame are shown in the following illustration. Determine the required strength of the column using the simplified analysis method. The column consists of a W12 × 79 section with a yield stress of 50 kips/in². The bay length is 25 ft and the beams consist of W21 × 62 sections with a yield stress of 50 kips/in². No intermediate bracing is provided to the column about either axis. The design story drift is limited to 1/240.

Illustration for Ex. 4.23

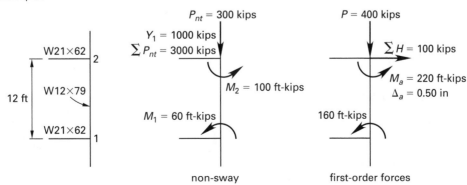

Solution

The sum of the gravity loads applied at the story, independent of loads from the upper stories is

$$Y_1 = 1000 \text{ kips}$$

The notional lateral load on the story, specified by LRFD Sec. C2.2(a)(3), is

$$\begin{aligned} N_1 &= 0.002Y_1 \\ &= (0.002)(1000 \text{ kips}) \\ &= 2 \text{ kips} \\ &< H \end{aligned}$$

The notional lateral load is not applicable.

The factored loads produce the first-order elastic forces, which are derived in Ex. 4.20 and are shown in the illustration. The interstory drift produced is

$$\Delta_a = 0.50 \text{ in}$$

For a design story drift of 1/240, the permissible deflection is

$$\begin{aligned} \Delta_p &= \frac{144 \text{ in}}{240} \\ &= 0.60 \text{ in} \end{aligned}$$

The lateral load required to produce the design story drift is

$$\begin{aligned} H_p &= \frac{(100 \text{ kips})\Delta_p}{\Delta_a} \\ &= \frac{(100 \text{ kips})(0.60 \text{ in})}{0.50 \text{ in}} \\ &= 120 \text{ kips} \end{aligned}$$

The ratio of the total story gravity load to the lateral load that produces the drift limit is

$$\begin{aligned} \frac{Y_1}{H_p} &= \frac{1000 \text{ kips}}{120 \text{ kips}} \\ &= 8.3 \end{aligned}$$

From the table in LRFD Part 2, the amplification factor is obtained as

$$B_2 = 1.03$$

The required forces are given by

$$\begin{aligned} M_r &= B_2 M \\ &= (1.03)(220 \text{ ft-kips}) \\ &= 227 \text{ ft-kips} \\ P_r &= B_2 P \\ &= (1.03)(400 \text{ kips}) \\ &= 412 \text{ kips} \end{aligned}$$

Combined Compression and Flexure

The adequacy of a member to sustain combined compression and flexure is determined by means of the interaction equations given in LRFD Sec. H1.1 as follows.

For $P_r/\phi_c P_n \geq 0.2$,

$$\frac{P_r}{\phi_c P_n} + \left(\frac{8}{9}\right)\left(\frac{M_{rx}}{\phi_b M_{nx}} + \frac{M_{ry}}{\phi_b M_{ny}}\right) \leq 1.0 \quad [LRFD \text{ } H1\text{-}1a]$$

This expression reduces to

$$pP_r + b_x M_{rx} + b_y M_{ry} \leq 1.0$$

For $P_r/\phi_c P_n < 0.2$,

$$\frac{P_r}{2\phi_c P_n} + \left(\frac{M_{rx}}{\phi_b M_{nx}} + \frac{M_{ry}}{\phi_b M_{ny}}\right) \leq 1.0 \quad [LRFD \text{ } H1\text{-}1b]$$

This expression reduces to

$$\frac{pP_r}{2} + \left(\frac{9}{8}\right)(b_x M_{rx} + b_y M_{ry}) \leq 1.0$$

Values of p, b_x, and b_y are tabulated in LRFD Table 6-1 for W shapes with a yield stress of 50 kips/in^2 and assuming a bending coefficient of $C_b = 1.0$.

Example 4.24

Determine the adequacy of the W12 × 79 column in the frame analyzed by the effective length method in Ex. 4.20. The beams consist of W21 × 62 sections with a length of 25 ft. All members have a yield stress of 50 kips/in^2 and no intermediate bracing is provided to the column about either axis.

Solution

The frame was analyzed for second-order effects in Ex. 4.20 using the effective length method. The required forces in the column are

$$\begin{aligned} P_r &= 414 \text{ kips} \\ M_{rx} &= 237 \text{ ft-kips} \\ M_{ry} &= 0 \text{ ft-kips} \end{aligned}$$

Calculate the effective column length.

At both joint 1 and joint 2, the stiffness ratio is

$$\begin{aligned} G_2 &= \frac{\sum \dfrac{I_c}{L_c}}{\sum \dfrac{I_g}{L_g}} \\ &= \frac{(2)\left(\dfrac{662 \text{ in}^4}{12 \text{ ft}}\right)}{\dfrac{1330 \text{ in}^4}{25 \text{ ft}}} \\ &= 2.07 \end{aligned}$$

From the alignment chart for sway frames (Fig. 4.9) the effective length factor about the x-axis of column 12 is

$$K_2 = 1.65$$

The effective length of the column about the x-axis is

$$K_2L = (1.65)(12 \text{ ft})$$
$$= 19.80 \text{ ft}$$

The effective length of the column about the y-axis is

$$KL_y = (1.0)(12 \text{ ft})$$
$$= 12 \text{ ft}$$

From LRFD Table 4-1, a W12 × 79 column has a value of

$$\frac{r_x}{r_y} = 1.75$$

The equivalent effective length about the major axis with respect to the y-axis is

$$KL_{y(\text{equiv})} = \frac{\dfrac{K_xL_x}{r_x}}{r_y}$$
$$= \frac{19.80 \text{ ft}}{1.75}$$
$$= 11.31 \text{ ft} \quad [\text{does not govern}]$$
$$< KL_y$$

The effective length about the minor axis governs.

Apply the interaction equation.

From LRFD Table 4-1, a W12 × 79 column with an effective length of $KL_y = 12$ ft has a design axial strength of

$$\phi_c P_n = 887 \text{ kips}$$
$$\frac{P_r}{\phi_c P_n} = \frac{414 \text{ kips}}{887 \text{ kips}}$$
$$= 0.467$$
$$> 0.2$$

The LRFD Eq. (H1-1a) applies, and the reduced form of LRFD Eq. (H1-1a) is

$$pP_r + b_x M_{rx} + b_y M_{ry} \le 1.0$$

Substituting in the left hand side of this expression and using the design parameters from LRFD Table 6-1 gives

$$(1.13 \times 10^{-3} \text{ kips}^{-1})(414 \text{ kips})$$
$$+ (2.02 \times 10^{-3} \text{ ft-kips}^{-1})(237 \text{ ft-kips}) + 0$$
$$= 0.95$$
$$< 1.0 \quad [\text{satisfactory}]$$

Column Base Plates

The design of column base plates is covered in LRFD Part 14 and LRFD Sec. J8. As shown in Fig. 4.11, the base plate is assumed to cantilever about axes a distance m or n from the edge of the plate. The required base plate thickness is given by the larger value obtained from

$$t_{\text{req}} = m\sqrt{\frac{2P_u}{0.9F_y BN}}$$

$$t_{\text{req}} = n\sqrt{\frac{2P_u}{0.9F_y BN}}$$

$$t_{\text{req}} = \lambda n'\sqrt{\frac{2P_u}{0.9F_y BN}}$$

$$n' = \frac{\sqrt{db_f}}{4}$$

λ may conservatively be taken as 1.0.

Figure 4.11 Column Base Plate

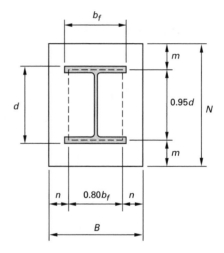

Example 4.25

A 19 in × 19 in grade A36 base plate is proposed for the W12×106 column of Ex. 4.16, which supports a factored load of 850 kips. Determine the minimum required base plate thickness.

Solution

From LRFD Table 1-1, the relevant dimensions are as follows.

$$d = 12.89 \text{ in}$$
$$b_f = 12.22 \text{ in}$$
$$\lambda n' = \frac{\lambda\sqrt{db_f}}{4}$$
$$= \frac{1.0\sqrt{(12.89 \text{ in})(12.22 \text{ in})}}{4}$$
$$= 3.14 \text{ in}$$

$$m = \frac{N - 0.95d}{2}$$

$$= \frac{19 \text{ in} - (0.95)(12.89 \text{ in})}{2}$$

$$= 3.38 \text{ in}$$

$$n = \frac{B - 0.8b_f}{2}$$

$$= \frac{19 \text{ in} - (0.8)(12.22 \text{ in})}{2}$$

$$= 4.61 \text{ in} \quad [\text{governs}]$$

$$t_{\text{req}} = n\sqrt{\frac{2P_u}{0.9F_y BN}}$$

$$= 4.61 \text{ in} \sqrt{\frac{(2)(850 \text{ kips})}{(0.9)\left(36 \ \dfrac{\text{kips}}{\text{in}^2}\right)(19 \text{ in})(19 \text{ in})}}$$

$$= 1.76 \text{ in}$$

5. PLASTIC DESIGN

Nomenclature

A	area of section	in^2
D	degree of indeterminacy of a structure	–
E	modulus of elasticity	kips/in^2
F_y	yield stress	kips/in^2
h	clear distance between flanges less the corner radius at each flange	in
H	horizontal force	kips
L	unbraced length	ft
L_{pd}	limiting laterally unbraced length for plastic analysis	ft or in
m_i	number of independent collapse mechanisms in a structure	–
M_s	bending moment produced by factored loads acting on the cut-back structure	ft-kips
M_1	smaller moment at end of unbraced length of beam	ft-kips
M_2	larger moment at end of unbraced length of beam	ft-kips
M_1/M_2	positive when moments cause reverse curvature and negative for single curvature	–
p	number of possible hinge locations in a structure	–
P	axial force	kips
P_u	required axial strength in compression	kips
P_y	member yield strength	kips
r	radius of gyration	in
t_w	web thickness	in
V	shear force	kips
w_u	factored uniformly distributed load	kips/ft

Symbols

ϕ_b	resistance factor for flexure	
ϕ_c	resistance factor for compression	

Statical Design Method

The factored loads are applied to the statically determinate cut-back structure, as shown in Fig. 4.12(a). The free moment diagram is drawn, as shown in Fig. 4.12(b). As shown in Fig. 4.12(c), the fixing moment line is superimposed to produce the partial collapse mechanism of Fig. 4.12(d). When the beam is of non-uniform section, a complete collapse mechanism is possible, as shown in Figs. 4.12(e) and 4.12(f).

Example 4.26

The factored loading, including the beam self-weight, acting on a three-span continuous beam is shown in the following illustration. Assuming that adequate lateral support is provided to the beam, determine the lightest adequate W12 shape using grade 50 steel.

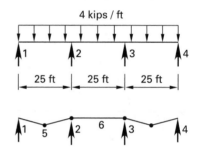

Solution

The free moment in each span is

$$M_s = \frac{w_u l^2}{8} = \frac{\left(4 \ \dfrac{\text{kips}}{\text{ft}}\right)(25 \text{ ft})^2}{8}$$

$$= 312 \text{ ft-kips}$$

Partial collapse occurs as shown in Fig. 4.12, with hinges forming in the end spans, and the required plastic moment of resistance is

$$M_p = 0.686 M_s$$

$$= 214 \text{ ft-kips}$$

The required plastic section modulus is

$$Z = \frac{M_p}{\phi F_y}$$

$$= \frac{(214 \text{ ft-kips})\left(12 \ \dfrac{\text{in}}{\text{ft}}\right)}{(0.9)\left(50 \ \dfrac{\text{kips}}{\text{in}^2}\right)}$$

$$= 57 \text{ in}^3$$

From LRFD Table 3-6, a W12 × 40 has a plastic section modulus of

$$Z = 57 \text{ in}^3 \quad [\text{satisfactory}]$$

Figure 4.12 *Statical Design Method*

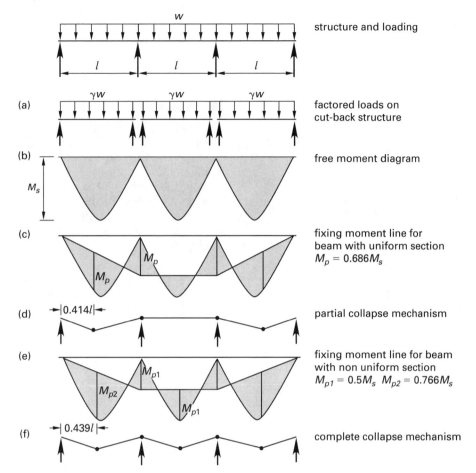

structure and loading

(a) factored loads on
cut-back structure

(b) free moment diagram

(c) fixing moment line for
beam with uniform section
$M_p = 0.686 M_s$

(d) partial collapse mechanism

(e) fixing moment line for beam
with non uniform section
$M_{p1} = 0.5 M_s$ $M_{p2} = 0.766 M_s$

(f) complete collapse mechanism

Beam Design Requirements

In accordance with LRFD App. 1 Sec. 1.2, plastic design is permitted only for steel with a yield stress not exceeding 65 kips/in², and LRFD App. 1 Sec. 1.4 requires that all members subjected to plastic hinge rotation be compact. The maximum unbraced length of a member adjacent to a plastic hinge is restricted by LRFD Eq. (A-1-7) to a value of

$$L_{pd} = \frac{\left(0.12 + (0.076)\left(\dfrac{M_1}{M_2}\right)\right) E r_y}{F_y}$$

Example 4.27

The three-span continuous beam of Ex. 4.26 is laterally braced at the midpoint of the central span and, in the end spans, at supports and at the locations of plastic hinges. Determine whether this bracing is adequate.

Solution

From LRFD Table 1-1, for a W12 × 40 the radius of gyration about the minor axis is

$$r_y = 1.94 \text{ in}$$

The relevant unbraced lengths in the end spans are

$$L_{15} = 0.414 L_{12} = (0.414)(25 \text{ ft})$$
$$= 10.35 \text{ ft}$$
$$L_{25} = 0.586 L_{12} = (0.586)(25 \text{ ft})$$
$$= 14.65 \text{ ft}$$

For section 15,

$$\frac{M_1}{M_2} = \frac{0}{M_p} = 0$$

The required unbraced length is given by LRFD Eq. (A-1-7) as

$$L_{pd} = \frac{\left(0.12 + (0.076)\left(\dfrac{M_1}{M_2}\right)\right) E r_y}{F_y}$$
$$= \frac{\left(0.12 \dfrac{\text{kips}}{\text{in}^2}\right)(1.94 \text{ in})\left(29{,}000 \dfrac{\text{kips}}{\text{in}^2}\right)}{50 \dfrac{\text{kips}}{\text{in}^2}}$$
$$= 135 \text{ in}$$
$$= 11.3 \text{ ft}$$
$$> L_{15} \quad \text{[satisfactory]}$$

For section 25,

$$\frac{M_1}{M_2} = +\frac{M_p}{M_p} \qquad \left[\begin{array}{c}\text{Moments cause}\\ \text{reverse curvature.}\end{array}\right]$$

$$= 1$$

The required unbraced length is given by LRFD Eq. (A-1-7) as

$$L_{pd} = \frac{\left(0.12 + (0.076)\left(\frac{M_1}{M_2}\right)\right)Er_y}{F_y}$$

$$= \frac{\begin{array}{c}\left(0.12\,\dfrac{\text{kips}}{\text{in}^2} + 0.076\,\dfrac{\text{kips}}{\text{in}^2}\right)(1.94\text{ in})\\[2mm] \times \left(29{,}000\,\dfrac{\text{kips}}{\text{in}^2}\right)\end{array}}{50\,\dfrac{\text{kips}}{\text{in}^2}}$$

$$= 220\text{ in}$$

$$= 18.4\text{ ft}$$

$$> L_{25} \quad \text{[satisfactory]}$$

The maximum moment in the central span is

$$M_6 = M_s - M_p$$

$$= 312\text{ ft-kips} - 214\text{ ft-kips}$$

$$= 98\text{ ft-kips}$$

The unbraced length in the central span is

$$L_{26} = 12.5\text{ ft}$$

For section 26,

$$\frac{M_1}{M_2} = +\frac{M_6}{M_p} \qquad \left[\begin{array}{c}\text{Moments cause}\\ \text{reverse curvature.}\end{array}\right]$$

$$= \frac{98\text{ ft-kips}}{214\text{ ft-kips}}$$

$$= 0.46$$

The required unbraced length is given by

$$L_{pd} = \frac{\left(0.12 + (0.076)\left(\frac{M_1}{M_2}\right)\right)Er_y}{F_y}$$

$$= \frac{\begin{array}{c}\left(0.12\,\dfrac{\text{kips}}{\text{in}^2} + \left(0.076\,\dfrac{\text{kips}}{\text{in}^2}\right)(0.46)\right)(1.94\text{ in})\\[2mm] \times \left(29{,}000\,\dfrac{\text{kips}}{\text{in}^2}\right)\end{array}}{50\,\dfrac{\text{kips}}{\text{in}^2}}$$

$$= 174\text{ in}$$

$$= 14.5\text{ ft}$$

$$> L_{26} \quad \text{[satisfactory]}$$

From LRFD Table 3-6, for a W12 × 40,

$$L_r = 21.1\text{ ft}$$

$$> L_{26}$$

$$\phi M_r = 135\text{ ft-kips}$$

$$> M_6\text{ ft-kips} \quad \text{[satisfactory]}$$

Mechanism Design Method

The mechanism design method uses the principle of virtual displacements to determine the required plastic moments in rigid frames. Plastic hinges may form at the point of application of a concentrated load, at the ends of members, and at the location of zero shear in a prismatic beam. In the case of two members meeting at a joint, a plastic hinge forms in the weaker member. In the case of three or more members meeting at a joint, plastic hinges may form at the ends of each of the members.

For the three-degree indeterminate frame fabricated from members of a uniform section, shown in Fig. 4.13, there are five possible locations of plastic hinges; these are shown in Fig. 4.13(a). The number of possible independent mechanisms is then

$$m_i = p - D$$

$$= 5 - 3$$

$$= 2$$

These independent mechanisms are the beam mechanism shown in Fig. 4.13(b) and the sway mechanism shown in Fig. 4.13(c). In addition, these may be combined to form the combined mechanism shown in Fig. 4.13(d).

Figure 4.13 *Mechanism Design Method*

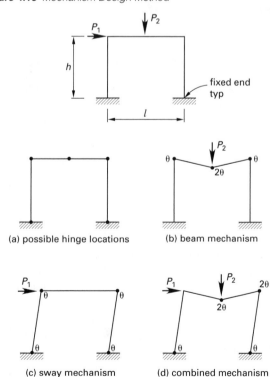

(a) possible hinge locations   (b) beam mechanism

(c) sway mechanism   (d) combined mechanism

Applying a virtual displacement to each of these mechanisms in turn and equating internal and external work yields three equations, from each of which a value of M_p may be obtained. The largest value of M_p governs. For the beam mechanism,

$$4M_p\theta = P_2\left(\frac{l}{2}\right)\theta$$

$$M_p = \frac{P_2 l}{8}$$

For the sway mechanism,

$$4M_p\theta = P_1 h\theta$$

$$M_p = \frac{P_1 h}{4}$$

For the combined mechanism,

$$6M_p\theta = P_2\left(\frac{l}{2}\right)\theta + P_1 h\theta$$

$$M_p = \frac{P_1 h + \dfrac{P_2 l}{2}}{6}$$

For the situation where $P_2 = 2P_1$ and $h = l$, the combined mechanism controls and

$$M_p = \frac{P_1 l}{3}$$

Example 4.28

The rigid frame shown in the following illustration is fabricated from members of a uniform section in grade 50 steel. For the factored loading indicated, ignoring the member self-weight and assuming adequate lateral support, determine the lightest adequate W shape.

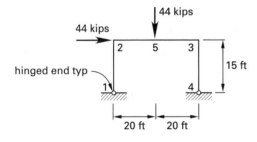

Solution

The three possible collapse mechanisms are shown in the following illustration.

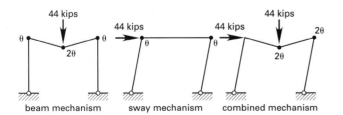

The beam mechanism gives

$$4M_p = (44\text{ kips})(20\text{ ft})$$

$$M_p = 220\text{ ft-kips}$$

The sway mechanism gives

$$2M_p = (44\text{ kips})(15\text{ ft})$$

$$M_p = 330\text{ ft-kips}$$

The combined mechanism gives

$$4M_p = 880\text{ ft-kips} + 660\text{ ft-kips}$$

$$M_p = 385\text{ ft-kips} \quad \text{[governs]}$$

From LRFD Table 3-6, a W21 × 50 has

$$\phi_b M_p = 413\text{ ft-kips}$$

$$> 385\text{ ft-kips} \quad \text{[satisfactory]}$$

[A W21 × 48 is non-compact and may not be used.]

Static Equilibrium Check

Mechanism methods lead to upper bounds on the collapse load. To confirm that the correct mechanism has been selected, it is necessary to check that the assumed plastic moment is not anywhere exceeded by constructing a moment diagram obtained by static equilibrium methods.

Example 4.29

Draw the bending moment diagram for the assumed collapse mechanism of Ex. 4.28.

Solution

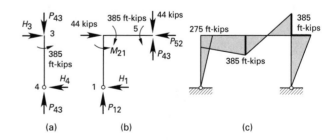

For member 34, as shown at (a) in the illustration,

$$H_4 = \frac{385\text{ ft-kips}}{15\text{ ft}}$$

$$= 25.67\text{ kips}$$

$$= H_3$$

$$= P_{52}$$

For member 125, as shown at (b) in the figure,

$$H_1 = 44 \text{ kips} - P_{52}$$
$$= 44 \text{ kips} - 25.67 \text{ kips}$$
$$= 18.33 \text{ kips}$$
$$P_{12} = M_p - (H_1)(h)$$
$$= \frac{385 \text{ ft-kips} - (18.33 \text{ kips})(15 \text{ ft})}{20 \text{ ft}}$$
$$= 5.50 \text{ kips}$$
$$M_{21} = (H_1)(h)$$
$$= (18.33 \text{ kips})(15 \text{ ft})$$
$$= 275 \text{ ft-kips}$$
$$P_{43} = 44 \text{ kips} - P_{12}$$
$$= 44 \text{ kips} - 5.50 \text{ kips}$$
$$= 38.50 \text{ kips}$$

The bending moment diagram is shown at (c) in the illustration. Because $M_p = 385$ ft-kips is not exceeded at any point in the frame, the combined mechanism is the correct failure mode.

Column Design Requirements

Flanges and webs of members subjected to combined flexure and compression shall be compact with width-thickness ratios not exceeding the values defined in LRFD Table B4.1. In addition, the webs of W-sections shall also comply with LRFD App. 1 Eqs. (A-1-1) and (A-1-2), which are as follows.

For $\dfrac{P_u}{\phi_b P_y} \leq 1.25$,

$$\frac{h}{t_w} \leq 3.76 \left(\frac{E}{F_y}\right)^{0.5} \left(1 - \frac{2.75 P_u}{\phi_b P_y}\right)$$

For $\dfrac{P_u}{\phi_b P_y} > 1.25$,

$$\frac{h}{t_w} \leq 1.12 \left(\frac{E}{F_y}\right)^{0.5} \left(2.33 - \frac{P_u}{\phi_b P_y}\right)$$
$$\geq 1.49 \left(\frac{E}{F_y}\right)^{0.5}$$

The member yield strength is

$$P_y = AF_y$$

The resistance factor for flexure is

$$\phi_b = 0.90$$

The maximum permitted slenderness ratio of a column is specified in LRFD App. 1 Sec. 1.6 as

$$\frac{L}{r} = 4.71 \left(\frac{E}{F_y}\right)^{0.5}$$
$$= 113 \quad [\text{for } F_y = 50 \text{ ksi}]$$

In accordance with LRFD App. 1 Sec. 1.5, the axial load in a column shall not exceed $0.85\phi_c A_g F_y$ for a braced frame or $0.75\phi_c A_g F_y$ for a sway frame. As for beams, the maximum unbraced length is controlled by LRFD Eq. (A-1-7), and for combined axial force and flexure, the interaction expressions of LRFD Eqs. (H1–1a) and (H1–1b) govern. In addition, LRFD Commentary Sec. 1.5 states that second-order effects may be neglected for low rise frames with small axial loads.

Example 4.30

Determine whether column 34 of the rigid frame in Ex. 4.28 is satisfactory. The column consists of a grade 50 W21 × 50 section and is laterally braced about its weak axis at 3.75 ft centers and at joint 3.

Solution

For the pinned connection at joint 4, LRFD Commentary Sec. C2 specifies a stiffness ratio of $G_4 = 10$.

At joint 3,

$$G_3 = \frac{\sum \dfrac{I_c}{L_c}}{\sum \dfrac{I_g}{L_g}} = \frac{\dfrac{I}{15}}{\dfrac{I}{40}}$$
$$= 2.7$$

From the alignment chart for sway frames, the effective length factor is

$$K_{34} = 2.2$$

From LRFD Table 1-1, a W21 × 50 has

$$A_g = 14.7 \text{ in}^2$$
$$r_y = 1.30 \text{ in}$$
$$r_x = 8.18 \text{ in}$$

The slenderness ratio about the x-axis is

$$\frac{K_{34}L_x}{r_x} = \frac{(2.2)(15 \text{ ft}) \left(12 \dfrac{\text{in}}{\text{ft}}\right)}{8.18 \text{ in}}$$
$$= 48.4 \quad [\text{governs for axial load}]$$
$$< 113 \quad [\text{satisfactory}]$$

The slenderness ratio about the y-axis is

$$\frac{K_{34}L_y}{r_y} = \frac{(1.0)(3.75 \text{ ft}) \left(12 \dfrac{\text{in}}{\text{ft}}\right)}{1.30 \text{ in}}$$
$$= 34.6$$

In accordance with LRFD App. 1 Sec. 1.5.2, the maximum axial load in the column of a sway frame is restricted to

$$P_{\max} = 0.75\phi_c A_g F_y$$
$$= (0.75)(0.90)\left(14.7 \text{ in}^2\right)\left(50 \frac{\text{kips}}{\text{in}^2}\right)$$
$$= 496 \text{ kips}$$
$$> P_{43} \quad [\text{satisfactory}]$$

From LRFD Table 3-6, for a W21 × 50,

$$L_p = 4.59 \text{ ft}$$
$$> 3.75 \text{ ft} \quad [\text{full plastic bending capacity available}]$$
$$\phi_b M_{nx} = \phi_b M_p$$
$$= 413 \text{ ft-kips}$$

From LRFD Table 4-22, for a $K_{34}L_x/r_x$ value of 48.4 the design stress for axial load is

$$\phi_c F_{cr} = 37.9 \text{ kips/in}^2$$

The design axial strength is

$$\phi_c P_n = \phi_c F_{cr} A_g$$
$$= \left(37.9 \frac{\text{kips}}{\text{in}^2}\right)\left(14.7 \text{ in}^2\right)$$
$$= 557 \text{ kips}$$
$$\frac{P_{34}}{\phi_c P_n} = \frac{38.50 \text{ kips}}{557 \text{ kips}}$$
$$= 0.07$$
$$< 0.20 \quad [\text{LRFD Eq. (H1–1b) governs.}]$$

Since secondary effects may be neglected, LRFD Eq. (H1-1b) reduces to

$$\frac{P_{34}}{2\phi_c P_n} + \frac{M_p}{\phi_b M_{nx}} \leq 1.0$$
$$= \frac{38.50 \text{ kips}}{(2)(557 \text{ kips})} + \frac{385 \text{ ft-kips}}{413 \text{ ft-kips}}$$
$$= 0.97$$
$$< 1.0 \quad [\text{satisfactory}]$$

6. DESIGN OF TENSION MEMBERS

Nomenclature

A_e	effective net area	in^2
A_g	gross area	in^2
A_n	net area	in^2
A_R	area of section required for fatigue loading	in^2
C_f	fatigue constant	–
d_b	nominal bolt diameter	in
d_h	specified hole diameter	in
$f_{\max}$	maximum tensile stress in member at service load	kips/in^2
$f_{\min}$	minimum stress in member at service load (compression negative)	kips/in^2
f_{SR}	actual stress range	kips/in^2
F_{SR}	allowable stress range for fatigue loading	kips/in^2
F_{TH}	threshold stress range	kips/in^2
F_u	specified minimum tensile strength	kips/in^2
g	transverse center-to-center spacing between fasteners (gage)	in
l	length of weld	in
N	number of stress range fluctuations	–
P_n	nominal axial strength	kips
P_u	required axial strength	kips
s	bolt spacing in direction of load	in
t	plate thickness	in
$T_{\max}$	maximum tensile force in member at service load	kips
$T_{\min}$	minimum force in member (compression negative) at service load	kips
$T_{\max} - T_{\min}$	force range	kips
U	shear lag factor used in calculating effective net area	–
w	plate width, distance between welds	in
$\overline{x}$	connection eccentricity	in

Plates in Tension

For yielding of the gross section, LRFD Sec. D2 gives

$$P_u = \phi_t P_n$$
$$= 0.9 F_y A_g$$

As shown in Fig. 4.14, the gross area is given by

$$A_g = wt$$

For tensile rupture at the connection, LRFD Sec. D2 gives

$$P_u = \phi_t P_n$$
$$= 0.75 F_u A_e$$

Figure 4.14 *Effective Net Area of Bolted Connection*

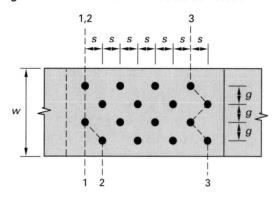

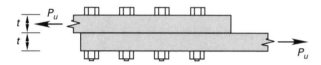

Figure 4.15 *Welded Connections for Plates*

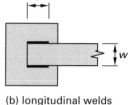

(a) transverse weld   (b) longitudinal welds

The effective net area of a bolted connection is shown in Fig. 4.14 and is defined in LRFD Sec. D3.2 as

$$A_e = t\,(w - 2d_h) \quad \text{[for Sec. 1–1]}$$

$$A_e = t\left(w - 3d_h + \frac{s^2}{4g}\right) \quad \text{[for Sec. 2–2]}$$

$$A_e = t\left(w - 4d_h + \frac{3s^2}{4g}\right) \quad \text{[for Sec. 3–3]}$$

$$\leq 0.85A_g \quad \text{[in accordance with LRFD Sec. J4.1]}$$

The specified hole diameter is defined in LRFD Sec. D3.2 and LRFD Table J3.3 as

$$d_h = d_b + \frac{1}{8}\ \text{in}$$

The effective net area of a welded connection is shown in Fig. 4.15 and is defined in LRFD Sec. D3.1. For the transverse welded connection shown in Fig. 4.15(a),

$$A_e = A_g$$

For the longitudinal welded connection shown in Fig. 4.15(b),

$$A_e = UA_g$$

The value of the shear lag factor is defined in LRFD Table D3.1 as

$$U = 1.00 \quad [l \geq 2w]$$
$$U = 0.87 \quad [2w > l \geq 1.5w]$$
$$U = 0.75 \quad [1.5w > l \geq w]$$

Example 4.31

As shown in the following illustration, two plates each $1/2$ in thick $\times$ 9 in wide are connected by three rows of bolts. The distance between rows is 3 in, the distance between bolts in a row is 3 in, and the center row of bolts is staggered. Determine the design axial strength of the plates in direct tension. The relevant properties of the plates are $F_y = 36$ kips/in², $F_u = 58$ kips/in², and specified hole diameter $d_h = 1.0$ in.

Solution

The gross area of each plate is given by

$$A_g = wt$$
$$= (9\ \text{in})(0.5\ \text{in})$$
$$= 4.5\ \text{in}^2$$

The design axial strength for yielding is

$$\phi_t P_n = 0.9F_y A_g$$
$$= (0.9)\left(36\ \frac{\text{kips}}{\text{in}^2}\right)\left(4.5\ \text{in}^2\right)$$
$$= 146\ \text{kips}$$

Illustration for Ex. 4.31

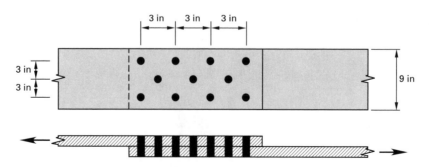

For a straight perpendicular fracture, the effective net area of the plate is given by

$$A_e = t(w - 2d_h)$$
$$= (0.5)\big((9 \text{ in}) - (2)(1.0 \text{ in})\big)$$
$$= 3.5 \text{ in}^2$$
$$< 0.85A_g \quad \text{[satisfactory]}$$

The design axial strength for tensile rupture is

$$\phi_t P_n = 0.75 F_u A_e$$
$$= (0.75)\left(58 \; \frac{\text{kips}}{\text{in}^2}\right)(3.5 \text{ in}^2)$$
$$= 152 \text{ kips}$$

For a staggered fracture, the effective net area of the plate is given by

$$A_e = t\left(w - 3d_h + \frac{2s^2}{4g}\right)$$
$$= (0.5 \text{ in})\left(9 \text{ in} - (3)(1.0 \text{ in}) + \frac{(2)(1.5 \text{ in})^2}{(4)(3 \text{ in})}\right)$$
$$= 3.19 \text{ in}^2$$

The corresponding design axial strength is

$$\phi_t P_n = 0.75 F_u A_e$$
$$= (0.75)\left(58 \; \frac{\text{kips}}{\text{in}^2}\right)(3.19 \text{ in}^2)$$
$$= 139 \text{ kips} \quad \text{[governs]}$$

Rolled Sections in Tension

To account for the effects of eccentricity and shear lag in rolled structural shapes connected by bolts through only part of their cross-sectional elements, the effective net area is given by LRFD Sec. D3 as

$$A_e = U A_n \qquad \textit{[LRFD D3-1]}$$

The value of the shear lag factor is defined in LRFD Table D3.1 case 2 as

$$U = 1 - \frac{\bar{x}}{l}$$

The length of the connection l is defined in LRFD Commentary Sec. D3.3 as the distance, parallel to the line of force, between the first and last fasteners in a line. The connection eccentricity $\bar{x}$ is defined as the distance from the connection plane to the centroid of the member resisting the connection force. In lieu of applying this expression for U, LRFD Table D3.1 permits the adoption of the following values for the shear lag factor.

$U = $ 0.90 for W, M, and S shapes and for T-sections with $b_f \geq 2d/3$, connected by the flange, with not fewer than three bolts in line in the direction of stress

$U = $ 0.85 for W, M, and S shapes and for T-sections with $b_f < 2d/3$ connected by the flange, with not fewer than three bolts in line in the direction of stress

$U = $ 0.70 for W, M, and S shapes and for T-sections connected by the web, with not less than four bolts in line in the direction of stress

$U = $ 0.80 for single angles with not less than four bolts in line in the direction of stress

$U = $ 0.60 for single angles with two or three bolts in line in the direction of stress

For a welded connection, when the axial force is transmitted only by transverse welds, as shown in Fig 4.16(a), the effective net area is given by LRFD Table D3.1 as

$$A_e = \text{area of directly connected elements}$$

Figure 4.16 *Welded Connections for Rolled Sections*

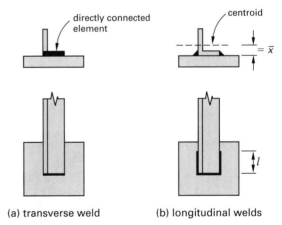

(a) transverse weld   (b) longitudinal welds

For a welded connection, when the axial force is transmitted only by longitudinal welds or in combination with transverse welds, as shown in Fig 4.16(b), the effective net area is given by LRFD Sec. D3 as

$$A_e = U A_g \qquad \textit{[LRFD D3-1]}$$

The value of the shear lag factor is defined in LRFD Table D3.1 case 2 as

$$U = 1 - \frac{\bar{x}}{l}$$

The length of the connection l is shown in Fig 4.16(b) and is defined in LRFD Commentary Sec. D3.3 as the length of the weld, parallel to the line of force.

Example 4.32

Assuming that the welds are adequate, determine the design axial strength of the grade A36 W12×65 member connected as shown in the following illustration.

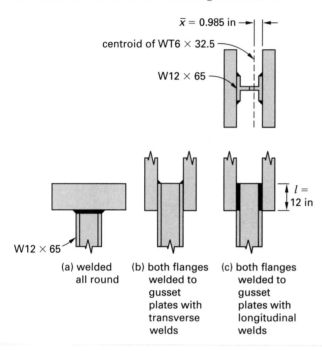

(a) welded all round

(b) both flanges welded to gusset plates with transverse welds

(c) both flanges welded to gusset plates with longitudinal welds

Solution

The relevant properties of the W12 × 65 are obtained from LRFD Table 1-1 and are

$$A_g = 19.1 \text{ in}^2$$
$$b_f = 12 \text{ in}$$
$$t_f = 0.605 \text{ in}$$

(a) The W12×65 is welded all around to the supporting member.

$$A_e = \text{area of directly connected elements}$$
$$= A_g$$

The design axial strength for rupture is

$$\phi_t P_n = 0.75 F_u A_e$$
$$= (0.75) \left(58 \; \frac{\text{kips}}{\text{in}^2} \right) (19.1 \text{ in}^2)$$
$$= 830 \text{ kips}$$

The design axial strength for yielding is

$$\phi_t P_n = 0.9 F_y A_g$$
$$= (0.9) \left(36 \; \frac{\text{kips}}{\text{in}^2} \right) (19.1 \text{ in}^2)$$
$$= 619 \text{ kips} \quad [\text{governs}]$$

(b) Both flanges are welded by transverse welds to gusset plates.

$$A_e = \text{area of directly connected elements}$$
$$= 2b_f t_f$$
$$= (2)(12 \text{ in})(0.605 \text{ in})$$
$$= 14.5 \text{ in}^2$$

The design axial strength for fracture is

$$\phi_t P_n = 0.75 F_u A_e$$
$$= (0.75) \left(58 \; \frac{\text{kips}}{\text{in}^2} \right) (14.5 \text{ in}^2)$$
$$= 631 \text{ kips}$$

The design axial strength for yielding is

$$\phi_t P_n = 619 \text{ kips} \quad [\text{governs}]$$

(c) Both flanges are welded by longitudinal welds to gusset plates.

In accordance with LRFD Commentary Sec. D3.3, the W section is treated as two WT sections as shown in the figure. The centroidal height of a WT6 × 32.5 cut from a W12 × 65 is obtained from LRFD Table 1-8 and is

$$\bar{x} = 0.985 \text{ in}$$

The value of the shear lag factor is defined in LRFD Table D3.1 case 2 as

$$U = 1 - \frac{\bar{x}}{l} = 1 - \frac{0.985 \text{ in}}{12 \text{ in}}$$
$$= 0.92$$

The effective net area is

$$A_e = U A_g = (0.92) \left(19.1 \text{ in}^2 \right)$$
$$= 17.6 \text{ in}^2$$

The design axial strength for fracture is

$$\phi_t P_n = 0.75 F_u A_e$$
$$= (0.75) \left(58 \; \frac{\text{kips}}{\text{in}^2} \right) (17.6 \text{ in}^2)$$
$$= 766 \text{ kips}$$

The design axial strength for yielding is

$$\phi_t P_n = 619 \text{ kips} \quad [\text{governs}]$$

Design for Fatigue

Fatigue failure is caused by fluctuations of tensile stress that cause crack propagation in the parent metal. Fatigue must be considered for tensile stresses, stress reversals, and shear when the number of loading cycles exceeds 20,000 and is based on the stress level at service loads.

The design procedure is given in LRFD App. 3 and consists of establishing the applicable loading condition, from LRFD Table A-3.1.

The applicable values of the fatigue constant, the threshold stress range, and the stress category are obtained from the table.

The stress range is defined as the magnitude of the change in stress due to the application or removal of the unfactored live load. Fatigue must be considered if the stress range in the member exceeds the threshold stress range. The actual stress range, at service level values, is given by

$$f_{SR} = f_{\max} - f_{\min}$$

Ten stress categories are defined in LRFD Table A-3.1. For stress categories A, B, B′, C, D, E, and E′, the design stress range in the member is given by LRFD Eq. (A-3-1) as

$$F_{SR} = \left(\frac{C_f}{N}\right)^{0.333}$$

For stress category F the design stress range in the member is given by LRFD Eq. (A-3-2).

$$F_{SR} = \left(\frac{C_f}{N}\right)^{0.167}$$

Example 4.33

A tie member in a steel truss consists of a pair of grade A36 5 in × 5 in by $^3/_8$ in angles fillet welded to a gusset plate. The force in the member, due to dead load only, is 90 kips tension. The additional force in the member, due to live load only, varies from a compression of 7 kips to a tension of 50 kips. During the design life of the structure, the live load may be applied 600,000 times. Determine whether fatigue effects must be considered.

Solution

From LRFD Table A-3.1, the loading condition of Sec. 4.1 is applicable and the relevant factors are

$$E = \text{stress category}$$

$$F_{SR} = \text{design stress range}$$
$$= \left(\frac{C_f}{N}\right)^{0.333} = \left(\frac{11 \times 10^8}{6 \times 10^5}\right)^{0.333}$$
$$= 12.21 \text{ kips/in}^2$$

The actual force range is

$$T_{\max} - T_{\min} = 50 \text{ kips} - (-7 \text{ kips})$$
$$= 57 \text{ kips}$$

The area of the tie is

$$A_s = 7.22 \text{ in}^2$$

The actual stress range is

$$f_{SR} = \frac{T_{\max} - T_{\min}}{A_s}$$
$$= \frac{57 \text{ kips}}{7.22 \text{ in}^2}$$
$$= 7.9 \text{ kips/in}^2$$
$$< F_{SR}$$

Fatigue effects need not be considered.

7. DESIGN OF BOLTED CONNECTIONS

Nomenclature

A_b	nominal unthreaded body area of bolt	in^2
C	coefficient for eccentrically loaded bolt and weld groups	–
d	nominal bolt diameter	in
d_m	moment arm between resultant tensile and compressive forces due to an eccentric force	in
D_u	a multiplier that reflects the ratio of the mean installed bolt tension to the specified minimum bolt pretension	–
f_v	computed shear stress	kips/in^2
F_{nt}	nominal tensile stress of bolt	kips/in^2
F'_{nt}	nominal tensile stress of a bolt subjected to combined shear and tension	kips/in^2
F_{nv}	nominal shear stress of bolt	kips/in^2
F_u	specified minimum tensile strength	kips/in^2
h_{sc}	modification factor for type of hole	–
k_s	slip-critical combined tension and shear coefficient	–
L_c	clear distance, in the direction of force, between the edge of the hole and the edge of the adjacent hole or edge of the material	in
n	number of bolts in a connection	–
n'	number of bolts above the neutral axis (in tension)	–
N	number of slip planes	–
N_b	number of bolts carrying strength level tension T_u	–
P_u	factored load on connection	kips
R_n	nominal strength	kips
s	center-to-center pitch of two consecutive bolts	in
t	thickness of connected part	in

Steel

T_b	minimum pre-tension force	kips/in^2
T_u	factored tensile force	kips/in^2

Symbols

μ	mean slip coefficient for the applicable surface
ϕ	resistance factor

Bearing-Type Bolts in Shear

The design strength in shear is given by LRFD Sec. J3.6 as

$$\phi R_n = \phi F_{nv} A_b$$
$$= 0.75 F_{nv} A_b$$

Values of the nominal shear stress F_{nv} are given in LRFD Table J3.2 for all types of bolts. Values of ϕR_n are given in LRFD Table 7-1.

Bearing-Type Bolts in Tension

The design strength in tension is given by LRFD Sec. J3.6 as

$$\phi R_n = \phi F_{nt} A_b$$
$$= 0.75 F_{nt} A_b$$

Values of the nominal tensile stress F_{nt} are given in LRFD Table J3.2 for all types of bolts. Values of ϕR_n are given in LRFD Table 7-2.

When a bearing-type bolt is subjected to combined shear and tension, the design strength in shear is unaffected, and the design strength in tension is reduced in accordance with LRFD Sec. J3.7.

Example 4.34

The connection analyzed in Ex. 4.31 consists of 11 grade A307 $^7/_8$ in diameter bolts. Determine the design shear strength of the bolts in the connection.

Solution

From LRFD Table 7-1, the design strength of the 11 bolts in shear is

$$\phi R_n = \phi F_{nv} A_b n$$
$$= \left(10.8 \ \frac{\text{kips}}{\text{bolt}} \right) (11 \text{ bolts})$$
$$= 119 \text{ kips}$$

Slip-Critical Bolts in Shear

The minimum pre-tension force T_b in a bolt is specified in LRFD Table J3.1.

Slip-critical bolts in standard holes or slots transverse to the direction of load are designed for slip at the serviceability limit state. Slip-critical bolts in oversized holes or slots parallel to the direction of load are designed for slip at the strength level limit state. For both cases, the required strength is determined using factored load combinations and the design strength is determined using the resistance factor appropriate to each case.

The design slip resistance, for both cases, is given by LRFD Eq. (J3-4) as

$$\phi R_n = \phi \mu D_u h_{sc} T_b N$$

The condition of the faying surface determines the frictional resistance developed at the connection. Values of the mean slip coefficient for two types of surface condition are given in LRFD Sec. J3.8. Class A surfaces consist of unpainted clean mill scale surfaces or blast-cleaned surfaces with class A coatings. The mean slip coefficient for this condition is

$$\mu = 0.35$$

Class B surfaces consist of unpainted blast-cleaned surfaces or blast-cleaned surfaces with class B coatings. The mean slip coefficient for this condition is

$$\mu = 0.50$$

The ratio of the mean installed bolt tension to the specified minimum bolt pretension is defined as

$$D_u = 1.13$$

The modification factor, h_{sc}, is 1.00 for standard sized holes, 0.85 for oversized or short-slotted holes, and 0.70 for long-slotted holes.

The resistance factor, ϕ, is 1.0 for slip as the serviceability limit state and 0.85 for the strength level limit state.

Values for the design slip-critical shear resistance ϕR_n for a class A faying surface and for standard, short-slotted, and long-slotted holes are given in LRFD Tables 7-3 and 7-4.

Slip-Critical Bolts in Tension

The design tensile strength of slip-critical bolts is independent of the pretension in the bolt.

The design tensile strength at factored loads is given by LRFD Sec. J3.6 as

$$\phi R_n = \phi F_{nt} A_b$$
$$= 0.75 F_{nt} A_b$$

Values of the nominal tensile stress F_{nt} are given in LRFD Table J3.2 for all types of bolts. Values of ϕR_n are given in LRFD Table 7-2.

When a slip-critical bolt is subjected to combined shear and tension, the design strength in tension is unaffected. However, in accordance with LRFD Sec. J3.9, the design resistance to shear is reduced by the factor $k_s = 1 - T_u/D_u T_b N_b$.

Example 4.35

The connection analyzed in Ex. 4.31 consists of 11 grade A490 $^7/_8$ in diameter slip-critical bolts. Determine the strength level design resistance to shear of the bolts in the connection. The bolts are in standard holes with a class A faying surface.

Solution

The connection may be designed for slip as the serviceability limit state. From LRFD Table 7-4, the strength level design strength of the 11 bolts in shear is

$$\phi R_n = (19.4 \text{ kips})(11 \text{ bolts})$$
$$= 213 \text{ kips}$$

Bolts in Bearing

When checking bearing of bolts in connected parts, both bearing-type bolts and slip-critical bolts are designed on the basis of ultimate load conditions. When deformation is a design consideration, the nominal bearing strength is specified by LRFD Sec. J3.10(a) as

$$R_n = 1.2 L_c t F_u \leq 2.4 dt F_u \quad \text{[LRFD J3-6a]}$$

The design bearing strength is

$$\phi R_n = 0.75 R_n$$

When deformation is not a design consideration,

$$R_n = 1.5 L_c t F_u \leq 3.0 dt F_u \quad \text{[LRFD J3-6b]}$$

Values of ϕR_n are tabulated in LRFD Tables 7-5 and 7-6.

Example 4.36

The connection analyzed in Ex. 4.31 consists of 11 grade A307 $^7/_8$ in diameter bolts. Determine the design bearing strength of the bolts in the connection if the edge distance is $L_c = 2.5$ in and $s = 3$ in.

Solution

From LRFD Table 7-6, the minimum edge distance for full bearing strength is

$$L_c = 2.25 \text{ in}$$
$$< 2.5 \text{ in provided}$$

The edge distance does not govern.

From LRFD Table 7-5, the design strength of the 11 bolts in bearing is

$$\phi R_n = \left(91.4 \, \frac{\frac{\text{kips}}{\text{in}}}{\text{bolt}} \right) (0.5 \text{ in})(11 \text{ bolts})$$
$$= 503 \text{ kips}$$

Bolt Group Eccentrically Loaded in Plane of Faying Surface

Eccentrically loaded bolt groups of the type shown in Fig. 4.17 may be conservatively designed by means of the elastic unit area method. The moment of inertia of the bolt group about the x-axis is

$$I_x = \sum y^2$$

Figure 4.17 *Eccentrically Loaded Bolt Group*

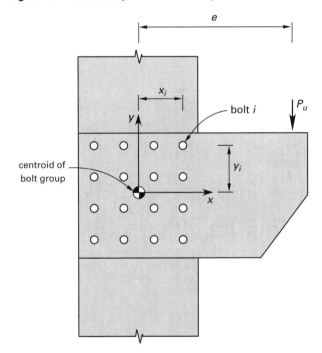

The moment of inertia of the bolt group about the y-axis is

$$I_y = \sum x^2$$

The polar moment of inertia of the bolt group about the centroid is

$$I_o = I_x + I_y$$

The vertical force on bolt i due to the applied load P_u is

$$V_P = \frac{P_u}{n}$$

The vertical force on bolt i due to the eccentricity e is

$$V_e = \frac{P_u e x_i}{I_o}$$

The horizontal force on bolt i due to the eccentricity e is

$$H_e = \frac{P_u e y_i}{I_o}$$

The resultant force on bolt i is

$$R = \sqrt{(V_P + V_e)^2 + (H_e)^2}$$

The instantaneous center of rotation method of analyzing eccentrically loaded bolt groups affords a more realistic estimate of a bolt group's capacity. LRFD Tables 7-7 through 7-14 provide a means of designing common bolt group patterns by this method.

Example 4.37

Determine the diameter of the A325 bearing-type bolts required in the bolted bracket shown in the following illustration. Use the elastic unit area method and compare with the instantaneous center of rotation method.

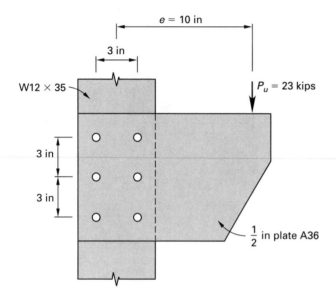

Solution

The geometric properties of the bolt group are obtained by applying the unit area method.

The moment of inertia about the x-axis is

$$I_x = \sum y^2 = (4)(3 \text{ in})^2 = 36 \text{ in}^4/\text{in}^2$$

The moment of inertia about the y-axis is

$$I_y = \sum x^2 = (6)(1.5 \text{ in})^2 = 13.5 \text{ in}^4/\text{in}^2$$

The polar moment of inertia about the centroid is

$$I_o = I_x + I_y = 49.5 \text{ in}^4/\text{in}^2$$

The top right bolt is the most heavily loaded, and the coexistent forces on this bolt are as follows.

- vertical force due to applied load

$$V_p = \frac{P_u}{n} = \frac{23 \text{ kips}}{6}$$
$$= 3.83 \text{ kips}$$

- vertical force due to eccentricity

$$V_e = \frac{P_u e x_i}{I_o} = \frac{(23 \text{ kips})(10 \text{ in})(1.5 \text{ in})}{49.5 \dfrac{\text{in}^4}{\text{in}^2}}$$
$$= 6.98 \text{ kips}$$

- horizontal force due to eccentricity

$$H_e = \frac{P_u e y_i}{I_o} = \frac{(23 \text{ kips})(10 \text{ in})(3 \text{ in})}{49.5 \dfrac{\text{in}^4}{\text{in}^2}}$$
$$= 13.94 \text{ kips}$$

- resultant force

$$R = \sqrt{(V_p + V_e)^2 + (H_e)^2}$$
$$= \sqrt{(3.83 \text{ kips} + 6.98 \text{ kips})^2 + (13.94 \text{ kips})^2}$$
$$= 17.6 \text{ kips}$$

Shear controls, and from LRFD Table 7-1 the design shear strength of a $7/8$ in diameter A325N bolt in a standard hole in single shear is

$$\phi R_n = 21.6 \text{ kips}$$
$$> 17.6 \text{ kips} \quad \text{[satisfactory]}$$

From LRFD Table 7-8, the coefficient C is given as 1.46, and the required design strength of an individual bolt, based on the instantaneous center of rotation method, is

$$\phi R_n = \frac{P_u}{C} = \frac{23 \text{ kips}}{1.46}$$
$$= 15.8 \text{ kips}$$

Shear controls, and from LRFD Table 7-1 the design shear strength of a $3/4$ in diameter A325N bolt in a standard hole in single shear is

$$\phi R_n = 15.9 \text{ kips}$$
$$> 15.8 \text{ kips} \quad \text{[satisfactory]}$$

Figure 4.18 *Bolt Group Eccentrically Loaded Normal to Faying Surface*

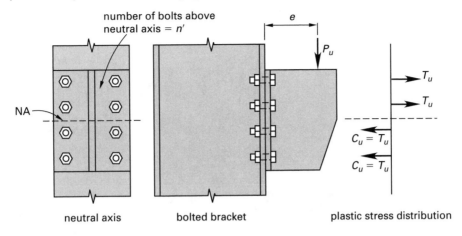

neutral axis   bolted bracket   plastic stress distribution

Bolt Group Eccentrically Loaded Normal to the Faying Surface

Eccentrically loaded bolt groups of the type shown in Fig. 4.18 may be conservatively designed by assuming that the neutral axis is located at the centroid of the bolt group and that a plastic stress distribution is produced in the bolts. The tensile force in each bolt above the neutral axis due to the eccentricity is given by

$$T_u = \frac{P_u e}{n' d_m}$$

The shear force in each bolt due to the applied load is given by

$$V_P = \frac{P_u}{n}$$

Example 4.38

Determine whether the $^7/_8$ in diameter A325N bearing-type bolts in the bolted bracket shown in the following illustration are adequate. Prying action may be neglected.

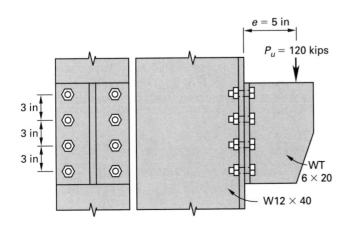

e = 5 in

P_u = 120 kips

3 in

3 in

3 in

WT 6 × 20

W12 × 40

Solution

The tensile force in each bolt above the neutral axis due to the eccentricity is given by

$$T_u = \frac{P_u e}{n' d_m} = \frac{(120 \text{ kips})(5 \text{ in})}{(4)(6 \text{ in})}$$
$$= 25 \text{ kips}$$

The shear force in each bolt due to the applied load is given by

$$V_P = \frac{P_u}{n} = \frac{120 \text{ kips}}{8}$$
$$= 15 \text{ kips}$$

The calculated shear stress on each bolt is

$$f_v = \frac{V_p}{A_b} = \frac{15 \text{ kips}}{0.601 \text{ in}^2}$$
$$= 25 \text{ kips/in}^2$$

The design shear stress for grade A325 bolts, with threads not excluded from the shear plane, is obtained from LRFD Table J3.2 as

$$\phi F_{nv} = (0.75) \left(48 \, \frac{\text{kips}}{\text{in}^2} \right)$$
$$= 36 \, \frac{\text{kips}}{\text{in}^2}$$
$$> f_v \quad [\text{satisfactory}]$$
$$f_v > 0.2 \phi F_{nv}$$

The design tensile stress for grade A325 bolts is obtained from LRFD Table J3.2 as

$$\phi F_{nt} = (0.75) \left(90 \, \frac{\text{kips}}{\text{in}^2} \right)$$
$$= 67.5 \, \frac{\text{kips}}{\text{in}^2}$$

The factored tensile stress in each $^7/_8$ in diameter bolt is

$$f_t = \frac{T_u}{A_b}$$
$$= \frac{25 \text{ kips}}{0.601 \text{ in}^2}$$
$$= 41.60 \ \frac{\text{kips}}{\text{in}^2}$$
$$< \phi F_{nt} \quad \text{[satisfactory]}$$
$$> 0.2\phi F_{nt}$$

Hence, it is necessary to investigate the effects of the combined shear and tensile stress. The nominal tensile stress F'_{nt} of a bolt, subjected to combined shear and tension, is given by LRFD Eq. (J3-3a) as

$$F'_{nt} = 1.3F_{nt} - \frac{f_v F_{nt}}{\phi F_{nv}}$$
$$= (1.3)\left(90 \ \frac{\text{kips}}{\text{in}^2}\right) - \frac{\left(25 \ \frac{\text{kips}}{\text{in}^2}\right)\left(90 \ \frac{\text{kips}}{\text{in}^2}\right)}{36 \ \frac{\text{kips}}{\text{in}^2}}$$
$$= 54.50 \text{ kips/in}^2$$

The design tensile stress $\phi F'_{nt}$ of a bolt, subjected to combined shear and tension, is given by LRFD Eq. (J3-2) as

$$\phi F'_{nt} = (0.75)\left(54.50 \ \frac{\text{kips}}{\text{in}^2}\right)$$
$$= 40.88 \ \frac{\text{kips}}{\text{in}^2}$$
$$< f_t \quad \text{[unsatisfactory]}$$

8. DESIGN OF WELDED CONNECTIONS

Nomenclature

a	coefficient for eccentrically loaded weld group	–
A_w	effective area of the weld	in^2
C	coefficient for eccentrically loaded weld group	–
D	number of sixteenths-of-an-inch in the weld size	–
F_{EXX}	classification of weld metal	
F_w	nominal strength of weld electrode	kips/in^2
k	coefficient for eccentrically loaded weld group	
l	characteristic length of weld group used in tabulated values of instantaneous center method	in
$\bar{l}$	total length of weld	in

R_n	nominal strength	–
R_{wl}	total nominal strength of longitudinally loaded fillet welds, as determined in accordance with LRFD Table J2.5	–
R_{wt}	total nominal strength of transversely loaded fillet welds, as determined in accordance with LRFD Table J2.5 without the amplification of the weld shear strength given by LRFD Eq. (J2-5)	–
w	fillet weld size	in

Symbols

θ	angle of inclination of loading measured from the weld longitudinal axis	
ϕ	resistance factor	

Weld Design Strength

The design strength of the weld metal is given by LRFD Sec. J2.4 as the product of the resistance factor ϕ, the nominal strength of the weld electrode F_w, and the effective area A_w. For a complete-penetration groove weld, the effective thickness is the thickness of the thinner part joined. For partial-penetration groove welds and flare groove welds, the effective throat thickness is given in LRFD Tables J2.1 and J2.2. For fillet welds made by the shielded metal arc process, the effective throat thickness, in accordance with LRFD Sec. J2.2, is $0.707w$.

The design strength of a $^1/_{16}$ in fillet weld per inch run of E70XX grade electrodes is then

$$q_u = \phi F_w A_w$$
$$= (0.75)(0.6)\left(70 F_{EXX} \ \frac{\text{kips}}{\text{in}^2}\right)(0.707)\left(\frac{1}{16} \text{ in}\right)$$
$$= 1.39 \text{ kips/in per } ^1/_{16} \text{ in}$$

The values of the resistance factors and the nominal strengths for other weld types are given in LRFD Table J2.5. Minimum sizes of fillet welds are given in LRFD Table J2.4.

The strength of linear weld groups, in which all the elements are in line or are parallel, may be analyzed by the method specified in LRFD Sec. J2.4a. This method accounts for the angle of inclination of the applied loading to the longitudinal axis of the weld. For an angle of inclination θ, the design strength in shear is given by LRFD Eq. (J2-4) as

$$\phi R_n = \phi F_w A_w$$

The nominal strength of the weld metal is given by LRFD Eq. (J2-5) as

$$F_w = 0.60 F_{EXX}(1.0 + 0.50\sin^{1.5}\theta)$$

The resistance factor is given by

$$\phi = 0.75$$

For concentrically loaded weld groups, with elements oriented both longitudinally or transversely to the direction of the applied load, the strength is determined as specified in LRFD Sec. J2.4. The combined strength of the weld group is given by the greater of

$$R_n = R_{wl} + R_{wt} \qquad \text{[LRFD J2.9a]}$$
$$R_n = 0.85R_{wl} + 1.5R_{wt} \qquad \text{[LRFD J2.9b]}$$

Example 4.39

The two grade A36 plates shown in the following illustration are connected by E70XX fillet welds as indicated. Determine the size of weld required to develop the full design axial strength of the $^5/_8$ in plate.

Solution

The design axial strength of the $^5/_8$ in plate is

$$P_u = \phi_t P_n = 0.9 F_y A_g$$
$$= (0.9)\left(36 \ \frac{\text{kips}}{\text{in}^2}\right)(3 \ \text{in})(0.625 \ \text{in})$$
$$= 60.75 \ \text{kips}$$

The total length of the longitudinally loaded weld is

$$\ell_{wl} = (2)(4 \ \text{in})$$
$$= 8 \ \text{in}$$

The total length of the transversely loaded weld is

$$\ell_{wt} = 3 \ \text{in}$$

The design shear capacity of a $^1/_4$ in fillet weld is

$$Q_w = Dq_u$$
$$= (4 \ \text{sixteenths})\left(1.39 \ \frac{\text{kips}}{\text{in}} \ \text{per} \ 1/16 \ \text{in}\right)$$
$$= 5.56 \ \text{kips/in}$$

Applying LRFD Eq. (J2-9a), the design strength of the connection is

$$\phi R_n = \phi(R_{wl} + R_{wt})$$
$$= \ell_{wl}Q_w + \ell_{wt}Q_w$$
$$= (8 \ \text{in})\left(5.56 \ \frac{\text{kips}}{\text{in}}\right) + (3 \ \text{in})\left(5.56 \ \frac{\text{kips}}{\text{in}}\right)$$
$$= 61.16 \ \text{kips}$$

Applying LRFD Eq. (J2-9b), the design strength of the connection is

$$\phi R_n = \phi(0.85R_{wl}) + \phi(1.5R_{wt})$$
$$= 0.85\ell_{wl}Q_w + 1.5\ell_{wt}Q_w$$
$$= (0.85)(8 \ \text{in})\left(5.56 \ \frac{\text{kips}}{\text{in}}\right)$$
$$\quad + (1.5)(3 \ \text{in})\left(5.56 \ \frac{\text{kips}}{\text{in}}\right)$$
$$= 62.83 \ \text{kips} \quad \text{[governs]}$$
$$> 61.16 \ \text{kips}$$
$$> P_u$$

From LRFD Table J2.4, the minimum size of fillet weld required for the $^1/_2$ plate is

$$w_{\min} = \frac{3}{16}$$
$$< 1/4 \ \text{in} \quad \text{[satisfactory]}$$

From LRFD Sec. J2.2b, the maximum size of fillet weld permitted at the edge of the $^5/_8$ in plate is

$$w_{\max} = \frac{5}{8} \ \text{in} - \frac{1}{16} \ \text{in}$$
$$= 9/16 \ \text{in}$$
$$> 1/4 \ \text{in} \quad \text{[satisfactory]}$$

A $^1/_4$ in fillet weld is adequate.

Illustration for Ex. 4.39

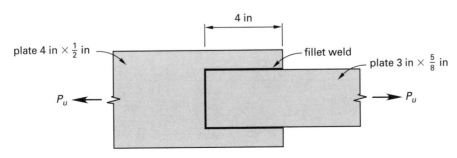

Weld Group Eccentrically Loaded in Plane of Faying Surface

Eccentrically loaded weld groups of the type shown in Fig. 4.19 may be conservatively designed by means of the elastic vector analysis technique assuming unit size of weld. The polar moment of inertia of the weld group about the centroid is

$$I_o = I_x + I_y$$

Figure 4.19 Eccentrically Loaded Weld Group

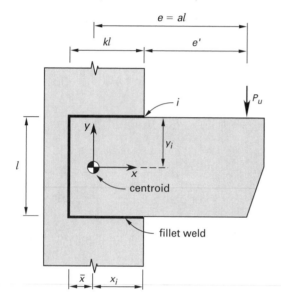

For a total length of weld $\bar{l}$, the vertical force per linear inch of weld due to the applied load P_u is

$$V_P = \frac{P_u}{\bar{l}}$$

The vertical force at point i due to the eccentricity e is

$$V_e = \frac{P_u e x_i}{I_o}$$

The horizontal force at point i due to the eccentricity e is

$$H_e = \frac{P_u e y_i}{I_o}$$

The resultant force at point i is

$$R = \sqrt{(V_p + V_e)^2 + (H_e)^2}$$

The instantaneous center of rotation method of analyzing eccentrically loaded weld groups affords a more realistic estimate of a weld group's capacity. LRFD Tables 8-4 through 8-11 provide a means of designing common weld group patterns by this method.

Example 4.40

Determine the size of E70XX fillet weld required in the welded bracket shown in the following illustration. Use the elastic unit area method and compare with the instantaneous center of rotation method.

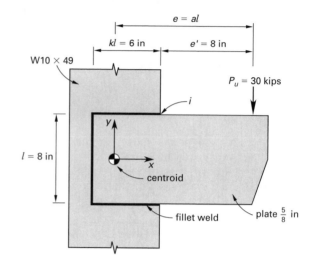

Solution

Assuming unit size of weld, the properties of the weld group are obtained by applying the elastic vector technique. The total length of the weld is

$$\bar{l} = l + 2kl$$
$$= 8 \text{ in} + (2)(6 \text{ in})$$
$$= 20 \text{ in}$$

The centroid location is given by LRFD Table 8-8, for a value of $k = 0.75$, as

$$\bar{x} = xl$$
$$= (0.225)(8 \text{ in})$$
$$= 1.8 \text{ in}$$

The moment of inertia about the x-axis is

$$I_x = \frac{l^3}{12} + (2)(kl)\left(\frac{l}{2}\right)^2$$
$$= \frac{(8 \text{ in})^3}{12} + (2)(6 \text{ in})(4 \text{ in})^2$$
$$= 235 \text{ in}^4/\text{in}$$

The moment of inertia about the y-axis is

$$I_y = \frac{(2)(kl)^3}{12} + (2)(kl)\left(\frac{kl}{2} - \bar{x}\right)^2 + l\bar{x}^2$$
$$= \frac{(2)(6 \text{ in})^3}{12} + (2)(6 \text{ in})(1.2 \text{ in})^2 + (8 \text{ in})(1.8 \text{ in})^2$$
$$= 79 \text{ in}^4/\text{in}$$

The polar moment of inertia is

$$I_o = I_x + I_y$$
$$= \frac{235 \text{ in}^4}{\text{in}} + \frac{79 \text{ in}^4}{\text{in}}$$
$$= 314 \text{ in}^4/\text{in}$$

The eccentricity of the applied load about the centroid of the weld profile is

$$e = e' + kl - \bar{x}$$
$$= 8 \text{ in} + 6 \text{ in} - 1.8 \text{ in}$$
$$= 12.2 \text{ in}$$

The top right corner of the weld profile is the most highly stressed, and the coexistent forces acting at this point in the x-direction and y-direction are as follows.

- vertical force due to applied load

$$V_P = \frac{P_u}{\bar{l}} = \frac{30 \text{ kips}}{20 \text{ in}}$$
$$= 1.5 \text{ kips/in}$$

- vertical force due to eccentricity

$$V_e = \frac{P_u e x_i}{I_o}$$
$$= \frac{(30 \text{ kips})(12.2 \text{ in})(4.2 \text{ in})}{314 \frac{\text{in}^4}{\text{in}}}$$
$$= 4.9 \text{ kips/in}$$

- horizontal force due to eccentricity

$$H_e = \frac{P_u e y_i}{I_o}$$
$$= \frac{(30 \text{ kips})(12.2 \text{ in})(4 \text{ in})}{314 \frac{\text{in}^4}{\text{in}}}$$
$$= 4.7 \text{ kips/in}$$

- resultant force

$$R = \sqrt{(V_p + V_e)^2 + (H_e)^2}$$
$$= \sqrt{(1.5 + 4.9)^2 + (4.7)^2}$$
$$= 7.9 \text{ kips/in}$$

The required fillet weld size per $^1/_{16}$ in is

$$D = \frac{R}{q_u}$$
$$= \frac{7.9 \text{ kips}}{1.39 \frac{\text{kips}}{\text{in}} \text{ per } \frac{1}{16} \text{ in}}$$
$$= 5.7 \text{ sixteenths}$$

Use a weld size of

$$w = 3/8 \text{ in}$$

The flange thickness of the W10 × 49 is

$$t_f = 0.560 \text{ in}$$

From LRFD Table J2.4, the minimum size of fillet weld is

$$w_{\min} = \frac{1}{4} \text{ in}$$
$$< 3/8 \text{ in} \quad [\text{satisfactory}]$$

From LRFD Sec. J2.2b, the maximum size of fillet weld for the $^5/_8$ in plate is

$$w_{\max} = \frac{5}{8} \text{ in} - \frac{1}{16} \text{ in}$$
$$> 3/8 \text{ in} \quad [\text{satisfactory}]$$

From LRFD Table 8-8, for values of $a = 1.53$ and $k = 0.75$, the coefficient C is given as 1.59, and the required fillet weld size per $^1/_{16}$ in, based on the instantaneous center of rotation method, is

$$D = \frac{P_u}{\phi C l}$$
$$= \frac{30 \text{ kips}}{(0.75)\left(1.59 \frac{\text{kips}}{\text{in}} \text{ per } \frac{1}{16} \text{ in}\right)(8 \text{ in})}$$
$$= 3.1 \text{ sixteenths}$$

Use a weld size of

$$w = 1/4 \text{ in}$$

Weld Group Eccentrically Loaded Normal to Faying Surface

Eccentrically loaded weld groups of the type shown in Fig. 4.20 may be conservatively designed by means of the elastic vector analysis technique assuming unit size of weld. For a total length of weld $\bar{l}$, the vertical force per linear inch of weld due to the applied load P_u is

$$V_P = \frac{P_u}{\bar{l}} = \frac{P_u}{2l}$$

Moment of inertia about the x-axis is

$$I_x = \frac{2l^3}{12} = \frac{l^3}{6}$$

The horizontal force at point i due to the eccentricity e is

$$H_e = \frac{P_u e y_i}{I_x} = \frac{3P_u e}{l^2}$$

The resultant force at point i is

$$R = \sqrt{(V_p)^2 + (H_e)^2}$$

The instantaneous center of rotation method of analyzing eccentrically loaded weld groups may also be used to determine a weld group's capacity. LRFD Table 8-4, with $k = 0$, provides a means of designing weld groups by this method.

Figure 4.20 *Weld Group Eccentrically Loaded Normal to Faying Surface*

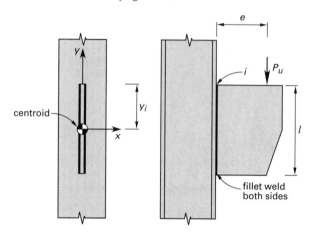

Example 4.41

Determine the size of E70XX fillet weld required in the welded gusset plate shown in the following illustration. Use the elastic unit area method and compare with the instantaneous center of rotation method.

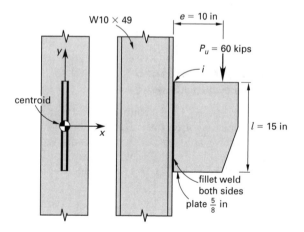

Solution

Assuming unit size of weld, the properties of the weld group are obtained by applying the elastic vector technique. The total length of the weld is

$$\bar{l} = 2l = 30 \text{ in}$$

The vertical force per linear inch of weld due to the applied load P_u is

$$V_P = \frac{P_u}{2l} = 2.0 \text{ kips/in}$$

Moment of inertia about the x-axis is

$$I_x = \frac{2l^3}{12} = 563 \text{ in}^3$$

The horizontal force at point i due to the eccentricity e is

$$\begin{aligned}
H_e &= \frac{P_u e y_i}{I_x} \\
&= \frac{(60 \text{ kips})(10 \text{ in})(7.5 \text{ in})}{563 \text{ in}^3} \\
&= 8.0 \text{ kips/in}
\end{aligned}$$

The resultant force at point i is

$$\begin{aligned}
R &= \sqrt{(V_p)^2 + (H_e)^2} \\
&= \sqrt{\left(2.0 \frac{\text{kips}}{\text{in}}\right)^2 + \left(8.0 \frac{\text{kips}}{\text{in}}\right)^2} \\
&= 8.2 \text{ kips/in}
\end{aligned}$$

The required fillet weld size per $1/16$ in is

$$\begin{aligned}
D &= \frac{R}{q_u} \\
&= \frac{8.2 \dfrac{\text{kips}}{\text{in}}}{1.39 \dfrac{\text{kips}}{\text{in}} \text{ per } \dfrac{1}{16} \text{ in}} \\
&= 5.9 \text{ sixteenths}
\end{aligned}$$

Use a weld size of

$$w = 3/8 \text{ in}$$

From LRFD Table 8-4, for values of $a = 0.67$ and $k = 0$, the coefficient C is given as 1.83, and the required fillet weld size per $1/16$ in, based on the instantaneous center of rotation method, is

$$\begin{aligned}
D &= \frac{P_u}{\phi C l} \\
&= \frac{60 \text{ kips}}{(0.75)\left(1.83 \dfrac{\text{kips}}{\text{in}} \text{ per } \dfrac{1}{16} \text{ in}\right)(15 \text{ in})} \\
&= 2.9 \text{ sixteenths}
\end{aligned}$$

Use $w_{\min} = 1/4$ in.

9. PLATE GIRDERS

Nomenclature

a	clear distance between transverse stiffeners	in

a_w ratio of web area to compression flange area   –

A_f compression flange area in^2

A_{pb} bearing area of stiffener after allowing for corner snip in^2

A_{sc} cross-sectional area of a stud shear connector in^2

A_{st} area of transverse stiffener in^2

A_T area of compression flange plus $^1/_6$ web in^2

A_w web area in^2

b_f flange width   in

b_{st} width of transverse stiffener   in

C_b bending coefficient   –

C_v shear coefficient for tension field action   –

d overall depth   in

D_s factor dependent on the type of transverse stiffener used   –

E_c modulus of elasticity of concrete $kips/in^2$

f'_c specified compressive strength of concrete $kips/in^2$

F_{cr} critical plate girder compression flange stress $kips/in^2$

F_{cr} critical column axial compression stress $kips/in^2$

F_r compressive residual stress in the flange $kips/in^2$

F_u minimum specified tensile strength of stud shear connector $kips/in^2$

F_{uv} required design shear strength of the stiffener-to-web weld $kips/in$

F_{yf} specified minimum yield stress of flange material $kips/in^2$

F_{yst} specified minimum yield stress of the stiffener material $kips/in^2$

F_{yw} specified minimum yield stress of the web material $kips/in^2$

h clear distance between flanges of a welded plate girder   in

I_{oy} moment of inertia of flange plus $^1/_6$ web referred to the y-axis in^4

I_{st} moment of inertia of transverse stiffener in^4

I_x moment of inertia referred to the x-axis in^4

j factor used to define moment of inertia of transverse stiffener   –

k_v web plate buckling coefficient   –

K effective length factor   –

l largest unbraced length along either flange at the point of load   in

l laterally unbraced length of column   in

L_b unbraced length of compression flange   in or ft

L_p maximum unbraced length for the limit state of yielding   ft

L_r maximum unbraced length for the limit state of inelastic lateral-torsional buckling   ft

M_n nominal flexural capacity   ft-kips

P_u factored end reaction   kips

q_u design strength of a $^1/_{16}$ in fillet weld per inch run of E70XX grade electrodes per $^1/_{16}$ $kips/in$ per $^1/_{16}$

R_g stud group coefficient   –

R_p stud position coefficient   –

r_{st} radius of gyration of bearing stiffener   in

r_t radius of gyration of compression flange plus $^1/_6$ web referred to the y-axis   in

R_{pg} plate girder flexural coefficient   –

R_u nominal bearing strength $kips/in^2$

S_x elastic section modulus referred to the x-axis in^3

t_f flange thickness   in

t_{st} stiffener thickness   in

t_w web thickness   in

V_n nominal shear capacity   kips

w fillet weld size   in

w unit weight of concrete lbf/ft^3

Symbols

λ slenderness parameter

λ_p limiting slenderness parameter for compact element

λ_r limiting slenderness parameter for noncompact element

ϕ_b resistance factor for flexure

ϕ_v resistance factor for shear

Girder Proportions

Doubly symmetric, nonhybrid beams are designed by LRFD Sec. F5 when the web depth-to-thickness ratio is

$$\frac{h}{t_w} > 5.70\sqrt{\frac{E}{F_y}}$$

$$> 162 \quad [\text{for } F_y = 36 \text{ kips/in}^2]$$

When this value of h/t_w is exceeded, the nominal flexural strength M_n is less than the plastic moment M_p.

In accordance with LRFD Sec. G2.2, and provided that the required shear strength does not exceed the design shear capacity given by LRFD Eq. G2.1, intermediate stiffeners are not required when

$$\frac{h}{t_w} \leq 2.46\sqrt{\frac{E}{F_y}}$$

For an unstiffened web, LRFD Sec. F13.2 requires that

$$\frac{h}{t_w} \leq 260$$

When intermediate stiffeners are provided, this limit may be exceeded. For $a/h > 1.5$,

$$\frac{h}{t_w} \le \frac{0.42E}{F_y} \qquad \text{[LRFD F13-4]}$$

$$\le 338 \quad [\text{for } F_y = 36 \text{ kips/in}^2]$$

For $a/h \le 1.5$,

$$\frac{h}{t_w} \le 11.7\sqrt{\frac{E}{F_{yf}}} \qquad \text{[LRFD F13-3]}$$

$$\le 332 \quad [\text{for } F_y = 36 \text{ kips/in}^2]$$

In accordance with LRFD Sec. G3.1, when $a/h > 3.0$ or $> (260t_w/h)^2$, design using tension field action is not permitted.

Example 4.42

For the welded plate girder of grade A36 steel shown in the following illustration, determine the maximum allowable spacing of intermediate stiffeners.

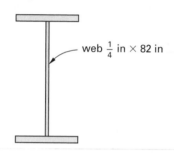

web $\frac{1}{4}$ in × 82 in

Solution

$$\frac{h}{t_w} = \frac{82 \text{ in}}{0.25 \text{ in}}$$
$$= 328$$
$$< 332$$

The maximum allowable stiffener spacing is

$$a = 1.5h = (1.5)(82 \text{ in})$$
$$= 123 \text{ in}$$

Design for Flexure

For doubly symmetric, nonhybrid beams with slender webs and values of $h/t_w > 5.7\sqrt{E/F_{yw}}$, the nominal flexural capacity is given by LRFD Sec. F5 as

$$M_n = S_x R_{pg} F_{cr} \qquad \text{[LRFD F5-2]}$$

$$R_{pg} = 1 - \frac{a_w\left(\frac{h}{t_w} - 5.7\sqrt{\frac{E}{F_y}}\right)}{1200 + 300a_w}$$
$$\qquad \text{[LRFD F5-6]}$$
$$\le 1.0$$

For the limit state of lateral-torsional buckling, the relevant parameters are

$$L_p = 1.1r_t\sqrt{\frac{E}{F_y}} \qquad \text{[LRFD F4-7]}$$

$$L_r = \pi r_t\sqrt{\frac{E}{0.7F_y}} \qquad \text{[LRFD F5-5]}$$

For an unbraced length of $L_b \le L_p$, the critical stress is given by

$$F_{cr} = F_{yf} \qquad \text{[LRFD F5-1]}$$

When $L_p < L_b \le L_r$, the critical stress is given by

$$F_{cr} = C_b F_y\left(1 - \frac{(0.3)(L_b - L_p)}{L_r - L_p}\right) \text{[LRFD F5-3]}$$
$$\le F_y$$

When $L_b > L_r$, the critical stress is given by

$$F_{cr} = \frac{C_b\pi^2 E}{\left(\frac{L_b}{r_t}\right)^2} \le F_y \qquad \text{[LRFD F5-4]}$$

For the limit state of flange local buckling, the relevant parameters are given by LRFD Sec. F5.3 as

$$\lambda = \frac{b_f}{2t_f}$$

$$\lambda_p = 0.38\sqrt{\frac{E}{F_y}} \qquad \text{[LRFD Table B4.1]}$$

$$= 11 \quad [\text{for } F_y = 36 \text{ kips/in}^2]$$

$$\lambda_r = 0.95\sqrt{\frac{E}{0.7F_y}}{k_c} \qquad \text{[LRFD Table B4.1]}$$

$$k_c = \frac{4}{\sqrt{\frac{h}{t_w}}}$$
$$\ge 0.35$$
$$\le 0.76$$

Flange local buckling does not occur in a compact flange with $\lambda \le \lambda_p$, and the critical compression flange stress is given by LRFD Sec. F5.3(a) as

$$F_{cr} = F_y$$

Inelastic local buckling of the flange occurs in a non-compact flange with $\lambda_p < \lambda \le \lambda_r$, and the critical compression flange stress is given by LRFD Sec. F5.3(b) as

$$F_{cr} = F_y\left(1 - \frac{0.3(\lambda - \lambda_p)}{\lambda_r - \lambda_p}\right)$$

Elastic local buckling of the flange occurs in a slender flange with $\lambda > \lambda_r$, and the critical compression flange stress is given by LRFD Sec. F5.3(c) as

$$F_{cr} = \frac{0.9Ek_c}{\lambda^2}$$

The design flexural capacity is given by LRFD Sec. F1 as

$$\phi_b M_n = 0.90 M_n$$

Example 4.43

Determine the design flexural capacity for the welded plate girder of grade A36 steel shown in the following illustration. Lateral support to the compression flange is provided at 10 ft centers, and $C_b = 1.0$.

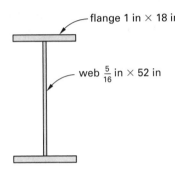

flange 1 in × 18 in

web $\frac{5}{16}$ in × 52 in

Solution

$$\frac{h}{t_w} = \frac{52 \text{ in}}{0.313 \text{ in}}$$
$$= 166$$
$$< 260 \quad \text{[stiffeners not mandatory]}$$
$$> 162 \quad \text{[LRFD Sec. F5 applies.]}$$

$$ht_w = (52 \text{ in})(0.313 \text{ in})$$
$$= 16.25 \text{ in}^2$$

$$A_f = b_f t_f = (18 \text{ in})(1.0 \text{ in})$$
$$= 18.00 \text{ in}^2$$

$$a_w = \frac{ht_w}{A_f} = \frac{16.25 \text{ in}^2}{18 \text{ in}^2}$$
$$= 0.903$$

The moment of inertia of the flange plus $^1/_6$ web about the y-axis is

$$I_{oy} = \frac{t_f b_f^3}{12} = \frac{(1.00 \text{ in})(18 \text{ in})^3}{12}$$
$$= 486 \text{ in}^4$$

$$A_T = A_f + \frac{ht_w}{6} = 18 \text{ in}^2 + \frac{16.25 \text{ in}^2}{6}$$
$$= 20.71 \text{ in}^2$$

$$r_t = \sqrt{\frac{I_{oy}}{A_T}} = \sqrt{\frac{486 \text{ in}^4}{20.71 \text{ in}^2}}$$
$$= 4.84 \text{ in}$$

The section modulus referred to the x-axis is

$$S_x = \frac{236{,}196 \text{ in}^4 - 207{,}250 \text{ in}^4}{27 \text{ in}}$$
$$= 1073 \text{ in}^3$$

For the limit state of lateral-torsional buckling,

$$L_p = 1.1 r_t \sqrt{\frac{E}{F_y}}$$
$$= (1.1)(4.84 \text{ in})\sqrt{806}$$
$$= 151 \text{ in}$$
$$> 120 \text{ in}$$

Hence,

$$F_{cr} = F_y$$

For the limit state of flange local buckling,

$$\lambda = \frac{b_f}{2t_f} = \frac{18 \text{ in}}{(2)(1 \text{ in})} \qquad \text{[LRFD Sec. 5.3]}$$
$$= 9$$
$$< \lambda_p = 11$$

Hence,

$$F_{cr} = F_y$$

$$R_{pg} = 1 - \frac{a_w \left(\dfrac{h}{t_w} - 5.7\sqrt{\dfrac{E}{F_y}} \right)}{1200 + 300 a_w}$$
$$= 1 - \frac{(0.903)(166 - 161.7)}{1200 + (300)(0.903)}$$
$$= 0.997$$

The design flexural capacity is given by LRFD Sec. F5.1 as

$$\phi_b M_n = \phi_b S_x R_{pg} F_{cr}$$
$$= (0.9)\left(1073 \text{ in}^3\right)(0.997)\left(36 \frac{\text{kips}}{\text{in}^2}\right)$$
$$= 34{,}660 \text{ in-kips}$$

Design for Shear Without Tension Field Action

For values of $h/t_w \leq 1.10\sqrt{(k_v E/F_y)}$, the nominal shear capacity, based on shear yielding of the stiffened or unstiffened web, is given by LRFD Sec. G2.1 as

$$V_n = 0.6 F_y A_w C_v \qquad \text{[LRFD G2-1]}$$

For values of $1.10\sqrt{(k_v E/F_y)} < h/t_w \leq 1.37\sqrt{(k_v E/F_y)}$ the nominal shear capacity, based on inelastic buckling of the web, is given by LRFD Sec. G2.1 as

$$V_n = 0.6 F_y A_w \left(\frac{1.10\sqrt{\dfrac{k_v E}{F_y}}}{\dfrac{h}{t_w}} \right)$$

For values of $h/t_w > 1.37\sqrt{(k_v E/F_{yw})}$, the nominal shear capacity, based on elastic buckling of the web, is given by LRFD Sec. G2.1 as

$$V_n = \frac{A_w (0.91 E k_v)}{\left(\dfrac{h}{t_w}\right)^2}$$

The web plate buckling coefficient is

$$k_v = 5 + \frac{5}{\left(\dfrac{a}{h}\right)^2}$$

$$= 5 \text{ when } a/h > 3$$

$$= 5 \text{ when } a/h > (260 t_w/h)^2$$

$$= 5 \text{ for unstiffened webs with } h/t_w < 260$$

LRFD Tables 3-16a and 3-17a provide values of $\phi_v V_n/A_w$ for a range of values of h/t_w and a/h.

Example 4.44

Determine the design shear capacity for the welded plate girder of Ex. 4.43 by using LRFD Sec. G2.1. Check the solution by using LRFD Table 3-16a.

Solution

From Ex. 4.43, for an unstiffened web,

$$A_w = dt_w$$
$$= (54 \text{ in})(0.313 \text{ in})$$
$$= 16.90 \text{ in}^2$$
$$k_v = 5$$
$$\frac{h}{t_w} = 166$$
$$> 1.37\sqrt{\frac{k_v E}{F_y}}$$
$$V_n = \frac{A_w (0.91 E k_v)}{\left(\dfrac{h}{t_w}\right)^2}$$
$$= \frac{(16.90 \text{ in}^2)(0.91)\left(29{,}000 \dfrac{\text{kips}}{\text{in}^2}\right)(5)}{166^2}$$
$$= 80.92 \text{ kips}$$

The design shear capacity is then

$$\phi_v V_n = 0.9 V_n$$
$$= (0.9)(80.92)$$
$$= 73 \text{ kips}$$

From LRFD Table 3-16a, for a value of $a/h > 3.0$ and a value of $h/t_w = 166$,

$$\phi_v V_n = 4.6 A_w$$
$$= (4.6)\left(16.90 \text{ in}^2\right)$$
$$= 78 \text{ kips}$$

Design for Shear with Tension Field Action

When tension field action is used, the nominal shear strength is determined in accordance with LRFD Sec. G3.2. LRFD Tables 3-16b and 3-17b, provide values of $\phi_v V_n/A_w$ for a range of values of h/t_w and a/h.

Tension field action is not permitted in end panels and when $a/h > 3.0$ or $> (260 t_w/h)^2$, in which case, in accordance with LRFD Sec. G2.1, the nominal shear strength is given by

$$V_n = 0.6 A_w F_y C_v \qquad \text{[LRFD G2-1]}$$

Example 4.45

The welded plate web girder of Ex. 4.43 has imposed factored loads of $M_u = 28{,}000$ in-kips and $V_u = 120$ kips. Intermediate stiffeners are provided at 100 in centers. Determine the design shear capacity.

Solution

From Ex. 4.43,

$$\frac{h}{t_w} = 166$$
$$\frac{a}{h} = \frac{100 \text{ in}}{52 \text{ in}}$$
$$= 1.92$$

From LRFD Table 3-16b, for a value of $a/h = 1.92$ and a value of $h/t_w = 166$,

$$\phi_v V_n = 11.5 A_w$$
$$= \left(11.5 \frac{\text{kips}}{\text{in}^2}\right)\left(16.90 \text{ in}^2\right)$$
$$= 194 \text{ kips}$$

Design of Intermediate Stiffeners

The required moment of inertia of a single stiffener about the face in contact with the web plate or of a pair of stiffener plates about the web center line is given by LRFD Sec. G2.2 as

$$I_{st} = at_w^3 j$$
$$j = \frac{2.5}{\left(\frac{a}{h}\right)^2} - 2 \qquad \text{[LRFD G2-6]}$$
$$\geq 0.5$$

When tension field action is used, the required stiffener area is given by LRFD Sec. G3.3 as

$$A_{st} = \left(\frac{F_y}{F_{yst}}\right)\left(0.15 D_s h t_w \left(1 - C_v\right)\left(\frac{V_u}{\phi_v V_n}\right) - 18 t_w^2\right)$$
$$\qquad \text{[LRFD G3-3]}$$
$$\geq 0$$

$D_s = 1.0$ for a pair of stiffeners

$\quad = 1.8$ for a single angle stiffener

$\quad = 2.4$ for a single plate stiffener

For values of $h/t_w = 1.1\sqrt{(k_v E/F_y)}$, the value of the shear coefficient is given by LRFD Sec. G2.1 as

$$C_v = 1.0$$

For values of $1.1\sqrt{(k_v E/F_y)} < h/t_w \leq 1.37\sqrt{(k_v E/F_y)}$, the shear coefficient is given by LRFD Sec. G2.1 as

$$C_v = \frac{1.1\sqrt{\dfrac{k_v E}{F_y}}}{\dfrac{h}{t_w}} \qquad \text{[LRFD G2-4]}$$

For values of $h/t_w > 1.37\sqrt{(k_v E/F_y)}$, the shear coefficient is given by LRFD Sec. G2.1 as

$$C_v = \frac{1.51 k_v E}{F_y \left(\dfrac{h}{t_w}\right)^2} \qquad \text{[LRFD G2-5]}$$

The maximum allowable width-to-thickness ratio of a stiffener plate is given by LRFD Sec. G3.3 as

$$\frac{b_{st}}{t_{st}} = 0.56\sqrt{\frac{E}{F_y}}$$
$$\qquad = 15.89 \text{ for } F_y = 36 \text{ kips/in}^2$$

The stiffener is connected to the web to resist a shear, in kips/in of

$$F_{uv} = 1.67 h \left(\frac{F_y}{340}\right)^{1.5}$$

As specified in LRFD Sec. G2.2, the weld used to attach the stiffener to the web shall terminate between four

times and six times the web thickness from the near toe of the web-to-flange weld.

Example 4.46

Design the intermediate stiffeners, using a pair of stiffener plates, for the plate web girder of Ex. 4.45, which utilizes tension field action.

Solution

For a value of $a/h = 1.92$, the moment of inertia factor is given by LRFD Sec. G2.2 as

$$j = \frac{2.5}{\left(\dfrac{a}{h}\right)^2} - 2 = \frac{2.5}{(1.92)^2} - 2$$
$$= 0.5 \text{ minimum}$$

The required moment of inertia of a pair of stiffener plates about the web center line is given by LRFD Sec. G2.2 as

$$I_{st} = at_w^3 j$$
$$= (100 \text{ in})(0.313 \text{ in})^3(0.5)$$
$$= 1.53 \text{ in}^4$$

For a pair of 4 in $\times$ $^1/_4$ in stiffener plates, the width-to-thickness ratio is

$$\frac{b_{st}}{t_{st}} = \frac{4 \text{ in}}{0.25 \text{ in}} = 16$$
$$\approx 0.56\sqrt{\frac{E}{F_y}} \quad \text{[satisfactory]}$$

In accordance with LRFD Sec. G2.2, the stiffener may terminate 2 in above the bottom flange.

The moment of inertia provided by the pair of plates is

$$I_{st} = \frac{t_{st}\left(2b_{st} + t_w\right)^3}{12}$$
$$= \frac{(0.25 \text{ in})(8.313 \text{ in})^3}{12}$$
$$= 11.97 \text{ in}^4$$
$$> 1.53 \text{ in}^4 \quad \text{[satisfactory]}$$

For a pair of stiffeners, the stiffener factor is given by LRFD Sec. G3.3 as

$$D = 1.0$$

The clear distance between the flanges is

$$h = 52 \text{ in}$$

Steel

The web thickness is

$$t_w = 0.313 \text{ in}$$

The panel aspect ratio is

$$\frac{a}{h} = \frac{100 \text{ in}}{52 \text{ in}}$$
$$= 1.92$$

Hence, the web plate buckling coefficient is given by LRFD Sec. G2.1(b)(ii) as

$$k_v = 5 + \frac{5}{\left(\frac{a}{h}\right)^2}$$
$$= 5 + \frac{5}{(1.92)^2}$$
$$= 6.36$$

The web height-to-thickness ratio is

$$\frac{h}{t_w} = \frac{52 \text{ in}}{0.313 \text{ in}}$$
$$= 166$$
$$> 1.37 \sqrt{\frac{k_v E}{F_y}}$$

Hence, the shear coefficient is given by LRFD Eq. (G2-5) as

$$C_v = \frac{1.51 k_v \dfrac{E}{F_y}}{\left(\dfrac{h}{t_w}\right)^2}$$

$$= \frac{(1.51)(6.36)\left(\dfrac{29{,}000 \ \dfrac{\text{kips}}{\text{in}^2}}{36 \ \dfrac{\text{kips}}{\text{in}^2}}\right)}{(166)^2}$$

$$= 0.281$$

From Ex. 4.45,

$$\frac{V_u}{\phi_v V_n} = \frac{120 \text{ kips}}{194 \text{ kips}}$$
$$= 0.62$$

The minimum required stiffener area is given by LRFD Equation (G3-3) as

$$A_{st} = \left(\frac{F_y}{F_{yst}}\right)\left(0.15 D_s h t_w (1 - C_v)\left(\frac{V_u}{\phi_v V_n}\right) - 18 t_w^2\right)$$

$$= 1.0 \left(\begin{array}{c} (0.15)(1.0)(52 \text{ in})(0.313 \text{ in})(0.719)(0.62) \\ - (18)(0.313 \text{ in})^2 \end{array} \right)$$

$$< 0.0$$

No specific area is required for the stiffener.

The web thickness is $5/16$ in, and the minimum allowable fillet weld size connecting the $1/4$ in stiffener to the web is given by LRFD Table J2.4 as

$$w_{\min} = 1/8 \text{ in}$$

The required design strength of the weld connecting the stiffener to the web is

$$F_{uv} = 1.67 h \left(\frac{F_y}{340}\right)^{1.5}$$

$$F_{uv} = (1.67)(52 \text{ in})\left(\frac{36 \ \dfrac{\text{kips}}{\text{in}^2}}{340}\right)^{1.5}$$

$$= 2.99 \text{ kips/in}$$

Providing four runs of $3/16$ in E70XX intermittent fillet welds 1.5 in long spaced at 8 in centers staggered on either side of the web and on either side of the stiffeners gives a design strength of

$$F_{uv} = \frac{(4)(3q_u)(1.5 \text{ in})}{8 \text{ in}}$$

$$= \frac{(4)(3)\left(1.39 \ \dfrac{\text{kips}}{\text{in}}\right)(1.5 \text{ in})}{8 \text{ in}}$$

$$= 3.13 \text{ kips/in}$$
$$> 2.99 \text{ kips/in} \quad \text{[satisfactory]}$$

Design of Bearing Stiffeners

Bearing stiffeners are required when the factored bearing reaction exceeds the design web buckling capacity given by LRFD Sec. J10.4. For flanges restrained against rotation, web sidesway buckling will occur when $(h/t_w)/(l/b_f) \le 2.3$. For flanges not restrained against rotation, sidesway web buckling will occur when $(h/t_w)/(l/b_f) \le 1.7$.

In accordance with LRFD Sec. J10.8, the stiffeners are designed as columns with a section composed of the two stiffener plates plus a strip of web having a width of $25 t_w$ at interior stiffeners and $12 t_w$ at end stiffeners. The effective length factor is $k = 0.75$. The stiffener-to-web fillet weld must be sufficient to transmit the reaction.

The nominal bearing strength is given by LRFD Sec. J7 as

$$R_n = 1.8 F_y A_{pb}$$

The design bearing strength is

$$\phi R_n = 0.75 R_n$$

Example 4.47

The welded plate web girder of Ex. 4.43 is provided with bearing stiffeners at each end consisting of a pair of $1/2$ in $\times$ 8 in plates of grade A36 steel. Determine the maximum factored reaction that may be applied to the girder.

Solution

Allowing for a 1 in corner snip to clear the weld, the nominal bearing area of the stiffener plates is

$$A_{pb} = 2t_{st}(b_{st} - 1)$$
$$= (2)(0.5 \text{ in})(7 \text{ in})$$
$$= 7 \text{ in}^2$$

The design bearing strength is given by LRFD Sec. J7 as

$$\phi R_n = (0.75)(1.8 F_y A_{pb})$$
$$= (0.75)(1.8)\left(36 \frac{\text{kips}}{\text{in}^2}\right)(7 \text{ in}^2)$$
$$= 340 \text{ kips}$$

The moment of inertia provided by the pair of plates is

$$I_{st} = \frac{t_{st}(2b_{st} + t_w)^3}{12}$$
$$= \frac{(0.5 \text{ in})(16.313 \text{ in})^3}{12}$$
$$= 181 \text{ in}^4$$

The effective area of the bearing stiffener is

$$A_{st} = 2t_{st}b_{st} + (12)(t_w)^2$$
$$= (2)(0.5 \text{ in})(8 \text{ in}) + (12)(0.313)^2$$
$$= 9.18 \text{ in}^2$$

The radius of gyration of the bearing stiffener is

$$r_{st} = \sqrt{\frac{I_{st}}{A_{st}}} = 4.44 \text{ in}$$

The slenderness ratio of the bearing stiffener is

$$\frac{Kl}{r_{st}} = \frac{(0.75)(52 \text{ in})}{4.44} = 8.78$$

From LRFD Table 4-22, the design axial stress is

$$\phi_c F_{cr} = 32.3 \text{ kips/in}^2$$

The maximum factored end reaction is

$$P_u = \phi_c F_{cr} A_{st}$$
$$= \left(32.3 \frac{\text{kips}}{\text{in}^2}\right)(9.18 \text{ in}^2)$$
$$= 297 \text{ kips} \quad [\text{governs}]$$

10. COMPOSITE BEAMS

Nomenclature

a	depth of compression block	in
a	distance between connectors	in
A_c	area of concrete slab within the effective width	in^2
A_s	cross-sectional area of structural steel	in^2
A_{sc}	cross-sectional area of a stud shear connector	in^2
b	effective concrete flange width	in
C_{con}	compressive force in slab at ultimate load	kips
d	depth of steel beam	in
d_s	diameter of stud shear connector	in
E_c	modulus of elasticity of concrete	kips/in^2
e_{mid-ht}	distance from the edge of the stud shank to the steel deck web, measured at mid-height of the deck rib, in the direction of maximum moment for a simply supported beam	–
f'_c	specified compressive strength of the concrete	kips/in^2
F_u	minimum specified tensile strength of stud shear connector	kips/in^2
F_y	specified minimum yield stress of the structural steel section	kips/in^2
h_r	nominal steel deck rib height	in
H_s	length of shear connector, not to exceed $(h_r + 3 \text{ in})$ in computations	in
I_{LB}	lower bound moment of inertia	in^4
L	span length	ft
M_n	nominal flexural strength of member	in-kips or ft-kips
n	number shear connectors between point of maximum positive moment and point of zero moment	–
N_r	number of studs in one rib at a beam intersection, not to exceed 3 in calculations	
Q_n	nominal shear strength of single shear connector	kips
R_g	stud group coefficient	–
R_p	stud position coefficient	–
s	beam spacing	ft or in
t_c	actual slab thickness	in
T_{stl}	tensile force in steel at ultimate load	kips
V'	total factored horizontal shear between point of maximum moment and point of zero moment	kips
w	unit weight of concrete	lbf/ft^3
w_r	average width of concrete rib	in

y moment arm between centroids of tensile force and compressive force in

Y_{con} distance from top of steel beam to top of concrete in

Y_1 distance from top of steel beam to plastic neutral axis in

Y_2 distance from top of steel beam to concrete flange force in

Symbols

ρ reduction factor for studs in ribbed steel deck

$\sum Q_n$ summation of Q_n between point of maximum moment and point of zero moment on either side kips

Section Properties

The composite beam shown in Fig. 4.21 consists of a concrete slab supported by a formed metal deck, with the slab acting compositely with a steel beam. In accordance with LRFD Sec. I3.1, the effective width of the concrete slab on either side of the beam centerline shall not exceed

- one-eighth of the beam span

- one-half of the beam spacing

- the distance to the edge of the slab

For the composite beam shown in Fig. 4.21, at the ultimate load the depth of the concrete stress block is less than the depth of the slab.[3] For this situation, the plastic neutral axis is located at the top of the steel beam and

$$Y_1 = 0$$

$$Y_2 = Y_{con} - \frac{a}{2}$$

When sufficient shear connectors are provided to ensure full composite action, the depth of the stress block is given by

$$a = \frac{F_y A_s}{0.85 f_c' b}$$

When insufficient shear connectors are provided to ensure full composite action, the depth of the stress block is given by

$$a = \frac{\sum Q_n}{0.85 f_c' b}$$

Using this value of the depth of the stress block to define an equivalent slab depth provides a lower bound on the actual moment of inertia based on elastic principles. LRFD Part 3, Table 3-20, provides values of I_{LB} for a range of values of Y_1 and Y_2. This moment of inertia is used to determine the deflection of the composite member.

Example 4.48

A simply supported composite beam consists of a 3 in concrete slab cast on a 3 in formed steel deck over a W21 × 50 grade 50 steel beam. The beams are spaced at 8 ft centers and span 30 ft; the slab consists of 4000 lbf/in² normal weight concrete. Determine the lower bound moment of inertia if full composite action is provided.

Solution

The effective width of the concrete slab is the lesser of

- $s = (8 \text{ ft})\left(12 \dfrac{\text{in}}{\text{ft}}\right)$
  $= 96 \text{ in}$

- $\dfrac{L}{4} = \dfrac{(30 \text{ ft})\left(12 \dfrac{\text{in}}{\text{ft}}\right)}{4}$
  $= 90 \text{ in}$ [governs]

Figure 4.21 *Fully Composite Beam Section Properties*

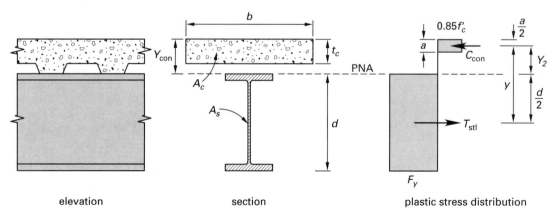

elevation   section   plastic stress distribution

For full composite action, the depth of the stress block is

$$a = \frac{F_y A_s}{0.85 f'_c b}$$

$$= \frac{\left(50 \dfrac{\text{kips}}{\text{in}^2}\right)(14.7 \text{ in}^2)}{(0.85)\left(4 \dfrac{\text{kips}}{\text{in}^2}\right)(90 \text{ in})}$$

$$= 2.40 \text{ in} \quad \text{[within the slab]}$$

The distance from the top of the steel beam to the line of action of the concrete slab force is

$$Y_2 = Y_{\text{con}} - \frac{a}{2}$$

$$= 3 \text{ in} + 3 \text{ in} - \frac{2.40 \text{ in}}{2}$$

$$= 4.80 \text{ in}$$

$$Y_1 = 0$$

From LRFD Part 3, Table 3-20,

$$I_{LB} = 2686 \text{ in}^4$$

Shear Connection

Shear connectors are provided to transfer the horizontal shear force across the interface. The nominal shear strengths Q_n of different types of shear connectors are given in LRFD Table 3-21. The required number of connectors may be uniformly distributed between the point of maximum moment and the support on either side, with the total horizontal shear being determined by the lesser value given by LRFD Eqs. (I3-1a) and (I3-1b) as

$$V' = 0.85 f'_c A_c \qquad \text{[LRFD I3-1a]}$$
$$V' = F_y A_s \qquad \text{[LRFD I3-1a]}$$

To provide complete shear connection and full composite action, the required number of connectors on either side of the point of maximum moment is given by

$$n = \frac{V'}{Q_n}$$

Figure 4.22 *Placement of Shear Connectors*

ribs perpendicular

ribs parallel

If a smaller number of connectors is provided, only partial composite action can be achieved, and the nominal flexural strength of the composite member is reduced. The number of shear connectors placed between a concentrated load and the nearest support shall be sufficient to develop the moment required at the load point.

The nominal strength of one stud shear connector embedded in a solid slab is given by LRFD Eq. (I3-3) as

$$Q_n = 0.5 A_{sc}(f'_c E_c)^{0.5}$$
$$\leq R_g R_p A_{sc} F_u$$

The minimum tensile strength of a Type B shear stud connector made from ASTM A108 material is

$$F_u = 65 \text{ kips/in}^2$$

The stud group coefficient for flat soffit, solid slabs with the stud welded directly to the girder flange is

$$R_g = 1.0$$

The stud position coefficient for flat soffit, solid slabs with the stud welded directly to the girder flange is

$$R_p = 1.0$$

The modulus of elasticity of concrete is given by

$$E_c = w^{1.5}(f'_c)^{0.5}$$

The unit weight of normal weight concrete is given by

$$w = 145 \text{ lbf/ft}^3$$
$$Q_n = 0.5 A_{sc}(w f'_c)^{0.75}$$

When the concrete is cast on a formed metal deck, the limitations imposed on the spacing and placement of shear connectors are given in LRFD Sec. I3.2c and are summarized in Fig. 4.22.

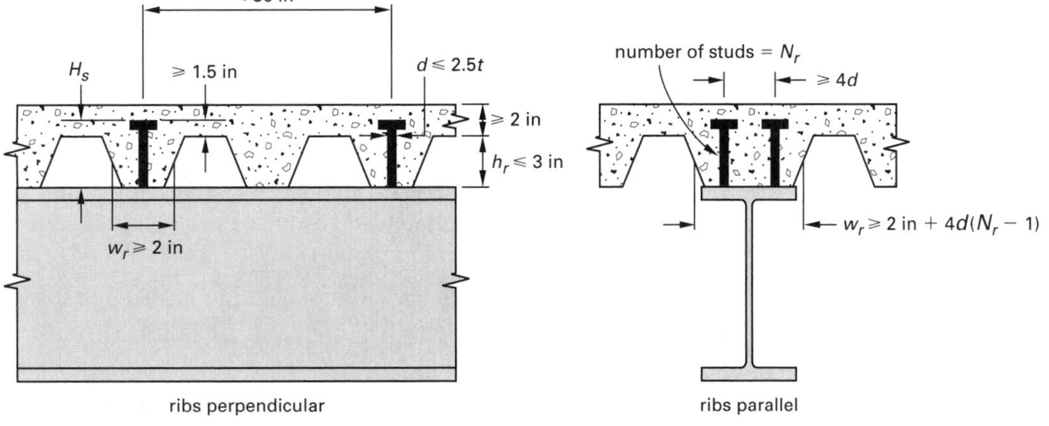

When the concrete slab is cast on a formed metal deck, the values of R_g and R_p are modified as detailed in LRFD Sec. I3.2d(3).

Deck Ribs Parallel to Steel Beam

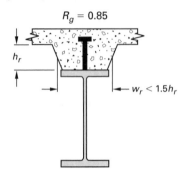
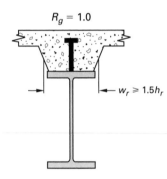
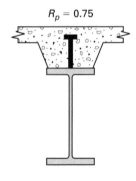

Figure 4.23 *Deck Ribs Parallel to Steel Beam, R_g and R_p Values*

When the deck ribs are parallel to the steel beam as shown in Fig. 4.23, the stud group coefficient for any number of studs welded in a row through the steel deck is given by

$$R_g = 1.0 \quad [\text{when } w_r \geq 1.5h_r]$$
$$R_g = 0.85 \quad [\text{when } w_r < 1.5h_r]$$

When the deck ribs are parallel to the steel beam as shown in Fig. 4.23, the stud position coefficient for studs welded through the steel deck is given by

$$R_p = 0.75$$

Deck Ribs Perpendicular to Steel Beam

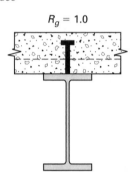
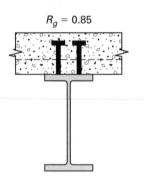
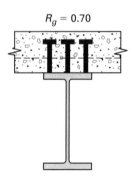

Figure 4.24 *Deck Ribs Perpendicular to Steel Beams, R_g Values*

When the deck ribs are perpendicular to the steel beam as shown in Fig. 4.24, the stud group coefficient for studs welded through the steel deck is given by

$$R_g = 1.0 \quad [\text{for one stud welded in a steel deck rib}]$$
$$R_g = 0.85 \quad [\text{for two studs welded in a steel deck rib}]$$
$$R_g = 0.70 \quad \begin{bmatrix} \text{for three or more studs} \\ \text{welded in a steel deck rib} \end{bmatrix}$$

Figure 4.25 *Deck Ribs Perpendicular to Steel Beams, R_p Values*

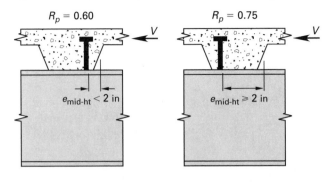

When the deck ribs are perpendicular to the steel beam as shown in Fig. 4.25, the stud position coefficient for studs welded through the steel deck is given by

$$R_p = 0.75 \quad \begin{bmatrix} \text{for studs welded in a steel deck rib} \\ \text{with } e_{\text{mid-ht}} \geq 2 \text{ in} \end{bmatrix}$$

$$R_p = 0.60 \quad \begin{bmatrix} \text{for studs welded in a steel deck rib} \\ \text{with } e_{\text{mid-ht}} < 2 \text{ in} \end{bmatrix}$$

The nominal strength of different stud diameters in 3 kips/in² and 4 kips/in² normal weight and light weight concrete is given in LRFD Table 3-21.

Example 4.49

Determine the number of ³/₄ in diameter stud shear connectors required in the composite beam of Ex. 4.48 to provide full composite action. The ribs of the formed steel deck are perpendicular to the steel beams with $h_r = 3$ in, $w_r = 3^1/2$ in, $H_s = 5$ in, and $N_r = 2$. The beam is loaded with a uniformly distributed load.

Solution

The total horizontal shear is given by

$$V' = F_y A_s$$
$$= \left(50 \frac{\text{kips}}{\text{in}^2}\right)(14.7 \text{ in}^2)$$
$$= 735 \text{ kips}$$

Two ³/₄ in diameter studs are located in each rib in the weak position.

The nominal shear strength of each ³/₄ in diameter stud is obtained from LRFD Table 3-21, as

$$Q_n = 14.6 \text{ kips}$$

The required number of studs in the beam is

$$2n = \frac{2V'}{Q_n} = \frac{(2)(735 \text{ kips})}{14.6 \text{ kips}}$$
$$= 100 \text{ studs}$$

Design for Flexure

LRFD Part 3, Table 3-19, provide values of ϕM_n for a range of values of Y_1, Y_2, and $\sum Q_n$. The value of $\sum Q_n$ is given by LRFD Sec. I3.2d as the least of

- $0.85 f_c' A_c$

- $F_y A_s$

- $n Q_n$

Because of redistribution of stresses at the ultimate load, the composite section is designed to support the total factored loads, due to all dead and live loads, for both shored and unshored construction. In addition, for unshored construction, the steel beam alone must be adequate to support all loads applied before the concrete has attained 75% of its required strength.

Example 4.50

Determine the design flexural strength of the composite beam of Ex. 4.49.

Solution

Because sufficient shear connectors are provided to ensure full composite action,

$$\sum Q_n = F_y A_s = 735 \text{ kips}$$

From Ex. 4.48,
$$Y_2 = 4.80 \text{ in}$$
$$Y_1 = 0$$

From LRFD Part 3, Table 3-19,

$$\phi M_n = 839 \text{ ft-kips}$$

References

1. American Institute of Steel Construction, Inc. *Steel Construction Manual*, 13th ed. 2005.

2. International Code Council. *International Building Code*. 2006.

3. Vogel, R. *LRFD-Composite Beam Design with Metal Deck*. Steel Committee of California. 1991.

PRACTICE PROBLEMS

1. The pair of shear legs shown in the following illustration consists of two nonstandard steel tubes 13 ft long, of 3.5 in outside diameter and 3 in inside diameter, pinned together at the top and inclined to each other at an angle of 45°. The yield stress of the tubes is $F_y = 36$ kips/in². The legs are laterally braced at the top and are pinned at the base. If the self-weight of the pipes is neglected, what is the maximum factored load the shear legs can lift?

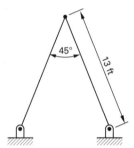

2. Both flanges of a W8 × 24 grade A36 steel section are each connected by six ³/₄ in diameter bolts to a steel bracket. A single row of three bolts is provided on each side of the beam web to both flanges, as shown in the following illustration. What is the capacity of the W section in direct tension?

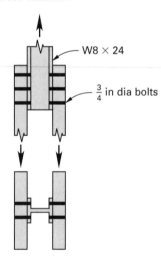

3. A fixed-ended steel beam is shown in the following illustration, with the factored loads indicated. The distributed load shown includes the beam self-weight. Full lateral support is provided to the beam, which is of grade 50 steel. What is the lightest W section beam that can support the factored loads?

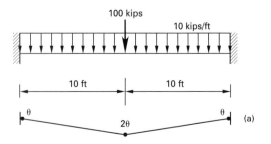

4. The simply supported composite beam shown in the following illustration consists of a 7¹/₂ in normal weight concrete slab cast on a W21 × 57 grade 50 steel beam. The beam forms part of a floor system with the beams spaced at 10 ft centers, and the concrete strength is 3000 lbf/in². The factored loads are indicated in the figure; these include the weight of the concrete slab and the self-weight of the steel beam. Seventeen stud shear connectors of ³/₄ in diameter are provided between sections 1 and 2. Is the number of connectors adequate?

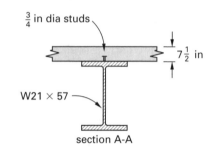

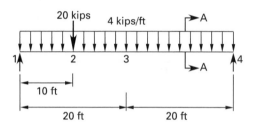

5. Determine the block shear design capacity of the welded connection shown in the following illustration. Both plates are of grade A36 steel and the plates are connected with E70XX fillet welds.

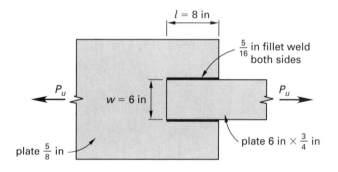

6. Determine the tensile design capacity of the fillet welds in the welded connection shown in Prob. 5.

7. Determine the tensile design capacity of the ³/₄ in plate in the welded connection shown in Prob. 5.

SOLUTIONS

1. The area of each pipe is

$$A = \pi \left(a^2 - b^2\right)$$
$$= \pi \left((1.75 \text{ in})^2 - (1.5 \text{ in})^2\right)$$
$$= 2.55 \text{ in}^2$$

The radius of gyration of each pipe is

$$r = \frac{\sqrt{a^2 + b^2}}{2} = 1.15 \text{ in}$$

The slenderness ratio of each pipe is given by

$$\frac{Kl}{r} = \frac{(1.0)(13 \text{ ft}) \left(12 \ \frac{\text{in}}{\text{ft}}\right)}{1.15 \text{ in}}$$
$$= 135.7$$
$$< 200 \quad [\text{satisfactory}]$$

From LRFD Table 4-22, the design axial stress is given by

$$\phi_c F_{cr} = 12.3 \text{ kips/in}^2$$

The design axial strength of each pipe is given by

$$\phi_c P_n = \phi_c F_{cr} A$$
$$= \left(12.3 \ \frac{\text{kips}}{\text{in}^2}\right) \left(2.55 \text{ in}^2\right)$$
$$= 31.37 \text{ kips}$$

The maximum factored load the pair of shear legs can lift is given by

$$P_u = 2\phi_c P_n \cos 22.5°$$
$$= 58 \text{ kips}$$

2. The relevant properties of the W8 × 24 are obtained from LRFD Table 1-1 and are

$$A_g = 7.08 \text{ in}^2$$
$$b_f = 6.5 \text{ in}^2$$
$$t_f = 0.40 \text{ in}$$
$$d = 7.93 \text{ in}$$

The hole diameter is

$$d_h = d_b + \frac{1}{8} \text{ in}$$
$$= \frac{3}{4} \text{ in} + \frac{1}{8} \text{ in}$$
$$= 0.875 \text{ in}$$

The ratio is

$$\frac{b_f}{d} = \frac{6.50 \text{ in}}{7.93 \text{ in}}$$
$$> 2/3$$

Three bolts are in line in the direction of stress, however, and hence from LRFD Sec. D3,

$$U = 0.90$$

The net area is given by

$$A_n = A_g - 4d_h t_f$$
$$= 7.08 \text{ in}^2 - (4)(0.875 \text{ in})(0.40 \text{ in})$$
$$= 5.68 \text{ in}^2$$

The design axial strength based on the gross section is

$$\phi_t P_n = 0.9 F_y A_g$$
$$= (0.9) \left(36 \ \frac{\text{kips}}{\text{in}^2}\right) \left(7.08 \text{ in}^2\right)$$
$$= 229 \text{ kips}$$

The design axial strength based on the net section is

$$\phi_t P_n = 0.75 U F_u A_n$$
$$= (0.75)(0.90) \left(58 \ \frac{\text{kips}}{\text{in}^2}\right) \left(5.68 \text{ in}^2\right)$$
$$= 222 \text{ kips} \quad [\text{governs}]$$

3. From the collapse mechanism shown in part (a) of the illustration, the required plastic moment of resistance is given by the expression

$$4M_p = (100 \text{ kips})(10 \text{ ft}) + \frac{(200 \text{ kips})(10 \text{ ft})}{2}$$
$$M_p = \frac{2000 \text{ ft-kips}}{4}$$
$$= 500 \text{ ft-kips}$$

From LRFD Table 3-2, a W24 × 55 has

$$\phi M_p = 503 \text{ ft-kips}$$
$$> 500 \text{ ft-kips} \quad [\text{satisfactory}]$$

4. The factored bending moment at section 2 is

Due to distributed load,

$$
\begin{aligned}
&= \frac{0.75wL^2}{8} \\
&= \frac{(0.75)\left(4\ \dfrac{\text{kips}}{\text{ft}}\right)(40\ \text{ft})^2}{8} \\
&= 600\ \text{ft-kips}
\end{aligned}
$$

Due to point load,

$$
\begin{aligned}
&= \frac{Wab}{L} \\
&= \frac{(20\ \text{kips})(10\ \text{ft})(30\ \text{ft})}{40\ \text{ft}} \\
&= 150\ \text{ft-kips}
\end{aligned}
$$

Total $M_2 = 750$ ft-kips

The effective width of the concrete slab is the lesser of

- $$
\begin{aligned}
s &= (10\ \text{ft})\left(12\ \frac{\text{in}}{\text{ft}}\right) \\
&= 120\ \text{in}
\end{aligned}
$$

- $$
\begin{aligned}
\frac{L}{4} &= \frac{(40\ \text{ft})\left(12\ \dfrac{\text{in}}{\text{ft}}\right)}{4} \\
&= 120\ \text{in}\quad [\text{governs}]
\end{aligned}
$$

The nominal shear strength of a $^3/_4$ in diameter stud shear connector in 3000 lbf/in^2 concrete is obtained from LRFD Table 3-21, as

$$
Q_n = 21.0\ \text{kips}
$$

The total horizontal shear transferred between sections 1 and 2 is given by

$$
\begin{aligned}
V_h &= \sum Q_n \\
&= nQ_n \\
&= (17\ \text{studs})\,(21.0\ \text{kips}) \\
&= 357\ \text{kips} \\
&< F_y A_s
\end{aligned}
$$

For this value of $\sum Q_n$, the plastic neutral axis lies below the top of the W21 × 57 a distance obtained from LRFD Table 3-19, as

$$
Y_1 = 1.84\ \text{in}
$$

The depth of the concrete stress block is given by

$$
\begin{aligned}
a &= \frac{\sum Q_n}{0.85 f'_c b} \\
&= \frac{357\ \text{kips}}{(0.85)\left(3\ \dfrac{\text{kips}}{\text{in}^2}\right)(120\ \text{in})} \\
&= 1.17\ \text{in}
\end{aligned}
$$

The distance between the top of the steel beam and the centroid of the concrete slab force is

$$
\begin{aligned}
Y_2 &= Y_{\text{con}} - \frac{a}{2} \\
&= 7.5\ \text{in} - \frac{1.17\ \text{in}}{2} \\
&= 6.92\ \text{in}
\end{aligned}
$$

For these values of Y_1 and Y_2, the design flexural strength at section 2 is obtained from LRFD Table 3-19, as

$$
\begin{aligned}
\phi M_n &= 829\ \text{ft-kips} \\
&> 750\ \text{ft-kips}\quad [\text{satisfactory}]
\end{aligned}
$$

The shear connectors provided are adequate.

5. From the illustration, the gross shear area is

$$
\begin{aligned}
A_{gv} &= (2)(8\ \text{in})(0.625\ \text{in}) \\
&= 10\ \text{in}^2 \\
&= A_{nv}
\end{aligned}
$$

From the illustration, the net tension area is

$$
\begin{aligned}
A_{nt} &= (6\ \text{in})(0.625\ \text{in}) \\
&= 3.75\ \text{in}^2
\end{aligned}
$$

The tensile stress is uniform and the reduction coefficient is

$$
U_{bs} = 1.0
$$

The rupture strength in tension is given by

$$
\begin{aligned}
U_{bs}F_u A_{nt} &= (1.0)\left(58\ \frac{\text{kips}}{\text{in}^2}\right)(3.75\ \text{in}^2) \\
&= 218\ \text{kips}
\end{aligned}
$$

The rupture strength in shear is given by

$$
\begin{aligned}
0.6 F_u A_{nv} &= (0.6)\left(58\ \frac{\text{kips}}{\text{in}^2}\right)(10\ \text{in}^2) \\
&= 348\ \text{kips}
\end{aligned}
$$

The yield strength in shear is given by

$$
\begin{aligned}
0.6 F_y A_{gv} &= (0.6)\left(36\ \frac{\text{kips}}{\text{in}^2}\right)(10\ \text{in}^2) \\
&= 216\ \text{kips}\quad [\text{governs}] \\
&< 0.6 F_u A_{nv}
\end{aligned}
$$

Hence, shear yielding governs and the resistance to block shear is given by LRFD Eq. (J4-5) as

$$\phi R_n = \phi(0.6F_y A_{gv} + U_{bs}F_u A_{nt})$$
$$= (0.75)(216 \text{ kips} + 218 \text{ kips})$$
$$= 326 \text{ kips}$$

6. The total length of longitudinally loaded weld is

$$\ell_{wl} = (2)(8 \text{ in})$$
$$= 16 \text{ in}$$

The design shear capacity of a $^5/_{16}$ in fillet weld is

$$Q_w = Dq_u$$
$$= (5 \text{ sixteenths})\left(1.39 \frac{\text{kips}}{\text{in}} \text{ per } \frac{1}{16} \text{ in}\right)$$
$$= 6.95 \text{ kips/in}$$

Applying LRFD Eq. (J2-5), the design strength of the weld is

$$\phi R_n = \phi R_{wl}$$
$$= \ell_{wl}Q_w$$
$$= (16 \text{ in})\left(6.95 \frac{\text{kips}}{\text{in}}\right)$$
$$= 111 \text{ kips}$$

7. The width of the $^3/_4$ in plate is

$$w = 6 \text{ in}$$

The length of the weld is

$$\ell = 8 \text{ in}$$
$$\frac{\ell}{w} = \frac{8 \text{ in}}{6 \text{ in}}$$
$$= 1.33$$

From LRFD Table D3.1, the reduction coefficient is given by

$$U = 0.75$$

The gross area of the $^3/_4$ in plate is given by

$$A_g = wt$$
$$= (6 \text{ in})(0.75 \text{ in})$$
$$= 4.5 \text{ in}^2$$
$$= A_n$$

The effective net area is given by LRFD Eq. (D3-1) as

$$A_e = UA_n$$
$$= (0.75)(4.5 \text{ in}^2)$$
$$= 3.38 \text{ in}^2$$

The corresponding design tensile capacity for tensile rupture is given by LRFD Section D2 as

$$\phi_t P_n = 0.75F_u A_e$$
$$= (0.75)\left(58 \frac{\text{kips}}{\text{in}^2}\right)(3.38 \text{ in}^2)$$
$$= 147 \text{ kips}$$

The design tensile capacity for yielding of the gross section is given by LRFD Section D2 as

$$\phi_t P_n = 0.9F_y A_g$$
$$= (0.9)\left(36 \frac{\text{kips}}{\text{in}^2}\right)(4.5 \text{ in}^2)$$
$$= 146 \text{ kips} \quad \text{[governs]}$$

Steel

5 Timber Design

1. Adjustment Factors 5-1
2. Design for Flexure 5-6
3. Design for Shear 5-8
4. Design for Compression 5-11
5. Design for Tension 5-16
6. Design of Connections 5-17
   Practice Problems 5-26
   Solutions 5-27

1. ADJUSTMENT FACTORS

Nomenclature

b	breadth of rectangular bending member	in
c	column parameter	–
C_b	bearing area factor	–
C_c	curvature factor for structural glued laminated members	–
C_D	load duration factor	–
C_F	size factor for sawn lumber	–
C_{fu}	flat use factor	–
C_H	shear stress adjustment factor	–
C_i	incising factor	–
C_L	beam stability factor	–
C_M	wet service factor	–
C_P	column stability factor	–
C_r	repetitive member factor for dimension lumber	–
C_t	temperature factor	–
C_V	volume factor for structural glued laminated timber	–
d	least dimension of rectangular compression member	in
E, E'	reference and adjusted modulus of elasticity	lbf/in²
E_{min}, E'_{min}	reference and adjusted modulus of elasticity for beam stability and column stability calculations	lbf/in²
F	ratio of F_{bE} to F_b^*	–
F'	ratio of F_{cE} to F_c^*	–
F_b^*	tabulated bending design value multiplied by all applicable adjustment factors except C_{fu}, C_V, and C_L	lbf/in²
F_b, F_b'	reference and adjusted design value	lbf/in²
F_{bE}	critical buckling design value for bending members	lbf/in²
F_c^*	tabulated compressive design value multiplied by all applicable adjustment factors except CP	lbf/in²
F_c, F_c'	reference and adjusted design value parallel to grain	lbf/in²
F_{cE}	critical buckling design value	lbf/in²

l_b	length of bearing parallel to the grain of the wood	in
l_e	effective span length of bending member	ft or in
l_e	effective length of compression member	ft or in
l_u	laterally unsupported length of beam	ft or in
L	span length of bending member	ft or in
R	radius of curvature of inside face of laminations	in
R_B	slenderness ratio of bending member	
t	thickness of laminations	in
x	species parameter for volume factor	–

Definitions and Terminology

A description of the wood products available and of the terminology used is as follows.

Decking consists of solid sawn lumber or glued laminated members 2–4 in nominal thickness and 4 in or more wide. For 2 in thicknesses, it is usually single tongue and groove, and for 3–4 in thicknesses it may be double tongue and groove.

Dimension lumber consists of solid sawn lumber members with 2–4 in nominal thickness and 2 in or more wide.

Dressed size refers to the dimensions of a lumber member after it has been surfaced with a planing machine. It is usually $^1/_2$–$^3/_4$ in less than nominal size.

Grade indicates the classification wood products are given with respect to strength in accordance with specific grading rules.

Joist is a lumber member with 2–4 in nominal thickness, and 5 in or wider. A joist is typically loaded on the narrow face and used as framing in floors and roofs.

Lumber is cut to size in the sawmill and surfaced in a planing machine, and is not further processed.

Mechanically graded lumber is dimension lumber that has been individually evaluated in a testing machine. Load is applied to the piece of lumber, the deflection is measured, and the modulus of elasticity is calculated. The strength characteristics of the lumber are directly related to the modulus of elasticity and can be determined. A visual check is also made on the lumber to detect visible flaws.

Nominal size is the term used to specify the undressed size of a lumber member. The finished size of a member after dressing is normally $^1/_2$–$^3/_4$ in smaller than the

Timber

original size. Thus, a 2 in nominal × 4 in nominal member has actual dimensions of $1^1/_2$ in × $3^1/_2$ in.

Structural glued laminated timber, or glulams, are built up from wood laminations bonded together with adhesives. The grain of all laminations is parallel to the length of the beam and the laminations are typically $1^1/_2$ in thick.

Timbers are lumber members of nominal 5 in × 5 in or larger.

Visually stress-graded lumber are lumber members that have been graded visually to detect flaws and defects and assess the inherent strength of the member.

Wood structural panels are manufactured from veneers or wood strands bonded together with adhesives. Examples are plywood, oriented strand board, and composite panels.

Adjustment of Design Values

The reference design values for sawn lumber and glued laminated members are tabulated in NDS SUPP[1] Tables 4A through 4F and 5A through 5D. These tabulated reference design values are applicable to normal conditions of use as defined in NDS Sec. 2.2. For other conditions of use, the tabulated values are multiplied by adjustment factors, specified in NDS Sec. 2.3, to determine the relevant adjusted design values. A summary of the adjustment factors follows, and the applicability of each to the reference design values is shown in Table 5.1.

Adjustment Factors Applicable to Sawn Lumber and Glued Laminated Members

Load Duration Factor, C_D

The load duration factor is applicable to all reference design values with the exception of compression-perpendicular-to-the-grain and modulus of elasticity. Values of the load duration factor are given in Table 5.2.

Wet Service Factor, C_M

When the moisture content of sawn lumber exceeds 19%, the adjustment factors given in NDS SUPP Tables 4A and 4B are applicable to visually graded dimension lumber, in NDS SUPP Table 4C to machine graded dimension lumber, in NDS SUPP Table 4D to visually graded timbers, in NDS SUPP Table 4E to decking, and in NDS SUPP Table 4F to non-North American visually graded dimension lumber. When the moisture content of glued laminated members exceeds 16%, the adjustment factors given in NDS SUPP Tables 5A through 5D are applicable.

Table 5.1 Applicability of Adjustment Factors

adjustment factor		F_b	F_t	F_v	$F_{c\perp}$	F_c	E	E_{min}
C_D	load duration	√	√	√	–	√	–	–
C_M	wet service	√	√	√	√	√	√	√
C_b	bearing area	–	–	–	√	–	–	–
C_L	beam stability[a]	√	–	–	–	–	–	–
C_P	column stability	–	–	–	–	√	–	–
C_t	temperature	√	√	√	√	√	√	√
C_i	incising[b]	√	√	√	√	√	√	√
C_F	size[c]	√	√	–	–	√	–	–
C_r	repetitive member[d]	√	–	–	–	–	–	–
C_{fu}	flat use[e]	√	–	–	–	–	–	–
C_V	volume[a]	√	–	–	–	–	–	–
C_c	curvature[f]	√	–	–	–	–	–	–

NOTE:
a. When applied to glued laminated members, only the lesser value of C_L or C_V is applicable.
b. Applies to incised sawn lumber
c. Applies to visually graded sawn lumber
d. Applies to dimension lumber
e. Applies to dimension lumber and glued laminated members
f. Applies to curved glued laminated members

Table 5.2 Load Duration Factors

design load	C_D
dead load	0.90
occupancy live load	1.00
snow load	1.15
construction load	1.25
wind or earthquake load	1.60
impact load	2.00

Bearing Area Factor, C_b

For bearings less than 6 in in length and not less than 3 in from the end of a member, the reference design values for compression perpendicular to the grain are modified by the adjustment factor C_b. This is specified in NDS Sec. 3.10.4 as

$$C_b = \frac{l_b + 0.375}{l_b} \qquad \text{[NDS 3.10-2]}$$

Beam Stability Factor, C_L

The beam stability factor is applicable to the reference bending design value for sawn lumber and glued laminated members. For glued laminated members, C_L is not applied simultaneously with the volume factor C_V, and the lesser of these two factors is applicable. The beam stability factor is given by NDS Sec. 3.3.3 as

$$C_L = \frac{1.0 + F}{1.9} - \sqrt{\left(\frac{1.0 + F}{1.9}\right)^2 - \frac{F}{0.95}}$$

[NDS 3.3-6]

The variables are defined as

$$F = \frac{F_{bE}}{F_b^*}$$

F_b^* = reference bending design value multiplied by all applicable adjustment factors except C_{fu}, C_V, and C_L

$$= F_b C_D C_M C_t C_i C_F C_r C_c$$

$\begin{bmatrix} C_F \text{ applies only to visually graded sawn} \\ \text{lumber, } C_r \text{ applies only to dimension lum-} \\ \text{ber, } C_c \text{ applies only to curved glued lam-} \\ \text{inated members, and } C_i \text{ applies only to} \\ \text{sawn lumber.} \end{bmatrix}$

F_{bE} = critical buckling design value

$$= \frac{1.20 E_{\min}'}{R_B^2}$$

$E_{\min}'$ = adjusted modulus of elasticity for stability calculations

$$= E_{\min} C_M C_t C_i \quad [C_i \text{ applies only to sawn lumber.}]$$

R_B = slenderness ratio

$$= \sqrt{\frac{l_e d}{b^2}} \qquad \text{[NDS 3.3-5]}$$

$$= \leq 50$$

As specified in NDS Sec. 3.3.3, $C_L = 1$ when the depth of the beam does not exceed its breadth or when continuous lateral restraint is provided to the compression edge of a beam with the ends restrained against rotation.

The effective span length l_e is determined in accordance with NDS Table 3.3.3. The value of l_e depends on the loading configuration and the distance between lateral restraints l_u. Typical values for l_e are given in Fig. 5.1.

In accordance with NDS Sec. 4.4.1, $C_L = 1.0$ when, based on nominal dimensions,

- $\dfrac{d}{b} \leq 2$

- $2 < \dfrac{d}{b} \leq 4$ and full depth bracing is provided at the ends of the member

- $4 < \dfrac{d}{b} \leq 5$ and the compression edge is continuously restrained

- $5 < \dfrac{d}{b} \leq 6$ and the compression edge is continuously restrained with full depth bracing provided at a maximum of 8 ft centers

- $6 < \dfrac{d}{b} \leq 7$ and both edges are continuously restrained

Column Stability Factor, C_P

The column stability factor is applicable to the reference compression design values parallel to the grain and is specified by NDS Sec. 3.7.1 as

$$C_p = \frac{1.0 + F'}{2c} - \sqrt{\left(\frac{1.0 + F'}{2c}\right)^2 - \frac{F'}{c}}$$

$$\text{[NDS 3.7-1]}$$

The variables are defined as

$$F' = \frac{F_{cE}}{F_c^*}$$

F_c^* = reference compression design value multiplied by all applicable adjustment factors except C_P

$$= F_c C_D C_M C_t C_i C_F$$

$\begin{bmatrix} C_F \text{ applies only to visually graded sawn lumber,} \\ \text{and } C_i \text{ applies only to sawn lumber.} \end{bmatrix}$

F_c = reference compression design value parallel to grain

F_{cE} = critical buckling design value

$$= \frac{0.822 E_{\min}'}{\left(\dfrac{l_e}{d}\right)^2}$$

$E_{\min}'$ = adjusted modulus of elasticity for stability calculations

$$= E_{\min} C_M C_t C_i \quad [C_i \text{ applies only to sawn lumber.}]$$

$E_{\min}$ = reference modulus of elasticity

c = column parameter

$$= 0.8 \quad \text{[for sawn lumber]}$$

$$= 0.9 \quad \text{[for glued laminated timber]}$$

Temperature Factor, C_t

The temperature factor is applicable to all reference design values for members exposed to sustained temperatures exceeding 100°F and is specified by NDS Sec. 2.3.3.

Example 5.1

The flexural stresses produced in a visually graded lumber member for various loading conditions are as follows.

dead load = 500 lbf/in^2

dead + floor live load = 1000 lbf/in^2

dead + floor live + snow + wind load = 1500 lbf/in^2

Determine the governing load combination.

Solution

By applying the appropriate load duration factor, the normalized stress for the dead load loading condition is

$$f_D = \frac{500 \, \dfrac{\text{lbf}}{\text{in}^2}}{C_D} = \frac{500 \, \dfrac{\text{lbf}}{\text{in}^2}}{0.9}$$

$$= 556 \text{ lbf/in}^2$$

Timber

Figure 5.1 *Typical Values of Effective Length, l_e*

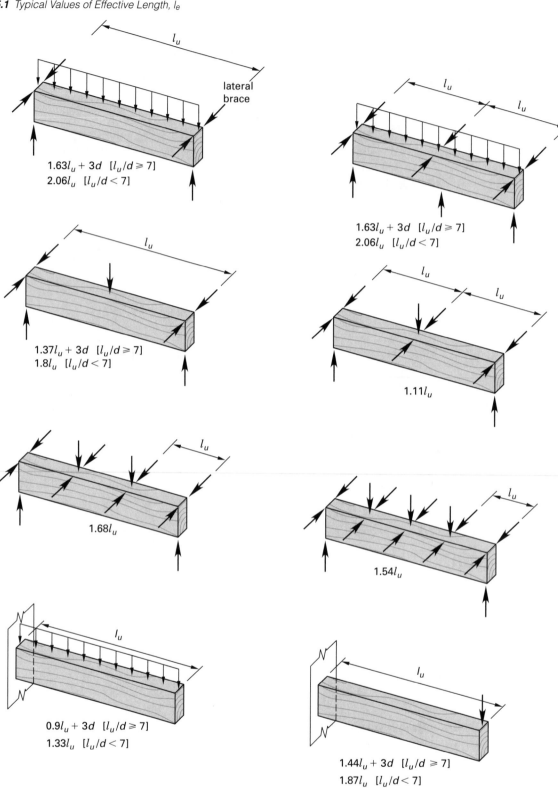

By applying the appropriate load duration factor, the normalized stress for the dead plus floor live load loading condition is

$$f_{D+L} = \frac{1000 \, \frac{\text{lbf}}{\text{in}^2}}{C_D} = \frac{1000 \, \frac{\text{lbf}}{\text{in}^2}}{1.0}$$
$$= 1000 \, \text{lbf/in}^2$$

The appropriate load combination for dead plus floor live plus snow plus wind is obtained from ASCE[2] Sec. 2.4.1 as

$$D + 0.75(L + S + W) = 500 \, \frac{\text{lbf}}{\text{in}^2} + (0.75) \left(1000 \, \frac{\text{lbf}}{\text{in}^2}\right)$$
$$= 1250 \, \text{lbf/in}^2$$

The normalized stress is obtained by applying the apropriate load duration factor to give

$$f_{D+L+S+W} = \frac{1250 \, \frac{\text{lbf}}{\text{in}^2}}{1.6}$$
$$= 781 \, \text{lbf/in}^2$$

The load combination dead plus live load governs.

Adjustment Factors Applicable to Sawn Lumber

Incising Factor, C_i

Values of the incising factor for a prescribed incising pattern are provided in NDS Sec. 4.3.8. These values are applicable to all design values of all sawn lumber.

Adjustment Factors Applicable to Visually Graded Sawn Lumber

Size Factor, C_F

The size factor is applicable to the design values for bending, tension, and compression. For sawn lumber 2 to 4 in thick, values of the size factor are given in NDS SUPP Tables 4A and 4B. For members exceeding 12 in depth and 5 in thickness, the size factor is specified in NDS Sec. 4.3.6 as

$$C_F = \left(\frac{12}{d}\right)^{1/9} \qquad \text{[NDS 4.3-1]}$$

Adjustment Factors Applicable to Dimension Lumber

Repetitive Member Factor, C_r

The repetitive member factor is applicable to the design value for bending when three or more sawn lumber elements not more than 4 in thick and spaced not more than 24 in apart are joined by a transverse load distributing element. The value of the repetitive member factor is given in NDS Sec. 4.3.9 as $C_r = 1.15$.

Adjustment Factors Applicable to Dimension Lumber and Glued Laminated Members

Flat Use Factor, C_{fu}

When wood members are loaded flatwise, flat use adjustment factors are applied to the bending stress. The adjustment factors given in NDS SUPP Tables 4A and 4B are applicable to visually graded dimension lumber, in NDS SUPP Table 4C to machine graded dimension lumber, and in NDS SUPP Tables 5A through 5D to glued laminated members.

Adjustment Factors Applicable to Glued Laminated Members

Volume Factor, C_V

The volume factor is applicable to the reference design value for bending and is not applied simultaneously with the beam stability factor C_L; the lesser of these two factors is applicable. The volume factor is defined in NDS Sec. 5.3.6 as

$$C_V = \left(\frac{1291.5}{bdL}\right)^{1/x} \qquad \text{[NDS 5.3-1]}$$

The variables are defined as

L = length of beam between points of zero moment, ft
b = beam width, in
d = beam depth, in
x = 20 [for Southern Pine]
   = 10 [for all other species]

Curvature Factor, C_c

To account for residual stresses in curved, glued laminated members, the curvature factor is specified in NDS Sec. 5.3.8 as

$$C_c = 1 - (2000) \left(\frac{t}{R}\right)^2 \qquad \text{[NDS 5.3-2]}$$

Example 5.2

A curved, glued laminated beam of combination 24F–V5 (24F–1.7E) western species with 1.5 in thick laminations has a radius of curvature of 30 ft, a width of $6^3/_4$ in, and a depth of 30 in. The beam has continuous lateral support, has a moisture content exceeding 16%, and is subjected to sustained temperatures between 100°F and 125°F. The governing loading combination is dead plus live load, and the span is 40 ft. The beam is simply supported, and all loading is uniformly distributed. Determine the allowable design values in bending, shear, and modulus of elasticity.

Solution

The reference design values for bending, shear, and modulus of elasticity are tabulated in NDS SUPP Table 5A and are

$$F_b = 2400 \text{ lbf/in}^2$$
$$F_v = 215 \text{ lbf/in}^2$$
$$E = 1.7 \times 10^6 \text{ lbf/in}^2$$
$$C_D = 1.0$$

The applicable adjustment factors for bending stress are as follows.

$$C_M = \text{wet service factor}$$
$$= 0.80 \quad [\text{NDS SUPP Table 5A}]$$

$$C_t = \text{temperature factor for wet}$$
$$\text{conditions}$$
$$= 0.7 \quad [\text{NDS Table 2.3.3}]$$

$$C_V = \text{volume factor}$$
$$= \left(\frac{1291.5}{bdL} \right)^{1/x}$$
$$= \left(\frac{1291.5 \text{ in}^2\text{-ft}}{(6.75)(30 \text{ in})(40 \text{ ft})} \right)^{1/10}$$
$$= 0.832$$

$$C_L = \text{stability factor}$$
$$= 1.0 \quad [\text{continuous lateral support}]$$
$$> \text{volume factor}$$

Hence, the volume factor governs.

$$C_c = \text{curvature factor}$$
$$= 1 - (2000) \left(\frac{t}{R} \right)^2 \qquad [\text{NDS Sec. 5.3.8}]$$
$$= 1 - (2000) \left(\frac{1.5 \text{ in}}{(30 \text{ ft}) \left(12 \frac{\text{in}}{\text{ft}} \right)} \right)^2$$
$$= 0.965$$

The adjusted bending stress is

$$F_b' = F_b C_M C_t C_V C_c$$
$$= \left(2400 \frac{\text{lbf}}{\text{in}^2} \right) (0.8)(0.7)(0.832)(0.965)$$
$$= 1079 \text{ lbf/in}^2$$

The applicable adjustment factors for shear stress are as follows.

$$C_M = \text{wet service factor}$$
$$= 0.875 \quad [\text{NDS SUPP Table 5A}]$$

$$C_t = \text{temperature factor for wet conditions}$$
$$= 0.7 \quad [\text{NDS Table 2.3.3}]$$

The adjusted shear stress is

$$F_v' = F_v C_M C_t$$
$$= \left(215 \frac{\text{lbf}}{\text{in}^2} \right) (0.875)(0.70)$$
$$= 132 \text{ lbf/in}^2$$

The applicable adjustment factors for modulus of elasticity are as follows.

$$C_M = \text{wet service factor}$$
$$= 0.833 \quad [\text{NDS SUPP Table 5A}]$$

$$C_t = \text{temperature factor for wet}$$
$$\text{or dry conditions}$$
$$= 0.9 \quad [\text{NDS Table 2.3.3}]$$

The adjusted modulus of elasticity is

$$E' = E C_M C_t$$
$$= \left(1.7 \times 10^6 \frac{\text{lbf}}{\text{in}^2} \right) (0.833)(0.9)$$
$$= 1.27 \times 10^6 \text{ lbf/in}^2$$

2. DESIGN FOR FLEXURE

General Requirements Applicable to Sawn Lumber and Glued Laminated Members

For all flexural members, in accordance with NDS Sec. 3.2.1, the beam span is taken as the clear span plus one-half the required bearing length at each end.

When the depth of a beam does not exceed its breadth or when continuous lateral restraint is provided to the compression edge of a beam with the ends restrained against rotation, the beam stability factor $C_L = 1.0$. For other situations, the value of C_L is calculated in accordance with NDS Sec. 3.3.3, and the effective span length l_e is determined in accordance with NDS Table 3.3.3. The value of l_e depends on the loading configuration and the distance between lateral restraints l_u. Typical values for l_e are given in Fig. 5.1.

Requirements Applicable to Visually Graded Sawn Lumber

For visually graded sawn lumber, both the stability factor C_L and the size factor C_F must be considered concurrently.

Example 5.3

A select structural Douglas Fir-Larch 4×12 beam is simply supported over a span of 20 ft. The governing load combination is a uniformly distributed dead plus live load, and the beam is laterally braced at midspan.

The beam is incised with the prescribed pattern specified in NDS Sec. 4.3.8. Determine the allowable design value in bending.

Solution

The reference design values for bending and modulus of elasticity for beam stability calculations are tabulated in NDS SUPP Table 4A and are

$$F_b = 1500 \text{ lbf/in}^2$$

$$E_{min} = 0.69 \times 10^6 \text{ lbf/in}^2$$

$$C_D = 1.0, \ C_M = 1.0, \ C_t = 1.0, \ C_r = 1.0$$

The applicable incising factor for modulus of elasticity is obtained from NDS Sec. 4.3.8 and is

$$C_i = 0.95$$

The adjusted modulus of elasticity for stability calculations is

$$E'_{min} = E_{min} C_i$$

$$= \left(0.69 \times 10^6 \ \frac{\text{lbf}}{\text{in}^2} \right) (0.95)$$

$$= 0.656 \times 10^6 \text{ lbf/in}^2$$

The distance between lateral restraints is

$$l_u = \frac{20 \text{ ft}}{2}$$

$$= 10 \text{ ft}$$

$$\frac{l_u}{d} = \frac{(10 \text{ ft}) \left(12 \ \frac{\text{in}}{\text{ft}} \right)}{11.25 \text{ in}}$$

$$= 10.7$$

$$> 7$$

For a uniformly distributed load and a l_u/d ratio > 7, the effective length is obtained from Fig. 5.1 as

$$l_e = 1.63 l_u + 3d$$

$$= (1.63)(10 \text{ ft}) \left(12 \ \frac{\text{in}}{\text{ft}} \right) + (3)(11.25 \text{ in})$$

$$= 229.4 \text{ in}$$

The slenderness ratio is given by NDS Sec. 3.3.3 as

$$R_B = \sqrt{\frac{l_e d}{b^2}}$$

$$= \sqrt{\frac{(229.4 \text{ in})(11.25 \text{ in})}{(3.5 \text{ in})^2}}$$

$$= 14.52$$

$$< 50 \quad \left[\begin{array}{c} \text{satisfies criteria} \\ \text{of NDS Sec. 3.3.3} \end{array} \right]$$

The critical buckling design value is

$$F_{bE} = \frac{1.20 E'_{min}}{R_B^2}$$

$$= \frac{(1.20) \left(0.656 \times 10^6 \ \dfrac{\text{lbf}}{\text{in}^2} \right)}{(14.52)^2}$$

$$= 3734 \text{ lbf/in}^2$$

The applicable incising factor for flexure is obtained from NDS Sec. 4.3.8 and is

$$C_i = 0.80$$

The applicable size factor is obtained from NDS SUPP Table 4A and is

$$C_F = 1.1$$

The reference flexural design value multiplied by all applicable adjustment factors except C_L is

$$F_b^* = F_b C_i C_F$$

$$= \left(1500 \ \frac{\text{lbf}}{\text{in}^2} \right) (0.80)(1.1)$$

$$= 1320 \text{ lbf/in}^2$$

$$F = \frac{F_{bE}}{F_b^*}$$

$$= \frac{3731 \ \dfrac{\text{lbf}}{\text{in}^2}}{1320 \ \dfrac{\text{lbf}}{\text{in}^2}}$$

$$= 2.83$$

The beam stability factor is given by NDS Sec. 3.3.3 as

$$C_L = \frac{1.0 + F}{1.9} - \sqrt{\left(\frac{1.0 + F}{1.9} \right)^2 - \frac{F}{0.95}}$$

$$= 0.974$$

The allowable flexural design value is

$$F_b' = C_L C_i C_F F_b$$

$$= (0.974)(0.80)(1.1) \left(1500 \ \frac{\text{lbf}}{\text{in}^2} \right)$$

$$= 1286 \text{ lbf/in}^2$$

Requirements Applicable to Glued Laminated Members

For glued laminated members, both the stability factor C_L and the volume factor C_V must be determined. Only the lesser of these two factors is applicable in determining the allowable design value in bending.

Example 5.4

A glued laminated $6^3/_4 \times 30$ beam of combination 24F–V5 (24F–1.7E) western species is simply supported over a span of 40 ft. The governing load combination is a uniformly distributed dead plus live load, and the beam is laterally braced at midspan. Determine the allowable design value in bending.

Solution

$$C_D = 1.0, C_M = 1.0, C_t = 1.0, C_c = 1.0$$
$$F_b = 2400 \text{ lbf/in}^2$$
$$E_{x(\text{min})} = 0.88 \times 10^6 \text{ lbf/in}^2$$
$$E_{y(\text{min})} = 0.78 \times 10^6 \text{ lbf/in}^2$$

From Ex. 5.2, the volume factor is

$$C_V = 0.832$$

The adjusted modulus of elasticity for stability calculations is

$$E'_{\text{min}} = E'_{y(\text{min})}$$
$$= 0.78 \times 10^6 \text{ lbf/in}^2$$

The distance between lateral restraints is

$$l_u = \frac{40 \text{ ft}}{2}$$
$$= 20 \text{ ft}$$

$$\frac{l_u}{d} = \frac{(20 \text{ ft}) \left(12 \dfrac{\text{in}}{\text{ft}} \right)}{30 \text{ in}}$$
$$= 8.0$$
$$> 7$$

For a uniformly distributed load and an l_u/d ratio > 7, the effective length is obtained from Fig. 5.1 as

$$l_e = 1.63 l_u + 3d$$
$$= (1.63)(20 \text{ ft}) \left(12 \dfrac{\text{in}}{\text{ft}} \right) + (3)(30 \text{ in})$$
$$= 481 \text{ in}$$

The slenderness ratio is given by NDS Eq. (3.3-5)

$$R_B = \sqrt{\frac{l_e d}{b^2}}$$
$$= \sqrt{\frac{(481 \text{ in})(30 \text{ in})}{(6.75 \text{ in})^2}}$$
$$= 17.80$$
$$< 50 \quad \text{[satisfies criteria of NDS Sec. 3.3.3]}$$

The critical buckling design value is

$$F_{bE} = \frac{1.20 E'_{\text{min}}}{R_B^2}$$
$$= \frac{(1.20) \left(0.78 \times 10^6 \dfrac{\text{lbf}}{\text{in}^2} \right)}{(17.80)^2}$$
$$= 2954 \text{ lbf/in}^2$$

No adjustment factors are to be applied to the reference flexural design value, and

$$F_b^* = F_b$$
$$= 2400 \text{ lbf/in}^2$$

$$F = \frac{F_{bE}}{F_b^*}$$
$$= \frac{2954 \dfrac{\text{lbf}}{\text{in}^2}}{2400 \dfrac{\text{lbf}}{\text{in}^2}}$$
$$= 1.23$$

The beam stability factor is given by NDS Sec. 3.3.3 as

$$C_L = \frac{1.0 + F}{1.9} - \sqrt{\left(\frac{1.0 + F}{1.9} \right)^2 - \frac{F}{0.95}}$$
$$= 0.89$$
$$> \text{volume factor derived in Ex. 5.2}$$

Hence, the volume factor governs. The allowable flexural design value is

$$F'_b = C_V F_b$$
$$= (0.832) \left(2400 \dfrac{\text{lbf}}{\text{in}^2} \right)$$
$$= 1997 \text{ lbf/in}^2$$

3. DESIGN FOR SHEAR

Nomenclature

d	depth of unnotched bending member	in
d_e	depth of member, less the distance from the unloaded edge of the member to the nearest edge of the nearest split ring or shear plate connector	in
d_e	depth of member, less the distance from the unloaded edge of the member to the center of the nearest bolt or lag screw	in

d_n	depth of member remaining at a notch	in
e	distance a notch extends past the inner edge of a support	in
f_v	actual shear stress parallel to grain	lbf/in^2
l_n	length of notch	in
V	shear force	lbf
V_r, V_r'	reference and adjusted design shear	lbf
x	distance from beam support face to load	in

General Requirements Applicable to Sawn Lumber and Glued Laminated Members

The shear stress in a rectangular beam is defined in NDS Sec. 3.4.2 as

$$f_v = \frac{1.5V}{bd} \qquad \text{[NDS 3.4-2]}$$

In determining the shear force on the member, uniformly distributed loads applied to the top of the beam within a distance from either support equal to the depth of the beam are ignored, as shown in Fig. 5.2.

Figure 5.2 *Shear Determination in a Beam*

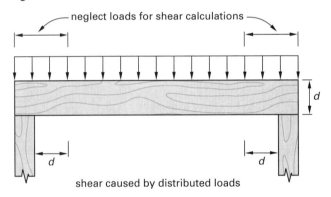

shear caused by distributed loads

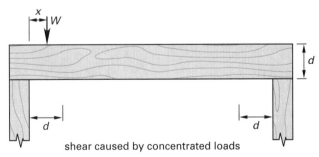

shear caused by concentrated loads

The procedure for a beam with concentrated loads is illustrated in Fig. 5.2. Concentrated loads within a distance from either support equal to the depth of the beam are multiplied by x/d, where x is the distance from the support to the load, to give an equivalent shear force.

Notches in a beam reduce the shear capacity, and NDS Sec. 3.2.3 imposes restrictions on their size and location, as shown in Fig. 5.3. The adjusted design shear at a

notch on the tension side of a beam is given by NDS Sec. 3.4.3.2 as

$$V_r' = \left(\frac{F_v' b d_n}{1.5} \right) \left(\frac{d_n}{d} \right)^2 \qquad \text{[NDS 3.4-3]}$$

The adjusted design shear at a notch on the compression side of a beam is given by NDS Sec. 3.4.3.2 as

$$V_r' = \left(\frac{F_v' b}{1.5} \right) \left(d - \left(\frac{d - d_n}{d_n} \right) e \right) \qquad \text{[NDS 3.4-5]}$$

For a connection less than five times the depth of the member from its end, as shown in Fig. 5.4, the adjusted design shear is given by NDS Sec. 3.4.3.3 as

$$V_r' = \left(\frac{F_v' b d_e}{1.5} \right) \left(\frac{d_e}{d} \right)^2 \qquad \text{[NDS 3.4-6]}$$

When the connection is at least five times the depth of the member from its end, the adjusted design shear is given by

$$V_r' = \frac{F_v' b d_e}{1.5} \qquad \text{[NDS 3.4-7]}$$

To facilitate the selection of glued laminated beam sections, tables are available[3] that provide shear and bending capacities of sections. For lumber joists, tables are available[4] that assist in the selection of a joist size for various span and live load combinations.

Example 5.5

A glued laminated $6^3/_4 \times 30$ beam of combination 24F–V5 western species is notched and loaded as shown in the following illustration. The beam has a moisture content exceeding 16% and is subjected to sustained temperatures between 100°F and 125°F. The governing load combination is dead plus live load. Determine the maximum allowable shear force at each support and at the hanger connection.

Solution

$$C_D = 1.0$$

From Ex. 5.2, the allowable shear design value is

$$F_v' = C_M C_t F_v$$

$$= 132 \text{ lbf/in}^2$$

At the left support and from NDS Eq. (3.4-3), the allowable shear force is

$$V_r' = \left(\frac{b d_n F_v'}{1.5} \right) \left(\frac{d_n}{d} \right)^2$$

$$= \left(\frac{(6.75 \text{ in})(27 \text{ in}) \left(132 \, \dfrac{\text{lbf}}{\text{in}^2} \right)}{1.5} \right) \left(\frac{27 \text{ in}}{30 \text{ in}} \right)^2$$

$$= 12{,}991 \text{ lbf}$$

Figure 5.3 *Notched Beams*

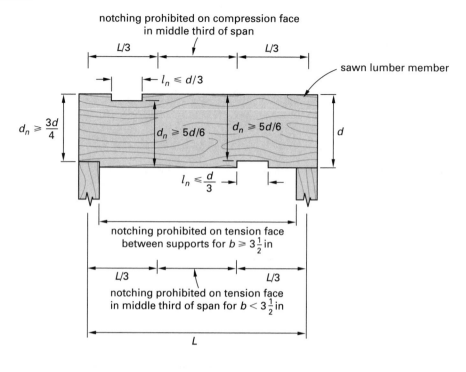

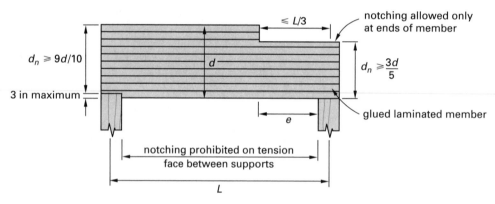

Figure 5.4 *Bolted Connections*

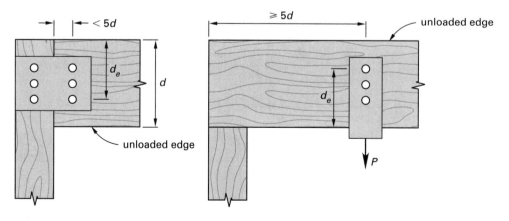

Illustration for Ex. 5.5

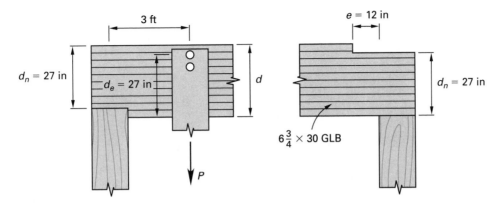

At the right support, $e < d_n$, and from NDS (Eq. 3.4-5), the allowable shear force is

$$V_r' = \frac{F_v'\left(b\left(d - \frac{e(d-d_n)}{d_n}\right)\right)}{1.5}$$

$$= \frac{\left(132 \ \frac{\text{lbf}}{\text{in}^2}\right)(6.75 \text{ in})\left(30 \text{ in} - \frac{(12 \text{ in})(3 \text{ in})}{27 \text{ in}}\right)}{1.5}$$

$$= 17{,}028 \text{ lbf}$$

The hanger connection is less than $5d$ from the end of the beam, and from NDS (Eq. 3.4-6) the allowable shear force is

$$V_r' = \left(\frac{bd_e F_v'}{1.5}\right)\left(\frac{d_e}{d}\right)^2$$

$$= \left(\frac{(6.75 \text{ in})(27 \text{ in})\left(132 \ \frac{\text{lbf}}{\text{in}^2}\right)}{1.5}\right)\left(\frac{27 \text{ in}}{30 \text{ in}}\right)^2$$

$$= 12{,}991 \text{ lbf}$$

Requirements Applicable to Sawn Lumber

The value of the incising factor for a prescribed incising pattern is provided in NDS Sec. 4.3.8.

Example 5.6

A select structural Douglas Fir-Larch 4 × 12 joist is simply supported over a span of 20 ft. The governing load combination is a uniformly distributed dead plus live load, and the joist is incised with the prescribed pattern specified in NDS Sec. 4.3.8. Determine the allowable design value in shear.

Solution

$$C_D = 1.0, \ C_M = 1.0, \ C_t = 1.0$$

The reference design value for shear is tabulated in NDS SUPP Table 4A and is

$$F_v = 180 \text{ lbf/in}^2$$

The applicable incising factor for shear is obtained from NDS Sec. 4.3.8 and is

$$C_i = 0.80$$

The allowable shear design value is

$$F_v' = C_i F_v = (0.8)\left(180 \ \frac{\text{lbf}}{\text{in}^2}\right)$$

$$= 144 \text{ lbf/in}^2$$

4. DESIGN FOR COMPRESSION

Nomenclature

A	area of cross section	in^2
C_{m1}	moment magnification factor for biaxial bending and axial compression, $1.0 - f_c/F_{cE1}$	–
C_{m2}	moment magnification factor for biaxial bending and axial compression, $1.0 - f_c/F_{cE2} - (f_{b1}/F_{bE})^2$	–
C_{m3}	moment magnification factor for axial compression and flexure with load applied to narrow face, $1.0 - f_c/F_{cE1}$	–
C_{m4}	moment magnification factor for axial compression and flexure with load applied to wide face, $1.0 - f_c/F_{cE2}$	–

Timber

C_{m5}	moment magnification factor for biaxial bending, $1.0 - (f_{b1}/F_{bE})^2$	–
d_1	dimension of wide face	in
d_2	dimension of narrow face	in
f_{b1}	actual edgewise bending stress for load applied to the narrow face	lbf/in^2
f_{b2}	actual flatwise bending stress for load applied to the wide face	lbf/in^2
f_c	actual compression stress parallel to grain	lbf/in^2
F_{bE}	critical buckling design value for bending member, $1.20E'_{min}/(R_B)^2$	lbf/in^2
F'_{b1}	allowable bending design value for load applied to the narrow face, including adjustment for slenderness ratio	lbf/in^2
F'_{b2}	allowable bending design value for load applied to the wide face, including adjustment for slenderness ratio	lbf/in^2
F'_c	allowable compression design value, including adjustment for largest slenderness ratio	lbf/in^2
F_{cE1}	critical buckling design value in plane of bending for load applied to the narrow face, $0.822E'_{min}/(l_{e1}/d_1)^2$	lbf/in^2
F_{cE2}	critical buckling design value in plane of bending for load applied to the wide face, $0.822E'_{min}/(l_{e2}/d_2)^2$	lbf/in^2
K_e	buckling length coefficient for compression members	–
l_1	distance between points of lateral support restraining buckling about the strong axis of compression member	ft or in
l_2	distance between points of lateral support restraining buckling about the weak axis of compression member	ft or in
l_{e1}	effective length between supports restraining buckling in plane of bending from load applied to narrow face of compression member, $K_e l_1$	ft or in
l_{e2}	effective length between supports restraining buckling in plane of bending from load applied to wide face of compression member, $K_e l_2$	ft or in
l_{e1}/d_1	slenderness ratio about the strong axis of compression member	–
l_{e2}/d_2	slenderness ratio about the weak axis of compression member	–
P	total concentrated load or total axial load	lbf or kips

Axial Load Only

Values of the buckling length coefficient K_e for compression members, for various restraint conditions, are given in NDS Table G1 and are summarized in Fig. 5.5. The effective length is defined in NDS Sec. 3.7.1.2 as

$$l_e = K_e l$$

The slenderness ratio is specified in NDS Sec. 3.7.1.4 as

$$\frac{l_e}{d} \leq 50$$

The applicable allowable compression design value F'_c is governed by the maximum slenderness ratio of a member, and the maximum allowable axial load on a column is given by

$$P_{al} = AF'_c$$

Figure 5.5 Buckling Length Coefficients

		K_e	
illus.	end conditions	theoretical	design
(a)	both ends pinned	1	1.00
(b)	both ends built in	0.5	0.65
(c)	one end pinned, one end built in	0.7	0.8
(d)	one end built in, one end free	2	2.10
(e)	one end built in, one end fixed against rotation but free to translate	1	1.20
(f)	one end pinned, one end fixed against rotation but free to translate	2	2.40

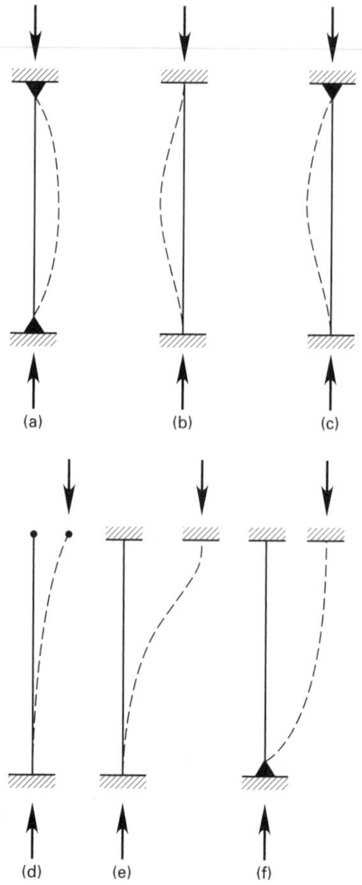

Example 5.7

The select structural 2×6 Douglas Fir-Larch top chord of a truss is loaded as shown in the following illustration. The governing load combination consists of dead plus live load, and the moisture content exceeds 19%. The chord is laterally braced at midlength about the weak axis, and the self-weight of the chord and bracing members may be neglected. Determine whether the member is adequate.

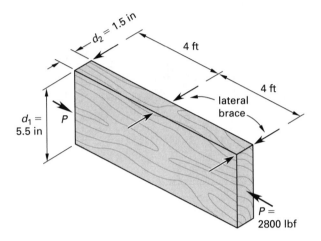

Solution

The reference design values for compression and modulus of elasticity are tabulated in NDS SUPP Table 4A and are

$$F_c = 1700 \text{ lbf/in}^2$$
$$E_{\min} = 0.69 \times 10^6 \text{ lbf/in}^2$$
$$C_D = 1.0, \ C_t = 1.0, \ C_i = 1.0$$

The applicable adjustment factors for compression and modulus of elasticity are as follows.

C_M = wet service factor from

NDS SUPP Table 4A

= 0.80 [compression member]

= 0.90 [modulus of elasticity]

C_F = size factor from

NDS SUPP Table 4A

= 1.1 [compression member]

From Fig. 5.5, the slenderness ratio about the strong axis is

$$\frac{K_e l_1}{d_1} = \frac{(1.0)(8 \text{ ft})\left(12 \ \frac{\text{in}}{\text{ft}}\right)}{5.5 \text{ in}}$$
$$= 17.46$$

From Fig. 5.5, the slenderness ratio about the weak axis is

$$\frac{K_e l_2}{d_2} = \frac{(1.0)(4 \text{ ft})\left(12 \ \frac{\text{in}}{\text{ft}}\right)}{1.5 \text{ in}}$$
$$= 32.00 \quad \text{[governs]}$$

The adjusted modulus of elasticity for stability calculations is

$$E'_{\min} = E_{\min} C_M = \left(0.69 \times 10^6 \ \frac{\text{lbf}}{\text{in}^2}\right)(0.9)$$
$$= 0.62 \times 10^6 \text{ lbf/in}^2$$

The reference compression design value multiplied by all applicable adjustment factors except C_P is given by

$$F_c^* = F_c C_M C_F = \left(1700 \ \frac{\text{lbf}}{\text{in}^2}\right)(0.8)(1.1)$$
$$= 1496 \text{ lbf/in}^2$$

The critical buckling design value is

$$F_{cE2} = \frac{0.822 E'_{\min}}{\left(\frac{l_{e2}}{d_2}\right)^2} = \frac{(0.822)\left(0.62 \times 10^6 \ \frac{\text{lbf}}{\text{in}^2}\right)}{(32.00)^2}$$
$$= 498 \text{ lbf/in}^2$$

The ratio of F_{cE2} to F_c^* is

$$F' = \frac{F_{cE2}}{F_c^*} = \frac{498 \ \frac{\text{lbf}}{\text{in}^2}}{1496 \ \frac{\text{lbf}}{\text{in}^2}}$$
$$= 0.333$$

The column parameter is obtained from NDS Sec. 3.7.1.5 as

$$c = 0.8 \quad \text{[for sawn lumber]}$$

The column stability factor is specified by NDS Sec. 3.7.1 as

$$C_P = \frac{1.0 + F'}{2c} - \sqrt{\left(\frac{1.0 + F'}{2c}\right)^2 - \frac{F'}{c}}$$
$$= \frac{1.333}{(2)(0.8)} - \sqrt{\left(\frac{1.333}{(2)(0.8)}\right)^2 - \frac{0.333}{0.8}}$$
$$= 0.31$$

The allowable compression design value parallel to grain is

$$F'_c = F_c C_M C_F C_P$$

$$= \left(1700 \ \frac{\text{lbf}}{\text{in}^2}\right)(0.8)(1.1)(0.31)$$

$$= 464 \ \text{lbf/in}^2$$

The actual compression stress on the chord is given by

$$f_c = \frac{P}{A} = \frac{2800 \ \text{lbf}}{8.25 \ \text{in}^2}$$

$$= 339 \ \text{lbf/in}^2$$

$$< F'_c$$

The chord is adequate.

Combined Axial Compression and Flexure

Members subjected to combined compression and flexural stresses due to axial and transverse loading must satisfy the interaction equations given in NDS Sec. 3.9.2 as

$$\left(\frac{f_c}{F'_c}\right)^2 + \frac{f_{b1}}{F'_{b1}C_{m1}} + \frac{f_{b2}}{F'_{b2}C_{m2}} \leq 1.00 \qquad [C3.9\text{-}3]$$

For bending load applied to the narrow face of the member and concentric axial compression load, the interaction equation reduces to

$$\left(\frac{f_c}{F'_c}\right)^2 + \frac{f_{b1}}{F'_{b1}C_{m3}} \leq 1.00$$

For bending load applied to the wide face of the member and concentric axial compression load, the equation reduces to

$$\left(\frac{f_c}{F'_c}\right)^2 + \frac{f_{b2}}{F'_{b2}C_{m4}} \leq 1.00$$

For bending loads applied to the narrow and wide faces of the member and no concentric axial load, the equation reduces to

$$\frac{f_{b1}}{F'_{b1}} + \frac{f_{b2}}{F'_{b2}C_{m5}} \leq 1.00$$

Example 5.8

The select structural 2 × 6 Douglas Fir-Larch top chord of a truss is loaded as shown in the following illustration. The governing load combination consists of dead plus live load, and the moisture content exceeds 19%. The chord is laterally braced at midlength about the weak axis, and the self-weight of the chord and bracing members may be neglected. Determine whether the member is adequate.

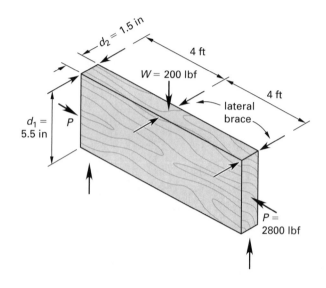

Solution

The reference design values for bending and modulus of elasticity are tabulated in NDS SUPP Table 4A and are

$$F_b = 1500 \ \text{lbf/in}^2$$

$$E_{\min} = 0.69 \times 10^6 \ \text{lbf/in}^2$$

$$C_D = 1.0, \ C_t = 1.0, \ C_i = 1.0, \ C_r = 1.0$$

The adjusted modulus of elasticity for stability calculations is

$$E'_{\min} = E_{\min} C_M$$

$$= 0.62 \times 10^6 \ \text{lbf/in}^2 \quad [\text{from Ex. 5.7}]$$

The distance between lateral restraints is

$$l_u = \frac{8 \ \text{ft}}{2}$$

$$= 4 \ \text{ft}$$

For a concentrated load at midspan and with lateral restraint at midspan, the effective length is obtained from Fig. 5.1 as

$$l_e = 1.11 l_u$$

$$= (1.11)(4 \ \text{ft})\left(12 \ \frac{\text{in}}{\text{ft}}\right)$$

$$= 53.28 \ \text{in}$$

The slenderness ratio is given by NDS Sec. 3.3.3 as

$$R_B = \sqrt{\frac{l_e d_1}{d_2^2}}$$

$$= \sqrt{\frac{(53.28 \text{ in}) (5.5 \text{ in})}{(1.5 \text{ in})^2}}$$

$$= 11.41$$

$$< 50 \quad \begin{bmatrix} \text{satisfies criteria of} \\ \text{NDS Sec. 3.3.3} \end{bmatrix}$$

The critical buckling design value is

$$F_{bE} = \frac{1.20 E'_{min}}{R_B^2}$$

$$= \frac{(1.20) \left(0.62 \times 10^6 \dfrac{\text{lbf}}{\text{in}^2} \right)}{(11.41)^2}$$

$$= 5715 \text{ lbf/in}^2$$

The applicable wet service factor for flexure is obtained from NDS SUPP Table 4A and is

$$C_M = 0.85$$

The applicable size factor for flexure is obtained from NDS SUPP Table 4A and is

$$C_F = 1.3$$

The reference flexural design value multiplied by all applicable adjustment factors except C_L is

$$F_b^* = F_b C_M C_F$$

$$= \left(1500 \dfrac{\text{lbf}}{\text{in}^2} \right) (0.85) (1.3)$$

$$= 1657 \text{ lbf/in}^2$$

$$F = \frac{F_{bE}}{F_b^*}$$

$$= \frac{5715 \dfrac{\text{lbf}}{\text{in}^2}}{1657 \dfrac{\text{lbf}}{\text{in}^2}}$$

$$= 3.45$$

The beam stability factor is given by NDS Sec. 3.3.3 as

$$C_L = \frac{1.0 + F}{1.9} - \sqrt{\left(\frac{1.0 + F}{1.9} \right)^2 - \frac{F}{0.95}}$$

$$= \frac{4.45}{1.9} - \sqrt{\left(\frac{4.45}{1.9} \right)^2 - \frac{3.45}{0.95}}$$

$$= 0.98$$

The allowable flexural design value for load applied to the narrow face is

$$F'_{b1} = F_b C_M C_L C_F$$

$$= \left(1500 \dfrac{\text{lbf}}{\text{in}^2} \right) (0.85) (0.98) (1.3)$$

$$= 1626 \text{ lbf/in}^2$$

The actual edgewise bending stress is

$$f_{b1} = \frac{WL}{4S}$$

$$= \frac{(200 \text{ lbf}) (8 \text{ ft}) \left(12 \dfrac{\text{in}}{\text{ft}} \right)}{(4) (7.56 \text{ in}^3)}$$

$$= 635 \text{ lbf/in}^2$$

From Ex. 5.7,

$$\frac{K_e l_1}{d_1} = 17.46$$

$$F'_c = 464 \text{ lbf/in}^2$$

$$f_c = 339 \text{ lbf/in}^2$$

The critical buckling design value, in the plane of bending, for load applied to the narrow face is

$$F_{cE1} = \frac{0.822 E'_{min}}{\left(\dfrac{l_{e1}}{d_1} \right)^2}$$

$$= \frac{(0.822) \left(0.62 \times 10^6 \dfrac{\text{lbf}}{\text{in}^2} \right)}{(17.46)^2}$$

$$= 1672 \text{ lbf/in}^2$$

The moment magnification factor for axial compression and flexure with load applied to the narrow face is

$$C_{m3} = 1.0 - \frac{f_c}{F_{cE1}}$$

$$= 1 - \frac{339 \dfrac{\text{lbf}}{\text{in}^2}}{1672 \dfrac{\text{lbf}}{\text{in}^2}}$$

$$= 0.797$$

The interaction equation for bending load applied to the narrow face of the member and concentric axial compression load is given in NDS Sec. 3.9.2 as

$$\left(\frac{f_c}{F'_c} \right)^2 + \frac{f_{b1}}{F'_{b1} C_{m3}} \leq 1.0$$

Timber

The left side of the expression is

$$\left(\frac{339 \frac{\text{lbf}}{\text{in}^2}}{464 \frac{\text{lbf}}{\text{in}^2}}\right)^2 + \frac{635 \frac{\text{lbf}}{\text{in}^2}}{\left(1626 \frac{\text{lbf}}{\text{in}^2}\right)(0.797)} = 0.535 + 0.490$$

$$= 1.025$$

$$\approx 1.0$$

The chord is adequate.

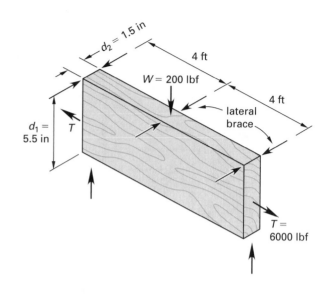

5. DESIGN FOR TENSION

Nomenclature

f_t	actual tension stress parallel to grain	lbf/in^2
F_b^*	reference bending design value multiplied by all applicable adjustment factors except C_L	lbf/in^2
F_b^{**}	reference bending design value multiplied by all applicable adjustment factors except C_V	lbf/in^2
F_t, F_t'	reference and adjusted tension design value parallel to grain	lbf/in^2
T	tensile force on member	lbf

Combined Axial Tension and Flexure

Members subjected to combined tension and flexural stresses due to axial and transverse loading must satisfy the two expressions given in NDS Sec. 3.9.1 as

$$\frac{f_t}{F_t'} + \frac{f_b}{F_b^*} \leq 1.0 \qquad \textit{[NDS 3.9-1]}$$

$$\frac{f_b - f_t}{F_b^{**}} \leq 1.0 \qquad \textit{[NDS 3.9-2]}$$

Example 5.9

The select structural 2 × 6 Douglas Fir-Larch bottom chord of a truss is loaded as shown in the following illustration. The governing load combination consists of dead plus live load, and the moisture content exceeds 19%. The chord is laterally braced at midlength about the weak axis, and the self-weight of the chord and bracing members may be neglected. Determine whether the member is adequate.

Solution

The reference design value for tension is tabulated in NDS SUPP Table 4A and is

$$F_t = 1000 \text{ lbf/in}^2$$

$$C_D = 1.0, \ C_t = 1.0, \ C_i = 1.0$$

The applicable adjustment factors for tension are as follows.

$$C_M = \text{wet service factor from}$$
$$\quad \text{NDS SUPP Table 4A}$$
$$= 1.00$$

$$C_F = \text{size factor from}$$
$$\quad \text{NDS SUPP Table 4A}$$
$$= 1.3$$

The allowable tension design value parallel to grain is

$$F_t' = F_t C_M C_F$$
$$= \left(1000 \ \frac{\text{lbf}}{\text{in}^2}\right)(1.0)(1.3)$$
$$= 1300 \text{ lbf/in}^2$$

The actual tension stress on the chord is given by

$$f_t = \frac{T}{A}$$
$$= \frac{6000 \text{ lbf}}{8.25 \text{ in}^2}$$
$$= 727 \text{ lbf/in}^2$$
$$< F_t' \quad \text{[satisfactory]}$$

From Ex. 5.8, the beam stability factor is

$$C_L = 0.98$$

The actual edgewise bending stress is

$$f_{b1} = 635 \text{ lbf/in}^2$$

The reference bending design value multiplied by all applicable adjustment factors except C_L is

$$F_b^* = F_b C_M C_F$$
$$= \left(1500 \; \frac{\text{lbf}}{\text{in}^2}\right)(0.85)(1.3)$$
$$= 1658 \text{ lbf/in}^2$$

The reference bending design value multiplied by all applicable adjustment factors except C_V is

$$F_b^{**} = F_b C_M C_L C_F$$
$$= \left(1500 \; \frac{\text{lbf}}{\text{in}^2}\right)(0.85)(0.98)(1.3)$$
$$= 1626 \text{ lbf/in}^2$$

Substituting in the two expressions given in NDS Sec. 3.9.1 gives

$$\frac{f_t}{F_t'} + \frac{f_{b1}}{F_b^*} = \frac{727 \; \dfrac{\text{lbf}}{\text{in}^2}}{1300 \; \dfrac{\text{lbf}}{\text{in}^2}} + \frac{635 \; \dfrac{\text{lbf}}{\text{in}^2}}{1658 \; \dfrac{\text{lbf}}{\text{in}^2}}$$
$$= 0.559 + 0.383$$
$$= 0.942$$
$$< 1.0 \quad \text{[satisfactory]}$$

$$\frac{f_{b1} - f_t}{F_b^{**}} = \frac{635 \; \dfrac{\text{lbf}}{\text{in}^2} - 727 \; \dfrac{\text{lbf}}{\text{in}^2}}{1626 \; \dfrac{\text{lbf}}{\text{in}^2}}$$
$$< 1.0 \quad \text{[satisfactory]}$$

The chord is adequate.

6. DESIGN OF CONNECTIONS

Nomenclature

a	center-to-center spacing between adjacent rows of fasteners	in
a_e	minimum edge distance with load parallel to grain	in
a_p	minimum end distance with load parallel to grain	in
a_q	minimum end distance with load perpendicular to grain	in
A	area of cross section	in^2
A_m	gross cross-sectional area of main wood member(s)	in^2
A_n	net area of member	in^2
A_s	sum of gross cross-sectional areas of side member(s)	in^2
C_d	penetration depth factor for connections	—
C_{di}	diaphragm factor for nailed connections	—
C_{eg}	end grain factor for connections	—
C_g	group action factor for connections	—
C_{st}	metal side plate factor for 4 in shear plate connections	—
C_{tn}	toe-nail factor for nailed connections	—
C_Δ	geometry factor for connections	—
d	pennyweight of nail or spike	—
d_e	effective depth of member at a connection	in
D	diameter	in
e_p	minimum edge distance unloaded edge	in
e_q	minimum edge distance loaded edge	in
E	length of tapered tip	in
g	gage of screw	—
l_m	length of bolt in wood main member	in
l_s	total length of bolt in wood side member(s)	in
L	length of nail	in
L	length of screw	in
n	number of fasteners in a row	—
N, N'	reference and adjusted lateral design value at an angle of α to the grain for a single split ring connector unit or shear plate connector unit	lbf
p	depth of fastener penetration into wood member	in
P, P'	reference and adjusted lateral design value parallel to grain for a single split ring connector unit or shear plate connector unit	lbf
Q, Q'	reference and adjusted lateral design value perpendicular to grain for a single split ring connector unit or shear plate connector unit	lbf
s	center-to-center spacing between adjacent fasteners in a row	in
S	unthreaded shank length	in
t_m	thickness of main member	in
t_s	thickness of side member	in
T	minimum thread length	in
W, W'	reference and adjusted withdrawal design value for fastener	lbf/in
Z_α'	allowable design value for lag screw with load applied at an angle α to the wood surface	lbf
Z, Z'	reference and adjusted lateral design value for a single fastener connection	lbf
$Z_\parallel$	reference lateral design value for a single bolt or lag screw connection with all wood members loaded parallel to grain	lbf

Timber

$Z_{m\perp}$ reference lateral design value for a single bolt or lag screw wood-to-wood connection with main member loaded perpendicular to grain and side member loaded parallel to grain lbf

$Z_{s\perp}$ reference lateral design value for a single bolt or lag screw wood-to-wood connection with main member loaded parallel to grain and side member loaded perpendicular to grain lbf

$Z_\perp$ reference lateral design value for a single bolt or lag screw, wood-to-wood, wood-to-metal, or wood-to-concrete connection with all wood member(s) loaded perpendicular to grain lbf

Symbols

α angle between wood surface and direction of applied load degree

γ load/slip modulus for a connection lbf/in

θ angle between direction of load and direction of grain (longitudinal axis of member) degree

Adjustment of Design Values

The reference design values for fasteners are given in NDS Parts 10 through 13. These design values are applicable to single fastener connections and normal conditions of use as defined in NDS Sec. 2.2. For other conditions of use, these values are multiplied by adjustment factors, specified in NDS Sec. 10.3, to determine the relevant design values. A summary of the adjustment factors follows, and the applicability of each to the nominal design values is shown in Table 5.3.

Load Duration Factor, C_D

With the exception of the impact load duration factor, values of the load duration factor given in Table 5.2 are applicable to connections.

Wet Service Factor, C_M

When the moisture content of the member exceeds 19%, the adjustment factors given in NDS Table 10.3.3 are applicable.

Temperature Factor, C_t

The temperature factor is applicable to all connectors and is specified by NDS Table 10.3.4.

Group Action Factor, C_g

The group action factors for various connection geometries and fastener types are given in NDS Tables 10.3.6A through 10.3.6D. This factor is dependent on the ratio of the area of the side members in a connection to the area of the main member, A_s/A_m. A_m and A_s are calculated by using gross areas without deduction for holes. When adjacent rows of fasteners are staggered, as shown in Fig. 5.6, the adjacent rows are considered a single row.

Table 5.3 *Adjustment Factors for Connections*

adjustment factor	bolts	lag screws		split rings and shear plates		screws		nails	
design value	Z	W	Z	P	Q	W	Z	W	Z
C_D load duration factor	√	√	√	√	√	√	√	√	√
C_M wet service factor	√	√	√	√	√	√	√	√	√
C_t temperature factor	√	√	√	√	√	√	√	√	√
C_g group action factor	√	–	√	√	√	–	–	–	–
C_Δ geometry factor	√	–	√	√	√	–	–	–	–
C_d penetration depth factor	–	–	√	√	√	–	√	–	√
C_{eg} end grain factor	–	√	√	–	–	–	√	–	√
C_{st} metal side plate factor	–	–	–	√	–	–	–	–	–
C_{di} diaphragm factor	–	–	–	–	–	–	–	–	√
C_{tn} toe-nail factor	–	–	–	–	–	–	–	√	√

NOTE: Z = lateral design value; W = withdrawal design value; P = parallel to grain design value; Q = perpendicular to grain design value

Figure 5.6 Staggered Fasteners

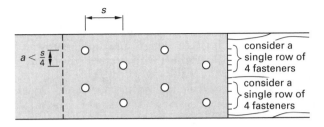

Geometry Factor, C_Δ

The geometry factor applies to bolts, lag screws, split rings, and shear plates. The factor is applied, in accordance with NDS Secs. 11.5.1 and 12.3.2, when end or edge distances or spacing is less than the specified minimum.

Penetration Depth Factor, C_d

The penetration depth factor applies to lag screws, split rings, shear plates, screws, and nails. The factor is applied in accordance with NDS Tables 12.2.3 and 11J through 11R when the penetration is less than the minimum specified.

End Grain Factor, C_{eg}

The end grain factor applies to lag screws, screws, and nails. The factor is applied in accordance with NDS Sec. 11.5.2 when the fastener is inserted in the end grain of a member.

Metal Side Plate Factor, C_{st}

The metal side plate factor is applicable to split rings and shear plates. The factor is applied in accordance with NDS Sec. 12.2.4 when metal side plates are used instead of wood side members.

Diaphragm Factor, C_{di}

The diaphragm factor applies to nails and spikes. The factor is applied in accordance with NDS Sec. 11.5.3 when the fasteners are used in diaphragm construction and $C_{di} = 1.1$.

Toe-Nail Factor, C_{tn}

The toe-nail factor applies to nails and spikes. The factor is applied in accordance with NDS Sec. 11.5.4 when toe-nailed connections are used and $C_{tn} = 0.83$ for lateral design values.

Example 5.10

A bolted connection in tension consists of a single row of eight $3/4$ in diameter bolts in two select structural 2×6 Douglas Fir-Larch members in single shear. The governing load combination consists of dead plus live load, and the moisture content exceeds 19%. The bolt spacing and end distance are 4 in. Determine the capacity of the connection.

Solution

From Ex. 5.9, the tension capacity of the members is

$$T = F_t' A_n$$
$$= F_t' \left(A - \left(D + \frac{1}{16} \right)(b) \right)$$
$$= \left(1300 \, \frac{\text{lbf}}{\text{in}^2} \right) (8.25 \text{ in}^2 - (0.813 \text{ in})(1.5 \text{ in}))$$
$$= 9140 \text{ lbf}$$

The nominal $3/4$ in diameter bolt design value for single shear is tabulated in NDS Table 11A as

$$Z_\parallel = 720 \text{ lbf}$$
$$C_D = 1.0, \; C_t = 1.0$$
$$A_s = A_m = 8.25 \text{ in}^2$$

The specified minimum end distance for the full bolt design value is specified in NDS Sec. 11.5.1 as

$$a_p = 7D$$
$$= (7)(0.75 \text{ in})$$
$$= 5.25 \text{ in}$$

The applicable adjustment factors for the bolts are as follows.

$$C_M = \text{wet service factor from}$$
$$\quad \text{NDS Table 10.3.3}$$
$$= 0.70$$
$$C_g = \text{group action factor from}$$
$$\quad \text{NDS Table 10.3.6A}$$
$$= 0.71$$
$$C_\Delta = \text{geometry factor from NDS Sec. 11.5.1}$$
$$= \frac{\text{actual end distance}}{\text{specified minimum end distance}}$$
$$= \frac{4 \text{ in}}{5.25 \text{ in}}$$
$$= 0.76$$

The allowable lateral design value for eight bolts is

$$T = n Z_\parallel C_M C_g C_\Delta$$
$$= (8)(720 \text{ lbf})(0.70)(0.71)(0.76)$$
$$= 2176 \text{ lbf} \quad \text{[governs]}$$

Bolted Connections

In accordance with NDS Sec. 11.1.2, bolt holes shall be $1/32$ in to $1/16$ in larger than the bolt diameter, and a metal washer or plate is required between the wood and the nut and bolt head. To ensure that the full design values of bolts are attained, spacing and edge and end distances are specified in NDS Sec. 11.5 and are illustrated in Fig. 5.7.

Reference design values for double shear connections are tabulated in NDS Table 11F for three sawn lumber members of identical species, in NDS Table 11G for a sawn lumber member with steel side plates, in NDS Table 11H for a glued laminated member with sawn lumber side members, and in NDS Table 11I for a glued laminated member with steel side plates.

Reference design values for single shear and symmetric double shear connections are specified in NDS Sec. 11.3 and are tabulated in NDS Table 11A for two sawn lumber members of identical species, in NDS Table 11B for a sawn lumber member with a steel side plate, in NDS Table 11C for a glued laminated member with sawn lumber side member, in NDS Table 11D for a glued laminated member with a steel side plate, and in NDS Table 11E for connections to concrete.

Example 5.11

Determine the minimum values for the dimensions A, B, C, and D, shown in the following illustration, that will allow the full design values to be applied to the $3/4$ diameter bolts. Determine the maximum tensile force T due to wind load that can be resisted by the connection. The $5^1/8 \times 12$ glued laminated member is of Douglas Fir-Larch species.

Solution

The reference $3/4$ in diameter bolt design value for double shear is tabulated in NDS Table 11I as

$$Z_\parallel = 3340 \text{ lbf}$$

$$C_M = 1.0, \ C_t = 1.0$$

$$A_s = 2 \text{ in}^2, \ A_m = 61.5 \text{ in}^2, \ \frac{A_m}{A_s} = 30.75$$

Figure 5.7 *Bolt Spacing Requirements for Full Design Values*

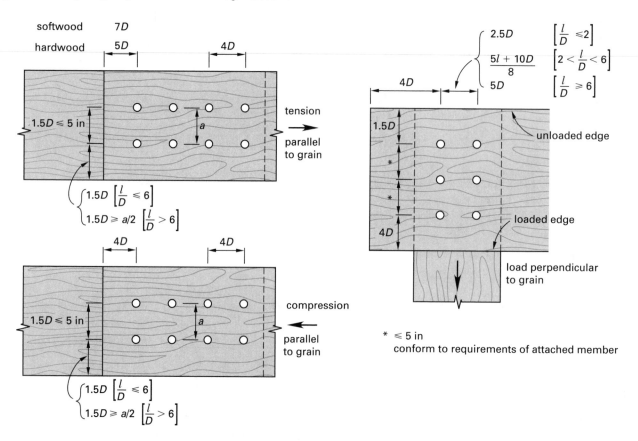

l = lesser of length of bolt in main member or total length of bolt in side member(s)
D = diameter of bolt

Illustration for Ex. 5.11

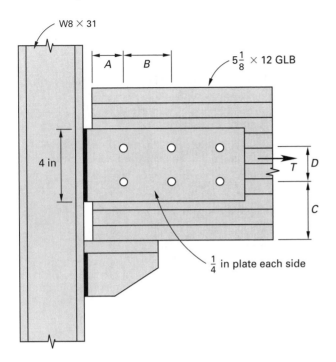

(not to scale)

The applicable adjustment factors for the bolts are as follows.

$$C_D = \text{load duration factor}$$
$$\text{from Table 5.2}$$
$$= 1.60$$

$$C_g = \text{group action factor}$$
$$\text{from NDS Table 10.3.6C}$$
$$= 0.99$$

$$C_\Delta = \text{geometry factor from}$$
$$\text{NDS Sec. 11.5.1}$$

$$= 1.0 \quad \left[\begin{array}{c} \text{All dimensions conform} \\ \text{to the specified minimums.} \end{array}\right]$$

The allowable lateral design value for six bolts is

$$T = nZ_\parallel C_D C_g C_\Delta$$
$$= (6)(3340 \text{ lbf})(1.60)(0.99)(1.00)$$
$$= 31{,}743 \text{ lbf}$$

The specified minimum end distance A for the full bolt design value is specified in NDS Table 11.5.1B as

$$a_p = 7D$$
$$= (7)(0.75 \text{ in})$$
$$= 5.25 \text{ in}$$

The specified minimum spacing B for the full bolt design value is specified in NDS Table 11.5.1C as

$$s = 4D$$
$$= (4)(0.75 \text{ in})$$
$$= 3.00 \text{ in}$$

The specified minimum spacing between rows D for the full bolt design value is specified in NDS Table 11.5.1D as

$$a = 1.5D$$
$$= (1.5)(0.75 \text{ in})$$
$$= 1.125 \text{ in}$$

The ratio of the length of the bolt in the main member to the bolt diameter is

$$\frac{l_m}{D} = \frac{5.125 \text{ in}}{0.75 \text{ in}}$$
$$= 6.83$$
$$> 6$$

Hence, the specified minimum edge distance C for the full bolt design value is specified in NDS Table 11.5.1A as the greater of

- $a_e = \dfrac{a}{2}$

$$= \frac{1.125 \text{ in}}{2}$$
$$= 0.563 \text{ in}$$

- $a_e = 1.5D$

$$= (1.5)(0.75 \text{ in})$$
$$= 1.125 \text{ in} \quad [\text{governs}]$$

Lag Screw Connections

Lateral Design Values in Side Grain

Minimum edge distances, end distances, spacing, and geometry factors are identical with those for bolts with a diameter equal to the shank diameter of the lag screw. As specified in NDS Table 11J, for full design values to be applicable, the depth of penetration (not including the length of the tapered tip) shall not be less than

$$p = 8D$$

The minimum allowable penetration is $4D$. When the penetration is between $4D$ and $8D$, the nominal design value is multiplied by the penetration factor, which is defined in NDS Table 11J as

$$C_d = \frac{p}{8D}$$
$$\leq 1.0$$

Reference design values for single shear connections are specified in NDS Sec. 11.3 and are tabulated in NDS Table 11J for connections with a wood side member and in NDS Table 11K for connections with a steel side plate.

Withdrawal Design Values in Side Grain

Minimum edge distance, end distance, and spacing are specified in NDS Table 11.5.1E and are

$$a_e = \text{edge distance} = 1.5D$$
$$a_p = \text{end distance} = 4D$$
$$s = \text{spacing} = 4D$$

Withdrawal design values in pounds per inch of thread penetration (not including the length of the tapered tip) are tabulated in NDS Table 11.2A.

Combined Lateral and Withdrawal Loads

When the load applied to a lag screw is at an angle α to the wood surface, the lag screw is subjected to combined lateral and withdrawal loading. The design value is determined by the Hankinson formula given by NDS Sec. 11.4.1 as

$$Z'_\alpha = \frac{W'pZ'}{W'p\cos^2\alpha + Z'\sin^2\alpha}$$

Example 5.12

A 3 in long, $^3/_8$ in diameter lag screw inserted into a Douglas Fir-Larch joist with a 10 gage steel side plate is subjected to a force inclined at an angle of 30° to the wood surface. Determine the maximum force that may be applied.

Solution

From NDS Table 11K, the nominal lateral design value for load applied parallel to the grain is

$$Z_\| = 220 \text{ lbf}$$
$$C_D = 1.0, \ C_M = 1.0, \ C_t = 1.0, \ C_g = 1.0, \ C_\Delta = 1.0$$

From NDS App. L2, the penetration into the main member of the screw shank, plus the threaded length, less the length of the tapered tip, is

$$p = S + T - E - t_s$$
$$= 1.0 \text{ in} + 1.781 \text{ in} - 0.134 \text{ in}$$
$$= 2.647 \text{ in}$$

From NDS Table 11K, the penetration factor is obtained as

$$C_d = \frac{p}{8D} = \frac{2.647 \text{ in}}{(8)(0.375 \text{ in})}$$
$$= 0.88$$

The adjusted lateral design value is

$$Z'_\| = Z_\| C_d = (220 \text{ lbf})(0.88)$$
$$= 194 \text{ lbf}$$

From NDS Table 11.3.2A, the specific gravity of the Douglas Fir-Larch joist is

$$G = 0.50$$

From NDS Table 11.2A, the nominal withdrawal design value is

$$W = 305 \text{ lbf/in}$$

The adjusted withdrawal design value is

$$W' = W$$
$$= 305 \text{ lbf/in}$$

From NDS App. L, the penetration into the main member of the threaded length, less the length of the tapered tip, is

$$p = T - E$$
$$= 1.781 \text{ in}$$

The maximum force that may be applied is determined by NDS Sec. 11.4.1 as

$$Z'_\alpha = \frac{W'pZ'_\|}{W'p\cos^2\alpha + Z'_\|\sin^2\alpha}$$
$$= \frac{\left(305 \ \frac{\text{lbf}}{\text{in}}\right)(1.781 \text{ in})(194 \text{ lbf})}{\left(305 \ \frac{\text{lbf}}{\text{in}}\right)(1.781 \text{ in})\left(\cos^2 30°\right) + (194 \text{ lbf})\left(\sin^2 30°\right)}$$
$$= 231 \text{ lbf}$$

Split Ring and Shear Plate Connections

Edge and end distances, spacing, and geometry factors C_Δ for various sizes of split ring and shear plate connectors are specified in NDS Table 12.3. When lag screws are used instead of bolts, nominal design values shall, where appropriate, be multiplied by the penetration depth factors specified in NDS Table 12.2.3 for various sizes of connectors and wood species. NDS Table 12.2.4 provides metal side plate factors C_{st} for 4 in shear plate connectors, loaded parallel to the grain, when metal side plates are substituted for wood side members. Group action factors C_g for 4 in split ring or shear plate connectors with wood side members are tabulated in NDS Table 10.3.6B. Group action factors C_g for 4 in shear plate connectors with steel side plates are tabulated in NDS Table 10.3.6D.

Reference design values for split ring connectors are provided in NDS Table 12.2A and for shear plate connectors in NDS Table 12.2B. When a load acts in the plane of the wood surface at an angle θ to the grain, the allowable design value is given by NDS Sec. 12.2.5 as

$$N' = \frac{P'Q'}{P'\sin^2\theta + Q'\cos^2\theta} \qquad \text{[NDS 12.2-1]}$$

Example 5.13

The Douglas Fir-Larch select structural members shown in the following illustration are connected with $2\frac{5}{8}$ in shear plate connectors. The governing load combination consists of dead plus live loads. The connector spacing and end distances are as shown. Determine the capacity of the connection.

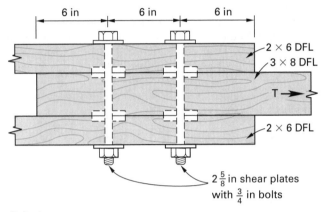

Solution

The reference $2\frac{5}{8}$ in shear plate design value for the $2\frac{1}{2}$ in thick main member of group B species with a connector on two faces is tabulated in NDS Table 12.2B as

$$P_{\text{main}} = 2860 \text{ lbf}$$

The nominal $2\frac{5}{8}$ in shear plate design value for a $1\frac{1}{2}$ in thick side member of group B species with a connector on one face is tabulated in NDS Table 12.2B as

$$P_{\text{side}} = 2670 \text{ lbf} \quad \text{[governs]}$$
$$C_D = 1.0, \ C_M = 1.0, \ C_t = 1.0$$
$$A_s = (2)\left(8.25 \text{ in}^2\right) = 16.5 \text{ in}^2$$
$$A_m = 18.13 \text{ in}^2$$
$$\frac{A_s}{A_m} = 0.91$$

The specified minimum spacing for the full shear plate design value is given in NDS Table 12.3 as

$$s = 6.75 \text{ in}$$

The applicable adjustment factors for the bolts are as follows.

$$C_g = \text{group action factor from NDS}$$

Table 10.3.6B for two shear plates

$$= 0.98$$

$$C_\Delta = \text{geometry factor from NDS}$$

Table 12.3 for a spacing of 6 in

$$= 0.5 + \frac{(0.5)(6 \text{ in} - 3.5 \text{ in})}{6.75 \text{ in} - 3.5 \text{ in}}$$

$$= 0.885$$

The allowable design value for four shear plates is

$$T = nP_{\text{side}}C_gC_\Delta$$
$$= (4)(2670 \text{ lbf})(0.98)(0.885)$$
$$= 9260 \text{ lbf}$$

Wood Screw Connections

Lateral Design Values in Side Grain

Recommended edge distances, end distances, and spacing are tabulated in NDS Commentary Table C11.1.4.7 for wood and steel side plates with and without prebored holes. Wood screws are not subject to the group action factor C_g.

As specified in NDS Table 11L, for full design values to be applicable, the depth of penetration shall not be less than

$$p = 10D$$

The minimum allowable penetration is $6D$. When the penetration is between $6D$ and $10D$, the reference design value is multiplied by the penetration factor, which is defined in NDS Table 11L as

$$C_d = \frac{p}{10D}$$
$$\leq 1.0$$

Reference design values for single shear connections are specified in NDS Sec. 11.3 and tabulated in NDS Table 11L for connections with a wood side member and in NDS Table 11M for connections with a steel side plate.

Withdrawal Design Values in Side Grain

Withdrawal design values in pounds per inch of thread penetration are tabulated in NDS Table 11.2B. The length of thread is specified in App. L as two-thirds the total screw length or four times the screw diameter, whichever is greater.

Combined Lateral and Withdrawal Loads

When the load applied to a wood screw is at an angle α to the wood surface, the wood screw is subjected to combined lateral and withdrawal loading, and the design value is determined by the Hankinson formula given by NDS Sec. 11.4.1 as

$$Z'_\alpha = \frac{W'pZ'}{W'p\cos^2\alpha + Z'\sin^2\alpha}$$

Example 5.14

A $7g$ steel strap is secured to a select structural Douglas Fir-Larch collector with ten $14g \times 3$ in wood screws. Edge and end distances and spacing are sufficient to prevent splitting of the wood. Determine the maximum tensile force T due to wind load that can be resisted by the connection.

Solution

The reference design value for single shear is tabulated in NDS Table 11M as

$$Z = 202 \text{ lbf}$$

$$C_M = 1.0, \ C_t = 1.0$$

The applicable adjustment factors for the screws are as follows.

$$C_D = \text{load duration factor from Table 5.2}$$

$$= 1.60$$

The penetration of the screw shank plus the threaded length is

$$p = L - t_s$$

$$= 3 \text{ in} - 0.179 \text{ in}$$

$$= 2.821 \text{ in}$$

This is greater than $10D$, and from NDS Table 11M,

$$C_d = \text{penetration depth factor}$$

$$= 1.0$$

The allowable lateral design value for ten screws is

$$T = nZC_D C_d$$

$$= (10)(202 \text{ lbf})(1.60)(1.0)$$

$$= 3232 \text{ lbf}$$

Connections with Nails and Spikes

Lateral Design Values in Side Grain

Recommended edge distances, end distances, and spacing are tabulated in NDS Commentary Table C11.1.5.6 for wood and steel side plates with and without pre-bored holes. Nails and spikes are not subject to the group action factor C_g.

As specified in NDS Table 11N, for full design values to be applicable, the depth of penetration shall not be less than

$$p = 10D$$

The minimum allowable penetration is $6D$. When the penetration is between $6D$ and $10D$, the reference design value is multiplied by the penetration factor, which is defined in NDS Table 11N as

$$C_d = \frac{p}{10D}$$

$$\leq 1.0$$

Reference design values for nails and spikes used in diaphragm construction shall be multiplied by the diaphragm factor $C_{di} = 1.1$.

Reference lateral design values for nails and spikes used in toe-nailed connections shall be multiplied by the toe-nail factor $C_{tn} = 0.83$.

Reference design values for single shear connections for two sawn lumber members of identical species are tabulated in NDS Table 11N.

Reference design values for single shear connections for a sawn lumber member with steel side plate are tabulated in NDS Table 11P.

The reference double shear value for a three-member sawn lumber connection is twice the lesser of the nominal design value for each shear plane. The minimum penetration into the side member shall be six times the connector diameter, or when the side member is at least $3/8$ in thick and $12d$ or smaller nails extend at least three diameters beyond the side member the nails shall be clinched.

Withdrawal Design Values in Side Grain

Reference withdrawal design values in pounds per inch of penetration are tabulated in NDS Table 11.2C. When toe-nailed connections are used, the reference design values shall be multiplied by the toe-nail factor $C_{tn} = 0.67$.

Combined Lateral and Withdrawal Loads

When the load applied to a nail or spike is at an angle α to the wood surface, the nail or spike is subjected to combined lateral and withdrawal loading, and in accordance with the NDS Sec. 11.4.2, the design value is determined by the interaction equation

$$Z'_\alpha = \frac{W'pZ'}{W'p \cos \alpha + Z' \sin \alpha}$$

Example 5.15

Determine the lateral design value for the 3 in long $10d$ common wire nail in the toe-nailed connection shown in the following illustration. Loading applied to the connection is due to wind load, and all members are Douglas Fir-Larch.

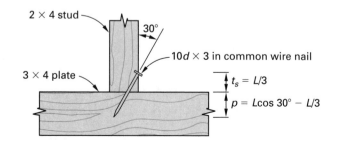

Solution

As specified in NDS Sec. 11.1.5, toe-nails are driven at an angle of 30° from the face of the member, with the point of penetration one-third the length of the nail from the member end. In accordance with NDS Commentary Sec. C11.1.5, the side member thickness is taken to be equal to this end distance and

$$t_s = \frac{L}{3} = \frac{3 \text{ in}}{3}$$
$$= 1 \text{ in}$$

The nominal design value for single shear is tabulated in NDS Table 11N as

$$Z = 118 \text{ lb}$$
$$C_M = 1.0, \ C_t = 1.0$$

The applicable adjustment factors for the nail are as follows.

- C_D = load duration factor from Table 5.2
  $= 1.60$

- The penetration of the nail into the main member, in accordance with NDS Commentary Sec. C11.1.5, is taken as the vertically projected length of the nail in the member and

$$p = L \cos 30° - \frac{L}{3}$$
$$= (3 \text{ in})(0.866) - 1.0 \text{ in}$$
$$= 1.60 \text{ in}$$

This is greater than $10D$, and from NDS Table 11N,

$$C_d = \text{penetration depth factor}$$
$$= 1.0$$

- C_{tn} = toe-nail factor from NDS Sec. 11.5.4
  $= 0.83$

The allowable lateral design value for the nail is

$$Z' = ZC_D C_d C_{tn}$$
$$= (118 \text{ lbf})(1.60)(1.0)(0.83)$$
$$= 157 \text{ lbf}$$

References

1. American Forest and Paper Association. *National Design Specification for Wood Construction ASD/LRFD: With Commentary and Supplement, (ANSI/AF&PA NDS-2005)*. 2005.

2. American Society of Civil Engineers. *Minimum Design Loads for Buildings and Other Structures, (ASCE/SEI 7-05)*. 2005.

3. American Plywood Association. *Glued Laminated Beam Design Tables*. 2007.

4. Western Wood Products Association. *Western Lumber Span Tables*. 2001.

Timber

PRACTICE PROBLEMS

1. The select structural 3 × 10 Douglas Fir-Larch rafter shown in the following illustration is notched over a supporting 3 in wall. Based on the bearing stress in the rafter, what is the maximum allowable reaction at the support caused by snow loading?

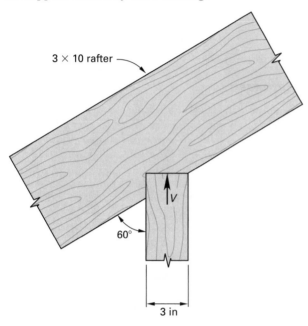

2. The select structural 4 × 10 Douglas Fir-Larch ledger shown in the following illustration supports a dead plus floor live load of 250 lbf/ft. Based on the $3/4$ in bolt design value in the ledger, what is the maximum allowable bolt spacing?

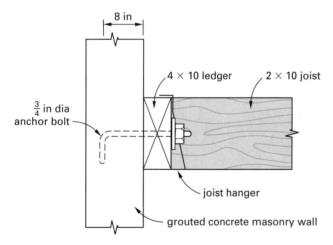

3. The floor system in an office building consists of select structural 2 × Douglas Fir-Larch joists at 16 in centers with $^{19}/_{32}$ plywood sheathing. Each joist supports a dead plus floor live load of 100 lbf/ft over a span of $L = 16$ ft. Floor live load is 50 lbf/ft^2, acceptable deflection due to live load is $\Delta_{ST} = L/360$, and acceptable deflection due to total load is $\Delta_T = L/240$. What is the depth of joist necessary to give acceptable stresses and deflection?

4. A select structural 6 × 6 Douglas Fir-Larch column is subjected to axial load due to dead plus floor live load. The column is 10 ft high and may be considered pin ended. What is the maximum load that may be applied?

SOLUTIONS

1. The reference design value for compressive bearing parallel to grain is tabulated in NDS SUPP Table 4A and is

$$F_c = 1700 \text{ lbf/in}^2$$

The applicable adjustment factors for compressive bearing parallel to grain are

$$C_t = 1.0$$
$$C_M = 1.0$$
$$C_F = 1.0$$
$$C_i = 1.0$$
$$C_D = \text{load duration factor}$$
$$\text{for snow load}$$
$$= 1.15 \qquad \text{[NDS Table 2.3.2]}$$

The adjusted compressive bearing design value parallel to grain is

$$F_c^* = C_t C_M C_F C_i C_D F_c$$
$$= \left(1700 \; \frac{\text{lbf}}{\text{in}^2}\right)(1.15)$$
$$= 1955 \text{ lbf/in}^2$$

The basic design value for compression perpendicular to grain is tabulated in NDS SUPP Table 4A and is

$$F_{c\perp} = 625 \text{ lbf/in}^2$$
$$C_M = 1.0, \; C_t = 1.0, \; C_i = 1.0$$

The applicable adjustment factor for compression perpendicular to grain is specified in NDS Sec. 3.10.4 as

$$C_b = \text{bearing area factor}$$
$$= \frac{l_b + 0.375}{l_b}$$
$$= \frac{3 \text{ in} + 0.375}{3 \text{ in}}$$
$$= 1.125$$

The adjusted compression design value perpendicular to grain is

$$F_{c\perp}' = F_{c\perp} C_b = \left(625 \; \frac{\text{lbf}}{\text{in}^2}\right)(1.125)$$
$$= 703 \text{ lbf/in}^2$$

The allowable bearing design value at an angle θ to the grain is given by NDS Sec. 3.10.3 as

$$F_\theta' = \frac{F_c^* F_{c\perp}'}{F_c^* \sin^2 \theta + F_{c\perp}' \cos^2 \theta}$$
$$= \frac{\left(1955 \; \frac{\text{lbf}}{\text{in}^2}\right)\left(703 \; \frac{\text{lbf}}{\text{in}^2}\right)}{\left(1955 \; \frac{\text{lbf}}{\text{in}^2}\right)(\sin^2 60°) + \left(703 \; \frac{\text{lbf}}{\text{in}^2}\right)(\cos^2 60°)}$$
$$= 837 \text{ lbf/in}^2$$

The allowable reaction at the support is

$$V = F_\theta' b l_b$$
$$= \left(837 \; \frac{\text{lbf}}{\text{in}^2}\right)(2.5 \text{ in})(3 \text{ in})$$
$$= 6278 \text{ lbf}$$

2. The nominal $^3/_4$ in diameter bolt design value, for single shear perpendicular to grain, into concrete is tabulated in NDS Table 11E as

$$Z_\perp = 900 \text{ lbf}$$
$$C_D = 1.0, \; C_M = 1.0, \; C_t = 1.0, \; C_g = 1.0$$

The applicable adjustment factor for the bolts is

$$C_\Delta = \text{geometry factor}$$
$$\text{from NDS Sec. 11.5.1}$$
$$= 1.0 \quad \begin{bmatrix} \text{All dimensions conform to} \\ \text{the specified minimums.} \end{bmatrix}$$

The allowable lateral design value is

$$Z_\perp' = Z_\perp C_\Delta = (900 \text{ lbf})(1.00)$$
$$= 900 \text{ lbf}$$

The maximum allowable bolt spacing is

$$s = \frac{900 \text{ lbf}}{250 \; \dfrac{\text{lbf}}{\text{ft}}}$$
$$= 3.6 \text{ ft}$$

3. The basic design values for bending and modulus of elasticity are tabulated in NDS SUPP Table 4A and are

$$F_b = 1500 \text{ lbf/in}^2$$
$$E = 1.9 \times 10^6 \text{ lbf/in}^2$$
$$C_D = 1.0, \; C_M = 1.0, \; C_L = 1.0, \; C_t = 1.0, \; C_i = 1.0$$

The applicable adjustment factors for bending stress are as follows.

$$C_F = \text{size factor}$$
$$= 1.10 \quad \begin{bmatrix} \text{NDS SUPP Table 4A,} \\ \text{assuming a 10 in joist} \end{bmatrix}$$
$$C_r = \text{repetitive member factor}$$
$$= 1.15 \quad \text{[NDS SUPP Table 4A]}$$

The adjusted bending stress is

$$F_b' = F_b C_F C_r$$
$$= \left(1500 \; \frac{\text{lbf}}{\text{in}^2}\right)(1.1)(1.15)$$
$$= 1898 \text{ lbf/in}^2$$

The applied moment on the joist is

$$M = \frac{wL^2}{8}$$

$$= \frac{\left(100\ \dfrac{\text{lbf}}{\text{ft}}\right)(16\ \text{ft})^2 \left(12\ \dfrac{\text{in}}{\text{ft}}\right)}{8}$$

$$= 38{,}400\ \text{in-lbf}$$

The required section modulus is

$$S_{xx} = \frac{M}{F_b'} = \frac{38{,}400\ \text{in-lbf}}{1898\ \dfrac{\text{lbf}}{\text{in}^2}}$$

$$= 20.24\ \text{in}^3$$

Hence, a 2×10 is adequate for acceptable stresses $[S_{xx} = 21.39\ \text{in}^3]$.

The adjusted modulus of elasticity is

$$E' = E$$

$$= 1.9 \times 10^6\ \text{lbf/in}^2$$

The floor live load is

$$w_L = \frac{\left(50\ \dfrac{\text{lbf}}{\text{ft}^2}\right)(16\ \text{in})}{12\ \dfrac{\text{in}}{\text{ft}}}$$

$$= 66.67\ \text{lbf/ft}$$

The floor dead load is

$$w_D = 100\ \frac{\text{lbf}}{\text{ft}} - 66.67\ \frac{\text{lbf}}{\text{ft}}$$

$$= 33.33\ \text{lbf/ft}$$

The required live load deflection is

$$\Delta_{ST} = \frac{L}{360} = \frac{(16\ \text{ft})\left(12\ \dfrac{\text{in}}{\text{ft}}\right)}{360}$$

$$= 0.53\ \text{in}$$

The corresponding required moment of inertia is

$$I_{xx} = \frac{5 w_L L^4}{384 E \Delta_{ST}}$$

$$= \frac{(5)\left(66.67\ \dfrac{\text{lbf}}{\text{ft}}\right)(16\ \text{ft})^4 \left(12\ \dfrac{\text{in}}{\text{ft}}\right)^3}{(384)\left(1.9 \times 10^6\right)(0.53\ \text{in})}$$

$$= 97.63\ \text{in}^4$$

Hence, a 2×10 is acceptable for live load deflection $[I_{xx} = 98.9\ \text{in}^4]$.

The required deflection for total load is given as

$$\Delta_T = \frac{L}{240} = \frac{(16\ \text{ft})\left(12\ \dfrac{\text{in}}{\text{ft}}\right)}{240}$$

$$= 0.80\ \text{in}$$

For long-term loads, NDS Sec. 3.5.2 specifies a creep factor K_{cr} of 1.5 for seasoned lumber. Hence, to determine the total deflection Δ_T, the applicable equivalent total load is

$$w_T = w_L + K_{cr} w_D$$

$$= 66.67\ \frac{\text{lbf}}{\text{ft}} + (1.5)\left(33.33\ \frac{\text{lbf}}{\text{ft}}\right)$$

$$= 116.67\ \text{lbf/ft}$$

The corresponding required moment of inertia is

$$I_{xx} = \frac{5 w_T L^4}{384 E \Delta_T}$$

$$= \frac{(5)\left(116.67\ \dfrac{\text{lbf}}{\text{ft}}\right)(16\ \text{ft})^4 \left(12\ \dfrac{\text{in}}{\text{ft}}\right)^3}{(384)\left(1.9 \times 10^6\ \dfrac{\text{lbf}}{\text{in}^2}\right)(0.80\ \text{in})}$$

$$= 113.18\ \text{in}^4 \quad [\text{governs}]$$

Hence, a 2×12 is necessary to control deflection $[I_{xx} = 178\ \text{in}^4]$.

4. The reference design values for compression and modulus of elasticity are tabulated in NDS SUPP Table 4D and are

$$F_c = 1150\ \text{lbf/in}^2$$

$$E_{\min} = 0.58 \times 10^6\ \text{lbf/in}^2$$

$$C_D = 1.0,\ C_M = 1.0,\ C_t = 1.0,\ C_i = 1.0$$

The applicable adjustment factor for compression is

$$C_F = \text{size factor from NDS SUPP Table 4D}$$

$$= 1.0$$

The slenderness ratio is

$$\frac{K_e l}{d} = \frac{(1.0)(10\ \text{ft})\left(12\ \dfrac{\text{in}}{\text{ft}}\right)}{5.5\ \text{in}}$$

$$= 21.82$$

The adjusted modulus of elasticity is

$$E_{\min}' = E_{\min}$$

$$= 0.58 \times 10^6\ \text{lbf/in}^2$$

The reference compression design value multiplied by all applicable adjustment factors except C_P is given by

$$F_c^* = F_c C_F = \left(1150 \ \frac{\text{lbf}}{\text{in}^2}\right)(1.0)$$
$$= 1150 \ \text{lbf/in}^2$$

The critical buckling design value is

$$F_{cE} = \frac{0.822 E'_{\text{min}}}{\left(\dfrac{l_e}{d}\right)^2}$$
$$= \frac{(0.822)\left(0.58 \times 10^6 \ \dfrac{\text{lbf}}{\text{in}^2}\right)}{(21.82)^2}$$
$$= 1001 \ \text{lbf/in}^2$$

The ratio of F_{cE} to F_c^* is

$$F' = \frac{F_{cE}}{F_c^*} = \frac{1001 \ \dfrac{\text{lbf}}{\text{in}^2}}{1150 \ \dfrac{\text{lbf}}{\text{in}^2}}$$
$$= 0.870$$

The column parameter is obtained from NDS Sec. 3.7.1.5 as

$$c = 0.8 \quad [\text{for sawn lumber}]$$

The column stability factor is specified by NDS Sec. 3.7.1 as

$$C_p = \frac{1.0 + F'}{2c} - \sqrt{\left(\frac{1.0 + F'}{2c}\right)^2 - \frac{F'}{c}}$$
$$= \frac{1.870}{(2)(0.8)} - \sqrt{\left(\frac{1.870}{(2)(0.8)}\right)^2 - \frac{0.870}{0.8}}$$
$$= 0.641$$

The allowable compression design value parallel to grain is

$$F_c' = F_c C_F C_P$$
$$= \left(1150 \ \frac{\text{lbf}}{\text{in}^2}\right)(1.0)(0.641)$$
$$= 737 \ \text{lbf/in}^2$$

The allowable load on the column is given by

$$P = F_c' A$$
$$= \left(737 \ \frac{\text{lbf}}{\text{in}^2}\right)(30.25 \ \text{in}^2)$$
$$= 22{,}294 \ \text{lbf}$$

6 Design of Reinforced Masonry

1. Design Principles 6-1
2. Design for Flexure 6-1
3. Design for Shear 6-5
4. Design of Masonry Columns 6-6
5. Design of Shear Walls 6-10
   Practice Problems 6-15
   Solutions 6-16

1. DESIGN PRINCIPLES

Nomenclature

D	dead load	lbf or kips
E	load effects of earthquake	–
E_m	modulus of elasticity of masonry in compression	lbf/in^2
E_s	modulus of elasticity of steel reinforcement	lbf/in^2
f'_m	specified masonry compressive strength	lbf/in^2
F	load effects of lateral pressure of liquids	–
H	load effects of lateral earth pressure	–
L	live load	lbf or kips
W	wind load	lbf or kips

General Requirements

Elastic analysis using service loads and permissible stresses is specified in BCRMS[1] Ch. 2 for the design of reinforced masonry.

IBC[2] Sec. 2107.1 adopts the allowable stress design method of BCRMS Ch. 2 with the exception of the load combinations of BCRMS Sec. 2.1.2.1 and the strength requirements of BCRMS Secs. 2.1.3.4 through 2.1.3.4.3. Hence, the applicable load combinations are obtained from ASCE[3] Sec. 2.4.

The modulus of elasticity of steel reinforcement is given by BCRMS Sec. 1.8.2.1 as

$$E_s = 29,000,000 \text{ lbf/in}^2$$

In accordance with BCRMS Sec. 1.8.2.2.1, the modulus of elasticity of concrete masonry may be based on the chord modulus of elasticity. This may be determined by means of a compression test, with the modulus being taken between 0.05 and 0.33 of the maximum compressive strength of the masonry prism. Alternatively the modulus may be derived from the expression

$$E_m = 900f'_m$$

2. DESIGN FOR FLEXURE

Nomenclature

A_s	area of tension reinforcement	in^2
b_w	width of beam	in
d	effective depth, distance from extreme compression fiber to centroid of tension reinforcement	in
d_b	diameter of reinforcement	in
f_b	calculated compressive stress in masonry due to flexure	lbf/in^2
f'_m	specified masonry compressive strength	lbf/in^2
f_s	calculated stress in reinforcement	lbf/in^2
f_y	yield strength of reinforcement	lbf/in^2
F_b	allowable compressive stress in masonry due to flexure	lbf/in^2
F_s	allowable stress in reinforcement	lbf/in^2
h	overall dimension of member	in
j	lever-arm factor, ratio of distance between centroid of flexural compression forces and centroid of tensile forces to effective depth, $1 - k/3$	–
k	neutral axis depth factor, $\sqrt{2\rho n + (\rho n)^2} - \rho n$	–
l	clear span length of beam	ft
l_c	distance between points of lateral support	ft
l_e	effective span	ft
M	applied moment	ft-kips or in-lbf
n	modular ratio, E_s/E_m	–

Symbols

ρ	tension reinforcement ratio, $A_s/b_w d$	

Beams with Tension Reinforcement

For grade 50 reinforcement, the allowable tensile stress and the allowable compressive stress are given by BCRMS Sec. 2.3.2 as

$$F_s = 20,000 \text{ lbf/in}^2$$

For grade 60 reinforcement, the allowable tensile stress and the allowable compressive stress are given by BCRMS Sec. 2.3.2 as

$$F_s = 24{,}000 \text{ lbf/in}^2$$

The allowable compressive stress in masonry due to flexure is given by BCRMS Sec. 2.3.3.2.2 as

$$F_b = 0.33 f'_m$$

The elastic design method, illustrated in Fig. 6.1, is used to calculate the stresses in a masonry beam under the action of the applied service loads and to ensure that these stresses do not exceed allowable values. From Fig. 6.1, the neutral axis depth factor may be derived[4] as

$$k = \sqrt{2\rho n + (\rho n)^2} - \rho n$$

Figure 6.1 *Reinforced Masonry Beam*

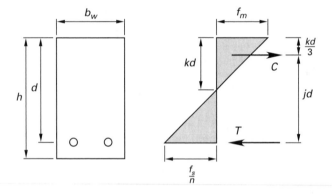

To facilitate the determination of k, Table A.2, in the appendix, tabulates values of k against ρn. In addition, the lever-arm factor is derived as

$$j = 1 - \frac{k}{3}$$

Figure 6.2 *Effective Span Length*

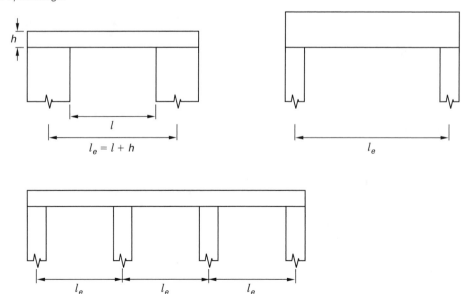

The stress in the reinforcement due to an applied moment M is

$$f_s = \frac{M}{A_s j d}$$

The stress in the masonry due to an applied moment M is

$$f_b = \frac{2M}{jk b_w d^2}$$

As an alternative to hand calculation, all parameters and stresses may be obtained with the aid of a calculator program.[5]

For simply supported beams, the effective span length l_e is defined in BCRMS Sec. 2.3.3.4 and is illustrated in Fig. 6.2 as clear span plus depth of member but not exceeding the distance between support centers. For continuous beams, BCRMS Sec. 2.3.3.4.2 defines the effective span as the distance between centers of supports; this is shown in Fig. 6.2.

In accordance with BCRMS Sec. 2.3.3.4.4, lateral support to the compression face of a beam shall be provided at a maximum spacing of

$$l_c = 32 b_w$$

Example 6.1

The 8 in solid grouted concrete block masonry beam shown in the illustration may be considered simply supported over an effective span of 15 ft. The masonry has a compressive strength of 1500 lbf/in² and a modulus of elasticity of 1,000,000 lbf/in²; tension reinforcement consists of two no. 6 grade 60 bars. The effective depth is 45 in, the overall depth is 48 in, and the beam is laterally braced at both ends. The 20 kips concentrated loads may be considered live loads. Determine whether the beam is adequate. The self-weight of the beam is 69 lbf/ft².

Illustration for Ex. 6.1

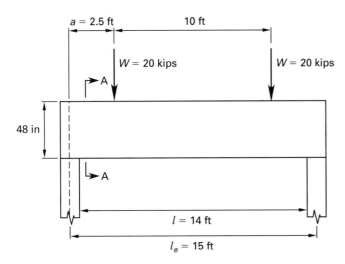

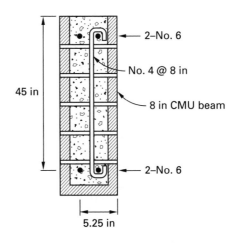

section A-A

Solution

The allowable stresses, in accordance with BCRMS Secs. 2.3.2 and 2.3.3, are

$$F_b = \frac{1}{3}f'_m = \left(\frac{1}{3}\right)\left(1500\ \frac{\text{lbf}}{\text{in}^2}\right)$$
$$= 500\ \text{lbf/in}^2$$

$$F_s = 24{,}000\ \text{lbf/in}^2$$

The beam self-weight is

$$w = \left(69\ \frac{\text{lbf}}{\text{ft}^2}\right)(4\ \text{ft})$$
$$= 276\ \text{lbf/ft}$$

At midspan, the bending moment produced by this self-weight is

$$M_s = \frac{wl^2}{8} = \frac{\left(276\ \dfrac{\text{lbf}}{\text{ft}}\right)(15\ \text{ft})^2}{(8)\left(1000\ \dfrac{\text{lbf}}{\text{kip}}\right)}$$
$$= 7.76\ \text{ft-kips}$$

At midspan, the bending moment produced by the concentrated loads is

$$M_c = Wa = \frac{(20\ \text{kips})(15\ \text{ft} - 10\ \text{ft})}{2}$$
$$= 50\ \text{ft-kips}$$

At midspan, the total moment is

$$M_y = M_s + M_c$$
$$= 7.76\ \text{ft-kips} + 50\ \text{ft-kips}$$
$$= 57.76\ \text{ft-kips}$$

The relevant parameters of the beam are

$$E_m = 1{,}000{,}000\ \text{lbf/in}^2$$

$$E_s = 29{,}000{,}000\ \text{lbf/in}^2$$

$$b_w = 7.63\ \text{in}$$

$$d = 45\ \text{in}$$

$$l_e = 15\ \text{ft}$$

$$A_s = 0.88\ \text{in}^2$$

$$\frac{l_e}{b_w} = \frac{(15\ \text{ft})\left(12\ \dfrac{\text{in}}{\text{ft}}\right)}{7.63\ \text{in}}$$
$$= 23.6$$
$$< 32 \quad \left[\begin{array}{c}\text{satisfies BCRMS}\\ \text{Sec. 2.3.3.4.4}\end{array}\right]$$

$$n = \frac{E_s}{E_m} = \frac{29{,}000{,}000\ \dfrac{\text{lbf}}{\text{in}^2}}{1{,}000{,}000\ \dfrac{\text{lbf}}{\text{in}^2}}$$
$$= 29$$

$$\rho = \frac{A_s}{b_w d} = \frac{0.88\ \text{in}^2}{(7.63\ \text{in})(45\ \text{in})}$$
$$= 0.00256$$

$$\rho n = (0.00256)(29)$$
$$= 0.0743$$

From Table A.2, in the appendix, the beam stresses, in accordance with BCRMS Sec. 2.3, are

$$k = \sqrt{2\rho n + (\rho n)^2} - \rho n$$
$$= 0.318$$

$$j = 1 - \frac{k}{3}$$
$$= 0.894$$

$$f_b = \frac{2M_y}{jkb_wd^2}$$

$$= \frac{(2)(57.76 \text{ ft-kips})\left(12\,\frac{\text{in}}{\text{ft}}\right)\left(1000\,\frac{\text{lbf}}{\text{kip}}\right)}{(0.894)(0.318)(7.63 \text{ in})(45 \text{ in})^2}$$

$$= 315 \text{ lbf/in}^2$$

$$< F_b \quad \text{[satisfactory]}$$

$$f_s = \frac{M_y}{jdA_s}$$

$$= \frac{(57.76 \text{ ft-kips})\left(12\,\frac{\text{in}}{\text{ft}}\right)\left(1000\,\frac{\text{lbf}}{\text{kip}}\right)}{(0.894)(45 \text{ in})(0.88 \text{ in}^2)}$$

$$= 19{,}580 \text{ lbf/in}^2$$

$$< F_s \quad \text{[satisfactory]}$$

The beam is adequate.

Biaxial Bending

The combined stresses produced by biaxial bending in members shall not exceed the allowable values.

Example 6.2

The masonry beam of Ex. 6.1, in addition to the vertical loads indicated, is subjected to a lateral force, due to wind, of 140 lbf/ft. Determine whether the beam is adequate.

Solution

At midspan, the bending moment produced by the wind load is

$$M_x = \frac{ql^2}{8}$$

$$= \frac{\left(140\,\frac{\text{lbf}}{\text{ft}}\right)(15 \text{ ft})^2}{(8)\left(1000\,\frac{\text{lbf}}{\text{kip}}\right)}$$

$$= 3.94 \text{ ft-kips}$$

The relevant parameters of the beam in the transverse direction are

$$b_w = 48 \text{ in}$$
$$d = 5.25 \text{ in}$$
$$l = 15 \text{ ft}$$
$$A_s = 0.88 \text{ in}^2$$
$$n = 29$$
$$\rho = \frac{A_s}{b_wd} = \frac{0.88 \text{ in}^2}{(48 \text{ in})(5.25 \text{ in})}$$
$$= 0.00349$$
$$\rho n = (0.00349)(29)$$
$$= 0.1013$$

From Table A.2, in the appendix, the beam stresses caused by the wind load, in accordance with BCRMS Sec. 2.3, are

$$k = \sqrt{2\rho n + (\rho n)^2} - \rho n = 0.360$$

$$j = 1 - \frac{k}{3} = 0.880$$

$$f_{bw} = \frac{2M_x}{jkb_wd^2}$$

$$= \frac{(2)(3.94 \text{ ft-kips})\left(12\,\frac{\text{in}}{\text{ft}}\right)\left(1000\,\frac{\text{lbf}}{\text{kip}}\right)}{(0.880)(0.360)(48 \text{ in})(5.25 \text{ in})^2}$$

$$= 225 \text{ lbf/in}^2$$

$$f_{sw} = \frac{M_x}{jdA_s}$$

$$= \frac{(3.94 \text{ ft-kips})\left(12\,\frac{\text{in}}{\text{ft}}\right)\left(1000\,\frac{\text{lbf}}{\text{kip}}\right)}{(0.880)(5.25 \text{ in})(0.88 \text{ in}^2)}$$

$$= 11{,}631 \text{ lbf/in}^2$$

From Ex. 6.1 the allowable stresses are

$$F_b = 500 \text{ lbf/in}^2$$

$$F_s = 24{,}000 \text{ lbf/in}^2$$

The bending moment produced by the beam self-weight is

$$M_s = 7.76 \text{ ft-kips}$$

The stresses produced in the masonry and in the reinforcement by the beam self-weight are

$$f_{bs} = \frac{2M_s}{jkb_wd^2}$$

$$= \frac{(2)(7.76 \text{ ft-kips})\left(12\,\frac{\text{in}}{\text{ft}}\right)\left(1000\,\frac{\text{lbf}}{\text{kip}}\right)}{(0.894)(0.318)(7.63 \text{ in})(45 \text{ in})^2}$$

$$= 42 \text{ lbf/in}^2$$

$$f_{ss} = \frac{M_s}{jdA_s}$$

$$= \frac{(7.76 \text{ ft-kips})\left(12\,\frac{\text{in}}{\text{ft}}\right)\left(1000\,\frac{\text{lbf}}{\text{kip}}\right)}{(0.894)(45 \text{ in})(0.88 \text{ in}^2)}$$

$$= 2630 \text{ lbf/in}^2$$

The bending moment produced by the live load is

$$M_c = 50 \text{ ft-kips}$$

The stresses produced in the masonry and in the reinforcement by the live load are

$$f_{bc} = \frac{2M_c}{jkb_w d^2}$$

$$= \frac{(2)(50 \text{ ft-kips})\left(12\ \dfrac{\text{in}}{\text{ft}}\right)\left(1000\ \dfrac{\text{lbf}}{\text{kip}}\right)}{(0.894)(0.318)(7.63 \text{ in})(45 \text{ in})^2}$$

$$= 273 \text{ lbf/in}^2$$

$$f_{sc} = \frac{M_c}{jdA_s}$$

$$= \frac{(50 \text{ ft-kips})\left(12\ \dfrac{\text{in}}{\text{ft}}\right)\left(1000\ \dfrac{\text{lbf}}{\text{kip}}\right)}{(0.894)(45 \text{ in})(0.88 \text{ in}^2)}$$

$$= 16{,}948 \text{ lbf/in}^2$$

Applying load combination 5 of ASCE Sec. 2.4.1 gives the combined stresses caused by the beam self-weight and the wind load as

$$f_b = f_{bs} + f_{bw}$$

$$= 42\ \frac{\text{lbf}}{\text{in}^2} + 225\ \frac{\text{lbf}}{\text{in}^2}$$

$$= 267 \text{ lbf/in}^2$$

$$< F_b \quad [\text{satisfactory}]$$

$$f_s = f_{ss} + f_{sw}$$

$$= 2630\ \frac{\text{lbf}}{\text{in}^2} + 11{,}631\ \frac{\text{lbf}}{\text{in}^2}$$

$$= 14{,}261 \text{ lbf/in}^2$$

$$< F_s \quad [\text{satisfactory}]$$

Applying load combination 6 of ASCE Sec. 2.4.1 gives the combined stresses caused by the beam self-weight, the wind load, and the live load as

$$f_b = f_{bs} + 0.75f_{bw} + 0.75f_{bc}$$

$$= 42\ \frac{\text{lbf}}{\text{in}^2} + (0.75)\left(225\ \frac{\text{lbf}}{\text{in}^2}\right) + (0.75)\left(273\ \frac{\text{lbf}}{\text{in}^2}\right)$$

$$= 416 \text{ lbf/in}^2$$

$$< F_b \quad [\text{satisfactory}]$$

$$f_s = f_{ss} + 0.75f_{sw} + 0.75f_{sc}$$

$$= 2630\ \frac{\text{lbf}}{\text{in}^2} + (0.75)\left(11{,}631\ \frac{\text{lbf}}{\text{in}^2}\right)$$

$$\quad + (0.75)\left(16{,}948\ \frac{\text{lbf}}{\text{in}^2}\right)$$

$$= 24{,}064 \text{ lbf/in}^2$$

$$> F_s \quad [\text{unsatisfactory}]$$

The beam is inadequate.

3. DESIGN FOR SHEAR

Nomenclature

A_v	area of shear reinforcement	in^2
f_v	calculated shear stress in masonry	lbf/in^2
F_v	allowable shear stress in masonry	lbf/in^2
s	spacing of shear reinforcement	in
V	design shear force	kips

Shear in Flexural Members

The allowable shear stress in a flexural member without shear reinforcement is given by BCRMS Sec. 2.3.5.2.2 as

$$F_v = \sqrt{f'_m} \qquad \text{[BCRMS 2-20]}$$

$$\le 50 \text{ lbf/in}^2$$

BCRMS Commentary Sec. 2.3.5 requires that when this value of the shear stress is exceeded, shear reinforcement be provided to carry the full shear load with any contribution from the masonry neglected. The shear stress in a flexural member with shear reinforcement designed to take the entire shear force is limited by BCRMS Sec. 2.3.5.2.3 to

$$F_v = 3\sqrt{f'_m} \qquad \text{[BCRMS 2-23]}$$

$$\le 150 \text{ lbf/in}^2$$

When necessary, the dimensions of the masonry member must be increased to conform to this requirement.

The area of shear reinforcement required when the allowable shear stress in the masonry is exceeded is given by BCRMS Sec. 2.3.5.3 as

$$\frac{A_v}{s} = \frac{V}{F_s d} \qquad \text{[BCRMS 2-26]}$$

The spacing of shear reinforcement shall not exceed the lesser of $d/2$ or 48 in.

In accordance with BCRMS Sec. 2.3.5.2.1, the shear stress in masonry may be determined from the following expression.

$$f_v = \frac{V}{b_w d} \qquad \text{[BCRMS 2-19]}$$

As specified in BCRMS Sec. 2.3.5.5, the maximum design shear may be calculated at a distance of $d/2$ from the face of the support, provided that no concentrated load occurs between the face of the support and a distance $d/2$ from the face.

Example 6.3

For the masonry beam of Ex. 6.1, determine whether the shear reinforcement provided is adequate.

Masonry

Solution

The allowable shear stress without shear reinforcement is given by BCRMS Eq. (2-20) as

$$F_v = \sqrt{f'_m}$$
$$= \sqrt{1500 \ \frac{\text{lbf}}{\text{in}^2}}$$
$$= 38.7 \ \text{lbf/in}^2$$
$$\leq 50 \ \text{lbf/in}^2 \quad \left[\begin{array}{c}\text{satisfies BCRMS} \\ \text{Sec. 2.3.5.2.2}\end{array}\right]$$

The shear stress with shear reinforcement provided to carry the total shear force is limited by BCRMS Eq. (2-23) to

$$F_v = 3\sqrt{f'_m}$$
$$= 3\sqrt{1500 \ \frac{\text{lbf}}{\text{in}^2}}$$
$$= 116 \ \text{lbf/in}^2$$
$$\leq 150 \ \text{lbf/in}^2 \quad \left[\begin{array}{c}\text{satisfies BCRMS} \\ \text{Sec. 2.3.5.2.3}\end{array}\right]$$

The shear force at a distance of $d/2$ from each support is given by

$$V = \frac{w\,(l - d)}{2} + W$$
$$= \frac{\left(0.276 \ \dfrac{\text{kips}}{\text{ft}}\right)(14 \ \text{ft} - 3.75 \ \text{ft})}{2} + 20 \ \text{kips}$$
$$= 21.42 \ \text{kips}$$

The shear stress at a distance of $d/2$ from each support is given by BCRMS Eq. (2-19) as

$$f_v = \frac{V}{b_w d}$$
$$= \frac{(21.42 \ \text{kips})\left(1000 \ \dfrac{\text{lbf}}{\text{kip}}\right)}{(7.63 \ \text{in})(45 \ \text{in})}$$
$$= 62.38 \ \text{lbf/in}^2$$
$$> 38.7 \ \text{lbf/in}^2 \quad \left[\begin{array}{c}\text{shear reinforcement required to} \\ \text{carry total shear force}\end{array}\right]$$
$$< 116 \ \text{lbf/in}^2 \quad \left[\begin{array}{c}\text{satisfies BCRMS} \\ \text{Sec. 2.3.5.2.3}\end{array}\right]$$

The minimum area of shear reinforcement required is given by BCRMS Eq. (2-26) as

$$\frac{A_v}{s} = \frac{V}{F_s d}$$
$$= \frac{(21.42 \ \text{kips})\left(12 \ \dfrac{\text{in}}{\text{ft}}\right)}{\left(24 \ \dfrac{\text{kips}}{\text{in}^2}\right)(45 \ \text{in})}$$
$$= 0.238 \ \text{in}^2/\text{ft}$$

The shear reinforcement provided supplies a value of

$$\frac{A_v}{s} = 0.300 \ \text{in}^2/\text{ft}$$
$$> 0.238 \ \text{in}^2/\text{ft} \quad [\text{satisfactory}]$$

The shear reinforcement provided is adequate.

4. DESIGN OF MASONRY COLUMNS

Nomenclature

a	distance between column reinforcement	in
A_n	net effective area of column	in^2
A_s	area of reinforcement	in^2
A_{st}	area of laterally tied longitudinal reinforcement	in^2
A_t	transformed area of column, $A_n\big(1 + (2n - 1)(\rho)\big)$	in^2
A_{ts}	transformed area of reinforcement, $A_s(2n - 1)$	in^2
b	width of section	in
f_a	calculated compressive stress due to axial load only	lbf/in^2
f_m	calculated stress in the masonry	lbf/in^2
F_a	allowable compressive stress due to axial load only	lbf/in^2
h	effective height of column	in
I_n	net effective moment of inertia of column	in^4
I_t	transformed moment of inertia of column	in^4
M	design bending moment	ft-kips
P	design axial load	kips
P_a	allowable compressive force due to axial load only	kips
P_m	allowable compressive force on the masonry due to axial load only	kips
P_s	allowable compressive force on the reinforcement due to axial load only	kips
r	radius of gyration	in
S_t	transformed section modulus of column	in^3

Axial Compression in Columns

The allowable compressive stress in an axially loaded reinforced masonry column is given by

$$F_a = \frac{P_a}{A_n}$$

For columns having an h/r ratio not greater than 99, the allowable axial load is given by BCRMS Sec. 2.3.3.2.1 as

$$P_a = \left(0.25 f'_m A_n + 0.65 A_{st} F_s\right)\left(1 - \left(\frac{h}{140r}\right)^2\right)$$

[BCRMS 2-17]

For columns having an h/r ratio greater than 99, the allowable axial load is given by BCRMS Sec. 2.3.3.2. as

$$P_a = (0.25 f'_m A_n + 0.65 A_{st} F_s) \left(\frac{70r}{h} \right)^2 \quad \text{[BCRMS 2-18]}$$

To allow for accidental eccentricities, BCRMS Sec. 2.1.6.3 requires that a column be designed for a minimum eccentricity equal to 0.1 times each side dimension. When actual eccentricity exceeds the minimum eccentricity, the actual eccentricity shall be used.

Limitations are imposed on column dimensions in BCRMS Secs. 2.1.6.1 and 2.1.6.2. The minimum nominal column dimension is 8 in, and the ratio between the effective height and the least nominal dimension shall not exceed 25.

Limitations are imposed on column reinforcement in BCRMS Sec. 2.1.6.4. Longitudinal reinforcement is limited to a maximum of 4% and a minimum of 0.25% of the net column area, with at least four bars being provided. The compressive stress in reinforcement is limited by BCRMS Sec. 2.3.2.2.2 to the lesser of $0.4f_y$ or 24,000 lbf/in^2.

In accordance with BCRMS Sec. 2.1.6.5, lateral ties for the confinement of longitudinal reinforcement shall be not less than $^1/_4$ in diameter. Lateral ties shall be placed at a spacing not exceeding the least of 16 longitudinal bar diameters, 48 lateral tie diameters, or the least cross-sectional dimension of the column. Lateral ties shall be arranged such that every corner and alternate longitudinal bar shall have support provided by the corner of a lateral tie, and no bar shall be farther than 6 in clear on each side from a supported bar.

Example 6.4

The nominal 16 in square solid grouted concrete block masonry column shown in the following illustration has a specified strength of 1500 lbf/in^2, has a modulus of elasticity of 1,000,000 lbf/in^2, and is reinforced with four no. 6 grade 60 bars. The column supports an axial load of 100 kips, has a height of 15 ft, and may be considered pinned at each end. The 100 kips axial load may be considered dead load. Neglecting accidental eccentricity, determine whether the column is adequate. The self-weight of the column is 180 lbf/ft^2.

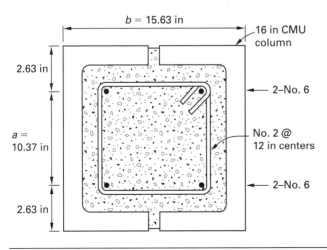

Solution

The relevant properties of the column are

$$b = \text{effective column width}$$
$$= 15.63 \text{ in}$$
$$h = \text{effective column height}$$
$$= 15 \text{ ft}$$
$$A_{st} = \text{reinforcement area}$$
$$= 1.76 \text{ in}^2$$
$$A_n = \text{effective column area}$$
$$= b^2$$
$$= (15.63 \text{ in})^2$$
$$= 244 \text{ in}^2$$
$$\rho = \frac{A_s}{A_n}$$
$$= \frac{1.76 \text{ in}^2}{244 \text{ in}^2}$$
$$= 0.0072$$
$$< 0.04$$
$$> 0.0025 \quad \begin{bmatrix} \text{satisfies BCRMS} \\ \text{Sec. 2.1.6.4} \end{bmatrix}$$
$$F_s = 0.4 f_y$$
$$= (0.4) \left(60 \frac{\text{kips}}{\text{in}^2} \right)$$
$$= 24 \text{ kips/in}^2$$

Including the self-weight of the column, the total load acting at the base of the column is

$$P = 100 \text{ kips} + \frac{(15 \text{ ft}) \left(180 \frac{\text{lbf}}{\text{ft}} \right)}{1000 \frac{\text{lbf}}{\text{kip}}}$$
$$= 102.7 \text{ kips}$$

The radius of gyration of the column is

$$r = \sqrt{\frac{I_n}{A_n}}$$
$$= 0.289b$$
$$= (0.289)(15.63 \text{ in})$$
$$= 4.52 \text{ in}$$

The slenderness ratio of the column is

$$\frac{h}{r} = \frac{(15 \text{ ft}) \left(12 \frac{\text{in}}{\text{ft}} \right)}{4.52 \text{ in}}$$
$$= 39.82$$
$$< 99 \quad \begin{bmatrix} \text{BCRMS Eq. (2-17)} \\ \text{is applicable.} \end{bmatrix}$$

The allowable column load is given by

$$P_a = (0.25f'_m A_n + 0.65A_{st}F_s)\left(1.0 - \left(\frac{h}{140r}\right)^2\right)$$

$$= \left(\begin{array}{c}(0.25)\left(1.5\,\dfrac{\text{kips}}{\text{in}^2}\right)(244\,\text{in}^2) \\[2mm] + (0.65)(1.76\,\text{in}^2)\left(24\,\dfrac{\text{kips}}{\text{in}^2}\right)\end{array}\right)$$

$$\times \left(1.0 - \left(\frac{39.82}{140}\right)^2\right)$$

$$= (91.5\,\text{kips} + 27.5\,\text{kips})(0.919)$$

$$= 84.1\,\text{kips} + 25.3\,\text{kips}$$

$$= P_m + P_s \quad \left[\begin{array}{c}\text{allowable loads on the} \\ \text{masonry and reinforcement}\end{array}\right]$$

$$= 109.4\,\text{kips}$$

$$> P \quad [\text{satisfactory}]$$

The column is adequate.

Combined Compression and Flexure

The allowable compressive stress in masonry due to combined axial load and flexure is given by BCRMS Sec. 2.3.3.2.2 as

$$F_b = \frac{1}{3}f'_m$$

In addition, the calculated compressive stress due to the axial load shall not exceed the allowable values given in BCRMS Sec. 2.2.3.1. For columns having an h/r ratio not greater than 99, this is

$$F_a = (0.25f'_m)\left(1.0 - \left(\frac{h}{140r}\right)^2\right)$$

[BCRMS 2-12]

For columns having an h/r ratio greater than 99, the allowable value is

$$F_a = (0.25f'_m)\left(\frac{70r}{h}\right)^2 \qquad \text{[BCRMS 2-13]}$$

When the axial load on the column causes a compressive stress larger than the tensile stress produced by the applied bending moment, the section is uncracked and stresses may be calculated by using the transformed section properties.[4] To allow for creep in the masonry,[6] the transformed reinforcement area is taken as $A_s(2n-1)$, and the resultant stresses at the extreme fibers of the section, as shown in Fig 6.3, are given by

$$f_m = f_a \pm f_b = \frac{P}{A_t} \pm \frac{M}{S_t}$$

Stress in the reinforcement is equal to $2n$ times the stress in the adjacent masonry.

Figure 6.3 *Uncracked Section Properties*

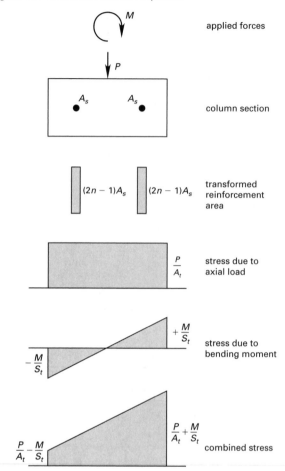

When the applied moment produces cracking in the section, the principle of superposition is no longer applicable. To determine the stresses on the section, the strain distribution over the section is estimated, and forces are determined as shown in Fig. 6.4. Internal forces on the section are compared with the applied loads, and the procedure is repeated until external and internal forces balance.

Example 6.5

The masonry column of Ex. 6.4, in addition to the vertical load supported, is subjected to a lateral force, due to wind, of 300 lbf/ft. Neglecting accidental eccentricity, determine whether the column is adequate.

Solution

From Ex. 6.4 and BCRMS Eq. (2-12), the allowable compressive stress in the masonry due to axial load is

$$F_a = (0.25f'_m)\left(1.0 - \left(\frac{h}{140r}\right)^2\right)$$

$$= \frac{P_m}{A_n} = \frac{(84.1\,\text{kips})\left(1000\,\dfrac{\text{lbf}}{\text{kip}}\right)}{244\,\text{in}^2}$$

$$= 345\,\text{lbf/in}^2$$

Figure 6.4 *Cracked Section Properties*

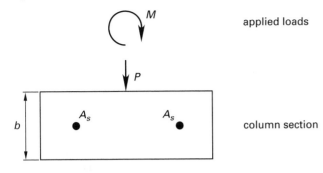

applied loads

column section

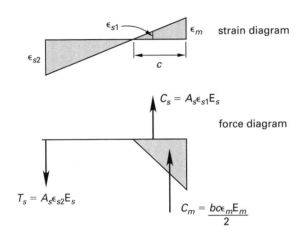

strain diagram

force diagram

Assuming that the section is not cracked, the transformed area of two no. 6 bars is

$$A_{st} = A_s(2n - 1)$$
$$= (0.88 \text{ in}^2)((2)(29) - 1)$$
$$= 50.2 \text{ in}^2$$

The transformed area of the column is

$$A_t = A_n(1 + (2n - 1)\rho)$$
$$= (244 \text{ in}^2)\left(1 + ((2)(29) - 1)(0.0072)\right)$$
$$= 344 \text{ in}^2$$

At midheight of the column, the stress in the masonry due to the axial load is

$$f_a = \frac{P}{A_t}$$
$$= \frac{(101.35 \text{ kips})\left(1000 \frac{\text{lbf}}{\text{kip}}\right)}{344 \text{ in}^2}$$
$$= 295 \text{ lbf/in}^2$$
$$< F_a \quad \begin{bmatrix} \text{BCRMS Sec. 2.3.3.2.2} \\ \text{is satisfied.} \end{bmatrix}$$

At midheight, the bending moment produced by the wind load is

$$M_w = \frac{ql^2}{8}$$
$$= \frac{\left(300 \frac{\text{lbf}}{\text{ft}}\right)(15 \text{ ft})^2}{(8)\left(1000 \frac{\text{lbf}}{\text{kip}}\right)}$$
$$= 8.44 \text{ ft-kips}$$

The column is pinned at the base, and the moment at midheight due to the accidental eccentricity is

$$M_e = (0.1)\left(\frac{Pb}{2}\right)$$
$$= (0.1)\left(\frac{(100 \text{ kips})(15.63 \text{ in})}{(2)\left(12 \frac{\text{in}}{\text{ft}}\right)}\right)$$
$$= 6.51 \text{ ft-kips}$$
$$< M_w$$

The bending moment produced by the wind load governs. The stresses in the extreme fibers of the column due to the wind moment are

$$f_b = \frac{M_w b}{2I_t} = \pm \frac{\begin{array}{c}(8.44 \text{ ft-kips})\left(12 \frac{\text{in}}{\text{ft}}\right) \\ \times \left(1000 \frac{\text{lbf}}{\text{kip}}\right)(15.63 \text{ in})\end{array}}{(2)(7660 \text{ in}^4)}$$
$$= \pm 103 \text{ lbf/in}^2$$
$$< f_a \quad \text{[The section is uncracked.]}$$

From Ex. 6.1, the allowable stresses produced by the bending moment are

$$F_b = 500 \text{ lbf/in}^2$$
$$F_s = 24{,}000 \text{ lbf/in}^2$$

Load combination 5 of ASCE Sec. 2.4.1 governs.

The maximum stress in the masonry due to combined axial and wind loading is

$$f_m = f_a + f_b$$
$$= 295 \frac{\text{lbf}}{\text{in}^2} + 103 \frac{\text{lbf}}{\text{in}^2}$$
$$= 398 \text{ lbf/in}^2$$
$$< F_b \text{ lbf/in}^2 \quad \text{[satisfactory]}$$

Masonry

The maximum compressive stress in the reinforcement due to combined axial and wind loading is given by

$$f_s = 2n \left(f_a + \frac{f_b a}{b} \right)$$

$$= (2)(29) \left(295 \; \frac{\text{lbf}}{\text{in}^2} + \frac{\left(103 \; \frac{\text{lbf}}{\text{in}^2} \right)(10.37 \text{ in})}{15.63 \text{ in}} \right)$$

$$= 21{,}074 \text{ lbf/in}^2$$

$$< F_s \text{ lbf/in}^2 \quad [\text{satisfactory}]$$

The column is adequate.

5. DESIGN OF SHEAR WALLS

Nomenclature

A_g	gross cross-sectional area of masonry	in^2
A_{sh}	area of horizontal reinforcement in shear wall	in^2
A_{sv}	area of vertical reinforcement in shear wall	in^2
d_v	actual depth of masonry in direction of shear	in
h	height of masonry shear wall	in
M	moment occurring simultaneously with V at the section under consideration	in-lbf
V	design shear force	lbf

Shear Wall Nominal Reinforcement Requirements

In accordance with IBC Sec. 2106.5.1, shear walls in seismic design category D and above designed to resist seismic forces, using the allowable stress method, shall be designed to resist 1.5 times the seismic forces calculated by IBC Ch. 16.

The reinforcement requirements for a shear wall depend on the shear force in the wall and on the seismic design category assigned to the structure. In accordance with BCRMS Sec. 1.14.2.2.5, shear reinforcement in special reinforced shear walls shall be anchored around vertical reinforcement with a standard hook. The minimum nominal reinforcement requirements for an intermediate reinforced wall assigned to seismic design category C are given in BCRMS Secs. 1.14.2.2.4 and 1.14.2.2.1 are shown in Fig. 6.5.

The minimum nominal reinforcement requirements for a special reinforced wall assigned to seismic design category D and above, for other than stack bond masonry, are given in BCRMS Sec. 1.14.6.3 and Sec. 1.14.2.2.5 and are shown in Fig. 6.6.

The minimum reinforcement requirements for a wall in stack bond masonry assigned to seismic design category E or F are given in BCRMS Secs. 1.14.7.3 and 1.14.6.3 and are shown in Fig. 6.7.

Figure 6.5 Reinforcement Details for Intermediate Reinforced Shear Wall in Seismic Design Category C

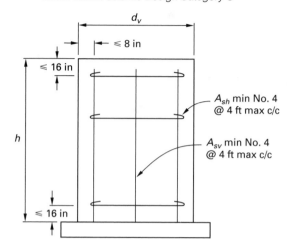

Figure 6.6 Reinforcement Details for Special Reinforced Shear Wall in Seismic Design Categories D, E, and F

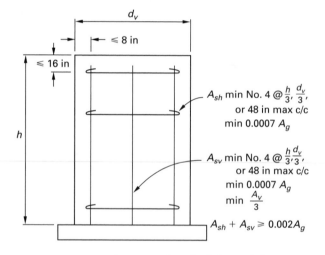

A_v = area of required shear reinforcement

Figure 6.7 Reinforcement Details for Stack Bond Shear Wall in Seismic Design Categories E and F

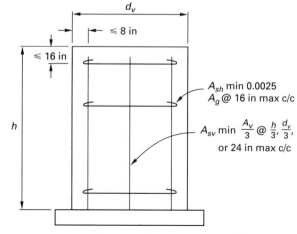

A_v = area of required shear reinforcement

Masonry Wall Without Shear Reinforcement

The allowable shear stress depends on the ratio M/Vd, M being the moment acting at the location where the applied shear force V is calculated. In a masonry wall without shear reinforcement and with $M/Vd < 1.0$, the allowable shear stress is given by BCRMS Sec. 2.3.5.2.2(b) as

$$F_v = \left(\frac{1}{3}\right)\left(4 - \frac{M}{Vd}\right)\sqrt{f'_m} \qquad \text{[BCRMS 2-21]}$$

$$\leq \left(80 - \frac{45M}{Vd}\right) \quad \text{[in lbf/in}^2\text{]}$$

In a masonry wall without shear reinforcement and with $M/Vd \geq 1.0$, the allowable shear stress is given by BCRMS Sec. 2.3.5.2.2(b) as

$$F_v = \sqrt{f'_m} \qquad \text{[BCRMS 2-22]}$$

$$\leq 35 \text{ lbf/in}^2$$

The shear stress in the masonry is determined from BCRMS Sec. 2.3.5.2.1 and BCRMS Commentary Sec. 2.3.5.3 as

$$f_v = \frac{V}{bd_v} \qquad \text{[BCRMS 2-19]}$$

For shear walls without shear reinforcement and with shear parallel to the plane of the wall, BCRMS Commentary Sec. 2.3.5.3 specifies the substitution of the overall depth of the wall d_v in place of the effective depth d. Similarly, for shear walls with horizontal shear reinforcement and with vertical reinforcement uniformly distributed along the depth of the wall, d_v may be substituted for d.

Example 6.6

The nominal 8 in solid grouted concrete block masonry shear wall shown in the following illustration has a specified strength of 1500 lbf/in² and a modulus of elasticity of 1,000,000 lbf/in². An in-plane wind load of 19 kips acts at the top of the wall, as shown, and this is the governing shear load. The wall is located in a structure assigned to seismic design category D. Determine the reinforcement required in the wall. Axial forces may be neglected.

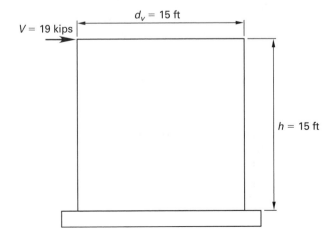

Solution

The allowable stresses, in accordance with BCRMS Secs. 2.3.2 and 2.3.3, are

$$F_b = \frac{1}{3}f'_m$$

$$= \left(\frac{1}{3}\right)\left(1500 \ \frac{\text{lbf}}{\text{in}^2}\right)$$

$$= 500 \text{ lbf/in}^2$$

$$F_s = 24{,}000 \text{ lbf/in}^2$$

The bending moment, produced by the wind load, at the base of the wall is

$$M = Vh = (19 \text{ kips})(15 \text{ ft})$$

$$= 285 \text{ ft-kips}$$

Assuming that two no. 6 reinforcing bars are located 4 in from each end of the wall, the relevant parameters of the wall are

$$b_w = 7.63 \text{ in}$$

$$d = 176 \text{ in}$$

$$A_s = 0.88 \text{ in}^2$$

$$n = 29$$

$$\rho = \frac{A_s}{b_w d} = \frac{0.88 \text{ in}^2}{(7.63 \text{ in})(176 \text{ in})}$$

$$= 0.000655$$

$$\rho n = (0.000655)(29)$$

$$= 0.0190$$

From Table A.2, in the appendix, the wall stresses caused by the wind load, in accordance with BCRMS Sec. 2.3, are

$$k = \sqrt{2\rho n + (\rho n)^2} - \rho n$$

$$= \sqrt{(2)(0.0190) + (0.0190)^2} - 0.0190$$

$$= 0.177$$

$$j = 1 - \frac{k}{3}$$

$$= 1 - \frac{0.177}{3}$$

$$= 0.941$$

$$f_b = \frac{2M}{jkb_w d^2}$$

$$= \frac{(2)(285 \text{ ft-kips})\left(12 \ \frac{\text{in}}{\text{ft}}\right)\left(1000 \ \frac{\text{lbf}}{\text{kip}}\right)}{(0.941)(0.177)(7.63 \text{ in})(176 \text{ in})^2}$$

$$= 174 \text{ lbf/in}^2$$

$$< 500 \text{ lbf/in}^2 \quad \text{[satisfactory]}$$

$$f_s = \frac{M}{jdA_s}$$

$$= \frac{(285 \text{ ft-kips})\left(12 \frac{\text{in}}{\text{ft}}\right)\left(1000 \frac{\text{lbf}}{\text{kip}}\right)}{(0.941)(176 \text{ in})(0.88 \text{ in}^2)}$$

$$= 23{,}466 \text{ lbf/in}^2$$

$$< 24{,}000 \text{ lbf/in}^2 \quad [\text{satisfactory}]$$

The flexural reinforcement provided is adequate.

The shear stress in the masonry wall is given by BCRMS Eq. (2-19) and BCRMS Commentary Sec. 2.3.5.3 as

$$f_v = \frac{V}{bd_v} = \frac{19{,}000 \text{ lbf}}{(7.63 \text{ in})(180 \text{ in})}$$

$$= 13.8 \text{ lbf/in}^2$$

The allowable stress is obtained by applying BCRMS Sec. 2.3.5.2.2.

$$\frac{M}{Vd} = \frac{Vh}{Vd}$$

$$= \frac{h}{d}$$

$$= \frac{(15 \text{ ft})\left(12 \frac{\text{in}}{\text{ft}}\right)}{176 \text{ in}}$$

$$= 1.02$$

$$> 1.0 \qquad \begin{bmatrix} \text{BCRMS Eq. (2-22)} \\ \text{is applicable.} \end{bmatrix}$$

Applying BCRMS Eq. (2-22), the allowable stress is the smaller value given by

- $F_v = \sqrt{f'_m}$

$$= \sqrt{1500 \frac{\text{lbf}}{\text{in}^2}}$$

$$= 38.7 \text{ lbf/in}^2$$

- $F_v = 35 \text{ lbf/in}^2 \quad [\text{governs}]$

$$> f_v \quad [\text{satisfactory}]$$

Hence, the masonry takes all the shear force, and nominal reinforcement is required as detailed in BCRMS Secs. 1.14.6.3 and 1.14.2.2.5 for a structure assigned to seismic design category D. The maximum allowable reinforcement spacing is 48 in, and the minimum specified horizontal and vertical reinforcement areas are

$$A_{sh} = A_{sv}$$

$$A_{sh} = 0.0007A_g$$

$$= (0.0007)(7.63 \text{ in})(12 \text{ in})$$

$$= 0.064 \text{ in}^2/\text{ft}$$

Providing no. 4 horizontal bars at 32 in on center gives a reinforcement area of

$$A_{sh} = 0.075 \text{ in}^2/\text{ft}$$

$$> 0.064 \text{ in}^2/\text{ft} \quad [\text{satisfactory}]$$

In addition to the flexural reinforcement of two no. 6 bars at each end of the wall, provide three no. 4 vertical bars at 48 in on center to give a combined vertical reinforcement of

$$A_{sv} = \frac{1.76 \text{ in}^2 + 0.60 \text{ in}^2}{15 \text{ ft}}$$

$$= 0.157 \text{ in}^2/\text{ft}$$

$$> 0.064 \text{ in}^2/\text{ft} \quad [\text{satisfactory}]$$

The sum of the horizontal and vertical reinforcement areas provided is

$$A_{sh} + A_{sv} = 0.075 \frac{\text{in}^2}{\text{ft}} + 0.157 \frac{\text{in}^2}{\text{ft}}$$

$$= 0.23 \text{ in}^2/\text{ft}$$

The required sum is

$$A_{sh} + A_{sv} = 0.002A_g$$

$$= (0.002)(7.63 \text{ in})\left(12 \frac{\text{in}}{\text{ft}}\right)$$

$$= 0.18 \text{ in}^2/\text{ft}$$

$$< 0.23 \text{ in}^2/\text{ft} \quad [\text{satisfactory}]$$

The required reinforcement details are shown in the following illustration.

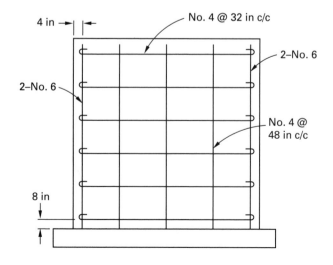

Masonry Wall with Shear Reinforcement

When the allowable shear stress of the masonry is exceeded, shear reinforcement shall be provided to carry the entire shear force without any contribution from the masonry. In a masonry wall with shear reinforcement designed to carry the entire shear force and with $M/Vd < 1.0$, the shear stress is limited by BCRMS Sec. 2.3.5.2.3(b) to

$$F_v = \left(\frac{1}{2}\right)\left(4 - \frac{M}{Vd}\right)\sqrt{f'_m} \qquad \text{[BCRMS 2-24]}$$

$$\leq \left(120 - \frac{45M}{Vd}\right) \quad \left[\text{in lbf/in}^2\right]$$

In a masonry wall with shear reinforcement designed to carry the entire shear force and with $M/Vd \geq 1.0$, the shear stress is limited by BCRMS Sec. 2.3.5.2.3(b) to

$$F_v = 1.5\sqrt{f'_m} \qquad \text{[BCRMS 2-25]}$$

$$\leq 75 \text{ lbf/in}^2$$

The area of shear reinforcement required when the allowable shear stress in the masonry is exceeded is given by BCRMS Sec. 2.3.5.3 as

$$\frac{A_v}{s} = \frac{V}{F_s d} \qquad \text{[BCRMS 2-26]}$$

The spacing of shear reinforcement shall not exceed the lesser of $d/2$ or 48 in. Reinforcement shall be provided perpendicular to the shear reinforcement and shall be at least equal to $A_v/3$. This perpendicular reinforcement shall be uniformly distributed and shall not exceed a spacing of 8 ft.

Example 6.7

The nominal 8 in solid grouted concrete block masonry shear wall shown in the following illustration has a specified strength of 1500 lbf/in² and a modulus of elasticity of 1,000,000 lbf/in². An in-plane wind load of 49.5 kips acts at the top of the wall, as shown, and this is the governing shear load. The wall is located in a structure assigned to seismic design category D and is laid in running board. Determine the reinforcement required in the wall.

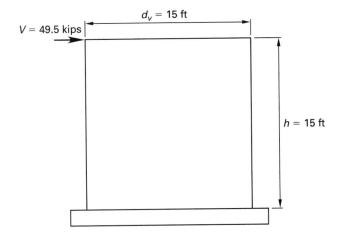

Solution

From Ex. 6.6, the allowable stresses are

$$F_b = 500 \text{ lbf/in}^2$$

$$F_s = 24,000 \text{ lbf/in}^2$$

The bending moment, produced by the wind load, at the base of the wall is

$$M = Vh$$
$$= (49.5 \text{ kips})(15 \text{ ft})$$
$$= 743 \text{ ft-kips}$$

Assuming that four no. 7 reinforcing bars are centered 8 in from each end of the wall, the relevant parameters of the wall are

$$b_w = 7.63 \text{ in}$$

$$d = 172 \text{ in}$$

$$A_s = 2.40 \text{ in}^2$$

$$n = 29$$

$$\rho = \frac{A_s}{b_w d}$$

$$= \frac{2.40 \text{ in}^2}{(7.63 \text{ in})(172 \text{ in})}$$

$$= 0.00183$$

$$\rho n = (0.00183)(29)$$

$$= 0.0530$$

From Table A.2, in the appendix, the wall stresses caused by the wind load, in accordance with BCRMS Sec. 2.3, are

$$k = \sqrt{2\rho n + (\rho n)^2} - \rho n$$
$$= 0.277$$

$$j = 1 - \frac{k}{3}$$
$$= 0.908$$

$$f_b = \frac{2M}{jkb_w d^2}$$

$$= \frac{(2)(743 \text{ ft-kips})\left(12 \frac{\text{in}}{\text{ft}}\right)\left(1000 \frac{\text{lbf}}{\text{kip}}\right)}{(0.908)(0.277)(7.63 \text{ in})(172 \text{ in})^2}$$

$$= 314 \text{ lbf/in}^2$$

$$< 500 \text{ lbf/in}^2 \quad \text{[satisfactory]}$$

$$f_s = \frac{M}{jd A_s}$$

$$= \frac{(743 \text{ ft-kips})\left(12 \frac{\text{in}}{\text{ft}}\right)\left(1000 \frac{\text{lbf}}{\text{kip}}\right)}{(0.908)(172 \text{ in})(2.40 \text{ in}^2)}$$

$$= 23,787 \text{ lbf/in}^2$$

$$< 24,000 \text{ lbf/in}^2 \quad \text{[satisfactory]}$$

The flexural reinforcement provided is adequate.

The shear stress in the masonry wall is given by BCRMS Eq. (2-19) and BCRMS Commentary Sec. 2.3.5.3 as

$$f_v = \frac{V}{bd_v}$$
$$= \frac{49{,}500 \text{ lbf}}{(7.63 \text{ in}) (180 \text{ in})}$$
$$= 36.0 \text{ lbf/in}^2$$

From Ex. 6.6,

$$\frac{M}{Vd} > 1.0 \quad \begin{bmatrix} \text{BCRMS Eq. (2-22)} \\ \text{is applicable.} \end{bmatrix}$$
$$F_v = 35 \text{ lbf/in}^2$$
$$< f_v$$

Hence, reinforcement is required to carry the entire shear force. In a masonry wall with shear reinforcement designed to carry the entire shear force and with $M/Vd > 1.0$, the shear stress for wind loading is limited by BCRMS Eq. (2-25) to the lesser of

- $F_v = 75 \text{ lbf/in}^2$
- $F_v = 1.5\sqrt{f_m'}$
  $$= 1.5\sqrt{1500 \ \frac{\text{lbf}}{\text{in}^2}}$$
  $$= 58 \text{ lbf/in}^2 \quad \text{[governs]}$$
  $$> f_v \quad \begin{bmatrix} \text{satisfies BCRMS} \\ \text{Sec. 2.3.5.2.3(b)} \end{bmatrix}$$

The minimum area of shear reinforcement required is given by BCRMS Eq. (2-26) as

$$\frac{A_v}{s} = \frac{V}{F_s d}$$
$$= \frac{(49.5 \text{ kips}) \left(12 \ \frac{\text{in}}{\text{ft}}\right)}{\left(24 \ \frac{\text{kips}}{\text{in}^2}\right) (172 \text{ in})}$$
$$= 0.144 \text{ in}^2/\text{ft}$$

Providing no. 4 horizontal bars at 16 in on center gives a reinforcement area of

$$A_{sh} = 0.150 \text{ in}^2/\text{ft}$$
$$> 0.144 \text{ in}^2/\text{ft} \quad \begin{bmatrix} \text{satisfies BCRMS} \\ \text{Sec. 2.3.5.3} \end{bmatrix}$$
$$> 0.0007 A_g \quad \begin{bmatrix} \text{satisfies BCRMS} \\ \text{Sec. 1.14.6.3} \end{bmatrix}$$

To comply with BCRMS Sec. 2.3.5.3.2, the vertical reinforcement shall not be less than

$$A_{sv} = \frac{A_v}{3}$$
$$= \frac{0.144 \ \frac{\text{in}^2}{\text{ft}}}{3}$$
$$= 0.048 \text{ in}^2/\text{ft}$$

Providing no. 4 vertical bars at 48 in on center gives a reinforcement area of

$$A_{sv} = 0.049 \text{ in}^2/\text{ft}$$
$$> 0.048 \text{ in}^2/\text{ft} \quad \text{[satisfactory]}$$

Including the flexural reinforcement, the total vertical reinforcement provided is

$$A_{sv} = 0.049 \ \frac{\text{in}^2}{\text{ft}} + \frac{4.8 \text{ in}^2}{15 \text{ ft}}$$
$$= 0.369 \text{ in}^2/\text{ft}$$
$$> 0.0007 A_g \quad \begin{bmatrix} \text{satisfies BCRMS} \\ \text{Sec. 1.14.6.3} \end{bmatrix}$$

The sum of the horizontal and vertical reinforcement areas provided is

$$A_{sh} + A_{sv} = 0.150 \ \frac{\text{in}^2}{\text{ft}} + 0.369 \ \frac{\text{in}^2}{\text{ft}}$$
$$= 0.519 \text{ in}^2/\text{ft}$$

The required sum, in accordance with BCRMS Sec. 1.14.6.3, is

$$A_{sh} + A_{sv} = 0.002 A_g$$
$$= (0.002)(7.63 \text{ in})(12 \text{ in})$$
$$= 0.18 \text{ in}^2/\text{ft}$$
$$< 0.519 \text{ in}^2/\text{ft} \quad \text{[satisfactory]}$$

The required reinforcement details are shown in the following illustration.

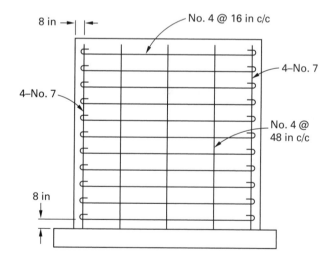

References

1. American Concrete Institute. *Building Code Requirements for Masonry Structures, (ACI 530-05/ ASCE 5-05/TMS 402-05)*. 2005.

2. International Code Council. *International Building Code*. 2006.

3. American Society of Civil Engineers. *Minimum Design Loads for Buildings and Other Structures, (ASCE/SEI 7-05)*. 2005.

4. Brandow, G.E., Ekwueme, C., and Hart, G.C. *Design of Reinforced Masonry Structures*. Concrete Masonry Association of California and Nevada. 2006.

5. Williams, A. *Structural Engineering License Review: Problems and Solutions*, 7th ed. 2010.

6. Concrete Masonry Association of California and Nevada. *Design of Masonry Walls to Resist In-Plane Forces*. 1999.

PRACTICE PROBLEMS

1. The nominal 8 in solid grouted concrete block masonry bearing wall shown in the following illustration has a specified strength of 1500 lbf/in² and a modulus of elasticity of 1,000,000 lbf/in². The wall supports an axial load, including its own weight, of 20 kips/ft, has a height of 15 ft, and may be considered pinned at the top and bottom. What is the reinforcement required in the wall? Ignore accidental eccentricity. The wall is not part of the lateral-force resisting system.

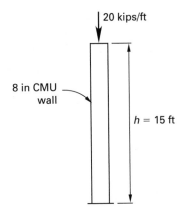

2. The nominal 8 in solid grouted concrete block masonry retaining wall shown in the following illustration has a specified strength of 1500 lbf/in² and a modulus of elasticity of 1,000,000 lbf/in². The reinforcement consists of no. 4 grade 60 bars at 16 in centers. The wall retains a soil with an equivalent fluid pressure of 30 lbf/ft²/ft, and the self-weight of the wall may be neglected. Is the wall adequate?

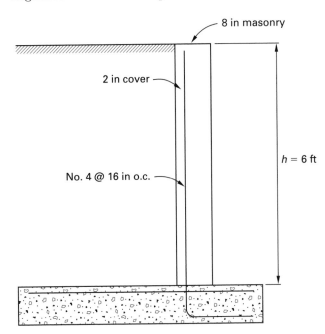

3. The 8 in solid grouted concrete block masonry beam shown in the following illustration may be considered simply supported over an effective span of 15 ft. The masonry has a compressive strength of 1500 lbf/in^2 and a modulus of elasticity of 1,000,000 lbf/in^2, and the reinforcement consists of two no. 7 grade 60 bars. The effective depth is 36 in, the overall depth is 40 in, and the beam is laterally braced at both ends. Is the beam adequate in flexure to support a uniformly distributed load, including its own weight, of 2000 lbf/ft?

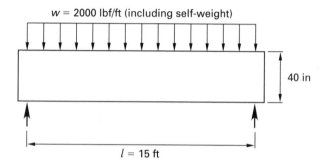

$w = 2000$ lbf/ft (including self-weight)

40 in

$l = 15$ ft

4. For the masonry beam of Prob. 3, what is the shear reinforcement required at each support?

SOLUTIONS

1. For axial loading, assuming the vertical reinforcement consists of no. 5 bars at 16 in centers and considering a 1 ft length of wall, the relevant parameters of the wall are

$$b = 12 \text{ in}$$

$$d_n = \text{nominal depth of wall}$$

$$= 7.63 \text{ in}$$

$$h = \text{effective column height}$$

$$= 15 \text{ ft}$$

$$A_s = \text{reinforcement area}$$

$$= 0.23 \text{ in}^2$$

$$A_n = \text{effective column area}$$

$$= bd_n$$

$$= (12 \text{ in})(7.63 \text{ in})$$

$$= 91.56 \text{ in}^2$$

$$\rho = \frac{A_s}{A_n}$$

$$= \frac{0.23 \text{ in}^2}{91.56 \text{ in}^2}$$

$$= 0.00251$$

$$< 0.04$$

$$> 0.0025 \quad \begin{bmatrix} \text{satisfies BCRMS} \\ \text{Sec. 2.1.6.4} \end{bmatrix}$$

$$F_s = 0.4f_y \quad [\text{BCRMS Sec. 2.3.2.2.2}]$$

$$= (0.4)\left(60 \ \frac{\text{kips}}{\text{in}^2}\right)$$

$$= 24 \text{ kips/in}^2$$

$$P = 20 \text{ kips}$$

The radius of gyration of the wall is

$$r = \sqrt{\frac{I_n}{A_n}}$$

$$= 0.289d_n$$

$$= (0.289)(7.63 \text{ in})$$

$$= 2.21 \text{ in}$$

The slenderness ratio of the wall is

$$\frac{h}{r} = \frac{(15 \text{ ft})\left(12 \ \frac{\text{in}}{\text{ft}}\right)}{2.21 \text{ in}}$$

$$= 81.63$$

$$< 99 \quad \begin{bmatrix} \text{BCRMS Eq. (2-17)} \\ \text{is applicable.} \end{bmatrix}$$

Ignoring the vertical reinforcement in conformity with BCRMS Sec. 2.3.2.2.1, the allowable wall load is given by BCRMS Eq. (2-12) as

$$P_a = (0.25 f'_m A_n)\left(1.0 - \left(\frac{h}{140r}\right)^2\right)$$

$$= (0.25)\left(1.5\ \frac{\text{kips}}{\text{in}^2}\right)(91.56\ \text{in}^2)\left(1.0 - \left(\frac{81.63}{140}\right)^2\right)$$

$$= (34.34\ \text{kips})(0.660)$$

$$= 22.66\ \text{kips}$$

$$> P \quad [\text{satisfactory}]$$

The wall is adequate for axial loading, and in accordance with BCRMS Sec. 1.14.5.2.3, the necessary minimum horizontal reinforcement is no. 4 bars at 48 in centers. The required reinforcement details are shown in the following illustration.

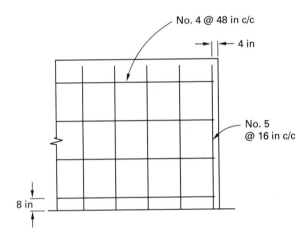

2. The allowable stresses, in accordance with BCRMS Secs. 2.3.2 and 2.3.3, are

$$F_b = \frac{1}{3} f'_m$$

$$= \left(\frac{1}{3}\right)\left(1500\ \frac{\text{lbf}}{\text{in}^2}\right)$$

$$= 500\ \text{lbf/in}^2$$

$$F_s = 24{,}000\ \text{lbf/in}^2$$

At the base of the wall, the bending moment produced in a 1 ft length of wall by the backfill is

$$M = \frac{qh^3}{6}$$

$$= \frac{\left(30\ \dfrac{\text{lbf}}{\text{ft}^2}\right)(6\ \text{ft})^3}{6}$$

$$= 1080\ \text{ft-lbf}$$

With vertical reinforcement consisting of no. 4 bars at 16 in centers located with 2 in cover to the earth face, the relevant parameters of the wall are

$$b_w = 12\ \text{in}$$

$$d = 7.63\ \text{in} - 2\ \text{in} - 0.25\ \text{in}$$

$$= 5.38\ \text{in}$$

$$A_s = 0.15\ \text{in}^2$$

$$n = 29$$

$$\rho = \frac{A_s}{b_w d}$$

$$= \frac{0.15\ \text{in}^2}{(12\ \text{in})(5.38\ \text{in})}$$

$$= 0.00232$$

$$\rho n = (0.00232)(29)$$

$$= 0.0674$$

From Table A.2, in the appendix, the stresses, in accordance with BCRMS Sec. 2.3, are

$$k = \sqrt{2\rho n + (\rho n)^2} - \rho n$$

$$= 0.306$$

$$j = 1 - \frac{k}{3}$$

$$= 0.898$$

$$f_b = \frac{2M}{jkb_w d^2}$$

$$= \frac{(2)(1080\ \text{ft-lbf})\left(12\ \dfrac{\text{in}}{\text{ft}}\right)}{(0.898)(0.306)(12\ \text{in})(5.38\ \text{in})^2}$$

$$= 272\ \text{lbf/in}^2$$

$$< F_b \quad [\text{satisfactory}]$$

$$f_s = \frac{M}{jdA_s}$$

$$= \frac{(1080\ \text{ft-lbf})\left(12\ \dfrac{\text{in}}{\text{ft}}\right)}{(0.898)(5.38\ \text{in})(0.15\ \text{in}^2)}$$

$$= 17{,}880\ \text{lbf/in}^2$$

$$< F_s \quad [\text{satisfactory}]$$

The wall is adequate.

3. The allowable stresses, in accordance with BCRMS Secs. 2.3.2 and 2.3.3, are

$$F_b = \frac{1}{3} f'_m$$

$$= \left(\frac{1}{3}\right)\left(1500\ \frac{\text{lbf}}{\text{in}^2}\right)$$

$$= 500\ \text{lbf/in}^2$$

$$F_s = 24{,}000\ \text{lbf/in}^2$$

At midspan, the bending moment produced by the distributed load is

$$M = \frac{wl^2}{8} = \frac{\left(2000 \; \frac{\text{lbf}}{\text{ft}}\right) (15 \; \text{ft})^2}{(8) \left(1000 \; \frac{\text{lbf}}{\text{kip}}\right)}$$

$$= 56.25 \; \text{ft-kips}$$

The relevant parameters of the beam are

$$E_m = 1{,}000{,}000 \; \frac{\text{lbf}}{\text{in}^2}$$

$$E_s = 29{,}000{,}000 \; \frac{\text{lbf}}{\text{in}^2}$$

$$b_w = 7.63 \; \text{in}$$

$$d = 36 \; \text{in}$$

$$l = 15 \; \text{ft}$$

$$A_s = 1.20 \; \text{in}^2$$

$$\frac{l}{b_w} = \frac{(15 \; \text{ft}) \left(12 \; \frac{\text{in}}{\text{ft}}\right)}{7.63 \; \text{in}}$$

$$= 23.6$$

$$< 32 \quad \begin{bmatrix} \text{satisfies BCRMS} \\ \text{Sec. 2.3.3.4.4} \end{bmatrix}$$

$$n = \frac{E_s}{E_m}$$

$$= \frac{29{,}000{,}000 \; \frac{\text{lbf}}{\text{in}^2}}{1{,}000{,}000 \; \frac{\text{lbf}}{\text{in}^2}}$$

$$= 29$$

$$\rho = \frac{A_s}{b_w d}$$

$$= \frac{1.20 \; \text{in}^2}{(7.63 \; \text{in}) (36 \; \text{in})}$$

$$= 0.00437$$

$$\rho n = (0.00437)(29)$$

$$= 0.127$$

From Table A.2, in the appendix, the beam stresses, in accordance with BCRMS Sec. 2.3, are

$$k = \sqrt{2\rho n + (\rho n)^2} - \rho n$$

$$= 0.393$$

$$j = 1 - \frac{k}{3}$$

$$= 0.869$$

$$f_b = \frac{2M}{jkb_w d^2}$$

$$= \frac{(2) (56.25 \; \text{ft-kips}) \left(12 \; \frac{\text{in}}{\text{ft}}\right) \left(1000 \; \frac{\text{lbf}}{\text{kip}}\right)}{(0.869) (0.393) (7.63 \; \text{in}) (36 \; \text{in})^2}$$

$$= 400 \; \text{lbf/in}^2$$

$$< F_b \quad \text{[satisfactory]}$$

$$f_s = \frac{M}{jdA_s}$$

$$= \frac{(56.25 \; \text{ft-kips}) \left(12 \; \frac{\text{in}}{\text{ft}}\right) \left(1000 \; \frac{\text{lbf}}{\text{kip}}\right)}{(0.869) (36 \; \text{in}) \left(1.20 \; \text{in}^2\right)}$$

$$= 17{,}980 \; \text{lbf/in}^2$$

$$< F_s \quad \text{[satisfactory]}$$

The beam is adequate in flexure.

4. The allowable shear stress without shear reinforcement is given by BCRMS Eq. (2-20) as

$$F_v = \sqrt{f'_m}$$

$$= \sqrt{1500 \; \frac{\text{lbf}}{\text{in}^2}}$$

$$= 38.7 \; \text{lbf/in}^2$$

$$\leq 50 \; \text{lbf/in}^2 \quad \begin{bmatrix} \text{satisfies BCRMS} \\ \text{Sec 2.3.5.2.2} \end{bmatrix}$$

The shear stress with shear reinforcement provided to carry the total shear force is limited by BCRMS Eq. (2-23) to

$$F_v = 3\sqrt{f'_m}$$

$$= 3\sqrt{1500}$$

$$= 116 \; \text{lbf/in}^2$$

$$\leq 150 \; \text{lbf/in}^2 \quad \begin{bmatrix} \text{satisfies BCRMS} \\ \text{Sec. 2.3.5.2.3} \end{bmatrix}$$

The shear force at a distance of $d/2$ from each support is given by

$$V = \frac{w(l-d)}{2}$$

$$= \frac{\left(2.0 \; \frac{\text{kips}}{\text{ft}}\right) (15 \; \text{ft} - 3 \; \text{ft})}{2}$$

$$= 12.0 \; \text{kips}$$

The shear stress at a distance of $d/2$ from each support is given by BCRMS Eq. (2-19) as

$$f_v = \frac{V}{b_w d}$$

$$= \frac{(12.0 \text{ kips}) \left(1000 \, \dfrac{\text{lbf}}{\text{kip}}\right)}{(7.63 \text{ in}) (36 \text{ in})}$$

$$= 43.69 \text{ lbf/in}^2$$

$$> 38.7 \text{ lbf/in}^2 \quad \left[\begin{array}{c} \text{shear reinforcement required to} \\ \text{carry total shear force} \end{array}\right]$$

$$< 116 \text{ lbf/in}^2 \quad \left[\begin{array}{c} \text{satisfies BCRMS} \\ \text{Sec. 2.3.5.2.3} \end{array}\right]$$

The minimum area of shear reinforcement required is given by BCRMS Eq. (2-26) as

$$\frac{A_v}{s} = \frac{V}{F_s d}$$

$$= \frac{(12.0 \text{ kips}) \left(12 \, \dfrac{\text{in}}{\text{ft}}\right)}{\left(24 \, \dfrac{\text{kips}}{\text{in}^2}\right) (36 \text{ in})}$$

$$= 0.167 \text{ in}^2/\text{ft}$$

Provide no. 3 stirrups at 8 in centers to give a value of

$$\frac{A_v}{s} = 0.165 \text{ in}^2/\text{ft}$$

$$\approx 0.167 \text{ in}^2/\text{ft} \quad [\text{satisfactory}]$$

Masonry

7 Seismic Design: International Building Code Lateral Force Procedure[a]

1. Equivalent Lateral Force Procedure 7-2
2. Vertical Distribution of Seismic Forces . . . 7-10
3. Diaphragm Loads 7-11
4. Story Drift 7-12
5. *P*-Delta Effects 7-13
6. Simplified Lateral Force Procedure 7-14
7. Seismic Load on an Element of a Structure . 7-18
   Practice Problems 7-20
   Solutions 7-21

Nomenclature

A_x	area of diaphragm immediately above the story	–
C_u	coefficient for upper limit on calculated period from ASCE Table 12.8-1	–
C_d	deflection amplification factor from ASCE Table 12.2-1	–
C_s	seismic response coefficient specified in ASCE Sec. 12.8.1	–
D	dead load applied to a structural element	lbf or kips
E	calculated seismic load on an element of a structure resulting from both horizontal and vertical earthquake induced forces as given by ASCE Eqs. (12.4-1) and (12.4-2)	lbf or kips
f_i	design seismic lateral force at level i	lbf or kips
F_a	short-period amplification factor	–
F_p	force on diaphragm	lbf or kips
F_v	long-period amplification factor	–
F_x	design seismic lateral force at level x as specified in ASCE Sec. 12.8.3	lbf or kips
g	acceleration due to gravity	32.2 ft/sec^2 or 386 in/sec^2
h_i	height above the base to level i	ft
h_n	height of the roof above the base, not including the height of penthouses or parapets	ft
h_{sx}	story height below level x	ft
h_x	height above the base to level x	ft
I_E	seismic importance factor	–
k	distribution exponent given in ASCE Sec. 12.8.3	–
k_i	stiffness of story i	lbf/in or kips/in
L	superimposed floor live load	lbf or kips
L_r	superimposed roof live load	lbf or kips
M_P	primary moment	kip-ft
M_S	secondary moment	kip-ft
N	number of stories	–
P_x	total unfactored vertical design load at and above level x	lbf or kips
Q_E	effect of horizontal seismic forces	lbf or kips
R	response modification coefficient for a specific structural system from ASCE Table 12.2-1	–
S	snow load applied to a structural element	lbf or kips
S_1	maximum considered response acceleration for a period of 1.0 sec	–
S_a	design spectral response acceleration	–
S_{DS}	design spectral response acceleration at a period of 0.2 sec	–
S_{D1}	design spectral response acceleration at a period of 1.0 sec	–
S_{MS}	modified spectral response acceleration at a period of 0.2 sec	–
S_{M1}	modified spectral response acceleration at a period of 1.0 sec	–
S_S	maximum considered response acceleration for a period of 0.2 sec	–
T	fundamental period of vibration, defined in ASCE Sec. 12.8.2	sec
T_a	approximate fundamental period of vibration determined using ASCE Sec. 12.8.2.1	sec
T_L	long-period transition period	sec
T_0	defined in ASCE Sec. 11.4.5 as $0.2S_{D1}/S_{DS}$	–
T_S	defined in ASCE Sec. 11.4.5 as S_{D1}/S_{DS}	–
V	total seismic base shear	lbf or kips
V_x	total shear force at level x	lbf or kips
V_S	design base shear	–

[a]The IBC[1] adopts by reference the American Society of Civil Engineers (ASCE)[2] standard for many of its code requirements. ASCE Ch. 14 and ASCE App. 11A are not adopted, and ASCE Secs. 12.3.1.1 and 17.5.4.2 are modified.

Seismic

V_Y base shear at formation of the collapse mechanism   –

w_i seismic dead load located at level i lbf or kips

w_p seismic dead load tributary to diaphragm   lbf or kips

w_x seismic dead load located at level x lbf or kips

W wind load applied to a structural element   lbf or kips

W effective seismic weight defined in ASCE Sec. 12.7.2   lbf or kips

$\sum F_i$ total shear force at level i lbf or kips

$\sum w_i$ total seismic dead load at level i and above   lbf or kips

Symbols

β ratio of shear demand to shear capacity for the story between levels x and x-1 as defined in ASCE Sec. 12.8.7   –

δ_x amplified horizontal deflection at level x, defined in ASCE Sec. 12.8.6   in

δ_{xe} horizontal deflection at level x, determined by an elastic analysis, as defined in ASCE Sec. 12.8.6   in

Δ design story drift, occurring simultaneously with the story shear V_x, defined in ASCE Sec. 12.8.6, and calculated using the amplification factor C_d in

Δ_A allowable story drift, defined in ASCE Table 12.12-1   in

Ω_0 overstrength factor tabulated in ASCE Table 12.2-1   –

θ stability coefficient defined in ASCE Sec. 12.8.7   –

1. EQUIVALENT LATERAL FORCE PROCEDURE

Determination of the seismic response of a structure depends on several factors, including ground motion parameters, site classification, site coefficient, adjusted response acceleration, design spectral response acceleration, importance factor, seismic design category, classification of the structural system, response modification coefficient, deflection amplification factor, overstrength factor, effective seismic weight, fundamental period of vibration, and seismic response coefficient. A summary of these factors follows.

Ground Motion Parameters

Ground motion parameters defined in ASCE Sec. 11.4.1 are values of the maximum considered ground acceleration that may be experienced at a specific location. As specified in ASCE Sec. 11.2, these are the most severe earthquake effects considered by the code. They are representative of a seismic event with a 2% probability of being exceeded in 50 yr[4] and have a recurrence interval of 2500 yr. Two values of the ground acceleration are required and these are designated S_S and S_1. S_S represents the 5% damped, maximum considered earthquake spectral response acceleration for a period of 0.2 sec for structures founded on rock (site classification B) and is applicable to short period structures. S_1 represents the 5% damped, maximum considered earthquake spectral response acceleration for a period of 1 sec for structures founded on rock and is applicable to structures with longer periods. Values of the ground accelerations S_S and S_1 are mapped in ASCE Fig. 22-1 through Fig. 22.14. The parameters are given as a percentage of the acceleration due to gravity.

Site Classification Characteristics

Site classification is defined in ASCE Sec. 11.4.2 and ASCE Table 20.3-1. Six different soil types are specified and range from site class A, which consists of hard rock, through site class F, which consists of peat, highly plastic clay, or collapsible soil. The soil profile may be determined on site from the average shear wave velocity in the top 100 ft of material. Alternatively, for site classification types C, D, or E, the classification may be made by measuring the standard penetration resistance or undrained shear strength of the material. An abbreviated listing of the site classifications is provided in Table 7.1.

Table 7.1 Site Classification Definitions

site classification	soil profile name	shear wave velocity (ft/sec)
A	hard rock	> 5000
B	rock	2500 to 5000
C	soft rock	1200 to 2500
D	stiff soil	600 to 1200
E	soft soil	< 600
F	–	–

Soil classification type A has the effect of reducing the ground response by 20%. Soil classification type E is defined as soft soil and has the effect of increasing the long period ground response by up to 350%. When soil parameters are unknown, in accordance with ASCE Sec. 11.4.2, soil classification type D may be assumed unless the building official determines that soil classification types E or F are likely to be present at the site.

Site Coefficients

Site coefficients are amplification factors applied to the maximum considered ground acceleration and are a function of the site classification. F_a is the short-period or acceleration-based amplification factor and is tabulated in ASCE Table 11.4-1. F_v is the long-period or velocity-based amplification factor and is tabulated in ASCE Table 11.4-2. ASCE Tables 11.4-1 and 11.4-2 are combined and reproduced in Table 7.2. Linear interpolation may be used to obtain intermediate values.

Adjusted Response Accelerations

The maximum considered ground accelerations must be adjusted by the site coefficients to allow for the site classification effects. ASCE Sec. 11.4.3 defines the modified spectral response accelerations at short periods and at a period of 1 sec as

$$S_{MS} = F_a S_S$$
$$S_{M1} = F_v S_1$$

Design Spectral Response Acceleration Parameters

The relevant design parameters are defined in ASCE Sec. 11.4.4 and are given by

S_{DS} = 5% damped design spectral response acceleration for a period of 0.2 sec

$$= \frac{2S_{MS}}{3}$$

S_{D1} = 5% damped design spectral response acceleration for a period of 1 sec

$$= \frac{2S_{M1}}{3}$$

Example 7.1

The two-story, reinforced concrete, moment-resisting frame shown in the following illustration is located on a site with a soil profile of stiff soil having a shear wave

velocity of 600 ft/sec. The 5% damped, maximum considered earthquake spectral response accelerations are obtained from the ASCE standard. They are $S_S = 1.5g$ and $S_1 = 0.7g$. Determine the 5% damped design spectral response accelerations S_{DS} and S_{D1}.

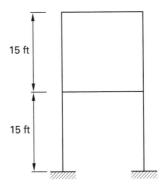

Solution

From ASCE Table 20.3-1 or from Table 7.1, the applicable site classification for stiff soil with a shear wave velocity of 600 ft/sec is site classification D. The site coefficients for this site classification and for the given values of the 5% damped, maximum considered earthquake spectral response accelerations are obtained from ASCE Tables 11.4-1 and 11.4-2 or from Table 7.2 as

$$F_a = 1.0$$
$$F_v = 1.5$$

Hence, the adjusted spectral response accelerations are given by ASCE Sec. 11.4.3 as

$$S_{MS} = F_a S_S$$
$$= (1.0)(1.5g)$$
$$= 1.5g$$
$$S_{M1} = F_v S_1$$
$$= (1.5)(0.7g)$$
$$= 1.05g$$

Table 7.2 *Site Coefficients F_a, Corresponding to S_S, and F_v, corresponding to S_1*

site classification	response acceleration S_S					response acceleration S_1				
	≤ 0.25	0.50	0.75	1.00	≥ 1.25	≤ 0.1	0.2	0.3	0.4	≥ 0.5
A	0.8	0.8	0.8	0.8	0.8	0.8	0.8	0.8	0.8	0.8
B	1.0	1.0	1.0	1.0	1.0	1.0	1.0	1.0	1.0	1.0
C	1.2	1.2	1.1	1.0	1.0	1.7	1.6	1.5	1.4	1.3
D	1.6	1.4	1.2	1.1	1.0	2.4	2.0	1.8	1.6	1.5
E	2.5	1.7	1.2	0.9	0.9	3.5	3.2	2.8	2.4	2.4
F	(a)	(a)	(a)	(a)	(a)	(a)	(a)	(a)	(a)	(a)

NOTE: (a) Site-specific geotechnical investigation and dynamic site response analysis required except for structures with $T \leq 0.5$ sec.

Seismic

The 5% damped design spectral response accelerations are given by ASCE Sec. 11.4.4 as

$$S_{DS} = \frac{2S_{MS}}{3}$$
$$= \frac{2(1.5g)}{3}$$
$$= 1.0g$$
$$S_{D1} = \frac{2S_{M1}}{3}$$
$$= \frac{2(1.05g)}{3}$$
$$= 0.7g$$

Occupancy Category and Importance Factors

In accordance with ASCE Sec. 1.5.1, each structure shall be assigned to an occupancy category, depending on the nature of its *occupancy*, with the corresponding *importance factor* indicated in ASCE Table 11.5-1. Table 7.3 lists the occupancy categories, seismic use groups, and seismic importance factors.

Table 7.3 *Occupancies and Importance Factors*

occupancy category	occupancy type	importance factor, I
I	low hazard structures	1.00
II	standard occupancy structures	1.00
III	assembly structures	1.25
IV	essential or hazardous structures	1.50

Category IV structures are those housing essential facilities that are required for post-earthquake recovery. Also included in category IV are structures containing substantial quantities of highly toxic substances that would endanger the safety of the public if released. Essential facilities are defined in ASCE Table 1-1 as hospitals, fire and police stations, emergency response centers, and buildings housing utilities and equipment required for these facilities. In order to ensure that category IV facilities remain functional after an upper level earthquake, an importance factor, I, of 1.5 is assigned to these facilities. This has the effect of increasing the design seismic forces by 50%, and raises the seismic level at which inelastic behavior occurs and the level at which the operation of essential facilities is compromised.

Category III structures are facilities that, if they failed, would become a substantial public hazard because of their high occupant load. These facilities are buildings where more than 300 people congregate in one area, schools with a capacity exceeding 250, colleges with a capacity exceeding 500, health care facilities with a capacity of 50 or more that do not have emergency treatment facilities, jails, and power stations. Also included are facilities containing explosive or toxic substances in a quantity not exceeding the exempt amounts in IBC Table 307.1(2). These structures are allocated a seismic importance factor, I, of 1.25.

Category II structures comprise standard occupancy structures and are allocated an importance factor of 1.00. Standard occupancy structures consist of residential, commercial, and office buildings.

Category I structures comprise low-hazard structures and are allocated an importance factor, I, of 1.00. Low-hazard structures consist of agricultural facilities, temporary facilities and minor storage facilities.

Determination of Seismic Design Category

Structures are assigned to a *seismic design category* based on their occupancy category and the design spectral response coefficients S_{DS} and S_{D1}. The seismic design category is defined in ASCE Sec. 11.6 and ASCE Tables 11.6-1 and 11.6-2, and establishes the design and detailing requirements necessary in a structure. The seismic design category is determined twice, first as a function of S_{DS} using ASCE Table 11.6-1, and then as a function of S_{D1} using ASCE Table 11.6-2. The most severe seismic design category governs. ASCE Table 11.6-1 and ASCE Table 11.6-2 are combined and reproduced in Table 7.4.

Six seismic design categories are defined, categories A through F, and these establish the design and detailing requirements necessary in a structure. Seismic design category A is applicable to structures in locations where anticipated ground movements are minimal. ASCE Sec. 11.7 specifies requirements to ensure the integrity of the structure in the event of a minor earthquake. Seismic design category B is applicable to structures in occupancy categories I, II, and III in regions of moderate seismicity. Seismic design category C is applicable to category IV structures in regions of moderate seismicity as well as structures in occupancy categories I, II, and III in regions of somewhat more severe seismicity. The use of some structural systems is restricted in this design category. Plain concrete and masonry structures are not permitted. Seismic design category D includes structures in occupancy categories I, II, III, and IV in regions of high seismicity, but not located close to a major active fault, as well as occupancy category IV structures in regions with less severe seismicity. In this design category some types of structural systems must be designed by dynamic analysis methods. Seismic design category E includes structures in occupancy categories I, II, and III located close to a major active fault. Seismic design category F includes occupancy category IV structures located close to a major active fault. In this design category restrictions are imposed on the use of structural systems and analysis methods.

Table 7.4 Seismic Design Categories

		occupancy category		
S_{DS}	S_{D1}	I or II	III	IV
$S_{DS} < 0.167g$	$S_{D1} < 0.067g$	A	A	A
$0.167g \leq S_{DS} < 0.33g$	$0.067g \leq S_{D1} < 0.133g$	B	B	C
$0.33g \leq S_{DS} < 0.50g$	$0.133g \leq S_{D1} < 0.20g$	C	C	D
$0.50g \leq S_{DS}$	$0.20g \leq S_{D1}$	D	D	D
MCE* acceleration at 1 sec period, $S_1 \geq 0.75g$		E	E	F

*MCE = maximum considered earthquake

Example 7.2

The two-story, reinforced concrete, moment-resisting frame analyzed in Ex. 7.1 is used as a residential building. Determine the applicable occupancy category, importance factor, and seismic design category.

Solution

The 5% damped design spectral response accelerations are obtained from Ex. 7.12 as

$$S_{DS} = 1.0g$$
$$S_{D1} = 0.7g$$

A residential building is classified as a standard occupancy structure. The applicable occupancy category is obtained from ASCE Table 11.5-1 or Table 7.3. The occupancy category is II. For occupancy category II, the seismic importance factor is obtained from ASCE Table 11.5-1 or Table 7.3 as

$$I = 1.00$$

The design spectral response acceleration at short periods is

$$S_{DS} = 1.0g$$
$$> 0.50g$$

For an occupancy category of II, the seismic design category for this acceleration is obtained from ASCE Table 11.6-1 or Table 7.4. The seismic design category is D.

The design spectral response acceleration at a period of 1 sec is

$$S_{D1} = 0.70g$$
$$> 0.20g$$

For an occupancy category of II, the seismic design category for this acceleration is obtained from ASCE Table 11.6-2 or Table 7.4. The category is D. Hence, the seismic design category for this building is D.

Classification of the Structural System

ASCE Sec. 12.2.1 and ASCE Table 12.2-1 detail six major categories of building types characterized by the method used to resist the lateral force. These categories consist of bearing walls, building frames, moment-resisting frames, dual systems with a special moment-resisting frame, dual systems with a reinforced concrete intermediate moment frame or a steel ordinary moment frame, and inverted pendulum structures.

A *bearing wall system* consists of shear walls or braced frames that provide support for the gravity loads and resist all lateral loads. A *building frame system* consists of shear walls or braced frames that resist all lateral loads, and a separate framework that provides support for gravity loads. *Moment-resisting frames* provide support for both lateral and gravity loads by flexural action. In a *dual system*, nonbearing walls or braced frames supply the primary resistance to lateral loads, with a moment frame providing primary support for gravity loads plus additional resistance to lateral loads. A *cantilevered column structure* consists of a building supported on column elements to produce an inverted pendulum structure.

Response Modification Coefficient

The structure *response modification coefficient*, R, is a measure of the ability of a specific structural system to resist lateral loads without collapse. ASCE Table 12.2-1 lists the different structural framing systems, with the height limitations, response modification coefficients, and deflection amplification factors for each. An abbreviated listing of structural systems, response modification coefficients, overstrength factors, and deflection amplification factors is provided in Table 7.5.

Deflection Amplification Factor

The *deflection amplification factor* is tabulated in ASCE Table 12.2-1 and in Table 7.5 and is given by

$$C_d = \frac{\delta_x}{\delta_{xe}}$$

After allowing for the occupancy importance factor, ASCE Eq. (12.8-15) gives the value of the actual displacement as

$$\delta_x = \frac{C_d \delta_{xe}}{I}$$

Table 7.5 *Response Modification Coefficients*

structural system	R	C_d	Ω_0
bearing wall			
light-framed walls with wood shear panels	6.5	4.0	3.0
special reinforced concrete shear walls	5.0	5.0	2.5
special reinforced masonry shear walls	5.0	3.5	2.5
building frame			
eccentrically braced frame, moment-resisting connections away from link beam	8.0	4.0	2.0
eccentrically braced frame, non-moment-resisting connections away from link beam	7.0	4.0	2.0
light-framed walls with wood shear panels	7.0	4.5	2.5
special reinforced concrete shear walls	6.0	5.0	2.5
special reinforced masonry shear walls	5.5	4.0	2.5
special steel concentrically braced frames	6.0	5.0	2.0
moment-resisting frame			
steel or concrete special moment-resisting frames	8.0	5.5	3.0
intermediate moment frames of reinforced concrete	5.0	4.5	3.0
ordinary moment frames of steel	3.5	3.0	3.0
dual system with special moment-resisting frame			
special reinforced concrete shear walls	7.0	5.5	2.5
special reinforced masonry shear walls	5.5	5.0	3.0
eccentrically braced frame	8.0	4.0	2.5
special steel concentrically braced frame	7.0	5.5	2.5
dual system with intermediate moment frames			
special reinforced concrete shear walls	6.5	5.0	2.5
special steel concentrically braced frames	6.0	5.0	2.5
intermediate reinforced masonry shear walls	3.5	3.0	3.0
cantilevered column			
steel special moment-resisting frames	2.5	2.5	1.25
ordinary moment frames of steel	1.25	1.25	1.25

Adapted from American Society of Civil Engineers, *Minimum Design Loads for Buildings and Other Structures*, Table 12.2-1.

Overstrength Factor

The *overstrength factor* is a measure of the actual strength of a structure compared to the design seismic force. Values of the overstrength factor for various building systems are tabulated in ASCE Table 12.2-1 and shown in Table 7.5. The overstrength factor is given by

$$\Omega_0 = \frac{V_Y}{V_S}$$

The system overstrength is produced by the following factors: conservative design methods, system redundancy, material overstrength, oversized members, application of load factors, and drift limitations controlling design.

Effective Seismic Weight

The *effective seismic weight*, W, as specified in ASCE Sec. 12.7.2, is the total dead load of the structure and the part of the service load that may be expected to be attached

to the building. The effective seismic weight consists of the following.

- 25% of the floor live load for storage and warehouse occupancies

- a minimum allowance of 10 lbf/ft^2 for moveable partitions

- flat roof snow loads exceeding 30 lbf/ft^2, which may be reduced by 80%

- the total weight of permanent equipment and fittings

Roof and floor live loads, except as noted above, are not included in the value of W.

Fundamental Period of Vibration

ASCE Secs. 12.8.2 and 12.8.2.1 provide three methods for determining the *fundamental period* of a structure.

From ASCE Eq. (12.8-7), the approximate fundamental period is given by

$$T_a = 0.028(h_n)^{0.8} \text{ for steel moment-resisting frames}$$

$$T_a = 0.016(h_n)^{0.9} \text{ for reinforced concrete moment-resisting frames}$$

$$T_a = 0.030(h_n)^{0.75} \text{ for eccentrically braced steel frames}$$

$$T_a = 0.020(h_n)^{0.75} \text{ for all other structural systems}$$

Alternatively, for moment-resisting frames not exceeding 12 stories in height and with a story height not less than 10 ft, the approximate fundamental period may be determined by ASCE Eq. (12.8-8) as

$$T_a = 0.1N$$

Example 7.3

A two-story, reinforced concrete, moment-resisting frame is shown in the following illustration. Calculate the natural period of vibration T_a.

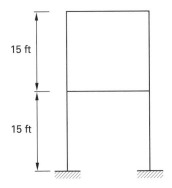

Solution

The number of stories is

$$N = 2$$
$$< 12$$

Then, for a moment-resisting frame, ASCE Eq. (12.8-8) specifies a value for a building period of

$$T_a = 0.1N$$
$$= \left(0.1 \frac{\text{sec}}{\text{story}}\right)(2 \text{ stories})$$
$$= 0.20 \text{ sec}$$

Example 7.4

For the two-story, reinforced concrete, moment-resisting frame analyzed in Ex. 7.3, calculate the fundamental period of vibration T_a by using ASCE Eq. (12.8-7).

Solution

For a reinforced concrete frame, the fundamental period is given by ASCE Eq. (12.8-7) as

$$T_a = (0.016)(30 \text{ ft})^{0.9}$$
$$= 0.342 \text{ sec}$$

Rayleigh Procedure

ASCE Sec. 12.8.2 permits the fundamental period to be determined by a "properly substantiated analysis." In accordance with NEHRP Commentary,[3] the *Rayleigh procedure* is an acceptable method, and the fundamental period is given by

$$T = 2\pi \sqrt{\frac{\sum w_i \delta_i^2}{g \sum f_i \delta_i}}$$

$$= 0.32 \sqrt{\frac{\sum w_i \delta_i^2}{\sum f_i \delta_i}}$$

The terms in this expression are illustrated in Fig. 7.1, where δ_i represents the elastic displacements due to a lateral force distribution f_i increasing approximately uniformly with height.

Figure 7.1 *Application of the Rayleigh Procedure*

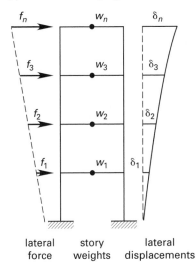

lateral force   story weights   lateral displacements

To allow for a possible underestimation of the stiffness of the structure, ASCE Sec. 12.8.2 specifies that the value of the natural period determined by this method may not exceed the value of

$$T = C_u T_a$$

Values of C_u are given in ASCE Table 12.8-1 and are shown in Table 7.6.

Table 7.6 *Coefficient for Upper Limit on the Calculated Period*

S_{D1}	≥ 0.40	0.30	0.20	0.15	≤ 0.10
C_u	1.4	1.4	1.5	1.6	1.7

Example 7.5

Using ASCE Sec. 12.8.2, determine the fundamental period of vibration of the two-story frame of Ex. 7.3, which is located in an area with a value for S_{D1} exceeding 0.4. The force system shown in the following illustration may be used; the effective seismic weight at each level and the total stiffness of each story are indicated.

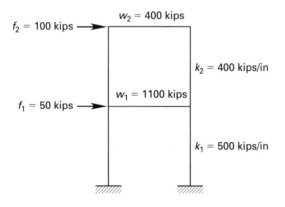

Solution

For the force system indicated, the displacements at each level are given by

$$\delta_1 = \frac{f_2 + f_1}{k_1}$$

$$= \frac{100 \text{ kips} + 50 \text{ kips}}{500 \, \frac{\text{kips}}{\text{in}}}$$

$$= 0.30 \text{ in}$$

$$\delta_2 = \frac{f_2}{k_2} + \delta_1$$

$$= \frac{100 \text{ kips}}{400 \, \frac{\text{kips}}{\text{in}}} + 0.30 \text{ in}$$

$$= 0.55 \text{ in}$$

The natural period is given by Rayleigh's procedure as

$$T = 0.32 \sqrt{\frac{\sum w_i \delta_i^2}{\sum f_i \delta_i}}$$

The relevant values are given in the table.

Rayleigh's procedure

level	w_i (kips)	f_i (kips)	δ_i (in)	$w_i\delta_i^2$ (kips-in^2)	$f_i\delta_i$ (in-kips)
2	400	100	0.55	121	55
1	1100	50	0.30	99	15
total	1500	–	–	220	70

$$T = \left(0.32 \, \frac{\text{sec}}{\sqrt{\text{in}}}\right) \sqrt{\frac{220 \text{ kips-in}^2}{70 \text{ in-kips}}}$$

$$= 0.567 \text{ sec}$$

In an area with a value for $S_{D1} > 0.4$, the value of the coefficient for the upper limit on the calculated period is obtained from ASCE Table 12.8-1 or Table 7.6 as

$$C_u = 1.4$$

Hence, the fundamental period, in accordance with ASCE Sec. 12.8.2, is limited to

$$T = 1.4 T_a$$

$$= (1.4)(0.2 \text{ sec})$$

$$= 0.28 \text{ sec}$$

$$< 0.567 \text{ sec}$$

Hence, use the maximum value of

$$T = 0.28 \text{ sec}$$

Alternatively, the value obtained for T_a in Ex. 7.4 may be used to give

$$T = (1.4)(0.342 \text{ sec})$$

$$= 0.479 \text{ sec}$$

General Procedure Response Spectrum

The *general procedure response spectrum* is defined in ASCE Sec. 11.4.5 and shown in ASCE Fig. 11.4-1. The response spectrum is reproduced in Fig. 7.2.

Figure 7.2 *Construction of ASCE Response Spectra*

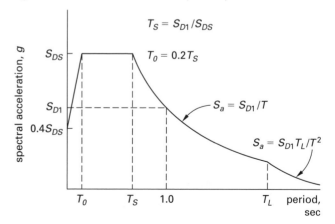

The response spectrum is constructed as shown using the following functions.

$$T_S = \frac{S_{D1}}{S_{DS}}$$

$$T_0 = \frac{0.2 S_{D1}}{S_{DS}}$$

For periods less than or equal to T_0, the design spectral response acceleration is given by ASCE Eq. (11.4-5) as

$$S_a = \frac{0.6 S_{DS} T}{T_0} + 0.4 S_{DS}$$

At $T = 0$,

$$S_a = 0.4 S_{DS}$$

For periods greater than or equal to T_0 and less than or equal to T_S, the design response acceleration is given by

$$S_a = S_{DS}$$

For periods greater than T_S and less than or equal to T_L, the design response acceleration is given by ASCE Eq. (11.4-6) as

$$S_a = \frac{S_{D1}}{T}$$

For periods greater than T_L, the design spectral response acceleration is given by ASCE Eq. (11.4-7) as

$$S_a = \frac{S_{D1} T_L}{T^2}$$

Values of T_L range from 4–16 sec, and are mapped in ASCE Fig. 22-15 through Fig. 22-20.

Seismic Response Coefficient

The *seismic response coefficient*, C_s, given in ASCE Sec. 12.8.1.1 represents the code design spectrum and is given by ASCE Eq. (12.8-3) for values of T not greater than T_L as

$$C_s = \frac{S_{D1} I}{RT}$$

For values of T greater than T_L, the seismic response coefficient is given by ASCE Eq. (12.8-4) as

$$C_s = \frac{S_{D1} T_L I}{RT^2}$$

The maximum value of the seismic response coefficient is given by ASCE Eq. (12.8-2) as

$$C_s = \frac{S_{DS} I}{R}$$

This latter expression controls for shorter periods up to approximately 1 sec. For longer periods, the expression provides conservative values.

In accordance with ASCE Eq. (12.8-5), the value of the seismic response coefficient shall not be taken less than

$$C_s = 0.01$$

For those structures for which the 1 sec spectral response value is $S_1 \geq 0.6g$, the minimum value of the seismic response coefficient is given by ASCE Eq. (12.8-6) as

$$C_s = \frac{0.5 S_1 I_E}{R}$$

Example 7.6

The two-story, reinforced concrete, special moment-resisting frame of Ex. 7.5 is used for a residential building. Calculate the seismic response coefficient by using the alternative value for the fundamental period determined in Ex. 7.5.

Solution

From previous examples, the relevant parameters are

$$S_{DS} = 1.0g$$
$$S_{D1} = 0.7g$$
$$I = 1.0$$
$$T = 0.479 \text{ sec}$$
$$T_S = 0.70 \text{ sec}$$

The value of the response modification coefficient for a special moment-resisting frame is obtained from ASCE Table 12.2-1 or Table 7.5 as

$$R = 8.0$$

The seismic response coefficient is given in ASCE Sec. 12.8.1.1 as

$$\begin{aligned} C_s &= \frac{S_{D1} I}{RT} \\ &= \frac{(0.7)(1.0)}{(8.0)(0.479 \text{ sec})} \\ &= 0.183 \end{aligned}$$

The maximum value of the seismic response coefficient is

$$\begin{aligned} C_s &= \frac{S_{DS} I}{R} \\ &= \frac{(1.0)(1.0)}{8.0} \\ &= 0.125 \quad \text{[governs]} \end{aligned}$$

This follows since $T < T_S$. The minimum values of C_s do not govern.

Seismic Base Shear

The *seismic base shear* is specified by ASCE Eq. (12.8-1) as

$$V = C_s W$$

Example 7.7

Calculate the seismic base shear for the two-story, reinforced concrete, moment-resisting frame of Ex. 7.6.

Solution

The value of the effective seismic weight was derived in Ex. 7.5 as

$$W = 1500 \text{ kips}$$

The value of the seismic response coefficient was derived in Ex. 7.6 as

$$C_s = 0.125$$

Hence, the base shear is given by ASCE Eq. (12.8-1) as

$$V = C_s W$$
$$= (0.125)(1500 \text{ kips})$$
$$= 188 \text{ kips}$$

Building Configuration Requirements

The static lateral force procedure is applicable to structures that satisfy prescribed conditions of regularity, occupancy, location, and height. A regular structure has mass, stiffness, and strength uniformly distributed over the height of the structure and is without irregular features that will produce stress concentrations. Vertical irregularities are defined in ASCE Table 12.3-2 and horizontal irregularities in ASCE Table 12.3-1. As defined in ASCE Table 12.6-1, a dynamic analysis is required for the following situations for structures in seismic design categories D, E, and F.

- occupancy category III and IV structures, not of light frame construction, with a fundamental period $T \geq 3.5 T_S$

- structures having soft story, mass, or geometric vertical irregularities, as defined in ASCE Table 12.3-2

- structures having torsional or extreme torsional horizontal irregularities as defined in ASCE Table 12.3-1

Occupancy category I and II structures of light-framed construction not exceeding three stories in height, or of any construction not exceeding two stories in height are exempted from these requirements.

Redundancy Factor

To improve the seismic performance of buildings in seismic design categories D, E, and F, it is advantageous to incorporate redundancy in the structure by providing multiple load resisting paths. The *redundancy factor*, ρ, is a factor that penalizes structures with relatively few lateral load-resisting elements and is specified in ASCE Sec. 12.3.4. The redundancy factor is fully defined in this text in the section "Seismic Load on an Element of a Structure."

2. VERTICAL DISTRIBUTION OF SEISMIC FORCES

The distribution of base shear over the height of a building is obtained from ASCE Sec. 12.8.3, and the design lateral force at level x is given by

$$F_x = \frac{V w_x h_x^k}{\sum w_i h_i^k}$$

The terms in this expression are illustrated in Fig. 7.3, where h_i represents the height above the base to any level i, h_x represents the height above the base to a specific level x, and $\sum w_i h_i^k$ represents the summation, over the whole structure, of the product of w_i and h_i^k. To allow for higher mode effects in long period buildings, when T has a value of 2.5 sec or more, the distribution exponent k is given by

$$k = 2$$

Figure 7.3 *Vertical Force Distribution*

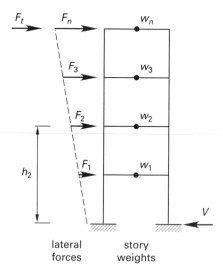

lateral forces   story weights

When T has a value not exceeding 0.5 sec, the distribution exponent is

$$k = 1$$

For intermediate values of T, a linear variation of k may be assumed.

Example 7.8

Determine the vertical force distribution for the two-story reinforced concrete, moment-resisting frame of Ex. 7.6.

Solution

The fundamental period was derived in Ex. 7.5 as

$$T = 0.479 \text{ sec}$$
$$< 0.5 \text{ sec}$$

The value of the distribution exponent factor is obtained from ASCE Sec. 12.8.3 as

$$k = 1.0$$

Hence, in accordance with ASCE Sec. 12.8.3, the expression for F_x reduces to

$$F_x = \frac{V w_x h_x}{\sum w_i h_i}$$

The effective seismic weights located at levels 1 and 2 are obtained from Ex. 7.5, and the relevant values are given in the following table.

Vertical Force Distribution

level	w_x (kips)	h_x (ft)	$w_x h_x$ (ft-kips)	F_x (kips)
2	400	30	12,000	79
1	1100	15	16,500	109
total	1500	–	28,500	188

From Ex. 7.7, the base shear is given by

$$V = 188 \text{ kips}$$

The design lateral force at level x is

$$F_x = \frac{V w_x h_x}{\sum w_i h_i}$$

$$= \frac{(188 \text{ kips}) w_x h_x}{28{,}500 \text{ ft-kips}}$$

$$= 0.00660 w_x h_x$$

The values of F_x are given in the previous table.

3. DIAPHRAGM LOADS

The load acting on a horizontal diaphragm is given by ASCE Sec. 12.10.1.1 as

$$F_{px} = \frac{w_{px} \sum F_i}{\sum w_i} \qquad \text{[ASCE 12.10-1]}$$

$$\geq 0.2 S_{DS} I w_{px}$$

$$\leq 0.4 S_{DS} I w_{px}$$

The terms in the expression are illustrated in Fig. 7.4, where $\sum F_i$ represents the total shear force at level i, $\sum w_i$ represents the total seismic weight at level i and above, and w_{px} represents the seismic weight tributary to the diaphragm at level x, not including walls parallel to the direction of the seismic load.

Figure 7.4 Diaphragm Loads

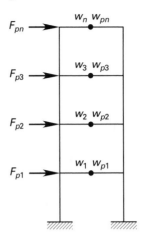

For a single-story structure, the expression reduces to

$$F_{px} = \frac{V w_{px}}{W}$$

$$= C_s w_{px}$$

Also, in a multistory structure, at the second-floor level,

$$\frac{F_t + \sum F_i}{\sum w_i} = \frac{V}{W}$$

$$= C_s$$

Example 7.9

Determine the diaphragm loads for the two-story reinforced concrete, moment-resisting frame of Ex. 7.8. The effective seismic weight tributary to the diaphragm at roof level is 300 kips and at the second-floor level is 600 kips.

Solution

From Ex. 7.1, the design response coefficient is

$$S_{DS} = 1.0g$$

The diaphragm load is given by ASCE Eq. (12.10-1) as

$$F_{px} = \frac{w_{px} \sum F_i}{\sum w_i}$$

level	w_i (kips)	$\sum w_i$ (kips)	F_i (kips)	$\sum F_i$ (kips)	$\dfrac{\sum F_i}{\sum w_i}$
2	400	400	79	79	0.198
1	1100	1500	109	188	0.125

level	max	min	w_{px} (kips)	F_{px} (kips)
2	0.40	0.20	300	60
1	0.40	0.20	600	120

The maximum value for the diaphragm load is given by ASCE Sec. 12.10.1.1 as

$$F_{p(\max)} = 0.4 S_{DS} I w_{px}$$
$$= (0.4)(1.0)(1.0) w_{px}$$
$$= 0.40 w_{px}$$

The minimum value for the diaphragm load is given by ASCE Sec. 12.10.1.1 as

$$F_{p(\min)} = 0.2 S_{DS} I w_{px}$$
$$= (0.2)(1.0)(1.0) w_{px}$$
$$= 0.20 w_{px} \quad \text{[governs at both levels]}$$

The relevant values are given in the previous table.

4. STORY DRIFT

Story drift is defined in ASCE Sec. 12.8.6 as the lateral displacement of one level of a multistory structure relative to the level below. The maximum allowable story drift Δ_a is given in ASCE Table 12.12-1 and is shown in Table 7.7.

To allow for inelastic deformations, drift is determined by using the deflection amplification factor C_d defined in Table 7.5, and the amplified deflection at level x is given by ASCE Eq. (12.8-15) as

$$\delta_x = \frac{C_d \delta_{xe}}{I}$$

Table 7.7 *Maximum Allowable Story Drift Δ_a*

| | occupancy category | | |
building type	I or II	III	IV
one-story buildings with fittings designed to accomodate drift	no limit	no limit	no limit
buildings other than masonry buildings of four stories or less with fittings designed to accommodate drift	$0.025 h_{sx}$	$0.020 h_{sx}$	$0.015 h_{sx}$
masonry cantilever shear wall buildings	$0.010 h_{sx}$	$0.010 h_{sx}$	$0.010 h_{sx}$
other masonry shear wall buildings	$0.007 h_{sx}$	$0.007 h_{sx}$	$0.007 h_{sx}$
all other buildings	$0.020 h_{sx}$	$0.015 h_{sx}$	$0.010 h_{sx}$

The term δ_{xe} represents the horizontal deflection at level x, determined by an elastic analysis using the code-prescribed design level forces. In accordance with ASCE Sec. 12.8.7, P-delta effects need not be included in the calculation of drift when the stability coefficient θ does not exceed 0.10.

For the calculation of drift, in accordance with ASCE Sec. 12.8.6.2, the full value of T, the fundamental period determined by using the Rayleigh procedure, may be used to determine the seismic base shear.

Example 7.10

Determine the drift in the bottom story of the two-story, reinforced concrete, moment-resisting frame of Ex. 7.8. The relevant details are shown in the following illustration. Fittings are designed to accommodate drift.

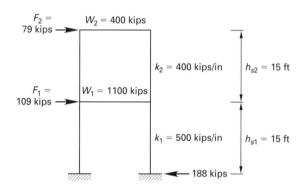

Solution

From Ex. 7.5, the fundamental period obtained by using the Rayleigh procedure is

$$T = 0.567 \text{ sec}$$

The seismic response coefficient is given in ASCE Sec. 12.8.1.1 as

$$C_s = \frac{S_{D1} I}{RT}$$
$$= \frac{(0.7)(1.0)}{(8.0)(0.567 \text{ sec})}$$
$$= 0.154$$

The maximum value of the seismic design coefficient is

$$C_s = \frac{S_{DS} I}{R}$$
$$= \frac{(1.0)(1.0)}{8.0}$$
$$= 0.125 \quad \text{[governs]}$$

Hence, the seismic base shear is given by ASCE Sec. 12.8.1 and is identical with the value calculated in Ex. 7.7 as

$$V = 188 \text{ kips}$$

In addition, the lateral forces are identical with the values calculated in Ex. 7.8.

For a moment-resisting frame, the amplification factor is obtained from ASCE Table 12.2-1 or Table 7.5 as

$$C_d = 5.5$$

From the lateral forces determined in Ex. 7.8, the drift in the bottom story is

$$\Delta_1 = C_d \delta_{xe}$$

$$= \frac{C_d \left(F_2 + F_1 \right)}{k_1}$$

$$= \frac{(5.5)(79 \text{ kips} + 109 \text{ kips})}{500 \ \dfrac{\text{kips}}{\text{in}}}$$

$$= 2.07 \text{ in}$$

In accordance with ASCE Table 12.2-1, the maximum allowable drift for a two-story structure in seismic use group I is

$$\Delta_a = 0.025 h_{s1}$$

$$= (0.025)(15 \text{ ft}) \left(12 \ \dfrac{\text{in}}{\text{ft}} \right)$$

$$= 4.50 \text{ in}$$

$$> 2.07 \text{ in}$$

Hence, the drift is acceptable.

5. *P*-DELTA EFFECTS

The *P-delta effects* are calculated by using the design level seismic forces and elastic displacements determined in accordance with ASCE Sec. 12.8.1. *P*-delta effects in a given story are a result of the secondary moments, caused by the eccentricity of the gravity loads above that story. The *secondary moment in a story* is defined as the product of the total dead load, floor live load, and snow load above the story multiplied by the elastic drift of that story. The *primary moment in a story* is defined as the seismic shear in the story multiplied by the height of the story.

The ratio of the secondary moment to primary moment is termed the *stability coefficient* and is given by ASCE Sec. 12.8.7 as

$$\theta = \frac{P_x \Delta}{V_x h_{sx} C_d}$$

$$= \frac{P_x \left(\delta_{xe} - \delta_{(x-1)e} \right)}{V_x h_{sx}}$$

The stability coefficient shall not exceed the value

$$\theta_{\max} = \frac{0.5}{\beta C_d}$$

$$\leq 0.25$$

The term β is the ratio of the shear demand to the shear capacity in a story and may conservatively be considered equal to 1.0. If the stability coefficient in any story exceeds 0.1, the effects of the secondary moments shall be included in the analysis of the whole structure. The revised story drift, allowing for *P*-delta effects, is obtained as the product of the calculated drift and the factor $1/(1 - \theta)$.

As shown in Fig. 7.5, with the designated lateral forces and story drift and with the combined dead load plus floor live load indicated by W_1 and the combined dead load plus roof snow load indicated by W_2, the primary moment in the second story of the frame is

$$M_{P2} = F_2 h_{s2}$$

Figure 7.5 *P-Delta Effects*

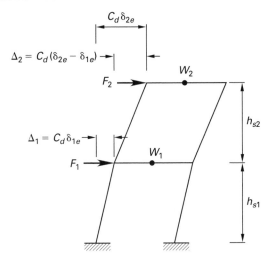

The secondary moment in the second story is

$$M_{S2} = \frac{P_2 \Delta_2}{C_d}$$

$$= W_2 \left(\delta_{2e} - \delta_{1e} \right)$$

The stability coefficient in the second story is

$$\theta_2 = \frac{M_{S2}}{M_{P2}}$$

The primary moment in the first story of the frame is

$$M_{P1} = \left(F_1 + F_2 \right) h_{s1}$$

The secondary moment in the first story is

$$M_{S1} = \frac{P_1 \Delta_1}{C_d}$$

$$= \left(W_2 + W_1 \right) \delta_{1e}$$

The stability coefficient in the first story is

$$\theta_1 = \frac{M_{S1}}{M_{P1}}$$

Seismic

Example 7.11

Determine the stability coefficient for the bottom story of the two-story reinforced concrete, moment-resisting frame of Ex. 7.10.

Solution

The drift in the bottom story is derived in Ex. 7.10 as

$$\Delta_1 = 2.07 \text{ in}$$

The primary moment in the bottom story is

$$M_{P1} = (F_2 + F_1) h_{s1}$$
$$= (79 \text{ kips} + 109 \text{ kips})(15 \text{ ft}) \left(12 \frac{\text{in}}{\text{ft}}\right)$$
$$= 33,840 \text{ in-kips}$$

The secondary moment is

$$M_{S1} = \frac{(W_2 + W_1) \Delta_1}{C_d}$$
$$= \frac{(1500 \text{ kips})(2.07 \text{ in})}{5.5}$$
$$= 565 \text{ in-kips}$$

The stability coefficient is

$$\theta_1 = \frac{M_{S1}}{M_{P1}}$$
$$= \frac{565 \text{ in-kips}}{33,840 \text{ in-kips}}$$
$$= 0.017$$
$$< 0.1 \quad \text{[Secondary moments need not be considered.]}$$

6. SIMPLIFIED LATERAL FORCE PROCEDURE

For some low-rise structures, ASCE Sec. 12.14 permits an alternative, conservative design method. The simplified method is applicable to a structure in which the following 12 limitations are met.

1. The structure does not exceed three stories in height.

2. The structure is assigned to occupancy category I or II.

3. The structure is situated in a location with a soil profile of site class A through D.

4. The lateral force resisting system is either a bearing wall system or a building frame system.

5. As shown in Fig. 7.6, the building shall have at least two lines of lateral resistance in each of two major axis directions.

6. As shown in Fig. 7.6, at least one line of resistance shall be provided on each side of the center of mass (CM) in each direction.

Figure 7.6 *Lines of Lateral Resistance*

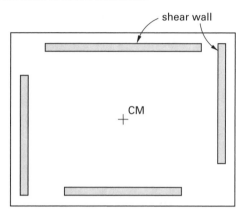

7. As shown in Fig. 7.7, for structures with flexible diaphragms, overhangs beyond the outside line of shear walls or braced frames shall satisfy ASCE Eq. (12.14-1). The distance, a, perpendicular to the forces being considered from the extreme edge of the diaphragm to the line of vertical resistance closest to that edge shall not exceed one-fifth the depth, d, of the diaphragm parallel to the forces being considered at the line of vertical resistance closest to the edge.

$$a \leq \frac{d}{5} \quad \text{[ASCE 12.14-1]}$$

Figure 7.7 *Flexible Diaphragm Overhang*

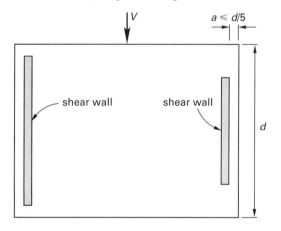

8. For buildings with a diaphragm that is nonflexible, the distance between the center of rigidity and the center of mass parallel to each major axis shall not exceed 15% of the greatest width of the diaphragm parallel to that axis. In addition, as shown in Fig. 7.8, the building layout shall satisfy both of the following equations.

$$\sum_{i=1}^{m} k_{1i}d_i^2 + \sum_{j=1}^{n} k_{2j}d_j^2$$

$$\geq 2.5\left(0.05 + \frac{e_1}{b_1}\right)b_1^2 \sum_{i=1}^{m} k_{1i}$$

[ASCE 12.14-2A]

$$\sum_{i=1}^{m} k_{1i}d_i^2 + \sum_{j=1}^{n} k_{2j}d_j^2$$

$$\geq 2.5\left(0.05 + \frac{e_2}{b_2}\right)b_2^2 \sum_{i=1}^{m} k_{1i}$$

[ASCE 12.14-2B]

The stiffness of wall i or braced frame i parallel to major axis 1 is

$$k = k_{1i}$$

The stiffness of wall j or braced frame j parallel to major axis 2 is

$$k = k_{2j}$$

The distance from the wall i or braced frame i to the center of rigidity, perpendicular to major axis 1 is

$$d = d_{1i}$$

The distance from the wall j or braced frame j to the center of rigidity, perpendicular to major axis 2 is

$$d = d_{2j}$$

The distance perpendicular to major axis 1 between the center of rigidity and the center of mass is

$$e = e_1$$

The width of the diaphragm perpendicular to major axis 1 is

$$b = b_1$$

The distance perpendicular to major axis 2 between the center of rigidity and the center of mass is

$$e = e_2$$

The width of the diaphragm perpendicular to major axis 2 is

$$b = b_2$$

m is the number of walls and braced frames resisting lateral force in direction 1.

n is the number of walls and braced frames resisting lateral force in direction 2.

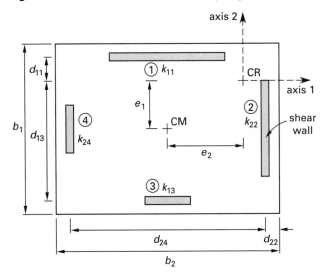

Figure 7.8 *Torsion Check for Nonflexible Diaphragms*

As shown in Fig. 7.9, these two equations need not be checked provided that the structure fulfills all of the following three conditions.

- the arrangement of walls or braced frames is symmetric about each major axis

- the distance between the two most separated lines of walls or braced frames is at least 90% of the dimension of the structure perpendicular to that axis direction

- the stiffness along each of the lines is at least 33% of the total stiffness in that axis direction

Figure 7.9 *Torsion Check Unnecessary*

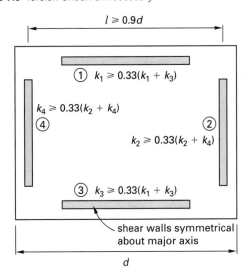

9. Lines of resistance of the lateral force-resisting system shall be oriented at angles of no more than 15° from alignment with the major orthogonal horizontal axes of the building.

10. The simplified design procedure shall be used for each major orthogonal horizontal axis direction of the building.

11. System irregularities caused by in plane or out of plane offsets of lateral force-resisting elements shall not be permitted, except in two-story buildings of light frame construction provided that the upper wall is designed for a factor of safety of 2.5 against overturning.

12. The lateral load resistance of any story shall not be less than 80% of the story above.

In applying the simplified design procedure, ASCE Sec. 12.14.3.1.1 indicates that the redundancy factor may be taken as

$$\rho = 1.0$$

In accordance with ASCE Sec. 12.14.3.2.1, the over-strength factor is

$$\Omega_0 = 2.5$$

When the simplified design procedure is used, ASCE Section 12.14.8.5 specifies that structural drift need not be calculated. If a drift value is required for design of cladding or to determine building separation, it may be assumed to be 1% of building height.

Example 7.12

The two-story structure shown in the following illustration is an office building located on a site with a soil classification type D. The structure consists of a bearing wall system with the reinforced concrete shear walls continuous through both stories, and the roof and floor diaphragms are rigid. Determine whether the simplified lateral force procedure is applicable.

Solution

The following hold true for the office building and are satisfactory.

- The occupancy category is II.

- It is situated on a site with a soil classification type D.

- It is two stories in height, which is less than three stories.

- The seismic force-resisting system is a bearing wall system.

- It has two lines of lateral resistance in each of the two major axis directions.

- One line of resistance is provided on each side of the center of mass in each direction.

- The lines of resistance of the lateral force-resisting system are parallel to the major orthogonal horizontal axes of the building.

Illustration for Ex. 7.12

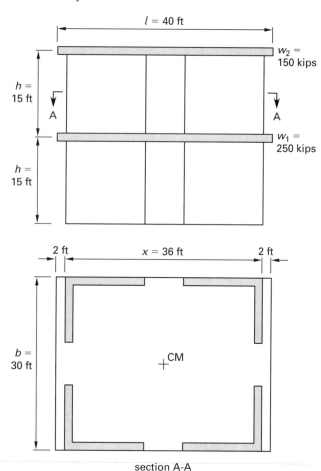

section A-A

- There are no irregularities caused by in plane or out of plane offsets of lateral force-resisting elements.

- The lateral load resistance is identical in both stories.

Since the building has rigid diaphragms, it is necessary to check out ASCE Eqs. 12.14-2A and 12.14-2B, or determine whether the limitations are such that doing so is not necessary. The relevant limitations are as follows.

The arrangement of the shear walls is symmetric about each major axis, and this is satisfactory.

The length of the building is

$$l = 40 \text{ ft}$$

The separation between the east and west shear walls is

$$x = 36 \text{ ft}$$

The ratio of separation of walls to length of building is

$$\frac{x}{l} = \frac{36 \text{ ft}}{40 \text{ ft}}$$
$$= 0.9 \quad [\text{satisfactory}]$$

The stiffness along each line of shear walls is 50% of the total stiffness in that axis direction. Therefore, it is not necessary to check out ASCE Eqs. 12.14-2A and 12.14-2B. The simplified lateral force procedure is applicable.

Simplified Determination of Seismic Base Shear

The simplified seismic base shear is given by ASCE Eq. (12.14-11) as

$$V = \left(\frac{FS_{DS}}{R}\right)W$$

The design spectral response acceleration at short periods is given by ASCE Sec. 12.14.8.1 as

$$S_{DS} = \frac{2F_a S_S}{3}$$

The 5% damped, maximum considered earthquake spectral response acceleration, for a period of 0.2 sec for structures founded on rock, is given by ASCE Sec. 11.4.1 with the limitation

$$S_S \leq 1.5g$$

The short-period site coefficient is obtained from ASCE Table 11.4-1, or may be taken as

$$F_a = 1.0 \quad \text{[rock sites]}$$
$$F_a = 1.4 \quad \text{[soil sites]}$$

ASCE Sec. 12.14.8.1 defines a rock site as having the height of the soil between the rock surface and the bottom of the building's foundations no greater than 10 ft.

The modification factor for building type is

$$F = 1.0 \quad \text{[one-story buildings]}$$
$$F = 1.1 \quad \text{[two-story buildings]}$$
$$F = 1.2 \quad \text{[three-story buildings]}$$

The effective seismic weight W, as specified in ASCE Sec. 12.14.8.1, includes the total dead load of the structure plus the following loads.

- 25% of the floor live load for storage and warehouse occupancies

- a minimum allowance of 10 lb/ft^2 for moveable partitions

- snow loads exceeding 30 lb/ft^2, which may be reduced by 80% when approved by the building official

- the total weight of permanent equipment and fittings

The response modification factor, R, and the limitations on the use of the various lateral force-resisting systems are given in ASCE Table 12.14-1 and are summarized in Table 7.8.

Table 7.8 *Design Factors for Simplified Lateral Force Procedure*

structural system	response modification coefficient R	limitations seismic design category		
		B	C	D, E
bearing wall				
special reinforced concrete shear walls	5	P	P	P
ordinary reinforced concrete shear walls	4	P	P	NP
special reinforced masonry shear walls	5	P	P	P
intermediate reinforced masonry shear walls	3.5	P	P	NP
ordinary reinforced masonry shear walls	2	P	NP	NP
light-framed walls with wood structural panels	6.5	P	P	P
building frame				
eccentrically braced frame, moment-resisting connections at column away from link	8	P	P	P
eccentrically braced frame, non-moment-resisting connections at column away from link	7	P	P	P
special steel concentrically braced frames	6	P	P	P
ordinary steel concentrically braced frames	3.25	P	P	P
special reinforced concrete shear walls	6	P	P	P
ordinary reinforced concrete shear walls	5	P	P	P
composite steel concentrically braced frames	5	P	P	P
special reinforced masonry shear walls	5.5	P	P	P
intermediate reinforced masonry shear walls	4	P	P	NP
light-framed walls with wood structural panels	7	P	P	P

P = permitted, NP = not permitted

Example 7.13

Determine the seismic base shear using the simplified procedure for the two-story special reinforced concrete bearing wall structure of Ex. 7.12. The 5% damped, maximum considered earthquake spectral response acceleration, for a period of 0.2 sec, is $S_S = 1.2g$. The effective seismic weight of the building is 400 kips.

Solution

The 5% damped, maximum considered earthquake spectral response acceleration for a period of 0.2 sec is given as

$$S_S = 1.2g$$

The short-period site coefficient for a soil site is given by ASCE Sec. 12.14.8.1 as

$$F_a = 1.4$$

The design spectral response acceleration at short periods is given by ASCE Sec. 12.14.8.1 as

$$S_{DS} = \frac{2F_a S_S}{3}$$
$$= (2)(1.4)\left(\frac{1.2g}{3}\right)$$
$$= 1.12g$$

For a bearing wall structure with special reinforced concrete shear walls, the response modification factor is obtained from ASCE Table 12.14-1, or from Table 7.8, as

$$R = 5$$

The modification factor for a two-story building is

$$F = 1.1$$

The effective seismic weight of the building is given as

$$W = 400 \text{ kips}$$

Therefore, the simplified base shear is given by ASCE Eq. (12.14-11) as

$$V = \left(\frac{FS_{DS}}{R}\right)W$$
$$= \left(\frac{(1.1)(1.12g)}{5}\right)(400 \text{ kips})$$
$$= 98.6 \text{ kips}$$

Simplified Vertical Distribution of Base Shear

When the simplified procedure is used to determine the seismic base shear, the forces at each level may be determined from ASCE Sec. 12.14.8.2 as

$$F_x = \frac{w_x V}{W} \qquad \text{[ASCE 12.14-12]}$$

Example 7.14

Determine the vertical force distribution by using the simplified procedure for the two-story, reinforced concrete, bearing wall structure of Ex. 7.13.

Solution

From Ex. 7.13, the following values are obtained.

$$V = 98.6 \text{ kips}$$
$$W = 400 \text{ kips}$$

The forces at each level are determined from ASCE Eq. (12.14-12) as

$$F_x = \frac{w_x V}{W}$$
$$= \frac{w_x(98.6 \text{ kips})}{400 \text{ kips}}$$
$$= 0.25w_x$$

The values of F_x are given in the following table.

level	w_x (kips)	F_x (kips)
2	150	37.0
1	250	61.6
total	400	98.6

Simplified Determination of Drift

In accordance with ASCE Sec. 12.14.8.5, when the simplified procedure is used to determine the seismic base shear, the design story drift in any story shall be taken as

$$\Delta_x = 0.01h_{sx}$$

Example 7.15

Using the simplified procedure, determine the drift in the bottom story of the two-story, reinforced concrete, bearing wall structure of Ex. 7.13.

Solution

The design story drift in the bottom story is given by

$$\Delta_1 = 0.01h_{s1}$$
$$= (0.01)(15)(12)$$
$$= 1.80 \text{ in}$$

7. SEISMIC LOAD ON AN ELEMENT OF A STRUCTURE

The *seismic load*, E, is a function of both horizontal and vertical earthquake-induced forces and is given by ASCE Sec. 12.4.2 as

$$E = \rho Q_E + 0.2S_{DS}D$$

The term Q_E is the lateral force produced by the calculated base shear V. The term $0.2S_{DS}D$ is the vertical force due to the effects of vertical acceleration. The redundancy factor ρ is a factor that penalizes structures with relatively few lateral load resisting elements and is defined by ASCE Sec. 12.3.4.

The seismic performance of a structure is enhanced by providing multiple lateral load-resisting elements. Therefore, the yield of one element will not result in an unstable condition that may lead to collapse of the structure. For this situation, the value of the redundancy factor is $\rho = 1.0$. When the yield of an element will result in an unstable condition, the value of the redundancy factor is $\rho = 1.3$.

In accordance with ASCE Sec. 12.3.4.1, the value of the redundancy factor is equal to 1.0 for the following.

- structures assigned to seismic design categories B and C

- drift calculation and P-delta effects

- design of nonstructural components

- design of nonbuilding structures that are not similar to buildings

- design of collector elements, splices, and their connections for which load combinations with overstrength factors are used

- design of members and connections for which load combinations with overstrength factors are required

- diaphragm loads determined using ASCE Eq. 12.10-1

- structures with damping systems designed in accordance with ASCE Ch. 18

In accordance with ASCE Sec. 12.3.4.2a, the value of the redundancy factor may be taken equal to 1.0, provided that each story resisting more than 35% of the base shear complies with the following.

- For a braced frame, the removal of an individual brace, or connection thereto, does not result in more than a 33% reduction in story strength, nor create an extreme torsional irregularity (horizontal structural irregularity type 1b).

- For a moment frame, loss of moment resistance at the beam to column connections at both ends of a single beam does not result in more than a 33% reduction in story strength, nor create an extreme torsional irregularity (horizontal structural irregularity type 1b).

- For a shear wall or a wall-pier system with a height to length ratio greater than 1.0, removal of a wall or pier, or collector connections thereto, does not result in more than a 33% reduction in story strength, nor create an extreme torsional irregularity (horizontal structural irregularity type 1b).

- For a cantilever column, loss of moment resistance at the base connections of any single cantilever column does not result in more than a 33% reduction in story strength, nor create an extreme torsional irregularity (horizontal structural irregularity type 1b).

- There are no requirements for all other structural systems.

In accordance with ASCE Sec. 12.3.4.2b, the value of the redundancy factor may be taken equal to 1.0 provided that the building is regular in plan at all levels with not less than two bays of lateral load-resisting perimeter framing on each side of the building in each orthogonal direction at each story resisting more than 35% of the base shear. The number of bays for a shear wall is calculated as the length of the shear wall divided by the story height. For light-framed construction, the number of bays for a shear wall is calculated as twice the length of the shear wall divided by the story height.

Example 7.16

Determine the redundancy factor for the moment-resisting framed structure shown. The stiffness of all frames is identical. The roof diaphragm is flexible.

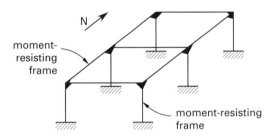

Solution

The building is regular in plan and this complies with ASCE Sec. 12.3.4.2b.

In the north-south direction, two bays of moment-resisting perimeter frames are provided on each side of the building. The frames on each perimeter resist 50% of the base shear. This complies with ASCE Sec. 12.3.4.2b.

In the east-west direction, only one bay of moment-resisting perimeter frames is provided on each side of the building. This does not comply with ASCE Sec. 12.3.4.2b.

In the north-south direction, removing one frame results in a reduction of shear strength of

$$\frac{1}{4} = 0.25$$

$$< 0.33$$

This complies with ASCE Sec. 12.3.4.2a.

In the east-west direction, removing one frame results in a reduction of shear strength of

$$\frac{1}{3} = 0.33$$

This complies with ASCE Sec. 12.3.4.2a.

The redundancy factor is given by ASCE Sec. 12.3.4 as

$$\rho = 1.0$$

Example 7.17

Determine the redundancy factor for the structure shown. The stiffness of all braced frames in the north-south direction is identical, and the stiffness of all moment-resisting frames in the east-west direction is identical. The roof diaphragm is flexible.

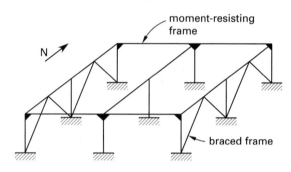

Solution

The building is regular in plan and this complies with ASCE Sec. 12.3.4.2b.

In the north-south direction, two bays of chevron bracing are provided on each side of the building. The frames on each perimeter resist 50% of the base shear. This complies with ASCE Sec. 12.3.4.2b.

Similarly, in the east-west direction, two bays of moment-resisting frames are provided on each side of the building. The frames on each perimeter resist 50% of the base shear. This complies with ASCE Sec. 12.3.4.2b.

The redundancy factor is given by ASCE Sec. 12.3.4 as

$$\rho = 1.0$$

References

1. International Code Council. *International Building Code*. 2006.

2. American Society of Civil Engineers. *Minimum Design Loads for Buildings and Other Structures, (ASCE/SEI 7-05)*. 2005.

3. Building Seismic Safety Council. *NEHRP Recommended Provisions for the Development of Seismic Regulations for New Buildings: Part 2, Commentary*. 2003.

PRACTICE PROBLEMS

1. A one-story industrial building is assigned to seismic design category D and is located in an area with a 0.2 sec acceleration coefficient of $S_{DS} = 1.0$ and a 1.0 sec acceleration coefficient of $S_{D1} = 0.7$. Details are shown in the illustration. The weight of the wood roof is 15 lbf/ft^2, and the weight of the masonry walls is 75 lbf/ft^2. The roof sheathing is $^{15}/_{32}$ in Structural I grade plywood. For north-south seismic loads, what is the spacing of 10d nails at the diaphragm boundaries?

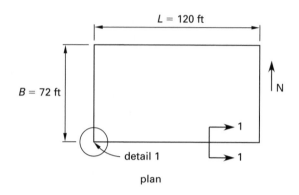

plan

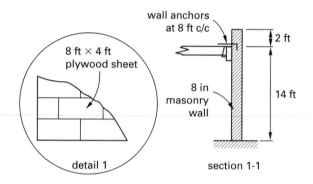

2. For the building of Prob. 1, wall anchors are provided to the masonry wall at 8 ft centers. What is the service level design force in each anchor?

SOLUTIONS

1. The relevant dead load tributary to the roof diaphragm in the north-south direction is due to the north and south wall and the roof dead load and is obtained as

$$\text{roof} = \left(15 \ \frac{\text{lbf}}{\text{ft}^2}\right)(72 \ \text{ft})$$

$$= 1080 \ \text{lbf/ft}$$

$$\text{north + south wall} = \frac{(2 \ \text{walls})\left(75 \ \frac{\text{lbf}}{\text{ft}^2}\right)(16 \ \text{ft})^2}{(2)(14 \ \text{ft})}$$

$$= 1371 \ \text{lbf/ft}$$

The total dead load tributary to the roof diaphragm is

$$w_{px} = \frac{\left(1080 \ \frac{\text{lbf}}{\text{ft}} + 1371 \ \frac{\text{lbf}}{\text{ft}}\right)(120 \ \text{ft})}{1000 \ \frac{\text{lbf}}{\text{kip}}}$$

$$= 294 \ \text{kips}$$

For a standard occupancy structure, the importance factor is

$$I = 1.0$$

For a bearing wall structure with reinforced masonry shear walls, the value of the response modification factor is obtained from Table 7.5 as

$$R = 5.0$$

For this type of structure, the maximum value of the seismic response coefficient controls, and the value of the seismic response coefficient is given by ASCE Sec. 12.8.1.1 as

$$C_s = \frac{S_{DS}I}{R}$$

$$= \frac{(1.0)(1.0)}{5}$$

$$= 0.2$$

In accordance with ASCE Sec. 12.10.1.1, the force acting on the roof diaphragm of a one-story building may be taken as

$$F_{px} = C_s w_{px}$$

$$= 0.2 w_{px}$$

The minimum allowable force acting on the roof diaphragm is given by ASCE Sec. 12.10.1.1 as

$$F_{px(\min)} = 0.2 S_{DS} I w_{px}$$

$$= (0.2)(1.0) w_{px}$$

$$= 0.2 w_{px}$$

Hence, the force on the diaphragm is given by

$$F_{px} = (0.2)(294 \ \text{kips})$$

$$= 59 \ \text{kips}$$

This value of F_{px} is at the strength level, and the equivalent service level value for design of the diaphragm using allowable stress design is given by ASCE Sec. 2.4.1 as

$$F'_{px} = 0.7 F_{px}$$

$$= (0.7)(59 \ \text{kips})$$

$$= 41 \ \text{kips}$$

The service level design unit shear along the diaphragm boundary is

$$q = \frac{F'_{px}}{2B}$$

$$= \frac{\left(1000 \ \frac{\text{lbf}}{\text{kip}}\right)(41 \ \text{kips})}{(2)(72 \ \text{ft})}$$

$$= 285 \ \text{lbf/ft}$$

The required nail spacing is obtained from IBC Table 2306.3.1 with a case 1 plywood layout applicable, all edges blocked, and 2 in framing. Using both $^{15}/_{32}$ in Structural I grade plywood and $10d$ nails with $1^1/_2$ in penetration, a nail spacing of 6 in at the diaphragm boundaries and 6 in at all other panel edges gives an allowable unit shear of

$$q_u = 320 \ \text{lbf/ft}$$

$$> 285 \ \text{lbf/ft} \quad [\text{satisfactory}]$$

2. The relevant weight of the element tributary to the wall anchors is obtained from Prob. 1 as

$$W_w = \frac{1371 \ \frac{\text{lbf}}{\text{ft}}}{2}$$

$$= 685.5 \ \text{lbf/ft}$$

The strength level seismic lateral force for a flexible roof diaphragm is given by ASCE Sec. 12.11.2.1 as

$$F_p = 0.80 I S_{DS} W_w$$

$$= (0.80)(1.0)(1.0)\left(685.5 \ \frac{\text{lbf}}{\text{ft}}\right)$$

$$= 548 \ \text{lbf/ft}$$

For concrete and masonry walls, ASCE Sec. 12.11.2 specifies a minimum value lateral seismic force of

$$F_{p(\min)} = 400 I S_{DS}$$

$$= (400)(1.0)(1.0)$$

$$= 400 \ \text{lbf/ft}$$

$$< 548 \ \text{lbf/ft}$$

Hence, the minimum lateral force of 400 lbf/ft does not govern; the service level design force on each wall anchor at a spacing of 8 ft is given by ASCE Sec. 2.4.1 as

$$P_s = (0.7)(8 \ \text{ft})\left(548 \ \frac{\text{lbf}}{\text{ft}}\right)$$

$$= 3069 \ \text{lbf}$$

Seismic

8 Design of Bridges

1. Design Loads 8-1
2. Reinforced Concrete Design 8-14
3. Prestressed Concrete Design 8-21
4. Structural Steel Design 8-33
5. Wood Structures 8-40
6. Seismic Design 8-43
   Practice Problems 8-53
   Solutions 8-53

1. DESIGN LOADS

Nomenclature

A	area of beam	in^2
b	width of beam	in
B_s	width of equivalent strip	in
d	depth of beam	in
D	dead load	kips
DC	dead load of components and attachments	kips
d_e	depth to the resultant of the tensile force	in
d_v	distance between the resultants of the tensile and compressive forces due to flexure	in
DW	dead load of wearing surface and utilities	kips
E	width of equivalent strip for a slab bridge	in
e_g	distance between the centers of gravity of the beam and the deck slab	in
EQ	earthquake load	lbf or kips
g	distribution factor for moment or shear	–
h	overall depth of member	in
I	dynamic factor	–
I	moment of inertia of beam	in^4
IM	dynamic load allowance	–
k_s	stiffness of equivalent strip	in^4
K_g	longitudinal stiffness parameter	in^4
L	span length	ft
L_1	modified span length	ft
LL	vehicular live load	kips
m	multiple presence factor	–
M	bending moment	ft-kips
M_D	dead load moment	ft-kips
M_L	live load moment	ft-kips
M_s	service load moment	ft-kips
M_u	factored design moment	ft-kips
n	ratio of the modulus of elasticity of the beam and the deck slab	–
N_b	number of beams	–
N_L	number of traffic lanes	
P	wheel load	kips

Q	factored force effect	–
R_n	nominal resistance capacity	ft-kips
S	spacing of beams	ft
t_s	deck slab thickness	in
V	shear force	kips
w	distributed load	kips/ft
w	roadway width between curbs	ft
w_D	dead load	kips/ft
w_L	lane width	ft
W	concentrated load	kips
W	edge-to-edge width of the bridge	ft
W_1	modified edge-to-edge width of the bridge	ft
W_L	live load	kips
W_n	nominal load	kips
W_s	service load	kips
W_u	factored design load	kips
x	distance from the center line of a stringer to the face of the stringer	in

Symbols

α	tabulated force coefficient for distributed loads	
γ	load factor	
η	load modifier	
ϕ	strength reduction factor	

Design Lanes

A bridge deck is divided into design lanes as defined in AASHTO[1] Sec. 3.6. For deck widths between 20 ft and 24 ft, two design lanes are specified, each equal to one-half the deck width. For all other deck widths, design lanes are defined as being 12 ft wide, with fractional parts of a lane discounted, and the number of design lanes is given by

$$N_L = \text{INT} \left(\frac{w}{12} \right)$$

INT is the integer part of the ratio.

Design lanes are positioned on the deck to produce the maximum effect. The determination of the number of design lanes is illustrated in Fig. 8.1.

Bridges

Figure 8.1 Design Traffic Lanes

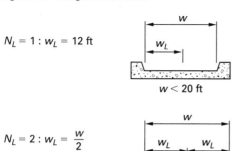

Example 8.1

For the bridge deck shown in the following illustration, determine the number of design lanes.

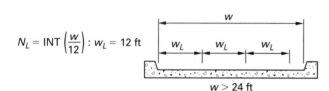
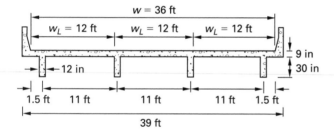

Solution

From AASHTO Sec. 3.6, the number of design lanes is given by

$$N_L = \frac{w}{12} = \frac{36 \text{ ft}}{12}$$
$$= 3 \text{ lanes}$$

Live Loads

The vehicular live loading for bridges is designated HL-93 and is specified in AASHTO Sec. 3.6.1.2. The loading consists of the most critical of the following two different loading types.

- a design lane load combined with a design truck

- a design lane load combined with a design tandem

As shown in Fig. 8.2, the design lane load consists of a load of 0.64 kips/ft uniformly distributed in the longitudinal direction. In considering the design lane load in the design of continuous spans, as many spans shall be loaded with the 0.64 kips/ft uniform load as is necessary

to produce the maximum effect. The design lane load is placed longitudinally only on those portions of the spans of a bridge to give the most critical effect. Transversely, the design lane load is uniformly distributed over a 10 ft width. The 10 ft loaded width is placed in the design lane to give the most critical effect without encroaching on the adjacent lane. A dynamic load allowance is not applied to the design lane load.

Figure 8.2 Design Lane Load

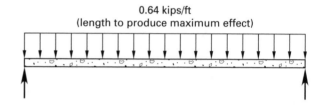

As shown in Fig. 8.3, the design truck load consists of three axles—the lead axle of 8 kips and the two following axles of 32 kips. The spacing between the two 32 kip axles is varied between 14 ft and 30 ft to produce the most critical effect. The transverse spacing of the wheels is 6 ft. Transversely, the design truck is positioned in a lane, as specified in AASHTO Sec. 3.6.1.3.1, so that the center of any wheel load is not closer than

- 1 ft from the face of a curb for the design of a deck overhang

- 2 ft from the edge of the design lane for all other components

A dynamic load allowance is applied to the design truck load.

Figure 8.3 Design Truck Load

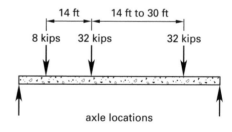

axle locations

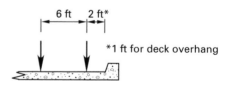

wheel locations

As shown in Fig. 8.4, the design tandem load consists of a pair of 25 kip axles spaced 4 ft apart. The transverse spacing of the wheels is 6 ft. Transversely, the design tandem is positioned in a lane in the same manner as the design truck. A dynamic load allowance is applied to the design tandem load.

Figure 8.4 Design Tandem Load

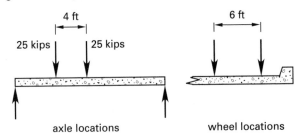

axle locations   wheel locations

In accordance with AASHTO Sec. 3.6.1.1.2, the number of loaded lanes is selected to produce the most critical effect. As specified in AASHTO Sec. 3.6.1.2.1, each lane under consideration shall be occupied by either the design truck or tandem combined with the lane load. Figure 8.5 shows the location of the design truck and the design lane load to produce the maximum positive moment at mid span of the end span of a three span continuous deck. The design lane load is placed on both end spans so as to produce the maximum effect. The design truck is placed with its central axle at the mid span of the end span.

Figure 8.5 Design Truck Positioned for Maximum Positive Moment at Point 5

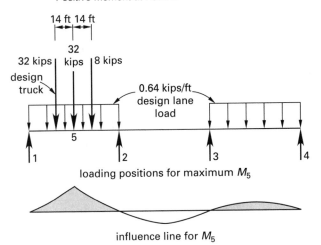

Similarly, Fig. 8.6 shows the design tandem positioned to produce the maximum positive moment at mid span of the end span of a three span continuous deck.

Figure 8.6 Design Tandem Positioned for Maximum Positive Moment at Point 5

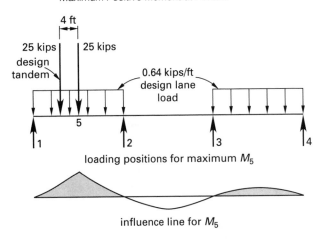

When several lanes are loaded, the force effect determined is multiplied by a multiple presence factor to account for the probability of simultaneous lane occupation by the full HL-93 design live load. The multiple presence factor, m, is given by AASHTO Table 3.6.1.1.2-1 as

- 1.2 for one loaded lane

- 1.0 for two loaded lanes

- 0.85 for three loaded lanes

- 0.65 for more than three loaded lanes

For the determination of maximum negative moments in a continuous deck, AASHTO Sec. 3.6.1.3.1 specifies that two design trucks may be located in each lane with a minimum distance of 50 ft between the lead axle of one truck and the rear axle of the other truck. The distance between the 32 kip axles of each truck is 14 ft. The two design trucks are placed in adjacent spans to produce maximum force effects. Axles that do not contribute to the negative moment are neglected. The truck loading is combined with the design lane load using patch loading to produce the maximum effect. The total combined moment is multiplied by a reduction factor of 0.9 to give the design moment. The same procedure is used to determine the reaction at interior piers.

Similarly, as shown in Fig. 8.7, two design tandems may be applied, spaced 26–40 ft apart, and combined with the design lane load. This represents the loading caused by "low-boy" type vehicles weighing in excess of 110 kips. The total combined moment obtained is the required design moment without multiplying by a reduction factor.

Bridges

Figure 8.7 *Design Lane Load and Two Design Tandems
Positioned for Maximum Moment at Support 3*

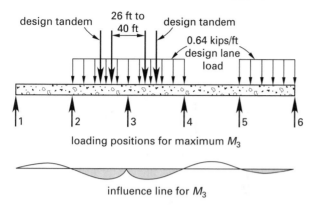

loading positions for maximum M_3

influence line for M_3

For continuous spans, influence lines may be used to determine the maximum effect, and these are available[2,3] for standard cases. For non-standard situations, several methods[4,5] may be used to determine the required influence lines.

Example 8.2

The four-span bridge shown in the following illustration has the superstructure analyzed in Ex. 8.1. Determine the maximum moment at support 2 produced by loading one design lane with the design lane load combined with the design truck. Neglect the multiple presence factor.

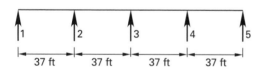

Solution

The locations of the design lane load to produce the maximum moment at support 2 are obtained[2] as shown in the following illustration. Span 34 is not loaded.

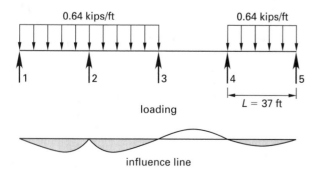

loading

influence line

The bending moment at support 2, produced by one design lane load, is given by[2]

$$M_2 = awL^2$$
$$= (0.1205)\left(0.64\ \frac{\text{kips}}{\text{ft}}\right)(37\ \text{ft})^2$$
$$= 106\ \text{ft-kips}$$

The axle locations of the two design trucks to produce the maximum moment at support 2 are obtained[2] as shown in the following illustration. The 8 kip lead axle on each truck is neglected.

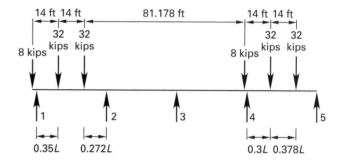

The bending moment at support 2, produced by the two design trucks, is given by[2]

$$M_2 = \sum \gamma W L$$
$$= \left(\begin{array}{c} 0.083 + 0.090 \\ + 0.0064 + 0.0051 \end{array}\right)(32\ \text{kips})(37\ \text{ft})$$
$$= 218\ \text{ft-kips}$$

The combined moment produced by the design lane load and the two standard trucks is

$$M_2 = 106\ \text{ft-kips} + 218\ \text{ft-kips}$$
$$= 324\ \text{ft-kips}$$

The design moment is given by AASHTO Sec. 3.6.1.3.1 as

$$M_2 = (0.9)(324\ \text{ft-kips})$$
$$= 292\ \text{ft-kips}$$

Dynamic Load Allowance

In accordance with AASHTO Sec. 3.6.2, an allowance for dynamic effects is applied to the static axle loads of the design truck and the design tandem. The dynamic load allowance, IM, is given by AASHTO Table 3.6.2.1-1 as

- 75% for deck joints for all limit states

- 15% for all other components for fatigue and fracture limit states

- 33% for all other components for all other limit states

The dynamic factor to be applied to the static load is

$$I = 1 + \frac{IM}{100}$$

The dynamic factor is not applied to

- the design lane load

- pedestrian loads

- centrifugal forces and braking forces

- retaining walls not subject to vertical loads from the superstructure

- foundation components that are entirely below ground level

- wood structures

The dynamic load allowance for culverts and other buried structures is given by AASHTO Eq. (3.6.2.2-1) as

$$IM = 33(1.0 - 0.125D_E)$$
$$\geq 0\%$$

D_E, the minimum depth of earth cover over the structure, is in ft.

Example 8.3

For the four-span bridge of Ex. 8.2, determine the maximum moment at support 2 produced by loading one design lane with the design lane load combined with the design truck. Include the effect of the dynamic load allowance.

Solution

The dynamic load allowance for the support moment is given by AASHTO Table 3.6.2.1-1 as

$$IM = 33\%$$

This is applied to the static axle loads of the design trucks to give the dynamic factor

$$I = 1 + \frac{IM}{100}$$
$$= 1 + \frac{33\%}{100}$$
$$= 1.33$$

The static moment at support 2 caused by the two design trucks is given by Ex. 8.2 as

$$M_2 = 218 \text{ ft-kips}$$

The moment at support 2 caused by the two design trucks, including the dynamic load allowance, is

$$M_2 = (1.33)(218 \text{ ft-kips})$$
$$= 290 \text{ ft-kips}$$

The combined moment produced by the design lane load and the two standard trucks, including the dynamic load allowance, is

$$M_2 = 106 \text{ ft-kips} + 290 \text{ ft-kips}$$
$$= 396 \text{ ft-kips}$$

The final design moment is given by AASHTO Sec. 3.6.1.3.1 as

$$M_2 = (0.9)(396 \text{ ft-kips})$$
$$= 356 \text{ ft-kips}$$

Distribution of Loads

In accordance with AASHTO Sec. 4.6.2.2.1 in the calculation of bending moments for T-beam bridges, permanent loads of and on the deck may be distributed uniformly to all beams.

The distribution of live load depends on the torsional stiffness of the bridge deck system and, if necessary, may be determined by several[6,7,8,9] analytical methods. In accordance with AASHTO Sec. 4.6.2.2, however, the distribution of live load may be calculated by empirical expressions, depending on the superstructure type and the stringer spacing.

The types of superstructure for which the distribution factor method may be used are illustrated in AASHTO Table 4.6.2.2.1-1. The method may be applied provided that the following conditions are met.

- a single lane of live loading is analyzed

- multiple lanes of live loading producing approximately the same force effect per lane are analyzed

- the deck width is constant

- the number of beams is not less than four (with some exceptions)

- beams are parallel and have approximately the same stiffness

- the roadway part of the overhang does not exceed 3 ft (with some exceptions)

- the curvature of the superstructure is less than the limit specified in AASHTO Sec. 4.4

Additional requirements are specified for each specific superstructure illustrated in AASHTO Table 4.6.2.2.1-1, and these are listed in AASHTO Table 4.6.2.2.2b-1.

A monolithic T-beam superstructure is listed as case (e) in AASHTO Table 4.6.2.2.1-1. For this type of bridge, the limitation on the beam spacing is

$$3.5 \text{ ft} \leq S \leq 16.0 \text{ ft}$$

The limitation on the deck slab thickness is

$$4.5 \text{ in} \leq t_s \leq 12.0 \text{ in}$$

The limitation on the superstructure span is

$$20 \text{ ft} \leq L \leq 240 \text{ ft}$$

The limitation on the number of beams is

$$N_b \geq 4$$

The limitation on the longitudinal stiffness parameter is

$$10,000 \text{ in}^4 \leq K_g \leq 7,000,000 \text{ in}^4$$

The longitudinal stiffness parameter is defined by AASHTO Eq. (4.6.2.2.1-1) as

$$K_g = n(I + Ae_g^2)$$

The ratio of the modulus of elasticity of the beam and the deck slab is defined by AASHTO Eq. (4.6.2.2.1-2) as

$$n = \frac{E_B}{E_D}$$

The moment of inertia of the beam is

$$I = \frac{bd^3}{12}$$

The area of the beam is

$$A = bd$$

The distance between the centers of gravity of the beam and the deck slab is

$$e_g = \frac{t_s + d}{2}$$

When these conditions are complied with, AASHTO Table 4.6.2.2.2b-1 gives the distribution factor for moment for one design lane loaded as

$$g_1 = 0.06 + \left(\frac{S}{14}\right)^{0.4} \left(\frac{S}{L}\right)^{0.3} \left(\frac{K_g}{12.0Lt_s^3}\right)^{0.1}$$

When two or more design lanes are loaded, the distribution factor for moment is

$$g_m = 0.075 + \left(\frac{S}{9.5}\right)^{0.6} \left(\frac{S}{L}\right)^{0.2} \left(\frac{K_g}{12.0Lt_s^3}\right)^{0.1}$$

In accordance with AASHTO Sec. 3.6.1.1.2, the multiple presence factors specified in AASHTO Table 3.6.1.1.2-1 are not applicable as these factors are already incorporated in the distribution factors. The dynamic load allowance must be applied to that portion of the bending moment produced by design trucks and design tandems.

These distribution factors are not applicable for the determination of bending moments in exterior beams. For exterior beams, with one lane loaded, and for interior beams in decks with less than four beams, the lever-rule method specified in AASHTO Sec. C4.6.2.2.1 may be used. For these analyses, both the multiple presence factor and the dynamic load allowance must be applied. Irrespective of the calculated moment, an exterior beam shall have a carrying capacity not less than that of an interior beam.

Example 8.4

For the four-span, concrete T-beam bridge of Ex. 8.2, determine the maximum live load moment for design of an interior beam at support 2. The ratio of the modulus of elasticity of the beam and the deck slab is $n = 1.0$.

Solution

From Ex. 8.3, the maximum live load moment produced at support 2 by loading one design lane with two design trucks, plus dynamic load allowance, and the design lane load is

$$M_2 = 356 \text{ ft-kips}$$

The ratio of the modulus of elasticity of the beam and the deck slab is given as

$$n = \frac{E_B}{E_D}$$
$$= 1.0$$

The moment of inertia of the beam is

$$I = \frac{bd^3}{12}$$
$$= \frac{(12 \text{ in})(30 \text{ in})^3}{12}$$
$$= 27,000 \text{ in}^4$$

The area of the beam is

$$A = bd$$
$$= (12 \text{ in})(30 \text{ in})$$
$$= 360 \text{ in}^2$$

The distance between the centers of gravity of the beam and the deck slab is

$$e_g = \frac{t_s + d}{2}$$
$$= \frac{9 \text{ in} + 30 \text{ in}}{2}$$
$$= 19.5 \text{ in}$$

The longitudinal stiffness parameter of the deck is defined by AASHTO Eq. (4.6.2.2.1-1) as

$$K_g = n(I + Ae_g^2)$$
$$= (1.0)\left(27{,}000 \text{ in}^4 + (360 \text{ in}^2)(19.5 \text{ in})^2\right)$$
$$= 163{,}890 \text{ in}^4 \quad \left[\begin{array}{c} \text{complies with} \\ \text{AASHTO Table 4.6.2.2b-1} \end{array}\right]$$
$$> 10{,}000 \text{ in}^4$$
$$< 7{,}000{,}000 \text{ in}^4$$

The beam spacing is

$$S = 11 \text{ ft} \quad \text{[complies with AASHTO Table 4.6.2.2b-1]}$$
$$> 3.5 \text{ ft}$$
$$< 16.0 \text{ ft}$$

The deck slab thickness is

$$t_s = 9 \text{ in} \quad \text{[complies with AASHTO Table 4.6.2.2b-1]}$$
$$> 4.5 \text{ in}$$
$$< 12.0 \text{ in}$$

The superstructure span is

$$L = 37 \text{ ft} \quad \text{[complies with AASHTO Table 4.6.2.2b-1]}$$
$$> 20 \text{ ft}$$
$$< 240 \text{ ft}$$

The number of beams in the deck is

$$N_b = 4 \quad \text{[complies with AASHTO Table 4.6.2.2b-1]}$$

Therefore, the configuration of the deck is in full conformity with the requirements of AASHTO Table 4.6.2.2b-1.

With one lane loaded, AASHTO Table 4.6.2.2b-1 gives the distribution factor for moment as

$$g_1 = 0.06 + \left(\frac{S}{14}\right)^{0.4}\left(\frac{S}{L}\right)^{0.3}\left(\frac{K_g}{12.0Lt_s^3}\right)^{0.1}$$
$$= 0.06 + \left(\frac{11 \text{ ft}}{14}\right)^{0.4}\left(\frac{11 \text{ ft}}{37 \text{ ft}}\right)^{0.3}$$
$$\times \left(\frac{163{,}890 \text{ in}^4}{(12.0)(37 \text{ ft})(9 \text{ in})^3}\right)^{0.1}$$
$$= 0.650$$

With two lanes loaded, as shown in the illustration, AASHTO Table 4.6.2.2b-1 gives the distribution factor for moment as

$$g_m = 0.075 + \left(\frac{S}{9.5}\right)^{0.6}\left(\frac{S}{L}\right)^{0.2}\left(\frac{K_g}{12.0Lt_s^3}\right)^{0.1}$$
$$= 0.075 + \left(\frac{11 \text{ ft}}{9.5}\right)^{0.6}\left(\frac{11 \text{ ft}}{37 \text{ ft}}\right)^{0.2}$$
$$\times \left(\frac{163{,}890 \text{ in}^4}{(12.0)(37 \text{ ft})(9 \text{ in})^3}\right)^{0.1}$$
$$= 0.875 \quad \text{[governs]}$$

The live load moment for the design of an interior beam at support 2 is

$$M_L = g_m M_2$$
$$= (0.875)(356 \text{ ft-kips})$$
$$= 312 \text{ ft-kips}$$

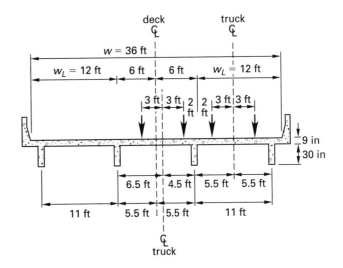

Shear Determination

The distribution factor method is also used to calculate design shear in interior beams. The distribution factors and the range of applicability are listed in AASHTO Table 4.6.2.3a-1. For a monolithic T-beam superstructure, the limitations on beam spacing, slab thickness, span length, and number of beams are identical with those for determining the distribution factor for moment. There is no requirement specified for the longitudinal stiffness parameter.

When these conditions are complied with, AASHTO Table 4.6.2.3a-1 gives the distribution factor for shear for one design lane loaded as

$$g_1 = 0.36 + \frac{S}{25}$$

When two or more design lanes are loaded, the distribution factor for shear is

$$g_m = 0.2 + \frac{S}{12} - \left(\frac{S}{35}\right)^{2.0}$$

In accordance with AASHTO Sec. 3.6.1.1.2, the multiple presence factors specified in AASHTO Table 3.6.1.1.2-1 are not applicable as these factors are already incorporated in the distribution factors. The dynamic load allowance must be applied to that portion of the shear produced by design trucks and design tandems.

These distribution factors are not applicable for the determination of shear in exterior beams. For exterior beams, with one lane loaded, and for interior beams in decks with less than four beams, the lever rule method specified in AASHTO Sec. C4.6.2.2.1 may be used. For these analyses, both the multiple presence factor and the dynamic load allowance must be applied. Irrespective of the calculated shear, an exterior beam shall not have less resistance than an interior beam.

The application of the lever rule is illustrated in Fig. 8.8 for the determination of the distribution factor for shear in the exterior girder of a T-beam superstructure. For one lane loaded, the center of one wheel of an axle of the design truck or the design tandem is located 2 ft from the edge of the design lane as specified in AASHTO Sec. 3.6.1.3.1. A notional hinge is introduced into the deck slab at the position of beam 2 and moments are taken about this hinge. The reaction at beam 1, in terms of one wheel load is

$$V_1 = \frac{P(4.5 \text{ ft} + 10.5 \text{ ft})}{11 \text{ ft}}$$
$$= 1.364P$$

The distribution factor for shear for one lane loaded with one axle is

$$g_1 = \frac{V_1}{2P}$$
$$= \frac{1.364P}{2P}$$
$$= 0.682$$

Applying the multiple presence factor for one lane loaded gives a distribution factor of

$$g = 1.2g_1$$
$$= (1.2)(0.682)$$
$$= 0.818$$

Figure 8.8 *Lever Rule for Shear in an Exterior Girder*

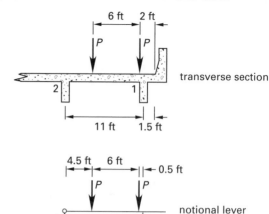

Example 8.5

For the four-span concrete T-beam bridge of Ex. 8.2, determine the live load shear V_{23} for design of an interior beam. Use the design truck load combined with the design lane load.

Solution

The influence line for V_{23} is shown in the following illustration.

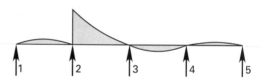

To produce the maximum value of V_{23}, the truck is positioned as shown in the following illustration.

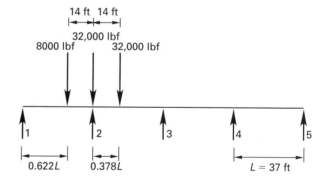

The shear at end 2 of span 23 produced by one standard truck is given by

$$V_{23} = \sum \gamma W$$
$$= \frac{(0.128)(8000 \text{ lbf})}{1000 \frac{\text{lbf}}{\text{kip}}} + \frac{(1.635)(32,000 \text{ lbf})}{1000 \frac{\text{lbf}}{\text{kip}}}$$
$$= 53.34 \text{ kips}$$

As determined in Ex. 8.4, the superstructure dimensions are within the allowable range of applicability.

AASHTO Table 4.6.2.2.3a-1 gives the distribution factor for shear for one design lane loaded as

$$g_1 = 0.36 + \frac{S}{25}$$
$$= 0.36 + \frac{11 \text{ ft}}{25}$$
$$= 0.800$$

When two or more design lanes are loaded, the distribution factor for shear is

$$g_m = 0.2 + \frac{S}{12} - \left(\frac{S}{35}\right)^{2.0}$$
$$= 0.2 + \frac{11 \text{ ft}}{12} - \left(\frac{11 \text{ ft}}{35 \text{ ft}}\right)^{2.0}$$
$$= 1.018 \quad [\text{governs}]$$

The dynamic load allowance for the shear is given by AASHTO Table 3.6.2.1-1 as

$$IM = 33\%$$

This is applied to the static axle load of the design truck to give the dynamic factor

$$I = 1 + \frac{IM}{100}$$
$$= 1 + \frac{33\%}{100}$$
$$= 1.33$$

The shear at support 2 caused by the design truck, including the dynamic load allowance, is

$$V_{23} = (1.33)(53.34 \text{ kips})$$
$$= 70.94 \text{ kips}$$

The design lane load is positioned in spans 12, 23, and 45 in order to produce the maximum shear at support 2. The shear produced by one design lane load is given by[2]

$$V_{23} = awL$$
$$= (0.6027)\left(0.64 \ \frac{\text{kips}}{\text{ft}}\right)(37 \text{ ft})$$
$$= 14.27 \text{ kips}$$

The combined shear produced by one lane of the design lane load and the standard truck, including the dynamic load allowance, is

$$V_{23} = 70.94 \text{ kips} + 14.27 \text{ kips}$$
$$= 85.21 \text{ kips}$$

The live load shear, produced by two loaded lanes, for the design of an interior beam at support 2, is

$$V_L = g_m V_{23}$$
$$= (1.018)(85.21 \text{ kips})$$
$$= 86.74 \text{ kips}$$

Design of Concrete Deck Slabs

The bending moments, caused by wheel loads, in concrete deck slabs supported by longitudinal stringers and transverse girders may be obtained by the methods proposed by Westergaard[10] and Pucher[11]. AASHTO Sec. 4.6.2.1 provides an equivalent strip method for the design of concrete deck slabs.

This method consists of dividing the deck into strips perpendicular to the supporting stringers and transverse girders. The principal features of the method are

- The extreme positive moment in any deck panel shall be applied to all positive moment regions.

- The extreme negative moment over any supporting component shall be applied to all negative moment regions.

- Where the deck slab spans primarily in the transverse direction, only the axles of the design truck or the design tandem shall be applied to the deck slab.

- Where the deck slab spans primarily in the longitudinal direction, and the span does not exceed 15 ft, only the axles of the design truck or the design tandem shall be applied to the deck slab.

- Where the deck slab spans primarily in the longitudinal direction, and the span exceeds 15 ft, the design truck combined with the design lane load or the design tandem combined with the design lane load shall be applied to the deck slab, and the provisions of AASHTO Sec. 4.6.2.3 shall apply.

- Where the deck slab spans primarily in the longitudinal direction, the width of the equivalent strip supporting an axle load shall not be taken greater than 40 in for open grids.

- Where the deck slab spans primarily in the transverse direction, the equivalent strip is not subject to width limits.

- Where the spacing of supporting components in the secondary direction exceeds 1.5 times the spacing in the primary direction, all of the wheel loads may be considered to be applied to the primary strip and distribution reinforcement complying with AASHTO Sec. 9.7.3.2 may be applied in the secondary direction.

- Where the spacing of supporting components in the secondary direction is less than 1.5 times the spacing in the primary direction, the deck shall be modeled as a system of intersecting strips.

- Wheel loads are distributed to the intersecting strips in proportion to their stiffnesses.

- The stiffness of a strip is specified as $k_s = EI_s/S^3$.

- Strips are treated as simply supported or continuous beams as appropriate with a span length equal to the center-to-center distance between the supporting components.

- Wheel loads may be modeled as concentrated loads or as patch loads whose length along the span is equal to the length of the tire contact area plus the depth of the deck slab.

- Both the multiple presence factor and the dynamic load allowance must be applied to the bending moments calculated.

- In lieu of determining the width of the equivalent strip, the moments may be obtained directly from AASHTO Table A4-1 and these values include an allowance for both the multiple presence factor and the dynamic load allowance.

AASHTO Table 4.6.2.1.3-1 defines the width of an equivalent strip. For cast-in-place deck slabs, the width, in inches, of both longitudinal and transverse strips for calculating positive moment is

$$B_s = 26 \text{ in} + 6.6(S \text{ ft})$$

The width of both longitudinal and transverse strips for calculating negative moment is

$$B_s = 48 \text{ in} + 3.0(S \text{ ft})$$

Example 8.6

For the four-span concrete T-beam bridge of Ex. 8.2, determine the maximum negative live load moment in the slab and the width of the equivalent strip. The layout of longitudinal and transverse girders is shown in the following illustration.

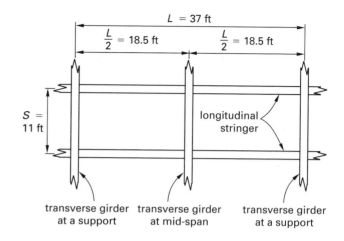

Solution

The aspect ratio of the slab is

$$AR = \frac{\dfrac{L}{2}}{S} = \frac{18.5 \text{ ft}}{11 \text{ ft}}$$
$$= 1.68$$
$$> 1.5$$

Therefore, all of the wheel loads may be considered to be applied to the primary strip in the transverse direction, and distribution reinforcement complying with AASHTO Sec. 9.7.3.2 may be applied in the secondary direction. Since the deck slab spans primarily in the transverse direction, only the axles of the design vehicle shall be applied to the deck slab.

The width, in inches, of the transverse strip for calculating negative moment is given by AASHTO Table 4.6.2.1.3-1 as

$$B_s = 48 \text{ in} + 3.0(S \text{ ft})$$
$$= \frac{48 \text{ in} + (3.0)(11 \text{ ft})}{12 \dfrac{\text{in}}{\text{ft}}}$$
$$= 6.75 \text{ ft}$$

The required moment may be determined from AASHTO Table A4-1. The span length of the transverse strip is

$$S = 11 \text{ ft}$$

The distance from the center line of a longitudinal stringer to the face of the stringer is

$$x = 6 \text{ in}$$

Therefore, from AASHTO Table A4-1, the maximum negative bending moment is

$$M_s = 7.38 \text{ ft-kips/ft}$$

Design of Slab-Type Bridges

The bending moments and shears in concrete slab-type decks, caused by axle loads, may be obtained by an equivalent strip method defined in AASHTO Sec. 4.6.2.3. The equivalent width of a longitudinal strip with two lines of wheels in one lane is given by AASHTO Eq. (4.6.2.3-1) as

$$E = 10.0 + 5.0(L_1 W_1)^{0.5}$$

The modified span length L_1 is equal to the lesser of the actual span length or 60 ft. The modified edge-to-edge width of the bridge W_1 is equal to the lesser of the actual width W, or 30 ft.

The equivalent width of a longitudinal strip with more than one lane loaded is given by AASHTO Eq. (4.6.2.3-2) as

$$E = 84.0 + 1.44(L_1 W_1)^{0.5}$$
$$\leq 12.0W/N_L$$

The modified span length L_1 is equal to the lesser of the actual span length, or 60 ft. The modified edge-to-edge width of the bridge W_1 is equal to the lesser of the actual width W, or 60 ft. The number of design lanes N_L is determined as specified in AASHTO Sec. 3.6.1.1.1.

An allowance for the multiple presence factor is included in the equivalent strip width. The dynamic load allowance must be applied to the bending moments calculated.

Example 8.7

A prestressed concrete slab bridge has a simply supported span of $L = 37$ ft. The overall width of the bridge is $W = 39$ ft, and the distance between curbs is $w = 36$ ft. Determine the width of the equivalent strip.

Solution

From AASHTO Sec. 3.6, the number of design lanes is

$$N_L = \frac{w}{12}$$
$$= \frac{36 \text{ ft}}{12 \frac{\text{ft}}{\text{lane}}}$$
$$= 3 \text{ lanes}$$

For one design lane loaded, the modified span length is equal to the lesser of the actual span length, or 60 ft, and

$$L_1 = L$$
$$= 37 \text{ ft}$$

The modified edge-to-edge width of the bridge is equal to the lesser of the actual width, or 30 ft, and

$$W_1 = 30 \text{ ft}$$

The equivalent width of a longitudinal strip is given by AASHTO Eq. (4.6.2.3-1) as

$$E = 10.0 + 5.0(L_1 W_1)^{0.5}$$
$$= 10.0 \text{ in} + (5.0)\big((37 \text{ ft})(30 \text{ ft})\big)^{0.5}$$
$$= 176.6 \text{ in}$$

For more than one design lane loaded, the modified span length is equal to the lesser of the actual span length, or 60 ft, and

$$L_1 = L$$
$$= 37 \text{ ft}$$

The modified edge-to-edge width of the bridge is equal to the lesser of the actual width, or 60 ft, and

$$W_1 = W$$
$$= 39 \text{ ft}$$

The equivalent width of a longitudinal strip is given by AASHTO Eq. (4.6.2.3-2) as

$$E = 84.0 + 1.44(L_1 W_1)^{0.5}$$
$$= 84.0 \text{ in} + (1.44)\big((37 \text{ ft})(39 \text{ ft})\big)^{0.5}$$
$$= 138.7 \text{ in}$$
$$\frac{12.0W}{N_L} = \frac{(12)(39 \text{ ft})}{3 \text{ lanes}}$$
$$= 156 \text{ in}$$
$$> E \quad [E = 138.7 \text{ in}]$$

The equivalent width for more than one design lane loaded governs, and

$$E = 138.7 \text{ in}$$

Combinations of Loads

The load and resistance factor design method presented in AASHTO Sec. 1.3.2, defines four limit states: the service limit state, the fatigue and fracture limit state, the strength limit state, and the extreme event limit state.

The *service limit state* governs the design of the structure under regular service conditions to ensure satisfactory stresses, deformations, and crack widths. Four service limit states are defined, with service I limit state comprising the load combination relating to the normal operational use of the bridge with a 55 mph wind, and all loads taken at their nominal values.

The *fatigue limit state* governs the design of the structure loaded with a single design truck for a given number of stress range cycles. The fracture limit state is defined as a set of material toughness requirements given in the AASHTO Materials Specifications.

The *strength limit state* ensures the structure's strength and structural integrity under the various load combinations imposed on the bridge during its design life. Five strength limit states are defined, with strength I limit state comprising the load combination relating to the normal vehicular use of the bridge without wind.

The *extreme event limit state* ensures the survival of the structure during a major earthquake or flood, or when subject to collision from a vessel, vehicle, or ice flow. Two extreme limit states are defined, with extreme event I limit state comprising the load combination that includes earthquake.

The factored load is influenced by the ductility of the components, the redundancy of the structure, and the operational importance of the bridge based on social or defense requirements. It is preferable for components to exhibit ductile behavior, as this provides warning of impending failure by large inelastic deformations. Brittle components are undesirable because failure occurs suddenly, with little or no warning, when the elastic limit is exceeded. For the strength limit state, the load modifier for ductility is given by AASHTO Sec. 1.3.3 as

$$\eta_D = 1.05 \quad \text{[non-ductile components]}$$
$$= 1.00 \quad \text{[conventional designs and details]}$$
$$\geq 0.95 \quad \left[\begin{array}{c}\text{components with}\\\text{ductility-enhancing features}\end{array}\right]$$

For all other limit states, the load modifier for ductility is given by AASHTO Sec. 1.3.3 as

$$\eta_D = 1.00$$

The component redundancy classification is based on the contribution of the component to the bridge safety. Major components, whose failure will cause collapse of the structure, are designated as failure-critical, and the associated structural system is designated nonredundant. Alternatively, components whose failure will not cause collapse of the structure are designated as nonfailure-critical, and the associated structural system is designated redundant. For the strength limit state, the load modifier for redundancy is given by AASHTO Sec. 1.3.4 as

$$\eta_R = 1.05 \quad \text{[nonredundant components]}$$
$$= 1.00 \quad \text{[conventional levels of redundancy]}$$
$$\geq 0.95 \quad \text{[exceptional levels of redundancy]}$$

For all other limit states, the load modifier for redundancy is given by AASHTO Sec. 1.3.4 as

$$\eta_R = 1.00$$

A bridge may be declared to be of operational importance based on survival or security reasons. For the strength limit state, the load modifier for operational importance is given by AASHTO Sec. 1.3.5 as

$$\eta_I = 1.05 \quad \text{[for important bridges]}$$
$$= 1.00 \quad \text{[for typical bridges]}$$
$$\geq 0.95 \quad \text{[for relatively less important bridges]}$$

For all other limit states, the load modifier for importance is given by AASHTO Sec. 1.3.5 as

$$\eta_I = 1.00$$

For loads where a maximum value is appropriate, the combined load modifier relating to ductility, redundancy, and operational importance is given by AASHTO Eq. (1.3.2.1-2) as

$$\eta_i = \eta_D \eta_R \eta_I$$
$$\geq 0.95$$

For loads where a minimum value is appropriate, the combined load modifier is given by AASHTO Eq. (1.3.2.1-3) as

$$\eta_i = \frac{1}{\eta_D \eta_R \eta_I}$$
$$\leq 1.0$$

The load factors applicable to permanent loads are listed in AASHTO Table 3.4.1-2 and are summarized in Table 8.1.

Table 8.1 *Load Factors for Permanent Loads*

type of load	load factor γ_p	
	max	min
components and attachments, DC	1.25	0.90
wearing surfaces and utilities, DW	1.5	0.65

The actual value of permanent loads may be less than or more than the nominal value, and both possibilities must be considered by using the maximum and minimum values given for the load factor.

Load combinations and load factors are listed in AASHTO Table 3.4.1-1, and those applicable to gravity and earthquake loads are summarized in Table 8.2.

Table 8.2 *Load Factors and Load Combinations*

load combination limit state	DC and DW	LL and IM	EQ
strength I	γ_p	1.75	–
extreme event I	γ_p	γ_{EQ}	1.00
service I	1.00	1.00	–
fatigue	–	0.75	–

The value of the load factor γ_p for the dead load of components and wearing surfaces is obtained from Table 8.1. The load factor γ_{EQ} in extreme event limit state I has traditionally been taken as 0.0. However, partial live load should be considered, and a reasonable value for the load factor is

$$\gamma_{EQ} = 0.50$$

The total factored force effect is given by AASHTO Eq. (3.4.1-1) as

$$Q = \sum \eta_i \gamma_i Q_i$$
$$\leq \phi R_n$$

Both positive and negative extremes must be considered for each load combination. For permanent loads, the load factor that produces the more critical effect is selected from Table 8.1. In strength I limit state, when the permanent loads produce a positive effect and the live loads a negative effect, the appropriate total factored force effect is

$$Q = 0.9DC + 0.65DW + 1.75(LL + IM)$$

In strength I limit state, when both the permanent loads and the live loads produce a negative effect, the appropriate total factored force effect is

$$Q = 1.25DC + 1.50DW + 1.75(LL + IM)$$

Example 8.8

For the four-span concrete T-beam bridge of Ex. 8.1, determine the strength I factored moment for design of an interior beam at support 2. Each concrete parapet has a weight of 0.5 kip/ft, and the parapets are constructed after the deck slab has cured.

Solution

The dead load acting on an interior beam consists of the beam self-weight, plus the applicable portion of the deck slab, plus the applicable portion of the two parapets. The dead load of a beam is

$$w_B = \left(0.15 \frac{\text{kip}}{\text{ft}^3}\right)(2.5 \text{ ft})(1 \text{ ft})$$
$$= 0.375 \text{ kip/ft}$$

The dead load of the applicable portion of the deck slab in accordance with AASHTO Sec. 4.6.2.2.1 is

$$w_S = \frac{\left(0.15 \frac{\text{kip}}{\text{ft}^3}\right)(39 \text{ ft})(0.75 \text{ ft})}{4 \text{ beams}}$$
$$= 1.097 \text{ kips/ft}$$

In accordance with AASHTO Sec. 4.6.2.2.1, the weights of the two concrete parapets are distributed equally to the four beams. Then, the applicable weight distributed to an interior beam is

$$w_P = \frac{\left(0.5 \frac{\text{kip}}{\text{ft}}\right)(2 \text{ parapets})}{4 \text{ beams}}$$
$$= 0.25 \text{ kip/ft}$$

The total dead load supported by an interior beam is

$$w_D = w_B + w_S + w_P$$
$$= 0.375 \frac{\text{kip}}{\text{ft}} + 1.097 \frac{\text{kips}}{\text{ft}} + 0.25 \frac{\text{kip}}{\text{ft}}$$
$$= 1.722 \text{ kips/ft}$$

The bending moment produced in an interior beam at support 2 by the uniformly distributed dead load is given by[2]

$$M_D = \alpha w_D L^2$$
$$= (0.1071)\left(1.722 \frac{\text{kips}}{\text{ft}}\right)(37 \text{ ft})^2$$
$$= 252 \text{ ft-kips}$$

The live load bending moment plus impact at support 2 is obtained from Ex. 8.4 as

$$M_L = 312 \text{ ft-kips}$$

The factored design moment for strength I limit state is given by AASHTO Eq. (3.4.1-1) and AASHTO Table 3.4.1-1 as

$$M_u = \eta_i(\gamma_p M_D + \gamma_{LL+IM} M_L)$$
$$= 1.0(1.25M_D + 1.75M_L)$$
$$= (1.0)\big((1.25)(252 \text{ ft-kips}) + (1.75)(312 \text{ ft-kips})\big)$$
$$= 861 \text{ ft-kips}$$

Critical Section for Shear

AASHTO Sec. 5.8.3.2 specifies that when the support reaction produces a compressive stress in a reinforced concrete beam, the critical section for shear is located at a distance from the support equal to the depth d_v. The depth d_v is defined in AASHTO Sec. 5.8.2.9 as the distance between the resultants of the tensile and compressive forces due to flexure. d_e is the depth to the resultant of the tensile force.

$$d_v \geq 0.9d_e$$
$$\geq 0.72h$$

Example 8.9

For the four-span concrete T-beam bridge of Ex. 8.1, determine the factored shear force, V_{23}, for design of an interior beam at support 2. The depth $d_v = 31.4$ in.

Solution

The live load shear force, including impact, on an interior beam at support 2 is obtained from Ex. 8.5 as

$$V_L = 86.74 \text{ kips}$$

The dead load supported by an interior beam is obtained from Ex. 8.8 as

$$w_D = 1.722 \text{ kips/ft}$$

The dead load shear at the support of an interior beam is given by[12]

$$V_s = \alpha w_D L$$
$$= (0.536) \left(1.722 \frac{\text{kips}}{\text{ft}} \right) (37 \text{ ft})$$
$$= 34.15 \text{ kips}$$

In accordance with AASHTO Sec. 5.8.3.2, the design shear for a distributed load may be determined at a distance d_v from the support and is given by

$$V_D = V_s - w_D d_v$$
$$= 34.15 \text{ kips} - \frac{\left(1.722 \frac{\text{kips}}{\text{ft}} \right) (31.4 \text{ in})}{12 \frac{\text{in}}{\text{ft}}}$$
$$= 29.64 \text{ kips}$$

The factored design shear for strength I limit state is given by AASHTO Eq. (3.4.1-1) and AASHTO Table 3.4.1-1 as

$$V_{23} = \eta_i (\gamma_p V_D + \gamma_{LL+IM} V_L)$$
$$= 1.0 (1.25 V_D + 1.75 V_L)$$
$$= (1.0) \big((1.25)(29.64 \text{ kips}) + (1.75)(86.74 \text{ kips}) \big)$$
$$= 189 \text{ kips}$$

Service Limit State

The service limit state governs stresses, deformations, and crack widths under regular service conditions. The service I limit state comprises the load combination relating to the normal operational use of a bridge with a 55 mph wind and all loads taken at their nominal values.

Example 8.10

For the four-span concrete T-beam bridge of Ex. 8.1, determine the service I design moment for an interior beam at support 2. Each concrete parapet has a weight of 0.5 kip/ft, and the parapets are constructed after the deck slab has cured. Wind effects may be neglected.

Solution

From Ex. 8.8, the bending moment at support 2 produced by the uniformly distributed dead load is

$$M_D = 252 \text{ ft-kips}$$

The live load bending moment plus impact at support 2 is obtained from Ex. 8.8 as

$$M_L = 312 \text{ ft-kips}$$

The service I design moment is given by AASHTO Sec. 3.4.1 as

$$M_s = M_D + M_L$$
$$= 252 \text{ ft-kips} + 312 \text{ ft-kips}$$
$$= 564 \text{ ft-kips}$$

2. REINFORCED CONCRETE DESIGN

Design for Flexure

Nomenclature

a	depth of equivalent rectangular stress block	in
$A_{\max}$	maximum area of tension reinforcement	in^2
A_s	area of tension reinforcement	in^2
A_{sk}	area of skin reinforcement per unit height in one side face	in^2/ft
b	width of compression face of member	in
b_w	web width	in
c	distance from extreme compression fiber to neutral axis	in
$\bar{c}$	distance from extreme tension fiber to centroid of tension reinforcement	in
d	distance from extreme compression fiber to centroid of tension reinforcement	in
d_b	diameter of bar	in
d_c	thickness of concrete cover measured from extreme tension fiber to center of nearest bar	in
f'_c	compressive strength of concrete	kips/in^2
f_f	allowable stress range	kips/in^2
$f_{\max}$	maximum stress in reinforcement	kips/in^2
$f_{\min}$	minimum stress in reinforcement	kips/in^2
f_r	modulus of rupture of concrete	kips/in^2
f_{ss}	calculated stress in tension reinforcement at service loads	kips/in^2
f_y	yield strength of reinforcement	kips/in^2
h	overall dimension of member	in
h_f	flange depth	in
$h_{\min}$	recommended minimum depth of superstructure	ft
I_g	moment of inertia of gross concrete section	in^4
K_u	design moment factor	lbf/in^2
l_a	lever arm for elastic design	in
M_{cr}	cracking moment	ft-kips
M_D	dead load moment	ft-kips

M_{max}	maximum moment	ft-kips
M_{min}	minimum design flexural strength	ft-kips
M_{mr}	maximum moment range	ft-kips
M_n	nominal flexural strength of a member	ft-kips
M_u	factored moment on the member	ft-kips
n	number of tensile reinforcing bars	–
s	spacing of reinforcement	in

Symbols

β_s	ratio of flexural strain at the extreme tension face to the strain of the centroid of the reinforcement layer nearest to the tension face	
β_1	compression zone factor	
ρ	ratio of tension reinforcement	
ρ_{max}	maximum allowable tension reinforcement ratio	
γ_e	exposure factor	
ϕ	strength reduction factor	
ω	tension reinforcement index	

Strength Design Method

The procedure specified in AASHTO Sec. 5.7 is similar to the procedure adopted in the ACI[13] building code. In addition, stresses at service load shall be limited to ensure satisfactory performance under service load conditions, and the requirements for deflection, cracking moment, flexural cracking, skin reinforcement, and fatigue must be satisfied.

Load Factor Design

When the depth of the equivalent stress block is not greater than the flange depth of a reinforced concrete T-beam, the section may be designed as a rectangular beam.

Example 8.11

For the four-span concrete T-beam bridge of Ex. 8.1, determine the tensile reinforcement required in an interior beam in the end span 12. The concrete strength is 4 kips/in^2, and the reinforcement consists of no. 9 grade 60 bars. Assume that the strength I factored moment is $M_u = 1216$ ft-kips.

Solution

The effective compression flange width is given by AASHTO Sec. 4.6.2.6.1 as the tributary width, which is

$$b = S$$
$$= (11 \text{ ft}) \left(12 \, \frac{\text{in}}{\text{ft}} \right)$$
$$= 132 \text{ in}$$

The factored design moment is given as

$$M_u = 1216 \text{ ft-kips}$$

Assuming that the stress block lies within the flange and the effective depth, d, is 34.6 in, the required tension reinforcement is determined from the principles of AASHTO Sec. 5.7. The design moment factor is

$$K_u = \frac{M_u}{bd^2}$$
$$= \frac{(1216 \text{ ft-kips}) \left(12 \, \dfrac{\text{in}}{\text{ft}} \right) \left(1000 \, \dfrac{\text{lbf}}{\text{kip}} \right)}{(132 \text{ in}) (34.6 \text{ in})^2}$$
$$= 92.3 \text{ lbf/in}^2$$

$$\frac{K_u}{f'_c} = \frac{92.3 \, \dfrac{\text{lbf}}{\text{in}^2}}{4000 \, \dfrac{\text{lbf}}{\text{in}^2}}$$
$$= 0.0231$$

From Table A.1, in the appendix, the corresponding tension reinforcement index is

$$\omega = 0.026$$
$$< 0.319\beta_1$$
$$= (0.319)(0.85)$$
$$= 0.271$$

Hence, the section is tension controlled, and $\phi = 0.9$.

The required reinforcement ratio is

$$\rho = \frac{\omega f'_c}{f_y}$$
$$= \frac{(0.026) \left(4 \, \dfrac{\text{kips}}{\text{in}^2} \right)}{60 \, \dfrac{\text{kips}}{\text{in}^2}}$$
$$= 0.00173$$

The reinforcement area required is

$$A_s = \rho bd$$
$$= (0.00173) (132 \text{ in}) (34.6 \text{ in})$$
$$= 7.90 \text{ in}^2$$

Bridges

Using eight no. 9 bars as shown in the following illustration, the reinforcement area provided is

$$A_s = 8 \text{ in}^2$$

$$> 7.90 \text{ in}^2 \quad \text{[satisfactory]}$$

$$\phi M_n = \frac{M_u(8 \text{ in}^2)}{7.90 \text{ in}^2} = \frac{(1216 \text{ ft-kips})(8 \text{ in}^2)}{7.90 \text{ in}^2}$$

$$= 1231 \text{ ft-kips}$$

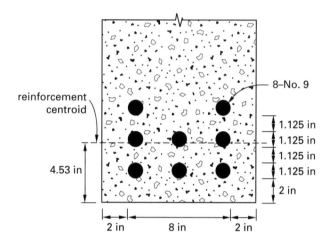

The height of the centroid of the tensile reinforcement is

$$\bar{c} = \frac{(3)(2.563 \text{ in} + 4.813 \text{ in}) + (2)(7.063 \text{ in})}{8}$$

$$= 4.53 \text{ in}$$

The effective depth is

$$d = h - \bar{c}$$

$$= 39 \text{ in} - 4.53 \text{ in}$$

$$= 34.47 \text{ in}$$

$$\approx 34.6 \text{ in} \quad \text{[assumed value of } d \text{ satisfactory]}$$

The stress block depth is

$$a = \frac{A_s f_y}{0.85 b f_c'}$$

$$= \frac{(8 \text{ in}^2)\left(60{,}000 \; \frac{\text{lbf}}{\text{in}^2}\right)}{(0.85)(132 \text{ in})\left(4000 \; \frac{\text{lbf}}{\text{in}^2}\right)}$$

$$= 1.07 \text{ in}$$

$$< h_f \quad \left[\begin{array}{c}\text{The stress block is contained} \\ \text{within the flange.}\end{array}\right]$$

Deflection Requirements

Deflections due to service live load plus impact are limited by AASHTO Sec. 2.5.2.6.2 to

$$\delta_{\max} = \frac{L}{800}$$

To achieve these limits, AASHTO Table 2.5.2.6.3-1 provides expressions for the determination of minimum superstructure depths. These are summarized in Table 8.3.

Table 8.3 *Recommended Minimum Depths*

superstructure type	minimum depth (ft)	
	simple spans	continuous spans
slabs spanning in direction of traffic	$1.2(L+10)/30$	$(L+10)/30 \geq 0.54$
T-beams	$0.070L$	$0.065L$
box girders	$0.060L$	$0.055L$

Actual deflections may be calculated in accordance with AASHTO Sec. 5.7.3.6.2, with the modulus of elasticity of normal weight concrete given by AASHTO Eq. (C5.4.2.4-1) as

$$E_c = 1820\sqrt{f_c'}$$

In determining deflections, the effective moment of inertia may be taken as the moment of inertia of the gross concrete section.

Example 8.12

Determine whether the deflection under live load of the four-span concrete T-beam bridge of Ex. 8.1 is satisfactory.

Solution

The recommended minimum depth of the T-beam superstructure, in accordance with AASHTO Table 2.5.2.6.3-1, is

$$h_{\min} = 0.065L$$

$$= (0.065)(37 \text{ ft})$$

$$= 2.4 \text{ ft}$$

The depth provided is

$$h = 3.25 \text{ ft}$$

$$> 2.4 \text{ ft} \quad \text{[satisfactory]}$$

Cracking Moment Requirements

The *cracking moment* is the moment that when applied to a reinforced concrete member, will produce cracking in the tension face of the member. In determining the cracking moment, AASHTO Sec. 5.7.3.6.2 allows the use of the gross section properties neglecting reinforcement. In the case of T-beam construction, it is appropriate to include the full width of the flange, tributary to the web, in determining the gross moment of inertia I_g. The modulus of rupture of normal weight concrete is given by AASHTO Sec. 5.4.2.6 as

$$f_r = 0.37\sqrt{f'_c}$$

When the neutral axis of the section is a distance $\bar{y}$ from the tension face, the cracking moment is given by AASHTO Sec. 5.7.3.3.2 as

$$M_{cr} = \frac{f_r I_g}{\bar{y}} \quad \text{[AASHTO 5.7.3.3.2-1]}$$

To prevent sudden tensile failure of a flexural member, AASHTO Sec. 5.7.3.3.2 requires the member to have a moment capacity at least equal to the lesser of

$$\phi M_n = 1.2 M_{cr}$$
$$\phi M_n = 1.33 M_u$$

Example 8.13

Determine whether the cracking moment of an interior beam in the end span 12 of the four-span concrete T-beam bridge of Ex. 8.1 is satisfactory.

Solution

Member Properties for Ex. 8.13

part	A (in^2)	y (in)	I (in^4)	Ay (in^3)	Ay^2 (in^4)
beams	360	15.0	27,000	5400	81,000
flange	1188	34.5	8019	40,986	1,414,017
total	1548	–	35,019	46,386	1,495,017

The gross moment of inertia of an interior beam is obtained as shown in the table. The height of the neutral axis of the section is

$$\bar{y} = \frac{\sum Ay}{\sum A} = \frac{46,386 \text{ in}^3}{1548 \text{ in}^2}$$
$$= 30 \text{ in}$$

$$I_g = \sum I + \sum Ay^2 - \bar{y}^2 \sum A$$
$$= 35,019 \text{ in}^4 + 1,495,017 \text{ in}^4$$
$$\quad - (1548 \text{ in}^2)(30 \text{ in})^2$$
$$= 136,836 \text{ in}^4$$

The modulus of rupture of the concrete is given by AASHTO Sec. 5.4.2.6 as

$$f_r = 0.37\sqrt{f'_c} = 0.37\sqrt{4 \frac{\text{kips}}{\text{in}^2}}$$
$$= 0.74 \text{ kips/in}^2$$

The cracking moment of an interior beam is given by AASHTO Eq. (5.7.3.3.2-1) as

$$M_{cr} = \frac{f_r I_g}{\bar{y}}$$
$$= \frac{\left(0.74 \frac{\text{kips}}{\text{in}^2}\right)(136,836 \text{ in}^4)}{(30 \text{ in})\left(12 \frac{\text{in}}{\text{ft}}\right)}$$
$$= 281 \text{ ft-kips}$$

In accordance with AASHTO Sec. 5.7.3.3.2, the minimum required factored design moment capacity is

$$M_{\min} = 1.2 M_{cr}$$
$$= (1.2)(281 \text{ ft-kips})$$
$$= 337 \text{ ft-kips} \quad \text{[governs]}$$
$$< 1.33 M_u$$

From Ex. 8.11, the ultimate moment of resistance of an interior beam is

$$\phi M_n = 1231 \text{ ft-kips}$$
$$> M_{\min} \quad \text{[satisfactory]}$$

Control of Flexural Cracking

To control flexural cracking of the concrete, the size and arrangement of tension reinforcement must be adjusted.

Two exposure conditions are defined in AASHTO Sec. 5.7.3.4. Class 1 exposure condition applies when cracks can be tolerated because of reduced concern for appearance or corrosion. Class 2 exposure condition applies when there is greater concern for appearance or corrosion.

The anticipated crack width depends on the following factors.

- the spacing, s, of reinforcement in the layer closest to the tension face

- the tensile stress, f_{ss}, in reinforcement at the service limit state

- the thickness of concrete cover, d_c, measured from the extreme tension fiber to center of reinforcement in the layer closest to the tension face

The exposure factor is defined as

$$\gamma_e = 1.00 \quad \text{[class 1 exposure conditions]}$$
$$\gamma_e = 0.75 \quad \text{[class 2 exposure conditions]}$$

The ratio of flexural strain at the extreme tension face to the strain at the centroid of the reinforcement layer nearest to the tension face is defined as

$$\beta_s = 1 + \frac{d_c}{0.7(h - d_c)}$$

The spacing of reinforcement in the layer closest to the tension face is given by AASHTO Eq. (5.7.3.4-1) as

$$s \le \frac{700\gamma_e}{\beta_s f_{ss}} - 2d_c$$

Example 8.14

For an interior beam in the end span 12 of the four-span concrete T-beam bridge of Ex. 8.1, determine the allowable spacing of tension reinforcement. Assume that the service I moment is $M_s = 639$ ft-kips.

Solution

The concrete cover measured to the center of the reinforcing bar closest to the tension face of the member is obtained from Ex. 8.11 as

$$d_c = 2.56 \text{ in}$$

The lever-arm for elastic design may conservatively be taken as

$$l_a = d - \frac{h_f}{2} = 34.47 \text{ in} - 4.5 \text{ in}$$
$$= 29.97 \text{ in}$$

The maximum service dead plus live load moment in an interior beam in the end span 12 is given as

$$M_s = 639 \text{ ft-kips}$$

The stress in the reinforcement is given by

$$f_s = \frac{M_s}{l_a A_s}$$
$$= \frac{(639 \text{ ft-kips})\left(12 \dfrac{\text{in}}{\text{ft}}\right)}{(29.97 \text{ in})\left(8 \text{ in}^2\right)}$$
$$= 31.98 \text{ kips/in}^2$$

The exposure factor for class 1 exposure conditions is given by AASHTO Sec. 5.7.3.4 as

$$\gamma_e = 1.00$$

The ratio of flexural strain at the extreme tension face to the strain at the centroid of the reinforcement layer nearest to the tension face is

$$\beta_s = 1 + \frac{d_c}{0.7(h - d_c)}$$
$$= 1 + \frac{2.56 \text{ in}}{(0.7)(39 \text{ in} - 2.56 \text{ in})}$$
$$= 1.10$$

The spacing of reinforcement in the layer closest to the tension face is given by AASHTO Eq. (5.7.3.4-1) as

$$s \le \frac{700\gamma_e}{\beta_s f_{ss}} - 2d_c$$
$$= \frac{(700)(1.00)}{(1.10)\left(31.98 \dfrac{\text{kips}}{\text{in}^2}\right)} - (2)(2.56 \text{ in})$$
$$= 14.8 \text{ in}$$

The spacing provided is

$$s = 2.3 \text{ in} \quad \text{[satisfactory]}$$

Longitudinal Skin Reinforcement

Longitudinal skin reinforcement is required, in the side faces of members exceeding 3 ft in effective depth, to control cracking. In accordance with AASHTO Sec. 5.7.3.4, skin reinforcement shall be provided over a distance of $d/2$ nearest the flexural tension reinforcement, and the area in each face, per foot of height, shall be not less than

$$A_{s(\min)} = (0.012)(d - 30) \quad [\text{in in}^2/\text{ft}]$$
$$\le A_s/4$$

The spacing of this reinforcement shall not exceed

$$s = \frac{d}{6}$$
$$\le 12 \text{ in}$$

Example 8.15

Determine the skin reinforcement required, in an interior beam, in the end span 12 of the four-span concrete T-beam bridge of Ex. 8.1.

Solution

The effective depth of the beam is

$$d = 34.47 \text{ in}$$
$$< 36 \text{ in}$$

Hence, in accordance with AASHTO Sec. 5.7.3.4, skin reinforcement is not required.

Fatigue Limits

Fatigue stress limits are defined in AASHTO Sec. 5.5.3 and depend on the stress in the reinforcement and the range of stress. The range between maximum and minimum stress levels must not exceed the value

$$f_f = 24 - 0.33f_{\min} \quad [AASHTO \ 5.5.3.2\text{-}1]$$

Stress levels are determined at the fatigue limit state load. In accordance with AASHTO Table 3.4.1-1, this consists of 75% of the design vehicle live load including dynamic load allowance. As specified in AASHTO Sec. 3.6.1.4, the design vehicle consists of the design truck with a constant spacing of 30 ft between the 32 kip axles. In accordance with AASHTO Sec. 3.6.1.3.1, axles that do not contribute to the maximum force under consideration are neglected.

Example 8.16

Determine whether the fatigue stress limits, in an interior beam, in the end span 12 of the four-span concrete T-beam bridge of Ex. 8.1 are satisfactory.

Solution

The maximum moment in an interior beam in span 12 caused by the design truck, plus the dynamic load allowance, is derived[2] as

$$M_{\max} = 443 \text{ ft-kips}$$

75% of the maximum moment is

$$M_{0.75,\max} = (0.75)(443 \text{ ft-kips})$$
$$= 332 \text{ ft-kips}$$

The maximum moment occurs at point x, a distance of 14 ft from support 1. The influence line for the bending moment at this section is shown in the following illustration.

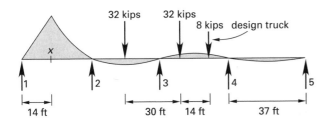

The location of the design truck to produce the minimum live load moment at point x is shown in the illustration. In accordance with AASHTO Sec. 3.6.1.3.1, the leading axle and the first 32 kips axle are ignored. Hence, the minimum moment caused by the design truck plus the dynamic load allowance is derived[2] as

$$M_{\min} = -45 \text{ ft-kips}$$

75% of the minimum moment is

$$M_{0.75,\min} = (0.75)(-45 \text{ ft-kips})$$
$$= -34 \text{ ft-kips}$$

The maximum moment range is

$$M_{mr} = M_{0.75,\max} - M_{0.75,\min}$$
$$= 332 \text{ ft-kips} - (-34 \text{ ft-kips})$$
$$= 366 \text{ ft-kips}$$

The lever arm for elastic design is obtained from Ex. 8.14 as

$$l_a = 29.97 \text{ in}$$

The actual stress range is

$$f_{f,\text{act}} = \frac{M_{mr}}{l_a A_s}$$
$$= \frac{(366 \text{ ft-kips})\left(12 \dfrac{\text{in}}{\text{ft}}\right)}{(29.97 \text{ in})(8 \text{ in}^2)}$$
$$= 18.31 \text{ kips/in}^2$$

The permanent dead load moment at section x is derived[2] as

$$M_D = 196 \text{ ft-kips}$$

The minimum live load moment resulting from the fatigue load combined with the permanent dead load moment is

$$M = M_D + M_{0.75,\min}$$
$$= 196 \text{ ft-kips} - 34 \text{ ft-kips}$$
$$= 162 \text{ ft-kips}$$

The corresponding stress is

$$f_{\min} = \frac{M}{l_a A_s}$$
$$= \frac{(162 \text{ ft-kips})\left(12 \dfrac{\text{in}}{\text{ft}}\right)}{(29.97 \text{ in})(8 \text{ in}^2)}$$
$$= 8.11 \text{ kips/in}^2$$

The range between maximum and minimum stress levels must not exceed the value given by AASHTO Eq. (5.5.3.2-1) as

$$f_f = 24 - 0.33 f_{\min}$$
$$= 24 - (0.33)\left(8.11 \frac{\text{kips}}{\text{in}^2}\right)$$
$$= 21.32 \text{ kips/in}^2$$
$$> f_{f,\text{act}}$$

The fatigue stress limits are satisfactory.

Bridges

Design for Shear

Nomenclature

A_s	area of tension reinforcement	in^2
A_v	area of shear reinforcement perpendicular to flexural tension reinforcement	in^2
b_v	web width	in
d_e	effective depth from extreme compression fiber to the centroid of the tensile force in the tensile reinforcement	in
d_v	effective shear depth	in
f_c'	specified compressive strength of concrete	kips/in^2
f_y	specified yield strength of reinforcing bars	kips/in^2
M_n	nominal flexural resistance	ft-kips
M_u	factored moment at the section	ft-kips
s	spacing of transverse reinforcement	in
V_c	nominal shear strength provided by concrete	kips
V_s	nominal shear strength provided by shear reinforcement	kips
v_u	average factored shear stress	kips/in^2
V_u	factored shear force at section	kips

Symbols

ρ_w	$A_s/b_w d$
ϕ	resistance factor

Design Methods

Two design methods are described in the *AASHTO LRFD Bridge Design Specifications*. For members in which the strain distribution is non linear, AASHTO Sec. 5.6.3 specifies the use of a strut-and-tie model. This method is applicable to pile caps and deep footings, and to members with abrupt changes in cross section. The traditional sectional model is applicable where engineering beam theory is valid, as is the case for typical bridge girders and slabs. The sectional model is specified in AASHTO Sec. 5.8.3. For nonprestressed concrete sections, not subjected to axial tension and with the minimum area of transverse reinforcement specified in AASHTO Eq. (5.8.2.5-1), a simplified procedure is permissible as specified in AASHTO Sec. 5.8.3.4.1.

Simplified Design Method

The nominal shear capacity of the concrete section is given by AASHTO Sec. 5.8.3.3 as

$$V_c = 0.0632 b_v d_v \sqrt{f_c'} \quad \textit{[AASHTO 5.8.3.3-3]}$$

The effective shear depth d_v is taken as the distance between the resultants of the tensile and compressive forces due to flexure.

The effective shear depth is given by AASHTO Eq. (C5.8.2.9-1) as

$$d_v = \frac{M_n}{A_s f_y}$$

The effective shear depth need not be taken to be less than the greater of $0.9d_e$ or $0.72h$. The effective web width b_v is taken as the minimum web width between the resultants of the tensile and compressive forces due to flexure.

The nominal shear capacity of vertical stirrups is given by AASHTO Sec. 5.8.3.3 as

$$V_s = \frac{A_v f_y d_v}{s} \quad \textit{[AASHTO 5.8.3.3-4]}$$

The shear stress on the concrete is calculated by AASHTO Eq. (5.8.2.9-1) as

$$v_u = \frac{V_u}{\phi b_v d_v}$$

For a value of v_u less than $0.125 f_c'$, AASHTO Sec. 5.8.2.7 limits the spacing of transverse reinforcement to the lesser of $0.8d_v$, or 24 in. When the value of v_u is not less than $0.125 f_c'$, the spacing is reduced to the lesser of $0.4d_v$, or 12 in.

A minimum area of transverse reinforcement is required to control diagonal cracking and this is specified by AASHTO Eq. (5.8.2.5-1) as

$$A_v = \frac{0.0316 \sqrt{f_c'} b_v s}{f_y}$$

The combined nominal shear resistance of the concrete section and the shear reinforcement is given by AASHTO Sec. 5.8.3.3 as the lesser of

$$V_n = V_c + V_s$$
$$V_n = 0.25 f_c' b_v d_v$$

The combined shear capacity of the concrete section and the shear reinforcement is

$$\phi V_n = \phi V_c + \phi V_s$$
$$\geq V_u$$

The resistance factor for normal weight concrete is given by AASHTO Sec. 5.5.4.2.1 as the following.

- For tension-controlled reinforced concrete sections,

$$\phi = 0.90$$

- For shear and torsion,

$$\phi = 0.90 \quad \text{[normal weight concrete]}$$
$$\phi = 0.70 \quad \text{[lightweight concrete]}$$

Example 8.17

For the four-span concrete T-beam bridge of Ex. 8.1, determine the shear reinforcement required in an interior beam at end 2 of span 23. The concrete strength is 4 kips/in², and the shear reinforcement consists of no. 4 grade 60 bars. The depth $d_v = 31.4$ in.

Solution

From Ex. 8.9, the factored shear at a distance d_v from the support is

$$V_{23} = 189 \text{ kips}$$

The shear strength provided by the concrete is given by

$$\phi V_c = 0.0632\phi b_v d_v \sqrt{f_c'}$$
$$= (0.0632)(0.90)(12 \text{ in})(31.4 \text{ in})\sqrt{4 \frac{\text{kips}}{\text{in}^2}}$$
$$= 42.87 \text{ kips}$$
$$< V_{23}$$

The factored shear force exceeds the shear strength of the concrete, and the shear strength required from shear reinforcement is given by

$$\phi V_s = V_{23} - \phi V_c$$
$$= 189 \text{ kips} - 42.87 \text{ kips}$$
$$= 146.13 \text{ kips}$$

The shear stress is given by AASHTO Eq. (5.8.2.9-1) as

$$v_u = \frac{V_{23}}{\phi b_v d_v}$$
$$= \frac{189 \text{ kips}}{(0.9)(12 \text{ in})(31.4 \text{ in})}$$
$$= 0.56 \text{ kips/in}^2$$
$$> 0.125 f_c'$$

Therefore, stirrups are required at a maximum spacing of 12 in. The area of shear reinforcement required is given by AASHTO Eq. (5.8.3.3-4) as

$$\frac{A_v}{s} = \frac{\phi V_s}{\phi d_v f_y}$$
$$= \frac{(146.13 \text{ kips})\left(12 \frac{\text{in}}{\text{ft}}\right)}{(0.90)(31.4 \text{ in})\left(60 \frac{\text{kips}}{\text{in}^2}\right)}$$
$$= 1.03 \text{ in}^2/\text{ft}$$

Shear reinforcement consisting of two arms of no. 4 bars at 4 in spacing provides a reinforcement area of

$$\frac{A_v}{s} = 1.2 \frac{\text{in}^2}{\text{ft}}$$
$$> 1.03 \text{ in}^2/\text{ft} \quad [\text{satisfactory}]$$

3. PRESTRESSED CONCRETE DESIGN

Design for Flexure

Nomenclature

a	depth of equivalent rectangular stress block	in
A	area of concrete section	in²
A_c	area of composite section	in²
A_{ps}	area of prestressing steel	in²
A_s	area of nonprestressed tension reinforcement	in²
b	width of compression face of member	in
c	distance from the extreme compression fiber to the neutral axis	in
d_e	effective depth from the extreme compression fiber to the centroid of the tensile force in the tensile reinforcement	in
d_p	distance from extreme compression fiber to centroid of prestressing tendons	in
d_s	distance from the extreme compression fiber to the centroid of nonprestressed reinforcement	in
d_t	distance from extreme compression fiber to centroid of extreme tensile reinforcement	in
e	eccentricity of prestressing force	in
f_{be}	bottom fiber stress at service load after allowance for all prestress losses	kips/in²
f_{bi}	bottom fiber stress immediately after prestress transfer and before time-dependent prestress losses	kips/in²
f_c'	specified compressive strength of concrete	kips/in²
f_{ci}'	compressive strength of concrete at time of prestress transfer	kips/in²
f_{cpe}	bottom fiber stress due only to effective prestressing force after allowance for all prestress losses	kips/in²
f_{pbt}	allowable stress in prestressing steel immediately prior to prestress transfer	kips/in²
f_{pe}	effective stress in prestressing steel after allowance for all prestress losses	kips/in²
f_{pj}	stress in the prestressing steel at jacking	kips/in²
f_{ps}	stress in prestressing steel at ultimate load	kips/in²
f_{pt}	stress in the prestressing steel immediately after transfer	kips/in²
f_{pu}	specified tensile strength of prestressing steel	kips/in²
f_{py}	specified yield strength of prestressing steel	kips/in²
f_r	modulus of rupture of concrete	kips/in²
f_s	stress in the tension reinforcement at nominal flexural resistance	kips/in²

f_{te}	top fiber stress at service loads after allowance for all prestress losses	kips/in^2
f_{ti}	top fiber stress immediately after prestress transfer and before time-dependent prestress losses	kips/in^2
f_y	specified yield strength of reinforcing bars	kips/in^2
h	overall depth of section	in
h_f	compression flange thickness	in
k	prestressing steel factor	–
L	span length	ft
M_{cr}	cracking moment	in-kips
M_D	bending moment due to superimposed dead load	in-kips
M_{DC}	bending moment due to superimposed dead load on composite section	in-kips
M_{dnc}	bending moment due to noncomposite dead load acting on the precast section, $(M_g + M_S)$	in-kips
M_g	bending moment due to self-weight of girder	in-kips
M_L	bending moment due to superimposed live load	in-kips
M_n	nominal flexural strength	in-kips
M_r	factored flexural resistance	in-kips
M_S	bending moment due to weight of deck slab	in-kips
M_u	factored moment	in-kips
P_e	force in prestressing steel at service loads after allowance for all losses	kips
P_i	force in prestressing steel immediately after prestress transfer	kips
S_c	section modulus of the composite section referred to the bottom fiber	in^3
S_{ci}	section modulus of the composite section referred to the interface of girder and slab	in^3
S_{nc}	section modulus of the noncomposite section referred to the bottom fiber	in^3
S_t	section modulus of the concrete section referred to the top fiber	in^3
y_b	height of centroid of the concrete section	in
y_s	height of centroid of the prestressing steel	in
w	distributed load	kips/ft

Symbols

β_1	compression zone factor	
ϵ_{cu}	failure strain of concrete in compression	
ϵ_t	net tensile strain in extreme tension steel at nominal resistance	
ϕ	strength reduction factor	
ϕ_w	reduction factor for slender members	

Conditions at Transfer

The allowable stresses in the concrete at transfer, in other than segmentally constructed bridges, are specified in AASHTO Sec. 5.9.4.1 and are

$$f_{ti} \geq -0.0948\sqrt{f'_{ci}}$$

$$\geq -0.2 \text{ kips/in}^2 \quad \text{[without bonded reinforcement]}$$

$$f_{ti} \geq -0.24\sqrt{f'_{ci}} \quad \text{[with bonded reinforcement]}$$

$$f_{bi} \leq 0.60 f'_{ci} \quad \text{[pre-tensioned members]}$$

$$f_{bi} \leq 0.60 f'_{ci} \quad \text{[post-tensioned members]}$$

In accordance with AASHTO Sec. 5.9.3, the maximum allowable stress in pre-tensioned tendons immediately prior to transfer is

$$f_{pbt} = 0.75 f_{pu} \quad \text{[low-relaxation strand]}$$

$$f_{pbt} = 0.70 f_{pu} \quad \text{[stress-relieved strand]}$$

The maximum allowable stress in post-tensioned tendons immediately after transfer is

$$f_{pt} = 0.70 f_{pu} \quad \text{[at the anchorage]}$$

$$f_{pt} = 0.74 f_{pu} \quad \left[\begin{array}{c}\text{elsewhere, low-} \\ \text{relaxation strand}\end{array}\right]$$

The maximum allowable stress at jacking is

$$f_{pj} = 0.90 f_{py}$$

Example 8.18

The post-tensioned girder shown in the following illustration is simply supported over a span of 100 ft and has the following properties.

A	S_t	S_b	y_b	f'_{ci}
800 in^2	14,700 in^3	15,600 in^3	37.8 in	4500 lbf/in^2

The concrete strength at transfer is $f'_{ci} = 4.5$ kips/in^2. The prestressing force immediately after transfer is 1000 kips, and the centroid of the tendons is 7 in above the bottom of the beam. Determine the actual and allowable stresses in the girder at midspan immediately after transfer if no bonded reinforcement is provided.

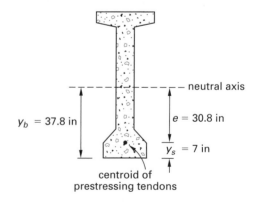

Solution

At midspan, the allowable tensile stress in the top fiber without bonded reinforcement is given by AASHTO Sec. 5.9.4.1 as

$$f_{ti} = -0.0948\sqrt{f'_{ci}} = -0.0948\sqrt{4.5 \; \frac{\text{kips}}{\text{in}^2}}$$

$$= -0.201 \; \text{kips/in}^2$$

Use the minimum allowable value of

$$f_{ti} = -0.200 \; \text{kips/in}^2$$

At midspan, the allowable compressive stress in the bottom fiber is given by AASHTO Sec. 5.9.4.1 as

$$f_{bi} = 0.60'_{ci} = (0.60)\left(4.5 \; \frac{\text{kips}}{\text{in}^2}\right)$$

$$= 2.70 \; \text{kips/in}^2$$

At midspan, the self-weight moment is

$$M_g = \frac{wL^2}{8}$$

$$= \frac{\left(150 \; \dfrac{\text{lbf}}{\text{ft}^3}\right)\left(\dfrac{800 \; \text{in}^2}{144 \; \dfrac{\text{in}^2}{\text{ft}^2}}\right)(100 \; \text{ft})^2\left(12 \; \dfrac{\text{in}}{\text{ft}}\right)}{(8)\left(1000 \; \dfrac{\text{lbf}}{\text{kip}}\right)}$$

$$= 12{,}500 \; \text{in-kips}$$

At midspan, the eccentricity of the prestressing force is

$$e = y_b - y_s$$

$$= 37.8 \; \text{in} - 7 \; \text{in}$$

$$= 30.8 \; \text{in}$$

At midspan, the actual stress in the top fiber is given by

$$f_{ti} = \frac{P_i}{A} - \frac{P_i e}{S_t} + \frac{M_g}{S_t}$$

$$= \frac{1000 \; \text{kips}}{800 \; \text{in}^2} - \frac{(1000 \; \text{kips})(30.8 \; \text{in})}{14{,}700 \; \text{in}^3}$$

$$+ \frac{12{,}500 \; \text{in-kips}}{14{,}700 \; \text{in}^3}$$

$$= +0.005 \; \text{kips/in}^2$$

$$> -0.20 \; \text{kips/in}^2 \quad \text{[satisfactory]}$$

At midspan, the actual compressive stress in the bottom fiber is given by

$$f_{bi} = \frac{P_i}{A} + \frac{P_i e}{S_b} - \frac{M_g}{S_b}$$

$$= \frac{1000 \; \text{kips}}{800 \; \text{in}^2} + \frac{(1000 \; \text{kips})(30.8 \; \text{in})}{15{,}600 \; \text{in}^3}$$

$$- \frac{12{,}500 \; \text{in-kips}}{15{,}600 \; \text{in}^3}$$

$$= 2.42 \; \text{kips/in}^2$$

$$< 2.70 \; \text{kips/in}^2 \quad \text{[satisfactory]}$$

Service Load Conditions

The allowable stresses in the concrete under service loads, in other than segmentally constructed bridges, after all prestressing losses have occurred are specified in AASHTO Sec. 5.9.4.2 as

$$f_{te} \leq 0.45 f'_c \quad \text{[for permanent load]}$$

$$f_{te} \leq 0.60 \phi_w f'_c \quad \text{[for permanent + transient loads]}$$

$$f_{te} \leq 0.40 f'_c \quad \left[\begin{array}{l}\text{for live load} \\ + \; (0.5)(\text{sustained load} + \text{prestress})\end{array}\right]$$

$$f_{be} \geq -0.19\sqrt{f'_c} \quad \left[\begin{array}{l}\text{with bonded prestressing tendons} \\ \text{or reinforcement and mild exposure}\end{array}\right]$$

$$f_{be} \geq -0.0948\sqrt{f'_c} \quad \left[\begin{array}{l}\text{with bonded prestressing} \\ \text{tendons or reinforcement} \\ \text{and severe exposure}\end{array}\right]$$

$$f_{be} \geq 0 \quad \text{[with unbonded prestressing tendons]}$$

In accordance with AASHTO Sec. 5.9.3, the maximum allowable stress in the tendons after all losses is

$$f_{pe} = 0.80 f_{py}$$

Example 8.19

The post-tensioned girder of Ex. 8.18 forms part of a composite deck, as shown in the following illustration, with girders located at 8 ft centers. The resulting composite section properties are tabulated as follows.

A_c	S_{ci}	S_c	f'_c (girder)
1250 in^2	45,400 in^3	21,200 in^3	6000 lbf/in^2

The concrete strength at 28 days is $f'_c = 6 \; \text{kips/in}^2$.

The prestressing force after all losses is 800 kips, and the losses occur before the deck slab is cast. Bending moment M_{DC} due to dead load imposed on the composite section is 3000 in-kips. Bending moment M_L due to live load plus impact is 16,250 in-kips. Determine the actual and allowable stresses in the girder at midspan if the girder is subject to mild exposure. Bonded reinforcement is provided at the bottom of the girder.

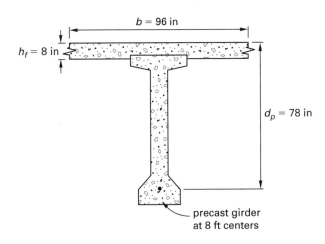

Solution

At midspan, the allowable tensile stress in the bottom fiber with bonded reinforcement is given by AASHTO Sec. 5.9.4.2 as

$$f_{be} = -0.19\sqrt{f_c'} = -0.19\sqrt{6\;\frac{\text{kips}}{\text{in}^2}}$$
$$= -0.465 \text{ kips/in}^2$$

At midspan, the allowable compressive stress in the top fiber is given by AASHTO Sec. 5.9.4.2 as

$$f_{te} = 0.45f_c' \quad \text{[for permanent loads]}$$
$$= (0.45)\left(6\;\frac{\text{kips}}{\text{in}^2}\right)$$
$$= 2.7 \text{ kips/in}^2$$

$$f_{te} = 0.60\phi_w f_c' \quad \text{[for permanent and transient loads]}$$
$$= (0.60)(1.0)\left(6\;\frac{\text{kips}}{\text{in}^2}\right)$$
$$= 3.6 \text{ kips/in}^2$$

$$f_{te} = 0.40f_c' \quad \left[\begin{array}{l}\text{for live load} \\ + (0.5)(\text{sustained load} + \text{prestress})\end{array}\right]$$
$$= (0.40)\left(6\;\frac{\text{kips}}{\text{in}^2}\right)$$
$$= 2.4 \text{ kips/in}^2$$

From Ex. 8.18, the midspan moment due to the self-weight of the girder is

$$M_g = 12{,}500 \text{ in-kips}$$

The resulting stresses in the girder are

$$f_{Gt} = \frac{M_g}{S_t} = \frac{12{,}500 \text{ in-kips}}{14{,}700 \text{ in}^3}$$
$$= 0.850 \text{ kips/in}^2$$

$$f_{Gb} = -\frac{M_g}{S_b} = -\frac{12{,}500 \text{ in-kips}}{15{,}600 \text{ in}^3}$$
$$= -0.801 \text{ kips/in}^2$$

At midspan, the moment due to the weight of the deck slab is

$$M_S = \frac{wL^2}{8}$$

$$= \frac{\left(150\;\frac{\text{lbf}}{\text{ft}^3}\right)\left(\dfrac{8 \text{ in}}{12\;\frac{\text{in}}{\text{ft}}}\right)(8 \text{ ft})(100 \text{ ft})^2\left(12\;\frac{\text{in}}{\text{ft}}\right)}{(8)\left(1000\;\frac{\text{lbf}}{\text{kip}}\right)}$$

$$= 12{,}000 \text{ in-kips}$$

The resulting stresses in the girder are

$$f_{St} = \frac{M_S}{S_t} = \frac{12{,}000 \text{ in-kips}}{14{,}700 \text{ in}^3}$$
$$= 0.816 \text{ kips/in}^2$$

$$f_{Sb} = -\frac{M_S}{S_b} = -\frac{12{,}000 \text{ in-kips}}{15{,}600 \text{ in}^3}$$
$$= -0.769 \text{ kips/in}^2$$

The resulting stresses in the girder due to the dead load imposed on the composite section are

$$f_{Dt} = \frac{M_{DC}}{S_{ci}} = \frac{3000 \text{ in-kips}}{45{,}400 \text{ in}^3}$$
$$= 0.066 \text{ kips/in}^2$$

$$f_{Db} = -\frac{M_{DC}}{S_c} = -\frac{3000 \text{ in-kips}}{21{,}200 \text{ in}^3}$$
$$= -0.142 \text{ kips/in}^2$$

The resulting stresses in the girder due to the live load imposed on the composite section are

$$f_{Lt} = \frac{M_L}{S_{ci}} = \frac{16{,}250 \text{ in-kips}}{45{,}400 \text{ in}^3}$$
$$= 0.358 \text{ kips/in}^2$$

$$f_{Lb} = -\frac{M_L}{S_c} = -\frac{16{,}250 \text{ in-kips}}{21{,}200 \text{ in}^3}$$
$$= -0.767 \text{ kips/in}^2$$

The stresses in the girder due to the effective prestressing force after all losses are

$$f_{Pt} = P_e\left(\frac{1}{A} - \frac{e}{S_t}\right)$$
$$= (800 \text{ kips})\left(\frac{1}{800 \text{ in}^2} - \frac{30.8 \text{ in}}{14{,}700 \text{ in}^3}\right)$$
$$= -0.676 \text{ kips/in}^2$$

$$f_{Pb} = P_e\left(\frac{1}{A} + \frac{e}{S_b}\right)$$
$$= (800 \text{ kips})\left(\frac{1}{800 \text{ in}^2} + \frac{30.8 \text{ in}}{15{,}600 \text{ in}^3}\right)$$
$$= 2.579 \text{ kips/in}^2$$

The final bottom fiber stress in the girder due to all loads is

$$f_b = f_{Gb} + f_{Sb} + f_{Db} + f_{Lb} + f_{Pb}$$
$$= -0.801\;\frac{\text{kips}}{\text{in}^2} + -0.769\;\frac{\text{kips}}{\text{in}^2} + -0.142\;\frac{\text{kips}}{\text{in}^2}$$
$$\quad + -0.767\;\frac{\text{kips}}{\text{in}^2} + 2.579\;\frac{\text{kips}}{\text{in}^2}$$
$$= 0.1 \text{ kips/in}^2$$
$$> -0.465 \text{ kips/in}^2 \quad \text{[satisfactory]}$$

The final top fiber stress in the girder due to all permanent and transient loads is

$$f_t = f_{Gt} + f_{St} + f_{Dt} + f_{Lt} + f_{Pt}$$
$$= 0.850 \, \frac{\text{kips}}{\text{in}^2} + 0.816 \, \frac{\text{kips}}{\text{in}^2} + 0.066 \, \frac{\text{kips}}{\text{in}^2}$$
$$+ 0.358 \, \frac{\text{kips}}{\text{in}^2} + -0.676 \, \frac{\text{kips}}{\text{in}^2}$$
$$= 1.414 \, \text{kips/in}^2$$
$$< 3.600 \, \text{kips/in}^2 \quad \text{[satisfactory]}$$

The final top fiber stress in the girder due to sustained loads is

$$f_t = f_{Gt} + f_{St} + f_{Dt} + f_{Pt}$$
$$= 0.850 \, \frac{\text{kips}}{\text{in}^2} + 0.816 \, \frac{\text{kips}}{\text{in}^2} + 0.066 \, \frac{\text{kips}}{\text{in}^2}$$
$$+ -0.676 \, \frac{\text{kips}}{\text{in}^2}$$
$$= 1.056 \, \text{kips/in}^2$$
$$< 2.700 \, \text{kips/in}^2 \quad \text{[satisfactory]}$$

The final top fiber stress in the girder due to live load plus (0.5)(sustained load + prestress) is

$$f_t = f_{Lt} + 0.5 \left(f_{Gt} + f_{St} + f_{Dt} + f_{Pt} \right)$$
$$= 0.358 \, \frac{\text{kips}}{\text{in}^2} + (0.5) \left(1.056 \, \frac{\text{kips}}{\text{in}^2} \right)$$
$$= 0.886 \, \text{kips/in}^2$$
$$< 2.400 \, \text{kips/in}^2 \quad \text{[satisfactory]}$$

Ultimate Load Conditions

Provided that the effective prestress in the tendons after losses, f_{pe}, is not less than half the tensile strength of the tendons, f_{pu}, the stress in bonded tendons at ultimate load, with non-prestressed reinforcement included in the member, is given by AASHTO Eq. (5.7.3.1.1-1) as

$$f_{ps} = f_{pu} \left(1 - \frac{kc}{d_p} \right)$$

This expression is based on the assumption that all of the prestressing steel is concentrated at a distance d_p from the extreme compression fiber. If this assumption is not justified, a method based on strain compatibility must be used.

The prestressing steel factor k is given by AASHTO Table C5.7.3.1.1-1 as

- 0.48 for type 2 high-strength bars with $f_{py}/f_{pu} = 0.80$

- 0.38 for stress-relieved strands and type 1 high-strength bars with $f_{py}/f_{pu} = 0.85$

- 0.28 for low-relaxation wire and strands with $f_{py}/f_{pu} = 0.90$

The compression zone factor β_1 given in AASHTO Sec. 5.7.2.2 is

- 0.85 for $f_c' \leq 4 \, \text{kips/in}^2$

- $0.85 - (f_c' - 4)/20$ for $4 \, \text{kips/in}^2 < f_c' \leq 8 \, \text{kips/in}^2$

- 0.65 minimum for $f_c' > 8 \, \text{kips/in}^2$

For a rectangular section, with nonprestressed tension reinforcement, the distance from the extreme compression fiber to the neutral axis is given by AASHTO Eq. (5.7.3.1.1-4) as

$$c = \frac{A_{ps} f_{pu} + A_s f_s}{0.85 f_c' \beta_1 b + \dfrac{k A_{ps} f_{pu}}{d_p}}$$

The previous expression is also applicable to a flanged section with the neutral axis within the flange.

f_y may replace f_s when, using f_y in the calculation, the resulting ratio c/d_s does not exceed 0.6. If c/d_s exceeds 0.6, strain compatibility shall be used to determine the stress in the mild steel tension reinforcement.

The depth of the equivalent rectangular stress block is given by AASHTO Sec. 5.7.2.2 as

$$a = \beta_1 c$$

The nominal flexural strength of a rectangular section is given by AASHTO Eq. (5.7.3.2.2-1) as

$$M_n = A_{ps} f_{ps} \left(d_p - \frac{a}{2} \right) + A_s f_s \left(d_s - \frac{a}{2} \right)$$

The factored flexural resistance is given by AASHTO Eq. (5.7.3.2.1-1) as

$$M_r = \phi M_n$$

The resistance factor for a tension-controlled prestressed concrete section is given by AASHTO Sec. 5.5.4.2.1 as

$$\phi = 1.0$$

Example 8.20

The area of the low-relaxation strand in the post-tensioned girder of Ex. 8.19 is 5.36 in^2, and the strand has a specified tensile strength of 270 kips/in^2. The 28 day compressive strength of the deck slab is 3 kips/in^2. Determine the maximum factored moment at midspan and the design flexural capacity of the composite section.

Solution

From Ex. 8.19, the total dead load moment on the composite section is

$$M_D = M_g + M_S + M_{DC}$$
$$= 12{,}500 \text{ in-kips} + 12{,}000 \text{ in-kips}$$
$$+ 3000 \text{ in-kips}$$
$$= 27{,}500 \text{ in-kips}$$

The live load moment plus impact is

$$M_L = 16{,}250 \text{ in-kips}$$

The strength I limit state moment is given by AASHTO Eq. (3.4.1-1) as

$$M_u = \gamma_p M_D + \gamma_{LL+IM} M_L$$
$$= (1.25)(27{,}500 \text{ in-kips})$$
$$+ (1.75)(16{,}250 \text{ in-kips})$$
$$= 62{,}813 \text{ in-kips}$$

The effective prestress in the tendons after all losses is obtained from Ex. 8.19 as

$$f_{pe} = \frac{P_e}{A_{ps}}$$
$$= \frac{800 \text{ kips}}{5.36 \text{ in}^2}$$
$$= 149 \text{ kips/in}^2$$
$$> 0.5 f_{pu}$$

Therefore, AASHTO Sec. 5.7.3.1 is applicable.

The compression zone factor for 3 kips/in^2 concrete is

$$\beta_1 = 0.85$$

The prestressing steel factor is given by AASHTO Table 5.7.3.1.1-1 as

$$k = 0.28 \quad \text{[for low-relaxation strand]}$$

Assuming that the neutral axis lies within the flange, for a section without non-prestressed tension reinforcement, the distance from the extreme compression fiber to the neutral axis is given by AASHTO Eq. (5.7.3.1.1-4) as

$$c = \frac{A_{ps} f_{pu}}{0.85 f'_c \beta_1 b + \dfrac{k A_{ps} f_{pu}}{d_p}}$$

$$= \frac{(5.36 \text{ in}^2)\left(270 \dfrac{\text{kips}}{\text{in}^2}\right)}{(0.85)\left(3 \dfrac{\text{kips}}{\text{in}^2}\right)(0.85)(96 \text{ in})}$$

$$+ \frac{(0.28)(5.36 \text{ in}^2)\left(270 \dfrac{\text{kips}}{\text{in}^2}\right)}{78 \text{ in}}$$

$$= 6.79 \text{ in}$$
$$< 8 \text{ in}$$

Therefore, the neutral axis does lie within the flange.

The depth of the equivalent rectangular stress block is given by AASHTO Sec. 5.7.2.2 as

$$a = \beta_1 c$$
$$= (0.85)(6.79 \text{ in})$$
$$= 5.8 \text{ in}$$

The stress in bonded tendons at ultimate load, is given by AASHTO Eq. (5.7.3.1.1-1) as

$$f_{ps} = f_{pu}\left(1 - \frac{kc}{d_p}\right)$$
$$= \left(270 \frac{\text{kips}}{\text{in}^2}\right)\left(1 - \frac{(0.28)(6.79 \text{ in})}{78 \text{ in}}\right)$$
$$= 263.42 \text{ kips/in}^2$$

The nominal flexural strength of the section is given by AASHTO Eq. (5.7.3.2.2-1) as

$$M_n = A_{ps} f_{ps}\left(d_p - \frac{a}{2}\right)$$
$$= (5.36 \text{ in}^2)\left(263.42 \frac{\text{kips}}{\text{in}^2}\right)\left(78 \text{ in} - \frac{5.8 \text{ in}}{2}\right)$$
$$= 106{,}036 \text{ in-kips}$$

The strain in the prestressing tendons at the nominal flexural strength is

$$\epsilon_t = \epsilon_{cu}\frac{d_p - c}{c}$$
$$= (0.003)\left(\frac{78 \text{ in} - 6.79 \text{ in}}{6.79 \text{ in}}\right)$$
$$= 0.031$$
$$> 0.005$$

Therefore, the section is tension controlled and the resistance factor is given by AASHTO Sec. 5.5.4.2.1 as

$$\phi = 1.0$$

The factored flexural resistance is

$$M_r = \phi M_n$$
$$= (1.0)(106{,}036 \text{ in-kips})$$
$$= 106{,}036 \text{ in-kips}$$
$$> M_u \quad \text{[satisfactory]}$$

Cracking Moment

The *cracking moment* is the external moment that, when applied to the member after all losses have occurred, will cause cracking in the bottom fiber. This cracking occurs when the stress in the bottom fiber exceeds the modulus of rupture, which is defined in AASHTO Sec. 5.4.2.6 for normal weight concrete as

$$f_r = 0.37\sqrt{f_c'}$$

For a composite section, the cracking moment is defined in AASHTO Sec. 5.7.3.3.2 as

$$M_{cr} = S_c\left(f_{cpe} + f_r\right) - M_{dnc}\left(\frac{S_c}{S_{nc}} - 1\right)$$
$$\geq S_c f_r$$

For noncomposite beams, S_{nc} is substituted for S_c in the previous expression. To prevent sudden tensile failure, AASHTO Sec. 5.7.3.3.2 requires that

$$\phi M_n \geq 1.2 M_{cr}$$

Example 8.21

Determine the cracking moment of the composite section of Ex. 8.19.

Solution

From Ex. 8.19, the bottom fiber stress due only to the effective prestressing force after allowance for all prestress losses is

$$f_{cpe} = 2.579 \text{ kips/in}^2$$

In addition, the bending moment due to the noncomposite dead load acting on the precast section is given by

$$M_{dnc} = M_g + M_S$$
$$= 12{,}500 \text{ in-kips} + 12{,}000 \text{ in-kips}$$
$$= 24{,}500 \text{ in-kips}$$

The modulus of rupture is given by AASHTO Sec. 5.4.2.6 as

$$f_r = 0.37\sqrt{f_c'}$$
$$= 0.37\sqrt{6 \frac{\text{kips}}{\text{in}^2}}$$
$$= 0.906 \text{ kips/in}^2$$

The cracking moment is given by AASHTO Sec. 5.7.3.3.2 as

$$M_{cr} = S_c\left(f_{cpe} + f_r\right) - M_{dnc}\left(\frac{S_c}{S_{nc}} - 1\right)$$
$$= \left(21{,}200 \text{ in}^3\right)\left(2.579 \frac{\text{kips}}{\text{in}^2} + 0.906 \frac{\text{kips}}{\text{in}^2}\right)$$
$$\quad - \left(24{,}500 \text{ in-kips}\right)\left(\left(\frac{21{,}200 \text{ in}^3}{15{,}600 \text{ in}^3}\right) - 1\right)$$
$$= 65{,}087 \text{ in-kips}$$
$$< \phi M_n / 1.2 \quad \begin{bmatrix} \text{satisfies AASHTO} \\ \text{Sec. 5.7.3.3.2} \end{bmatrix}$$

Design for Shear

Nomenclature

A_v	area of shear reinforcement	in^2
b_v	web width	in
d_e	effective depth from the extreme compression fiber to the centroid of the tensile force in the tensile reinforcement	in
d_p	distance from the extreme compression fiber to the centroid of the prestressing tendons	in
d_v	effective shear depth	in
f_c'	specified compressive strength of concrete	kips/in^2
f_{cpe}	compressive stress in the concrete, due to the final prestressing force only, at the bottom fiber of the section	lbf/in^2
f_d	tensile stress at bottom fiber of precast member due to unfactored dead load acting on the precast member	lbf/in^2
f_{pc}	compressive stress in the concrete, due to the final prestressing force and applied loads resisted by precast member, at the centroid of the composite section	lbf/in^2
f_r	modulus of rupture	kips/in^2
f_y	specified yield strength of reinforcing bars	kips/in^2
g	drape of the prestressing cable	in
h	depth of section	in
M_{cre}	moment causing flexural cracking at section due to externally applied loads	in-lbf or in-kips
M_d	moment due to unfactored dead load	in-kips
M_{dnc}	total unfactored dead load moment acting on the precast member	in-kips
M_{max}	maximum factored moment at section due to externally applied loads	in-lbf or in-kips
M_u	factored moment at the section due to total factored loads	in-kips
s	longitudinal spacing of shear reinforcement	in
S	section modulus at the centroid of the composite section	in^3
S_c	section modulus at the bottom of the composite member	in^3
S_{ci}	section modulus at the interface of the composite member	in^3
S_{nc}	section modulus at the bottom of the precast member	in^3
V_c	nominal shear strength provided by concrete	kips
V_{ci}	nominal shear strength provided by concrete when diagonal cracking results from combined shear and moment	kips
V_{cw}	nominal shear strength provided by concrete when diagonal cracking results from excessive principal tensile stress in the web	kips

V_d	shear force at section due to unfactored dead load	kips
V_i	factored shear force at section due to externally applied loads occurring simultaneously with $M_{\max}$	kips
V_p	vertical component of effective prestress force at section	kips
V_s	nominal shear strength provided by shear reinforcement	kips
V_u	factored shear force at section	kips
y_s	height of cable above beam soffit	in
z	half the beam length	ft

Symbols

θ	angle of inclination of diagonal compressive stresses	degree
ϕ	strength reduction factor	

Ultimate Load Design for Shear

The simplified procedure is permissible for prestressed concrete sections that are not subjected to significant axial tension and with the minimum amount of transverse reinforcement specified in AASHTO Eq. (5.8.2.5-1).

The nominal shear capacity of the concrete is provided by the lesser value of V_{ci} or V_{cw} given by AASHTO Sec. 5.8.3.4.3. For flexural-shear cracking, the nominal shear capacity is given by AASHTO Eq. (5.8.3.4.3-1).[14]

$$V_{ci} = 0.02 b_v d_v \sqrt{f'_c} + V_d + \frac{V_i M_{cre}}{M_{\max}}$$
$$\geq 0.06 b_v d_v \sqrt{f'_c}$$

The moment causing flexural cracking at the section due to externally applied loads is defined in AASHTO Eq. (5.8.3.4.3-2) as

$$M_{cre} = S_c \left(f_r + f_{cpe} - \frac{M_{dnc}}{S_{nc}} \right)$$

For web-shear cracking, the nominal shear capacity is given by AASHTO Eq. (5.8.3.4.3-3).[14]

$$V_{cw} = b_v d_v \left(0.06 \sqrt{f'_c} + 0.3 f_{pc} \right) + V_p$$

In a composite member, f_{pc} is the resultant compressive stress at the centroid of the composite section, or at the junction of the web and flange when the centroid lies within the flange, due to both the final prestresses and moments resisted by the precast member acting alone.

The effective shear depth d_v to the centroid of the prestressing steel need not be taken as less than the greater of

$$d_v = 0.72h$$
$$d_v = 0.9 d_e$$

The nominal shear capacity of vertical shear reinforcement is given by AASHTO Eq. (5.8.3.3-4) as

$$V_s = A_v f_y \left(\frac{d_v}{s} \right) \quad \text{[for } V_{ci} < V_{cw}]$$
$$= A_v f_y \left(\frac{d_v \cot \theta}{s} \right) \quad \text{[for } V_{ci} > V_{cw}]$$
$$\cot \theta = 1.0 + \frac{3 f_{pc}}{\sqrt{f'_c}}$$
$$\leq 1.8$$

For a composite section constructed in two stages, some of the dead load is resisted by the precast section and the remainder by the composite section. In AASHTO Eq. (5.8.3.4.3-1), V_d is the total shear force due to unfactored dead loads acting on the precast section, plus the unfactored superimposed dead load acting on the composite section. V_i is the factored shear force at a section caused by the externally applied loads occurring simultaneously with the maximum factored moment at the section, $M_{\max}$. In accordance with AASHTO Sec. C5.8.3.4.3, where V_u and M_u represent the factored shear and moment at the section, these values may be taken as

$$V_i = V_u - V_d$$
$$M_{\max} = M_u - M_d$$

The effective shear depth d_v is taken as the distance between the resultants of the tensile and compressive forces due to flexure. The effective shear depth is given by AASHTO Eq. (C5.8.2.9-1) as

$$d_v = \frac{M_n}{A_s f_y + A_{ps} f_{ps}}$$

The effective shear depth need not be taken to be less than the greater of $0.9 d_e$ or $0.72h$. The effective web width b_v is taken as the minimum web width between the resultants of the tensile and compressive forces due to flexure.

In AASHTO Eq. (5.8.3.4.3-2), the stress in the bottom fiber due to unfactored dead load acting on the precast section only is given by

$$f_d = \frac{M_{dnc}}{S_{nc}}$$

To account for the effects of differential shrinkage and thermal gradients, AASHTO Eq. (5.8.3.4.3-2) uses a reduced value of the modulus of rupture, and this is given by AASHTO Sec. 5.4.2.6 as

$$f_r = 0.20 \sqrt{f'_c}$$

The combined nominal shear resistance of the concrete section and the shear reinforcement is given by AASHTO Sec. 5.8.3.3 as the lesser of

$$V_n = V_c + V_s$$
$$V_n = 0.25 f'_c b_v d_v$$

The combined shear capacity of the concrete section and the shear reinforcement is

$$\phi V_n = \phi V_c + \phi V_s$$
$$\geq V_u$$

The resistance factor for normal weight concrete is given by AASHTO Sec. 5.5.4.2.1 as

$$\phi = 0.90$$

The shear stress on the concrete is calculated by AASHTO Eq. (5.8.2.9-1) as

$$v_u = \frac{V_u - \phi V_p}{\phi b_v d_v}$$

For a value of v_u less than $0.125 f_c'$, AASHTO Sec. 5.8.2.7 limits the spacing of transverse reinforcement to the lesser of $0.8d_v$ or 24 in. When the value of v_u is not less than $0.125 f_c'$, the spacing is reduced to the lesser of $0.4d_v$ or 12 in.

When the support reaction produces a compressive stress in the member, AASHTO Sec. 5.8.3.2 specifies that the critical section for shear may be taken at a distance from the support equal to the effective shear depth d_v.

Example 8.22

The tendon centroid of the post-tensioned girder of Ex. 8.18 is parabolic in shape, as shown in the following illustration. At section A-A, the unfactored shear and moment due to live load plus impact are 58 kips and 2340 in-kips. The moment of inertia of the precast section is $I = 589,680 \text{ in}^4$. Determine the required spacing of no. 3 grade 60 stirrups.

Solution

The equation of the parabolic cable profile is

$$y = \frac{gx^2}{z^2}$$

At section A-A, the rise of the cable is given by

$$y = \frac{(30 \text{ in}) (600 \text{ in} - 61.2 \text{ in})^2}{(600 \text{ in})^2}$$
$$= 24.2 \text{ in}$$

At section A-A, the effective depth of the prestressing cable referred to the composite section is

$$d_p = h - y - y_s$$
$$= 85 \text{ in} - 24.2 \text{ in} - 7 \text{ in}$$
$$= 53.8 \text{ in}$$
$$= d_e \quad \left[\begin{array}{l}\text{effective depth from the extreme compression} \\ \text{fiber to the centroid of the tensile force}\end{array}\right]$$

From Ex. 8.20, the depth of the stress block at midspan of the composite section is

$$a = 5.8 \text{ in}$$

The value of a may be conservatively taken as the stress block depth at section A-A and the effective shear depth is

$$d_v = d_e - \frac{a}{2}$$
$$= 53.8 \text{ in} - \frac{5.8 \text{ in}}{2}$$
$$= 50.9 \text{ in}$$

The effective shear depth need not be taken to be less than the greater of

$$0.9d_e = (0.9)(53.8 \text{ in})$$
$$= 48.4 \text{ in}$$
$$0.72h = (0.72)(85 \text{ in})$$
$$= 61.2 \text{ in} \quad \text{[governs]}$$

Therefore, as specified by AASHTO Sec. 5.8.3.2, the critical section for shear is located a distance of 61.2 in from the support.

Illustration for Ex. 8.22

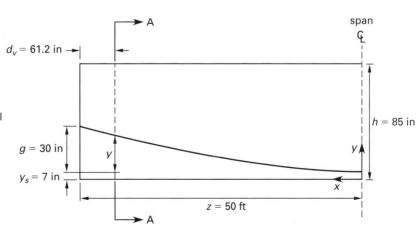

At section A-A, the cable eccentricity referred to the precast section is given by

$$e = y_b - y_s - y = 37.8 \text{ in} - 7 \text{ in} - 24.2 \text{ in}$$
$$= 6.6 \text{ in}$$

At section A-A, the slope of the cable is given by

$$\frac{dy}{dx} = \frac{2gx}{z^2} = \frac{(2)(30 \text{ in})(538.8 \text{ in})}{(600 \text{ in})^2}$$
$$= 0.0898$$

The vertical component of the final effective prestressing force at section A-A is

$$V_P = P_e\left(\frac{dy}{dx}\right) = (800 \text{ kips})(0.0898)$$
$$= 72 \text{ kips}$$

The centroid of the composite section is at a height of

$$y_{cb} = 53.2 \text{ in}$$

The section modulus of the precast section at the critical section for shear is

$$S = \frac{I}{y_{cb} - y_b} = \frac{589,680 \text{ in}^4}{53.2 \text{ in} - 37.8 \text{ in}}$$
$$= 38,290 \text{ in}^3$$

At section A-A, the stress in the concrete, at the centroid of the composite section, due to the final prestressing force only is

$$f_p = P_e\left(\frac{1}{A} - \frac{e}{S}\right)$$
$$= (800 \text{ kips})\left(\frac{1}{800 \text{ in}^2} - \frac{6.6 \text{ in}}{38,290 \text{ in}^3}\right)$$
$$= 0.862 \text{ kips/in}^2$$

At section A-A, the bending moment due to the girder self-weight is

$$M = M_g\left(1 - \left(\frac{x}{z}\right)^2\right)$$
$$= (12,500 \text{ in-kips})\left(1 - \left(\frac{538.8 \text{ in}}{600 \text{ in}}\right)^2\right)$$
$$= 2420 \text{ in-kips}$$

At section A-A, the stress in the concrete at the centroid of the composite section due to the girder self-weight is

$$f_G = \frac{M}{S} = \frac{2420 \text{ in-kips}}{38,290 \text{ in}^3}$$
$$= 0.063 \text{ kips/in}^2$$

At section A-A, the bending moment due to the weight of the deck slab is

$$M = M_s\left(1 - \left(\frac{x}{z}\right)^2\right)$$
$$= (12,000 \text{ in-kips})\left(1 - \left(\frac{538.8 \text{ in}}{600 \text{ in}}\right)^2\right)$$
$$= 2323 \text{ in-kips}$$

At section A-A, the stress in the concrete at the centroid of the composite section due to the weight of the deck slab is

$$f_s = \frac{M}{S} = \frac{2323 \text{ in-kips}}{38,290 \text{ in}^3}$$
$$= 0.061 \text{ kips/in}^2$$

At section A-A, the compressive stress at the centroid of the composite section due to final prestress and the bending moments resisted by the precast member acting alone is

$$f_{pc} = f_p + f_G + f_s$$
$$= 0.862 \frac{\text{kips}}{\text{in}^2} + 0.063 \frac{\text{kips}}{\text{in}^2} + 0.061 \frac{\text{kips}}{\text{in}^2}$$
$$= 0.986 \text{ kips/in}^2$$

The nominal web-shear capacity is given by AASHTO Eq. (5.8.3.4.3-3) as

$$V_{cw} = b_v d_v \left(0.06\sqrt{f_c'} + 0.3 f_{pc}\right) + V_p$$
$$= (7 \text{ in})(61.2 \text{ in})\left(\begin{array}{c}(0.06)\left(6\frac{\text{kips}}{\text{in}^2}\right)^{0.5} \\ + (0.3)\left(0.986\frac{\text{kips}}{\text{in}^2}\right)\end{array}\right)$$
$$+ 72 \text{ kips}$$
$$= 262 \text{ kips}$$

The compressive stress in the bottom fiber of the precast member at section A-A due to the final prestressing force is

$$f_{cpe} = P_e\left(\frac{1}{A} + \frac{e}{S_{nc}}\right)$$
$$= (800 \text{ kips})\left(\frac{1}{800 \text{ in}^2} + \frac{6.6 \text{ in}}{15,600 \text{ in}^3}\right)$$
$$= 1.338 \text{ kips/in}^2$$

At section A-A, the stress in the concrete at the bottom fiber due to the dead load imposed on the precast section is, from Ex. 8.19,

$$f_d = f_{Gb+Sb}\left(1 - \left(\frac{x}{z}\right)^2\right)$$
$$= \left(1.570\frac{\text{kips}}{\text{in}^2}\right)\left(1 - \left(\frac{538.8 \text{ in}}{600 \text{ in}}\right)^2\right)$$
$$= 0.304 \text{ kips/in}^2 \quad [\text{tension}]$$

The moment causing flexural cracking at the section due to externally applied loads is given by AASHTO Eq. (5.8.3.4.3-2) as

$$M_{cre} = S_c\left(f_r + f_{cpe} - f_d\right)$$

$$= \left(21{,}200 \text{ in}^3\right)\left(\begin{array}{c} 0.20\sqrt{6\ \dfrac{\text{kips}}{\text{in}^2}} + 1.338\ \dfrac{\text{kips}}{\text{in}^2} \\ -0.304\ \dfrac{\text{kips}}{\text{in}^2} \end{array}\right)$$

$$= 32{,}307 \text{ in-kips}$$

$$> 2340 \text{ in-kips} \quad \left[\begin{array}{c} \text{the given unfactored applied} \\ \text{moment at section A-A} \end{array}\right]$$

Hence, flexural-shear cracking does not occur at section A-A; the web-shear capacity controls with the nominal shear strength given by

$$V_c = V_{cw}$$
$$= 262 \text{ kips}$$

The design shear capacity is

$$\phi V_c = (0.9)(262 \text{ kips})$$
$$= 236 \text{ kips}$$

At section A-A, the shear force due to the girder self-weight is

$$V_G = w_G x$$

$$= \dfrac{\left(0.150\ \dfrac{\text{kip}}{\text{ft}^3}\right)\left(\dfrac{800 \text{ in}^2}{144\ \dfrac{\text{in}^2}{\text{ft}^2}}\right)(538.8 \text{ in})}{12\ \dfrac{\text{in}}{\text{ft}}}$$

$$= 37.4 \text{ kips}$$

At section A-A, the shear force due to the self-weight of the slab is

$$V_S = w_S x$$

$$= \dfrac{\left(0.150\ \dfrac{\text{kip}}{\text{ft}^3}\right)\left(\dfrac{64 \text{ in}}{12\ \dfrac{\text{in}}{\text{ft}}}\right)(538.8 \text{ in})}{12\ \dfrac{\text{in}}{\text{ft}}}$$

$$= 36.0 \text{ kips}$$

At section A-A, the shear force due to the dead load imposed on the composite section is

$$V_{DC} = w_{DC} x = \dfrac{\left(0.20\ \dfrac{\text{kip}}{\text{ft}}\right)(538.8 \text{ in})}{12\ \dfrac{\text{in}}{\text{ft}}}$$

$$= 9.0 \text{ kips}$$

At section A-A, the total dead load shear force on the composite section is

$$V_D = V_G + V_S + V_{DC}$$
$$= 37.4 \text{ kips} + 36.0 \text{ kips} + 9.0 \text{ kips}$$
$$= 82.4 \text{ kips}$$

The live load shear plus impact is given as

$$V_L = 58 \text{ kips}$$

The strength I limit state shear force is given by AASHTO Eq. (3.4.1-1) as

$$V_u = \gamma_p V_D + \gamma_{LL+IM} V_L$$
$$= (1.25)(82.4 \text{ kips}) + (1.75)(58 \text{ kips})$$
$$= 205 \text{ kips}$$
$$< \phi V_c$$

Hence, a minimum area of shear reinforcement is required and is given by AASHTO Eq. (5.8.2.5-1) as

$$\dfrac{A_v}{s} = 0.0316\left(\dfrac{\sqrt{f'_c}\,b_v}{f_y}\right)$$

$$= (0.0316)\left(\dfrac{\left(\sqrt{6\ \dfrac{\text{kips}}{\text{in}^2}}\right)(7 \text{ in})}{60\ \dfrac{\text{kips}}{\text{in}^2}}\right)$$

$$= 0.0090 \text{ in}^2/\text{in}$$

The shear stress on the concrete is calculated by AASHTO Eq. (5.8.2.9-1) as

$$v_u = \dfrac{V_u - \phi V_p}{\phi b_v d_v}$$

$$= \dfrac{203 \text{ kips} - (0.9)(72 \text{ kips})}{(0.9)(7 \text{ in})(61.2 \text{ in})}$$

$$= 0.358 \text{ kips/in}^2$$
$$< 0.125 f'_c \quad (0.75 \text{ kips/in}^2)$$

Therefore, AASHTO Sec. 5.8.2.7 limits the spacing of transverse reinforcement to the lesser of

$$s = 0.8 d_v$$
$$= (0.8)(61.2 \text{ in})$$
$$= 49 \text{ in}$$
$$s = 24 \text{ in} \quad [\text{governs}]$$

Providing no. 3 stirrups at the maximum permitted spacing of 24 in gives a value of

$$\dfrac{A_v}{s} = \dfrac{0.22 \text{ in}^2}{24 \text{ in}}$$
$$= 0.0092 \text{ in}^2/\text{in} \quad [\text{satisfactory}]$$

Prestress Losses

Nomenclature

A_g	area of precast girder	in^2
E_{ci}	modulus of elasticity of concrete at time of initial prestress	$kips/in^2$
E_p	modulus of elasticity of prestressing steel	$kips/in^2$
Δf_{pES}	loss of prestress due to elastic shortening	$kips/in^2$
f_{cgp}	compressive stress at centroid of prestressing steel due to prestress and self-weight of girder at transfer	$kips/in^2$
f_{pj}	stress in the prestressing steel at jacking	$kips/in^2$
g	drape of prestressing steel	in
I_g	moment of inertia of precast girder	in^4
K	wobble friction coefficient per foot of prestressing tendon	–
l_{px}	distance from free end of cable to section under consideration	ft
M_g	bending moment due to self-weight of precast member	in-kips
N	number of identical prestressing tendons	–
n_i	modular ratio at transfer	
P_{ES}	loss of prestress force due to elastic shortening	kips
P_i	force in prestressing steel immediately after transfer	kips
P_o	force in prestressing steel at anchorage	kips
R	radius of curvature of tendon profile	ft

Symbols

α	angular change of tendon profile from jacking end to any point x	radians
Δf_{pF}	prestress loss due to friction	$kips/in^2$
Δf_{pLT}	long-term prestress loss due to creep and shrinkage of concrete and relaxation of steel	$kips/in^2$
μ	curvature friction coefficient	–

Friction Losses

Friction losses are determined by AASHTO Sec. 5.9.5.2.2 from the expression

$$\Delta f_{pF} = f_{pj} \exp(Kx + \mu\alpha)$$

When $(Kx + \mu\alpha)$ is not greater than 0.3, this may be approximated to

$$f_{pj} = \Delta f_{pF} \left(1 - Kx - \mu\alpha\right)$$

Values of the wobble and friction coefficients are given in AASHTO Table 5.9.5.2.2b-1 and for prestressing strand are

$$K = 0.0002$$

$$\mu = 0.15 \text{ to } 0.25$$

Example 8.23

The beam of Ex. 8.22 is post-tensioned with low-relaxation strands with a total area of 5.36 in^2, a yield strength of 243 $kips/in^2$, and a tensile strength of 270 $kips/in^2$. The strands are located in 4 cables. The centroid of the prestressing steel is parabolic in shape and is stressed simultaneously from both ends with a jacking force P_o of 1036 kips. The value of the wobble friction coefficient is 0.0002/ft, and the curvature friction coefficient is 0.25. Determine the force in the prestressing steel at midspan of the member before elastic losses.

Solution

The nominal radius of the profile of the prestressing steel is

$$R = \frac{z^2}{2g} = \frac{(50 \text{ ft})^2}{(2)(2.5 \text{ ft})}$$
$$= 500 \text{ ft}$$

The length along the curve from the jacking end to midspan is

$$l_{px} = z + \frac{g^2}{3z} = 50 \text{ ft} + \frac{(2.5 \text{ ft})^2}{(3)(50 \text{ ft})}$$
$$= 50.04 \text{ ft}$$

The angular change of the cable profile over this length is

$$\alpha = \frac{l_{px}}{R} = \frac{50.04 \text{ ft}}{500 \text{ ft}}$$
$$= 0.100 \text{ radians}$$

$$(Kl_{px} + \mu\alpha) = (0.0002)(50.04 \text{ ft})$$
$$+ (0.25)(0.100 \text{ radians})$$
$$= 0.035$$
$$< 0.3$$

The cable force at midspan is given by

$$P_x = P_o \left(1 - Kl_{px} - \mu\alpha\right)$$
$$= (1036 \text{ kips})(1 - 0.035)$$
$$= 1000 \text{ kips}$$

Elastic Shortening

Losses occur due to the elastic shortening of the concrete. The concrete stress at the level of the centroid of the prestressing steel after elastic shortening is

$$f_{cgp} = P_i \left(\frac{1}{A_g} + \frac{e^2}{I_g}\right) - \frac{eM_g}{I_g}$$

AASHTO Sec. C5.4.2.4 specifies that the modulus of elasticity of normal weight concrete at transfer is

$$E_{ci} = 1820\sqrt{f'_{ci}}$$

The modulus of elasticity of prestressing strand is given as

$$E_p = 28{,}500 \text{ kips/in}^2$$

The modular ratio at transfer is

$$n_i = \frac{E_p}{E_{ci}}$$

For a pre-tensioned member, the loss of prestress due to elastic shortening is given by AASHTO Sec. 5.9.5.2.3a as

$$\Delta f_{pES} = n_i f_{cgp}$$

For a post-tensioned member, the loss of prestress is given by AASHTO Sec. 5.9.5.2.3b.

$$\Delta f_{pES} = \frac{N-1}{2N} n_i f_{cgp}$$

Example 8.24

The post-tensioned beam of Ex. 8.22 has a concrete strength at transfer of 4.5 kips/in². The initial force at midspan, after friction losses and before allowance for elastic shortening, is 1000 kips. The moment of inertia of the girder is 589,680 in⁴. Determine the loss of prestress due to elastic shortening.

Solution

From AASHTO Eq. (C5.4.2.4-1), the modulus of elasticity of the concrete at transfer is

$$\begin{aligned}
E_{ci} &= 1820\sqrt{f'_{ci}} \\
&= 1820\sqrt{4.5 \ \frac{\text{kips}}{\text{in}^2}} \\
&= 3861 \text{ kips/in}^2
\end{aligned}$$

The modular ratio at transfer is

$$\begin{aligned}
n_i &= \frac{E_s}{E_{ci}} = \frac{28{,}500 \ \dfrac{\text{kips}}{\text{in}^2}}{3861 \ \dfrac{\text{kips}}{\text{in}^2}} \\
&= 7.38
\end{aligned}$$

Assuming a 3% loss due to elastic shortening, the initial prestressing force at midspan is

$$P_i = 970 \text{ kips}$$

The compressive stress at the centroid of the prestressing steel immediately after transfer is

$$\begin{aligned}
f_{cgp} &= P_i \left(\frac{1}{A_g} + \frac{e^2}{I_g} \right) - \frac{e M_g}{I_g} \\
&= (970 \text{ kips}) \left(\left(\frac{1}{800 \text{ in}^2} \right) + \frac{(30.8 \text{ in})^2}{589{,}680 \text{ in}^4} \right) \\
&\quad - \frac{(30.8 \text{ in})(12{,}500 \text{ in-kips})}{589{,}680 \text{ in}^4} \\
&= 2.120 \text{ kips/in}^2
\end{aligned}$$

For a post-tensioned member, the loss of prestress is given by AASHTO Sec. 5.9.5.2.3b as

$$\begin{aligned}
\Delta f_{pES} &= \frac{N-1}{2N} n_i f_{cgp} = \frac{(3)(7.38)\left(2.120 \ \dfrac{\text{kips}}{\text{in}^2}\right)}{(2)(4)} \\
&= 5.87 \text{ kips/in}^2
\end{aligned}$$

The loss of prestressing force is

$$\begin{aligned}
P_{ES} &= A_{ps}\Delta f_{pES} = \left(5.36 \text{ in}^2\right)\left(5.87 \ \frac{\text{kips}}{\text{in}^2}\right) \\
&= 31.45 \text{ kips} \\
&\approx 30 \text{ kips} \quad \left[\begin{array}{l}\text{Assumed value is} \\ \text{sufficiently accurate}\end{array}\right]
\end{aligned}$$

Estimated Time-Dependent Losses

An estimate of time-dependent losses for members of usual design, with normal prestress levels, using normal weight concrete, and exposed to average exposure conditions may be obtained from AASHTO Table 5.9.5.3-1.

Example 8.25

Determine the average estimated time-dependent losses for the post-tensioned beam of Ex. 8.22. Low-relaxation strand is used, and all losses occur before the deck slab is cast.

Solution

From AASHTO Table 5.9.5.3-1, the estimated losses for a double T-section with low-relaxation strand are

$$\begin{aligned}
\Delta f_{pLT} &= 33\left(1 - \frac{0.15(f'_c - 6)}{6}\right) + 6\text{PPR} - 8 \\
&= 33\left(1 - \frac{0.15\left(6 \ \dfrac{\text{kips}}{\text{in}^2} - 6\right)}{6}\right) + 6 - 8 \\
&= 31 \text{ kips/in}^2
\end{aligned}$$

4. STRUCTURAL STEEL DESIGN

Design for Flexure

Nomenclature

A_{rb}	area of bottom layer of longitudinal reinforcement within the effective concrete deck width	in²
A_{rt}	area of top layer of longitudinal reinforcement within the effective concrete deck width	in²
A_s	cross-sectional area of structural steel	in²
b_c	width of the compression flange of the steel beam	in
b_s	effective concrete flange width	in or ft

Bridges

b_t width of the tension flange of the steel beam   in

C compressive force in slab at ultimate load   kips

d depth of steel beam   in

d_c distance from the plastic neutral axis to the midthickness of the compression flange used to compute the plastic moment   in

d_c distance from the plastic neutral axis to the midthickness of the web flange used to compute the plastic moment   in

d_{rb} distance from the plastic neutral axis to the centerline of the bottom layer of longitudinal concrete deck reinforcement used to compute the plastic moment   in

d_{rt} distance from the plastic neutral axis to the centerline of the top layer of longitudinal concrete deck reinforcement used to compute the plastic moment   in

d_t distance from the plastic neutral axis to the midthickness of the tension flange used to compute the plastic moment   in

D depth of the web of the steel beam   in

D_p distance from the top of slab to the plastic neutral axis   in

D_t total depth of composite section   in

f'_c specified compressive strength of the concrete   kips/in^2

F_y specified minimum yield strength of the structural steel section   kips/in^2

F_{yc} specified yield strength of the compression flange of the steel beam   kips/in^2

F_{yrb} specified yield strength of the bottom layer of longitudinal deck reinforcement   kips/in^2

F_{yrt} specified yield strength of the top layer of longitudinal deck reinforcement   kips/in^2

F_{yt} specified yield strength of the tension flange of the steel beam   kips/in^2

F_{yw} specified yield strength of the web of the steel beam   kips/in^2

L span length   ft

M_n nominal flexural resistance of the composite beam   in-kips or ft-kips

M_p full plastic moment of the member   in-kips or ft-kips

M_u moment due to factored loads   in-kips or ft-kips

P_c plastic force in the compression flange of the steel beam   kips

P_{rb} plastic force in the bottom layer of longitudinal deck reinforcement   kips

P_{rt} plastic force in the top layer of longitudinal deck reinforcement   kips

P_s plastic force in the full depth of the concrete deck   kips

P_t plastic force in the tension flange of the steel beam   kips

P_w plastic force in the web of the steel beam   kips

S beam spacing   ft or in

t_s slab thickness   in

y moment arm between centroids of tensile force and compressive force   in

Symbols

γ load factor

ϕ_f resistance factor for flexure

Strength Design Method

The strength design of a composite member is detailed in AASHTO Sec. 6.10.7 and AASHTO App. D6. As shown in Fig. 8.9, for positive bending with the plastic neutral axis (PNA) within the concrete slab, AASHTO Table D6.1-1 case V provides expressions for the depth of the plastic neutral axis and for the plastic moment of resistance. The depth of the plastic neutral axis is

$$D_p = t_s \frac{P_{rb} + P_c + P_w + P_t - P_{rt}}{P_s}$$

The plastic moment of resistance is

$$M_p = \frac{D_p^2 P_s}{2t_s} + P_{rb}d_{rb} + P_c d_c + P_w d_w + P_t d_t + P_{rt}d_{rt}$$

Figure 8.9 Determination of M_p

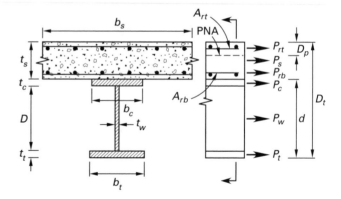

The plastic force in the bottom layer of longitudinal deck reinforcement is given by AASHTO Sec. D6.1 as

$$P_{rb} = F_{yrb}A_{rb}$$

The plastic force in the top layer of longitudinal deck reinforcement is given by AASHTO Sec. D6.1 as

$$P_{rt} = F_{yrt}A_{rt}$$

The plastic force in the top flange of the steel beam is given by AASHTO Sec. D6.1 as

$$P_c = F_{yc}b_c t_c$$

The plastic force in the bottom flange of the steel beam is given by AASHTO Sec. D6.1 as

$$P_t = F_{yt}b_tt_t$$

The plastic force in the web of the steel beam is given by AASHTO Sec. D6.1 as

$$P_w = F_{yw}Dt_w$$

The plastic compressive force in the full depth of the concrete deck is given by AASHTO Sec. D6.1 as

$$P_s = 0.85f'_cb_st_s$$

In the derivation of the expressions, concrete in tension is neglected. The plastic force in the portion of the concrete slab that is in compression is based on a magnitude of the compressive stress equal to $0.85f'_c$.

The forces in the longitudinal reinforcement in the slab may be conservatively neglected. The plastic moment of resistance of the composite section may then be determined as shown in Fig. 8.10.

For positive bending where the top flange of the steel beam is encased in concrete or anchored to the deck slab by shear connectors, the flange is considered continuously braced. Lateral bending stresses are then considered equal to zero. For this condition at the strength limit state, the moment due to factored loads shall satisfy the expression

$$M_u \leq \phi_f M_n$$

The resistance factor for flexure is given by AASHTO Sec. 6.5.4.2 as

$$\phi_f = 1.0$$

For $D_p \leq 0.1D_t$ the nominal flexural resistance of the section is given by AASHTO Eq. (6.10.7.1.2-1) as

$$M_n = M_p$$

For $D_p > 0.1D_t$ the nominal flexural resistance of the section is given by AASHTO Eq. (6.10.7.1.2-2) as

$$M_n = M_p \left(1.07 - \frac{0.7D_p}{D_t}\right)$$

For the composite beam shown in Fig. 8.10, the effective width of the concrete slab is given by AASHTO Sec. 4.6.2.6.1 as the tributary width, which is

$$b_s = S$$

Figure 8.10 *Fully Composite Beam Ultimate Strength*

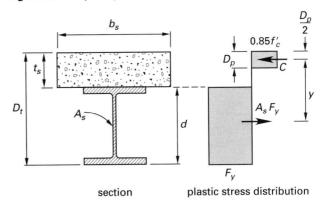

section   plastic stress distribution

Figure 8.10 shows conditions at the strength limit state when the depth of the compression zone at the ultimate load is less than the depth of the slab. In accordance with AASHTO Sec. D6.1, the depth of the stress block is given by

$$D_p = \frac{F_yA_s}{0.85f'_cb_s}$$

The distance between the centroids of the compressive force in the slab and the tensile force in the girder is

$$y = \frac{d}{2} + t_s - \frac{D_p}{2}$$

The plastic moment capacity in bending is given by

$$M_p = F_yA_sy$$

Example 8.26

The simply supported composite beam shown in the following illustration consists of an 8 in concrete slab cast on W36 × 194 grade A50 steel beams with adequate shear connection. The beams are spaced at 8 ft centers and span 100 ft; the slab consists of 4.5 kips/in² normal weight concrete. Bending moment M_{DC} due to dead load imposed on the composite section is 250 ft-kips. Bending moment M_L due to live load plus impact is 1354 ft-kips. Determine the maximum factored applied moment and the maximum flexural strength of the composite section in bending.

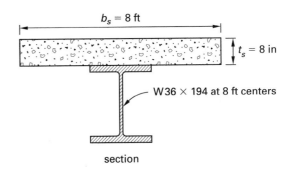

section

Solution

The W36 × 194 girder has the properties tabulated as follows.

A_s	d	t_w	t_f	D	I
57 in²	36.5 in	0.765 in	1.26 in	33.98 in	12,100 in⁴

The effective width of the concrete slab is given by AASHTO Sec. 4.6.2.6.1 as the tributary width, which is

$$b_s = S = (8 \text{ ft}) \left(12 \frac{\text{in}}{\text{ft}} \right)$$
$$= 96 \text{ in}$$

Assuming that the compression zone is contained within the slab, the depth of the stress block is given by AASHTO Sec. D6.1 as

$$D_p = \frac{F_y A_s}{0.85 f'_c b_s}$$
$$= \frac{\left(50 \frac{\text{kips}}{\text{in}^2} \right) (57 \text{ in}^2)}{(0.85) \left(4.5 \frac{\text{kips}}{\text{in}^2} \right) (96 \text{ in})}$$
$$= 7.76 \text{ in}$$

Hence, the compression zone is located within the slab. The distance between the centroids of the compressive force in the slab and the tensile force in the girder is

$$y = \frac{d}{2} + t_s - \frac{D_p}{2}$$
$$= \frac{36.5 \text{ in}}{2} + 8 \text{ in} - \frac{7.76 \text{ in}}{2}$$
$$= 22.37 \text{ in}$$

The plastic moment capacity in bending is given by

$$M_p = F_y A_s y$$
$$= \frac{\left(50 \frac{\text{kips}}{\text{in}^2} \right) (57 \text{ in}^2) (22.37 \text{ in})}{12 \frac{\text{in}}{\text{ft}}}$$
$$= 5313 \text{ ft-kips}$$
$$\frac{D_p}{D_t} = \frac{7.76 \text{ in}}{44.5 \text{ in}} = 0.174$$
$$> 0.1$$

Therefore, the nominal flexural resistance of the section is given by AASHTO Eq. (6.10.7.1.2-2) as

$$M_n = \left(1.07 - \frac{0.7 D_p}{D_t} \right) M_p$$
$$= \left(1.07 - (0.7) \left(\frac{7.76 \text{ in}}{44.5 \text{ in}} \right) \right) (5313 \text{ ft-kips})$$
$$= 5038 \text{ ft-kips}$$

At midspan, the self-weight moment is

$$M_g = \frac{wL^2}{8}$$
$$= \frac{\left(0.194 \frac{\text{kip}}{\text{ft}} \right) (100 \text{ ft})^2}{8}$$
$$= 243 \text{ ft-kips}$$

From Ex. 8.19, the moment due to the weight of the deck slab is

$$M_S = 1000 \text{ ft-kips}$$

The total dead load moment on the composite section is

$$M_D = M_g + M_S + M_{DC}$$
$$= 243 \text{ ft-kips} + 1000 \text{ ft-kips} + 250 \text{ ft-kips}$$
$$= 1493 \text{ ft-kips}$$

The live load moment plus impact is

$$M_L = 1354 \text{ ft-kips}$$

The factored applied moment is given by AASHTO Eq. (3.4.1-1) as

$$M_u = \gamma_P M_D + \gamma_{LL+IM} M_L$$
$$= (1.25)(1493 \text{ ft-kips}) + (1.75)(1354 \text{ ft-kips})$$
$$= 4236 \text{ ft-kips}$$

Design for Shear

Nomenclature

C	web buckling coefficient	–
D	depth of the web of the steel beam	in
E	modulus of elasticity of the steel beam	kips/in²
t_w	web thickness	in
V_n	nominal shear strength	kips
V_p	shear yielding strength of the web	kips
V_r	factored shear resistance	kips
V_u	factored applied shear force	kips

Symbols

γ	load factor	
ϕ_v	resistance factor for shear	

Strength Design Method

AASHTO Sec. 6.10.9.2 defines the nominal shear strength of a girder with unstiffened web as

$$V_n = CV_p \qquad \textit{[AASHTO 6.10.9.2-1]}$$

The plastic shear force of the web is given by

$$V_p = 0.58F_{yw}Dt_w \qquad \textit{[AASHTO 6.10.9.2-2]}$$

For values of $D/t_w \leq 1.12\sqrt{5E/F_{yw}}$, the web buckling coefficient is defined by AASHTO Eq. (6.10.9.3.2-4) as

$$C = 1.0$$

The factored shear resistance is given by AASHTO Eq. (6.12.1.2.3-1) as

$$V_r = \phi_v V_n$$

The resistance factor for shear is given by AASHTO Sec. 6.5.4.2 as

$$\phi_v = 1.0$$

Example 8.27

The simply supported composite beam of Ex. 8.26 is subjected to a support reaction V_L of 58 kips due to live load plus impact and a support reaction V_{DC} of 10 kips due to dead load imposed on the composite section. Determine whether the section is adequate.

Solution

The web buckling coefficient for the composite section is

$$C = 1.0$$

The nominal shear strength of the composite section is given by AASHTO Sec. 6.10.9.2 as

$$
\begin{aligned}
V_n = CV_p &= 0.58F_{yw}Dt_w \\
&= (0.58)\left(50 \ \frac{\text{kips}}{\text{in}^2}\right)(33.97 \ \text{in})(0.765 \ \text{in}) \\
&= 754 \ \text{kips}
\end{aligned}
$$

The factored shear resistance is

$$
\begin{aligned}
V_r = \phi_v V_n \\
&= (1.0)(754 \ \text{kips}) \\
&= 754 \ \text{kips}
\end{aligned}
$$

The support reaction due to the self-weight of the girder is

$$V_g = \frac{w_g L}{2} = \frac{\left(0.194 \ \frac{\text{kip}}{\text{ft}}\right)(100 \ \text{ft})}{2}$$
$$= 9.7 \ \text{kips}$$

The support reaction due to the weight of the slab is

$$V_S = \frac{w_S L}{2}$$

$$= \frac{\left(0.150 \ \frac{\text{kip}}{\text{ft}^3}\right)\left(\dfrac{64 \ \text{in}}{12 \ \frac{\text{in}}{\text{ft}}}\right)(100 \ \text{ft})}{2}$$

$$= 40 \ \text{kips}$$

The total dead load support reaction is

$$
\begin{aligned}
V_D = V_g + V_S + V_{DC} \\
&= 9.7 \ \text{kips} + 40 \ \text{kips} + 10 \ \text{kips} \\
&= 59.7 \ \text{kips}
\end{aligned}
$$

The live load reaction plus impact is

$$V_L = 58 \ \text{kips}$$

The factored applied reaction is given by AASHTO Eq. (3.4.1-1) as

$$
\begin{aligned}
V_u = \gamma_p V_D + \gamma_{LL+IM} V_L \\
&= (1.25)(59.7 \ \text{kips}) + (1.75)(58 \ \text{kips}) \\
&= 176 \ \text{kips} \\
&< V_r \quad \text{[The section is adequate.]}
\end{aligned}
$$

Shear Connection

Nomenclature

A_{ct}	transformed area of concrete slab	in^2
A_s	area of steel beam	in^2
A_{sc}	cross-sectional area of stud shear connector	in^2
b_s	effective width of the concrete deck	in
b_t	transformed width of concrete slab	in
d	diameter of stud shear connector	in
D	depth of the web of the steel beam	in
E	modulus of elasticity of the steel beam	kips/in^2
E_c	modulus of elasticity of concrete	kips/in^2
f_c'	specified 28 day compressive strength of concrete	kips/in^2
F_u	specified tensile strength of the steel beam	kips/in^2
F_y	specified yield strength of the steel beam	kips/in^2
H	stud height	in
I	moment of inertia of transformed composite section	in^4
I	impact factor	–
L	span length	ft
L	impact equation length	ft
m	modular ratio	–
n	number of shear connectors between point of maximum positive moment and point of zero moment	–
n	number of shear connectors in a cross-section	–
N	number of cycles	–
p	connector spacing	in

P total shear force at interface at ultimate limit state — kips

Q moment of transformed compressive concrete area about neutral axis — in^3

S_u ultimate strength of shear connector — lbf

V_r factored shear resistance — kips

V_r range of shear force due to live load plus impact — kips

V_{sr} range of horizontal shear at interface — kips/in

w unit weight of concrete — lbf/ft^3

y' distance from slab center to neutral axis of transformed composite — in

Z_r allowable range of shear for a welded stud — lbf

Symbols

α stress cycle factor

γ load factor

ϕ reduction factor

ϕ_v resistance factor for shear

General

Shear connectors are designed for fatigue and are checked for ultimate strength. Fatigue stresses are caused by the range of shear produced on the connector by live load plus impact and are calculated by using elastic design principles. The ultimate strength of the connectors must be adequate to develop the lesser of the strength of the steel girder or the ultimate strength of the concrete slab.

Fatigue Strength

The elastic design properties of the composite section are determined by using the transformed width of the concrete slab. As shown in Fig. 8.11, the transformed width is given by

$$b_t = \frac{b}{m}$$

The value of the modular ratio for short-term loads is given by AASHTO Sec. C6.10.1.1.1b as

$$m = 10 \quad [\text{for } f'_c = 2.4 - 2.8]$$
$$m = 9 \quad [\text{for } f'_c = 2.9 - 3.5]$$
$$m = 8 \quad [\text{for } f'_c = 3.6 - 4.5]$$
$$m = 7 \quad [\text{for } f'_c = 4.6 - 5.9]$$

The transformed area of the concrete slab is

$$A_{ct} = b_t t_s$$

The statical moment of the transformed concrete area about the neutral axis of the composite section is

$$Q = y' A_{ct}$$

For a straight girder, the range of horizontal shear at the interface is given by AASHTO Eq. (6.10.10.1.2-3) as

$$V_{sr} = \frac{V_f Q}{I}$$

The range of shear force V_f due to live load plus impact is the difference between the maximum and minimum applied shear under the fatigue load combination. For a ratio of height to diameter H/d of not less than 4, the allowable range of shear for a welded stud is given by

$$Z_r = \alpha d^2 \qquad \textit{[AASHTO 6.10.10.2-1]}$$
$$\geq 5.5 d^2 / 2$$

The stress cycle factor is given by AASHTO Eq. (6.10.10.2-2) as

$$\alpha = 34.5 - 4.28 \log N$$
$$\alpha = 13 \quad [\text{for } 100{,}000 \text{ cycles}]$$
$$\alpha = 10 \quad [\text{for } 500{,}000 \text{ cycles}]$$
$$\alpha = 7.53 \quad [\text{for } 2{,}000{,}000 \text{ cycles}]$$

The required connector pitch is given by AASHTO Eq. (6.10.10.1.2-1) as

$$p = \frac{n Z_r}{V_{sr}}$$
$$\leq 24 \text{ in}$$
$$\geq 6d$$

Example 8.28

For the composite beam of Ex. 8.27, determine the required spacing of $^3/_4$ in diameter stud shear connectors at the support for 2,000,000 stress cycles.

Figure 8.11 *Composite Section Properties*

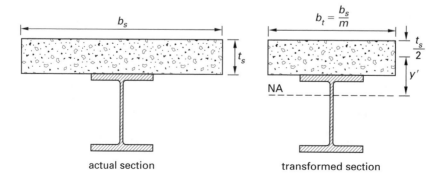

actual section   transformed section

Solution

The compressive strength of the concrete slab is 4.5 kips/in^2, and the corresponding modular ratio is

$$m = 8$$

The transformed area of the concrete slab is

$$A_{ct} = \frac{b_s t_s}{m} = \frac{(96 \text{ in})(8 \text{ in})}{8}$$

$$= 96 \text{ in}^2$$

The moment of inertia of the transformed section is derived as shown in the table.

Member Properties for Ex. 8.28

part	A (in^2)	y (in)	I (in^4)	Ay (in^3)	Ay^2 (in^4)
girder	57	18.25	12,100	1040	18,985
slab	96	40.49	512	3887	157,385
total	153	–	12,612	4927	176,370

The height of the neutral axis of the transformed section is

$$\bar{y} = \frac{\sum Ay}{\sum A} = \frac{4927 \text{ in}^3}{153 \text{ in}^2}$$

$$= 32.2 \text{ in}$$

The moment of inertia of the transformed section is

$$I = \sum I + \sum Ay^2 - \bar{y}^2 \sum A$$
$$= 12.612 \text{ in}^4 + 176,370 \text{ in}^4 - (32.2 \text{ in})^2(153 \text{ in}^2)$$
$$= 30,345 \text{ in}^4$$

The statical moment of the transformed slab about the neutral axis is

$$Q = y'A_{ct} = \left(d + \frac{t_s}{2} - \bar{y}\right) A_{ct}$$
$$= \left(36.5 \text{ in} + \frac{8 \text{ in}}{2} - 32.2 \text{ in}\right) (96 \text{ in}^2)$$
$$= 797 \text{ in}^3$$

$$\frac{Q}{I} = \frac{797 \text{ in}^3}{30,345 \text{ in}^4}$$
$$= 0.0263 \text{ in}^{-1}$$

The transformed section is shown in the following illustration.

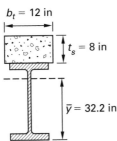

transformed composite section

The allowable range of shear for a $^3/_4$ in diameter welded stud subjected to 2,000,000 stress cycles is given by AASHTO Eq. (6.10.10.2-1) as

$$Z_r = \alpha d^2 = \left(7.53 \frac{\text{kips}}{\text{in}^2}\right)(0.75 \text{ in})^2$$
$$= 4.24 \text{ kips}$$

From Ex. 8.27, the maximum shear force at the support due to live load plus impact is

$$V_{\max} = 58 \text{ kips}$$

From AASHTO Table 3.4.1-1, the shear force at the support for the fatigue limit state is

$$V' = 0.75 V_{\max}$$
$$= (0.75)(58 \text{ kips})$$
$$= 43.5 \text{ kips}$$

The minimum shear force at the support is

$$V_{\min} = 0 \text{ kips}$$

The range of shear force at the support for the fatigue limit state is

$$V_f = V' - V_{\min}$$
$$= 43.5 \text{ kips} - 0 \text{ kips}$$
$$= 43.5 \text{ kips}$$

The range of horizontal shear at the support at the interface is given by AASHTO Eq. (6.10.10.1.2-3) as

$$V_{sr} = \frac{V_f Q}{I} = (43.5 \text{ kips})\left(0.0263 \text{ in}^{-1}\right)$$
$$= 1.14 \text{ kips/in}$$

With two studs per row, the required stud spacing is

$$p = \frac{n Z_r}{V_{sr}} = \frac{(2)(4.24 \text{ kips})}{1.14 \frac{\text{kips}}{\text{in}}}$$
$$= 7.4 \text{ in}$$

Ultimate Strength

To provide adequate connection at the interface at ultimate load, the number of connectors required on each side of the point of maximum moment is given by AASHTO Sec. 6.10.10.4.1 as

$$n = \frac{P}{\phi_{sc} Q_n}$$

For a straight girder, the total shear force at the interface at the ultimate limit state is given in AASHTO Sec. 6.10.10.4.2 as the lesser of

$$P = A_s F_y \quad \text{[girder governs]}$$
$$P = 0.85 f_c' b_s t_s \quad \text{[deck slab governs]}$$

The reduction factor is given by AASHTO Sec. 6.5.4.2 as

$$\phi_{sc} = 0.85$$

The nominal strength of a welded stud shear connector is given by AASHTO Eq. (6.10.10.4.3-1) as

$$Q_n = 0.5 A_{sc}\sqrt{f_c' E_c}$$
$$\leq A_{sc} F_u$$

The modulus of elasticity of normal weight concrete is given by AASHTO Eq. (C5.4.2.4-1) as

$$E_c = 1820\sqrt{f_c'}$$

Example 8.29

For the composite beam of Ex. 8.27, determine the required number of $^3/_4$ in diameter stud shear connectors to provide adequate connection at the ultimate load. The shear connectors have a tensile strength of 60 kips/in².

Solution

The compressive strength of the concrete slab is 4.5 kips/in², and the corresponding modulus of elasticity is given by AASHTO Eq. (C5.4.2.4-1) as

$$E_c = 1820\sqrt{f_c'}$$
$$= 1820\sqrt{4.5\ \frac{\text{kips}}{\text{in}^2}}$$
$$= 3860\ \text{kips/in}^2$$

The ultimate strength of a $^3/_4$ in diameter welded stud shear connector is given by AASHTO Eq. (6.10.10.4.3-1) as

$$Q_n = 0.5 A_{sc}\sqrt{f_c' E_c}$$
$$= (0.5)(0.44\ \text{in}^2)\sqrt{\left(4.5\ \frac{\text{kips}}{\text{in}^2}\right)\left(3860\ \frac{\text{kips}}{\text{in}^2}\right)}$$
$$= 29\ \text{kips}$$

Maximum strength is

$$Q_{n(\text{max})} = 60 A_{sc}$$
$$= (60)(0.44)$$
$$= 26.40\ \text{kips} \quad [\text{governs}]$$

The nominal shear force at the interface is governed by the girder, and the total number of connectors required on the girder is

$$2n = \frac{2 A_s F_y}{\phi_{sc} Q_n} = \frac{(2)(57\ \text{in}^2)\left(50\ \dfrac{\text{kips}}{\text{in}^2}\right)}{(0.85)(26.40\ \text{kips})}$$
$$= 254$$

5. WOOD STRUCTURES

Basic Design Values and Adjustment Factors

Nomenclature

a	species parameter for volume factor	–
A	ratio of F_{bE} to F_b	–
b	breadth of rectangular bending member	in
C_d	deck factor	–
C_{fu}	flat use factor	–
C_F	size factor for sawn lumber	–
C_i	incising factor	–
C_{KF}	format conversion factor	–
C_L	beam stability factor	–
C_M	wet service factor	–
C_V	volume factor for structural glued laminated timber	–
C_λ	time effect factor	–
d	depth of member	in
E, E_o	reference and adjusted modulus of elasticity	kips/in²
F_b	reference bending design value multiplied by all applicable adjustment factors	kips/in²
F_{bo}	reference bending design value	kips/in²
F_{bE}	critical buckling design value for bending members	kips/in²
F_v	adjusted design value of wood in shear	kips/in²
F_{vo}	reference design value of wood in shear	kips/in²
K_{bE}	Euler buckling coefficient for beams	–
L_e	effective bending member length	ft or in
L_u	laterally unsupported bending member length	ft or in
L	span length of bending member	ft or in
M_n	nominal flexural resistance	in-kips
M_r	factored flexural resistance	in-kips
R_B	slenderness ratio of bending member	–
S	section modulus	in³
V_n	nominal shear resistance	kips
V_r	factored shear resistance	kips

Symbols

ϕ	resistance factor	

Reference Design Values

The reference design values for sawn lumber are given in AASHTO Tables 8.4.1.1.4-1, 8.4.1.1.4-2, and 8.4.1.1.4-3. The reference design values for glued laminated timber are given in AASHTO Tables 8.4.1.2.3-1 and 8.4.1.2.3-2. These tabulated design values are applicable to normal conditions of use as defined in AASHTO Sec. C8.4.1. For other conditions of use, the tabulated values are multiplied by adjustment factors, specified in AASHTO Sec. 8.4.4, to determine the corresponding adjusted design values. In accordance with AASHTO Sec. 8.4.4.1, the adjusted design value in bending is

$$F_b = F_{bo} C_{KF} C_M (C_F \text{ or } C_V) C_{fu} C_i C_d C_\lambda$$

[AASHTO 8.4.4.1-1]

Adjustment Factors

The *time effect factor*, C_λ, given in AASHTO Sec. 8.4.4.9 is applicable to all reference design values with the exception of the modulus of elasticity. Values of the time effect factor are given in Table 8.4 [AASHTO Table 8.4.4.9-1].

Table 8.4 Time Effect Factor, C_λ

limit state	C_λ
strength I	0.8
strength II	1.0
strength III	1.0
strength IV (permanent)	0.6
extreme event I	1.0

The *wet service factor*, C_M, given in AASHTO Table 8.4.4.3-1, is applicable to sawn lumber when the moisture content exceeds 19%. Values of the wet service factor are given in Table 8.5.

Table 8.5 Wet Service Factors, C_M, for Sawn Lumber

design function	$F_{bo}C_F > 1.15$	F_{vo}
members not exceeding 4 in thickness	0.85	0.97
members exceeding 4 in thickness	1.00	1.00

When the moisture content of a glued laminated member exceeds 16%, the adjustment factor given in AASHTO Table 8.4.4.3-2 is applicable. Values of the wet service factor are given in Table 8.6.

Table 8.6 Wet Service Factor, C_M, for Glued Laminated Members

design function	F_{bo}	F_{vo}
wet service factor	0.80	0.875

The *beam stability factor* is applicable to the tabulated bending reference design value for sawn lumber and glued laminated members. For glued laminated members, C_L is not applied simultaneously with the volume factor, C_V, and the lesser of these two values is applicable.

The beam stability factor is given by AASHTO Sec. 8.6.2 as

$$C_L = \frac{1.0 + A}{1.9} - \sqrt{\left(\frac{1.0 + A}{1.9}\right)^2 - \frac{A}{0.95}}$$

[AASHTO 8.6.2-2]

The variables are defined as

$$A = \frac{F_{bE}}{F_b}$$
[AASHTO 8.6.2-3]

F_b = reference bending design value multiplied by all applicable adjustment factors

$\quad = F_{bo}C_{KF}C_MC_{fu}C_iC_dC_\lambda$ [C_F applies only to visually graded sawn lumber.]

F_{bE} = critical buckling design value

$\quad = \dfrac{K_{bE}E}{R_B^2}$ [AASHTO 8.6.2-4]

K_{bE} = Euler buckling coefficient

$\quad = 1.10$ $\begin{bmatrix} \text{for glued laminated timber and} \\ \text{machine stress rated lumber} \end{bmatrix}$

$\quad = 0.76$ [for visually graded lumber]

E = allowable modulus of elasticity

$\quad = E_oC_MC_i$ [AASHTO 8.4.4.1-6]

R_B = slenderness ratio

$\quad = \sqrt{\dfrac{L_e d}{b^2}}$ [AASHTO 8.6.2-5]

$\quad \leq 50$

The term L_e is the effective length of a bending member and is defined in AASHTO Sec. 8.6.2 and tabulated in Table 8.7.

Table 8.7 Effective Length, L_e

member dimensions	L_e
$\dfrac{L_u}{d} < 7$	$2.06L_u$
$7 \leq \dfrac{L_u}{d} \leq 14.3$	$1.63L_u + 3d$
$\dfrac{L_u}{d} > 14.3$	$1.84L_u$

In accordance with AASHTO Sec. 8.6.2, $C_L = 1.0$ when

- $\dfrac{d}{b} \leq 1.0$

- the compression edge is continuously restrained

The *size factor*, C_F, is applicable to sawn lumber and to glued laminated members with load applied parallel to the wide face of the laminations. For sawn lumber 2–4 in thick, values of the size factor are given in AASHTO Table 8.4.4.4-1. For members exceeding 12 in depth and 5 in thickness, the size factor is

$$C_F = \left(\frac{12}{d}\right)^{1/9}$$
[AASHTO 8.4.4.4-2]

The *volume factor*, C_V, is applicable to the reference design value for bending of glued laminated members and is not applied simultaneously with the *beam stability factor*, C_L; the lesser of these two factors is applicable.

Bridges

The volume factor is defined in AASHTO Sec. 8.4.4.5 as

$$C_V = \left(\frac{1291.5}{bdL} \right)^a \quad \text{[AASHTO 8.4.4.5-1]}$$
$$\leq 1.0$$

The variables are defined as

L = length of beam between points of zero moment, ft

b = beam width, in

d = beam depth, in

$a = 0.05$ [for Southern Pine]

  $= 0.10$ [for all other species]

The *flat-use factor*, C_{fu}, is applicable to dimension lumber with the load applied to the wide face. Values of the flat-use factor are given in AASHTO Table 8.4.4.6-1. Flat-use factors for glued laminated members, with the load applied parallel to the wide faces of the laminations, are given in AASHTO Table 8.4.4.6-2. The flat-use factor is not applied to dimension lumber graded as decking, as the design values already incorporate the appropriate factor.

Values of the *incising factor*, C_i, for a prescribed incising pattern are given in AASHTO Table 8.4.4.7-1. The prescribed pattern consists of incisions parallel to the grain a maximum depth of 0.4 in, a maximum length of 3/8 in, and a density of incisions of up to $1100/\text{ft}^2$.

The *deck factor*, C_d, is applied to mechanically laminated decks. Values of the deck factor for stressed wood, spike-laminated, and nail-laminated decks are given in AASHTO Table 8.4.4.8-1.

The format conversion factor, C_{KF}, is used to ensure that load and resistance factor design will result in the same size members as allowable stress design. AASHTO Sec. 8.4.4.2 gives the format conversion factor for all loading conditions except compression perpendicular to the grain, as

$$C_{KF} = \frac{2.5}{\phi}$$

Resistance factors are tabulated in AASHTO Sec. 8.5.2.2 and these include

$$\phi = 0.85 \text{ for flexure}$$
$$\phi = 0.75 \text{ for shear}$$

The factored flexural resistance is given by AASHTO Eq. (8.6.1-1) as

$$M_r = \phi M_n$$

The nominal flexural resistance is given by AASHTO Eq. (8.6.2-1) as

$$M_n = F_b S C_L$$

For shear, the corresponding values are

$$V_r = \phi V_n \quad \text{[AASHTO 8.7-1]}$$
$$V_n = \frac{F_v bd}{1.5} \quad \text{[AASHTO 8.7-2]}$$

Example 8.30

The bridge superstructure shown in the following illustration is simply supported over a span of 35 ft. The moisture content exceeds 16%. Determine the nominal resistance values in bending and shear for the strength I limit state for the glued laminated girders.

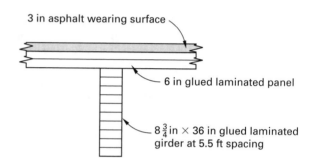

Solution

The reference design bending stress for a Southern Pine 24F-V4 glued laminated stringer is obtained from AASHTO Table 8.4.1.2.3-1 and is

$$F_{bo} = 2.4 \text{ kips/in}^2$$

The applicable adjustment factors for bending stress are as follows.

C_λ = time effect factor for the strength I limit state from Table 8.4.4.9-1

  $= 0.8$

C_M = wet service factor from AASHTO Table 8.4.4.3-2

  $= 0.80$

C_L = beam stability factor from AASHTO Sec. 8.6.2

  $= 1.0$ $\begin{bmatrix} \text{compression face of the girder} \\ \text{fully supported laterally} \end{bmatrix}$

C_V = volume factor given by AASHTO Eq. (8.4.4.5-1)

  $= \left(\frac{1291.5}{bdL} \right)^a$

  $= \left(\frac{1291.5 \text{ in}^2\text{-ft}}{(8.75 \text{ in})(36.0 \text{ in})(35 \text{ ft})} \right)^{0.10}$

  $= 0.81$ [governs]

  $< C_L$

In accordance with AASHTO Eq. (8.4.4.1-1), the adjusted design bending stress is

$$F_b = F_{bo}C_\lambda C_M C_V C_{KF}$$
$$= \left(2.4 \ \frac{\text{kips}}{\text{in}^2}\right)(0.8)(0.80)(0.81)\left(\frac{2.5}{0.85}\right)$$
$$= 3.66 \ \text{kips/in}^2$$

The nominal flexural resistance is given by AASHTO Eq. (8.6.2-1) as

$$M_n = F_b S C_L = (3.66 \ \text{kips})(1890 \ \text{in}^3)(1.0)$$
$$= 6917 \ \text{in-kips}$$

The reference design shear stress for a Southern Pine 24F-V4 glued laminated stringer is obtained from AASHTO Table 8.4.1.2.3-1 and is

$$F_{vo} = 0.21 \ \text{kips/in}^2$$

The applicable adjustment factors for shear stress are as follows.

C_λ = time effect factor for the strength I limit state from Table 8.4
= 0.8
C_M = wet service factor from AASHTO Table 8.4.4.3-2
= 0.875

In accordance with AASHTO Eq. (8.4.4.1-2), the adjusted design shear stress is

$$F_v = F_{vo}C_\lambda C_M C_{KF}$$
$$= \left(0.21 \ \frac{\text{kips}}{\text{in}^2}\right)(0.8)(0.875)\left(\frac{2.5}{0.75}\right)$$
$$= 0.49 \ \text{kips/in}^2$$

Design Requirements for Flexure

In accordance with AASHTO Sec. 13.6.1.2, the beam span is taken as the clear span plus one-half the required bearing length at each end. AASHTO Sec. 8.6.2 specifies that when the depth of a beam does not exceed its breadth or when continuous lateral restraint is provided to the compression edge of a beam with the ends restrained against rotation, the beam stability factor $C_L = 1.0$. For other situations, the value of C_L is calculated in accordance with AASHTO Sec. 8.6.2, and the effective span length, L_e, is determined in accordance with AASHTO Sec. 8.6.2. For visually graded sawn lumber, both the stability factor C_L and the size factor C_F must be considered concurrently. For glued laminated members, both the stability factor C_L and the volume factor C_V must be determined. Only the lesser of these two factors is applicable in determining the allowable design value in bending. In accordance

with AASHTO Sec. 3.6.2, impact need not be considered in timber structures.

Design Requirements for Shear

Shear shall be investigated at a distance from the support equal to the depth of the beam. The governing *shear force V for vehicle loads* is determined by placing the live load to produce the maximum shear at a distance from the support given by the lesser of

- $3d$

- $L/4$

In accordance with AASHTO Sec. 3.6.2, impact need not be considered in timber structures.

6. SEISMIC DESIGN

Nomenclature

A_S	peak seismic ground acceleration coefficient modified by zero period site factor from AASHTO 3.10.4.2	–
C_{sm}	seismic response coefficient specified in AASHTO Sec. 3.10.4.2	–
D	dead load applied to a structural element	lbf or kips
F_a	site factor for short-period range of acceleration response spectrum from AASHTO Sec. 3.10.3.2	–
F_{pga}	site factor at zero-period on acceleration response spectrum from AASHTO Sec. 3.10.3.2	–
F_v	site factor for long-period range of acceleration response spectrum from AASHTO Sec. 3.10.3.2	–
g	acceleration due to gravity	32.2 ft/sec² or 386 in/sec²
IC	importance category	–
K	total lateral stiffness of bridge	lbf/in or kips/in
L	length of bridge deck	ft
M_P	primary moment	ft-kips
N	minimum support length for girders	in
$p_e(x)$	intensity of the equivalent static seismic loading used to calculate the period in AASHTO Sec. C4.7.4.3.2b	lbf/in or kips/in
PGA	peak seismic ground acceleration coefficient on rock (site class B) from AASHTO Sec. 3.10.2.1	–

Bridges

p_o	assumed uniform loading used to calculate the period in AASHTO Sec. C.4.7.4.3.2b	lbf/ft or kips/ft
R	response modification factor from AASHTO Sec. 3.10.7.1	–
S_1	horizontal response spectral acceleration coefficient at 1.0 sec period on rock (site class B) from AASHTO Sec. 3.10.2.1	–
SC	site class from AASHTO Sec. 3.10.3.1	–
S_{DS}	horizontal response spectral acceleration coefficient at 0.2 sec period modified by short-period site factor from AASHTO Sec. 3.10.4.2	–
S_{D1}	horizontal response spectral acceleration coefficient at 1.0 sec modified by long-period site factor from AASHTO Sec. 3.10.4.2	–
SPZ	seismic performance zone	–
S_S	horizontal response spectral acceleration coefficient at 0.2 sec period on rock (site class B) from AASHTO Sec. 3.10.2.1	–
T_0	reference period used to define shape of acceleration response spectrum from AASHTO Sec. 3.10.4.2	–
T_m	fundamental period of vibration, defined in AASHTO Sec. C4.7.4.3.2b	sec
T_S	corner period at which acceleration response spectrum changes from being independent of period to being inversely proportional to period from AASHTO Sec. 3.10.4.2 sec	–
$v_e(x)$	static displacement profile resulting from applied load p_e used in AASHTO Sec. 4.7.4.3.2c	in
$v_s(x)$	static displacement profile resulting from applied load p_o used in AASHTO Sec. C.4.7.4.3.2b	in
$w(x)$	dead weight of bridge super-structure and tributary substructure per unit length	lbf/ft or kips/ft
W	total weight of bridge super-structure and tributary substructure	lbf or kips

Symbols

α	coefficient used to calculate period of the bridge in AASHTO Sec. C4.7.4.3.2b	ft^2

β	coefficient used to calculate period of the bridge in AASHTO Sec. C4.7.4.3.2b	ft-kips
γ	coefficient used to calculate period of the bridge in AASHTO Sec. C4.7.4.3.2b	ft^2-kips

Analysis Procedures

To determine the seismic response of the structure, several factors must be considered. These factors include the ground motion parameters, site class, fundamental period, and response modification factors. Selection of the design procedure depends on the type of bridge, the importance category, and the seismic zone. Four analysis procedures[15,16] are presented in AASHTO Sec. 4.7.4.3.1 and are shown in Table 8.8.

Table 8.8 Analysis Procedures

procedure	method
UL	uniform load elastic
SM	single-mode elastic
MM	multimode elastic
TH	time history

The uniform load and single-mode procedures both assume that the seismic response of a bridge can be represented by a single mode of vibration and are suitable for hand computation. The multimode and time history procedures account for higher modes of vibration and require analysis by computer.

Acceleration Coefficients

The acceleration coefficients PGA, S_S, and S_1 are defined in AASHTO Sec. 3.10.4.2 and shown in AASHTO Figs. 3.10.2.1-1 to 3.10.2.1-21. These are an estimate of the site-dependent design ground acceleration expressed as a percentage of the gravity constant g. The acceleration coefficients are also available on the CD-ROM provided with the AASHTO specification. The acceleration coefficients correspond to ground acceleration values with a recurrence interval of 1000 yr, which gives a 7% probability of being exceeded in a 75 yr period. This is termed the design earthquake.

Example 8.31

The two-span bridge shown in the following illustration is located at 33.70° north and −117.50° west on a non-essential route. The central circular column is fixed at the top and bottom. The soil profile at the site consists of a stiff soil with a shear wave velocity of 700 ft/sec. The relevant criteria are (a) column moment of inertia $I_c = 60$ ft^4, (b) column modulus of elasticity $E_c = 450{,}000$ kips/ft^2, (c) column height $h_c = 30$ ft, (d) weight of the superstructure and tributary substructure $w = 10$ kips/ft, (e) superstructure moment of inertia $I_s = 4000$ ft^4, and (f) superstructure modulus of elasticity $E_s = 450{,}000$ kips/ft^2. Determine the applicable acceleration coefficients.

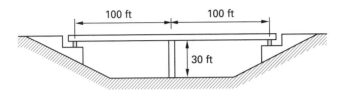

Solution

From AASHTO Figs. 3.10.2.1-4 to 3.10.2.1-6, the applicable acceleration coefficients are

$$PGA = 0.61$$
$$S_S = 1.45$$
$$S_1 = 0.52$$

Importance Category

The importance category is defined in AASHTO Sec. 3.10.5, and three categories are specified: critical bridges, essential bridges, and other bridges. The importance category of a bridge is determined on the basis of social and security requirements. An importance category of *critical* is assigned to bridges that must remain functional immediately after a 2500 year return period earthquake. An importance category of *essential* is assigned to bridges that must remain functional immediately after the design earthquake. An importance category of *other* is assigned to non-essential bridges.

Example 8.32

Determine the importance category for the bridge of Ex. 8.31.

Solution

From AASHTO Sec. 3.10.5, for a bridge on a non-essential route, the importance category is

$$IC = other$$

Site Class

Six soil profile types are identified in AASHTO Table 3.10.3.1-1. Table 8.9 gives a summary of the soil profile types.

Table 8.9 *Site Classes*

site class	soil profile name	shear wave velocity
A	hard rock	> 5000
B	rock	2500–5000
C	soft rock	1200–2500
D	stiff soil	600–1200
E	soft soil	< 600
F	(*)	(*)

Note: (*) consists of peat or high plasticity clay requiring a site-specific geotechnical investigation.

Example 8.33

Determine the site class for the bridge of Ex. 8.31.

Solution

From AASHTO Sec. 3.10.3.1 and Table 8.9, the relevant site class for a stiff soil is D.

Site Factors

Site factors are amplification factors applied to the ground accelerations and are a function of the site class. Site factor F_{pga} corresponds to PGA, F_a corresponds to S_S and F_v corresponds to S_1. Site Class B is the reference site category and has a site factor of 1.0. The site factors generally increase as the soil profile becomes softer (in going from site class A to E). The factors also decrease as the ground motion level increases due to the nonlinear behavior of the soil. Site factors F_{pga}, F_a, and F_v are specified in AASHTO Tables 3.10.3.2-1, 3.10.3.2-2, and 3.10.3.2-3 and are summarized in Table 8.10. Linear interpolation may be used to obtain intermediate values.

Table 8.10 *Site Factors (F_{pga} corresponding to PGA; F_a corresponding to S_S; F_v corresponding to S_1)*

site class	ground acceleration, PGA					ground acceleration, S_S					ground acceleration, S_1				
	≤ 0.1	0.2	0.3	0.4	≥ 0.5	≤ 0.25	0.50	0.75	1.00	≥ 1.25	≤ 0.1	0.2	0.3	0.4	≥ 0.5
A	0.8	0.8	0.8	0.8	0.8	0.8	0.8	0.8	0.8	0.8	0.8	0.8	0.8	0.8	0.8
B	1.0	1.0	1.0	1.0	1.0	1.0	1.0	1.0	1.0	1.0	1.0	1.0	1.0	1.0	1.0
C	1.2	1.2	1.1	1.0	1.0	1.2	1.2	1.1	1.0	1.0	1.7	1.6	1.5	1.4	1.3
D	1.6	1.4	1.2	1.1	1.0	1.6	1.4	1.2	1.1	1.0	2.4	2.0	1.8	1.6	1.5
E	2.5	1.7	1.2	0.9	(*)	2.5	1.7	1.2	0.9	(*)	3.5	3.2	2.8	2.4	(*)
F	(*)	(*)	(*)	(*)	(*)	(*)	(*)	(*)	(*)	(*)	(*)	(*)	(*)	(*)	(*)

Note: (*) Site-specific geotechnical investigation and dynamic site response analysis is required.

Example 8.34

Determine the site factors for the bridge of Ex. 8.31.

Solution

From Ex. 8.31, the ground motion parameters are PGA = 0.61 g, $S_S = 1.45$ g, and $S_1 = 0.52$ g. From Ex. 8.33, the site class at the location of the bridge is SC = D. From Table 8.10 the site factors are

$$F_{\text{pga}} = 1.0$$
$$F_a = 1.0$$
$$F_v = 1.5$$

Adjusted Response Parameters

As specified in AASHTO Sec. 3.10.4.2, the ground motion parameters are modified by the site factors to allow for the site class effects. Therefore, the adjusted response parameters are

$$A_S = F_{\text{pga}}(\text{PGA})$$
$$S_{DS} = F_a S_S$$
$$S_{D1} = F_v S_1$$

Example 8.35

Determine the adjusted response parameters for the bridge of Ex. 8.31.

Solution

From Ex. 8.31, the ground motion parameters are PGA = 0.61 g, $S_S = 1.45$ g, and $S_1 = 0.52$ g. From Ex. 8.34, the site factors are $F_{\text{pga}} = 1.0$, $F_a = 1.0$, and $F_v = 1.5$. From AASHTO Sec. 3.10.4.2, the adjusted response parameters are

$$A_s = F_{\text{pga}}(\text{PGA}) = (1.0)(0.61 \text{ g})$$
$$= 0.61 \text{ g}$$
$$S_{DS} = F_a S_S = (1.0)(1.45 \text{ g})$$
$$= 1.45 \text{ g}$$
$$S_{D1} = F_v S_1 = (1.5)(0.52 \text{ g})$$
$$= 0.78 \text{ g}$$

Design Response Spectrum

The adopted design response spectrum is given by AASHTO Fig. 3.10.4.1-1 and is shown in Fig. 8.12. The spectrum for a specific location is a graph of the elastic seismic response coefficient, C_{sm}, over a range of periods of vibration, T_m.

Figure 8.12 *Design Response Spectrum*

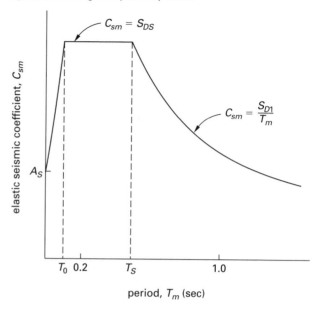

Elastic Seismic Response Coefficient

As shown in Fig. 8.12, the design response spectrum is composed of three segments demarcated by the periods of vibration.

$$T_m = 0$$
$$T_0 = 0.2 T_S$$
$$T_S = \frac{S_{D1}}{S_{DS}}$$

The values of the elastic seismic response coefficient are determined using equations given in AASHTO Sec. 3.10.4.2, which are summarized in Table 8.11.

Table 8.11 *Elastic Seismic Response Coefficient Equations*

period, T_m	elastic seismic response coefficient, C_{sm}
$T_m \leq T_0$	$A_S + (S_{DS} - A_S)(T_m/T_0)$
$T_0 < T_m \leq T_S$	S_{DS}
$T_S < T_m$	S_{D1}/T_m

Example 8.36

Determine the reference periods used to define the shape of the response spectrum for the bridge of Ex. 8.31.

Solution

From Ex. 8.35, the adjusted response parameters are $S_{DS} = 1.45$ g and $S_{D1} = 0.78$ g. From AASHTO Sec. 3.10.4.2, the reference periods are

$$T_S = \frac{S_{D1}}{S_{DS}} = \frac{0.78 \text{ g}}{1.45 \text{ g}}$$
$$= 0.538 \text{ sec}$$
$$T_0 = 0.2 T_S = (0.2)(0.538 \text{ sec})$$
$$= 0.108 \text{ sec}$$

Seismic Performance Zone

The seismic performance zone (SPZ) is a function of the acceleration coefficient S_{D1} and is defined in AASHTO Sec. 3.10.6. The four categories are shown in Table 8.12; these determine the necessary requirements for selection of the design procedure, minimum support lengths, and substructure design details.

Table 8.12 Seismic Performance Zones

acceleration coefficient, S_{D1}	seismic zone
$S_{D1} \le 0.15$	1
$0.15 < S_{D1} \le 0.30$	2
$0.30 < S_{D1} \le 0.50$	3
$0.50 < S_{D1}$	4

Example 8.37

Determine the seismic zone for the bridge of Ex. 8.31.

Solution

From AASHTO Sec. 3.10.6, for a value of the acceleration coefficient exceeding 0.50, the relevant seismic performance zone is

$$\text{SPZ} = 4$$

Selection of Analysis Procedure

In accordance with AASHTO Sec. 4.7.4.3, the analysis procedure selected depends on the seismic zone, importance category, and on the bridge regularity. This information is summarized in Table 8.13. A *regular bridge* is defined as having fewer than seven spans with no abrupt changes in weight, stiffness, or geometry. An *irregular bridge* does not satisfy the definition of a regular bridge, and in this type of structure, the higher modes of vibration significantly affect the seismic response. A detailed seismic analysis is not required for single-span bridges. Minimum support lengths are required, however, to accommodate the maximum inelastic displacement, in accordance with AASHTO Sec. 4.7.4.4.

A seismic analysis is not required for bridges in seismic zone 1.

Table 8.13 Selection of Analysis Procedure for Multispan Bridges

seismic zone	other bridges regular	other bridges irregular	essential bridges regular	essential bridges irregular	critical bridges regular	critical bridges irregular
2	SM/UL	SM	SM/UL	MM	MM	MM
3	SM/UL	MM	MM	MM	MM	TH
4	SM/UL	MM	MM	MM	TH	TH

Note: UL = uniform load elastic method; SM = single-mode elastic method; MM = multimode elastic method; TH = time history method.

Example 8.38

Determine the required analysis procedure for the bridge of Ex. 8.31.

Solution

From Ex. 8.37, the seismic performance zone is 4. From AASHTO Sec. 4.7.4.3, for a regular bridge in seismic zone 4 with an importance category of "other," the required analysis procedure is UL or SM.

The Uniform Load Elastic Method

The uniform load elastic method is defined in AASHTO Sec. 4.7.4.3.2c as being suitable for regular bridges that respond principally in their fundamental mode. The method may be used for both transverse and longitudinal earthquake motions. The seven stages in the procedure are as follows.

1. Calculate the maximum lateral displacement $v_{s(\max)}$ due to a uniform unit load p_o as shown in Fig. 8.13 for a transverse load. The uniform load is resisted by the lateral stiffness of the superstructure and by the stiffness of the central column. The abutments are assumed to be rigid and to provide a pinned end restraint at each end of the superstructure. The maximum displacement and the corresponding reaction V_o in the column may be determined by the virtual work method.[17]

2. The bridge transverse stiffness is given by

$$K = \frac{p_o L}{v_{s(\max)}} \quad \textit{[AASHTO C4.7.4.3.2c-1]}$$

3. The total weight of the bridge superstructure and tributary substructure is

$$W = \int w(x)\,dx \quad \textit{[AASHTO C4.7.4.3.2c-2]}$$

Figure 8.13 Transverse Displacement Due to Unit Transverse Load

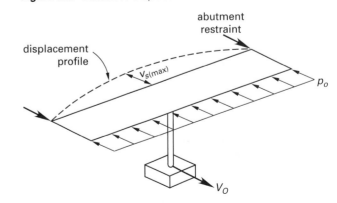

4. The fundamental period of the bridge is given by

$$T_m = 2\pi\sqrt{\frac{W}{gK}} \quad \text{[AASHTO C4.7.4.3.2c-3]}$$

5. The governing elastic seismic response coefficient, C_{sm}, is determined from AASHTO Sec. 3.10.4.2.

6. The uniform equivalent static seismic load is

$$p_e = \frac{C_{sm}W}{L} \quad \text{[AASHTO C4.7.4.3.2c-4]}$$

7. Apply p_e to the bridge as shown in Fig. 8.14 and determine the member forces due to the seismic load.

Figure 8.14 *Equivalent Static Seismic Load Applied to Bridge*

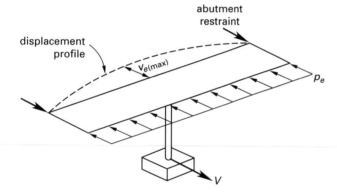

Example 8.39

Using the uniform load method, determine the elastic seismic design moment in the column, in the transverse direction, for the bridge of Ex. 8.31.

Solution

The stiffness of the column, fixed at the top and bottom, is given by

$$K_c = \frac{12E_cI_c}{h_c^3}$$

$$= \frac{(12)\left(450,000 \; \dfrac{\text{kips}}{\text{ft}^2}\right)(60 \; \text{ft}^4)}{(30 \; \text{ft})^3}$$

$$= 12,000 \; \text{kips/ft}$$

The transverse reaction in the column due to a uniform unit transverse load p_o on the superstructure is

$$V_o = v_{s(\text{max})}K_c$$

$$= v_{s(\text{max})}(12,000 \; \text{kips/ft})$$

The maximum transverse displacement of the superstructure alone due to a uniform unit transverse load p_o is

$$\delta_p = \frac{5p_oL^4}{384E_sI_s}$$

$$= \frac{(5)\left(1.0 \; \dfrac{\text{kip}}{\text{ft}}\right)(200 \; \text{ft})^4}{(384)\left(450,000 \; \dfrac{\text{kips}}{\text{ft}^2}\right)(4000 \; \text{ft}^4)}$$

$$= 0.0116 \; \text{ft}$$

The maximum transverse displacement of the superstructure due to the column reaction V_o is

$$\delta_v = -\frac{V_oL^3}{48E_sI_s}$$

$$= -\frac{v_{s(\text{max})}\left(12,000 \; \dfrac{\text{kips}}{\text{ft}}\right)L^3}{48E_sI_s}$$

$$= -\frac{v_{s(\text{max})}\left(12,000 \; \dfrac{\text{kips}}{\text{ft}}\right)(200 \; \text{ft})^3}{(48)\left(450,000 \; \dfrac{\text{kips}}{\text{ft}^2}\right)(4000 \; \text{ft}^4)}$$

$$= -v_{s(\text{max})}(1.1111)$$

The maximum transverse displacement of the superstructure due to p_o and V_o combined is

$$v_{s(\text{max})} = \delta_p + \delta_v$$

$$= 0.0116 \; \text{ft} - v_{s(\text{max})}(1.1111)$$

$$= 0.00548 \; \text{ft}$$

The bridge transverse stiffness is given by AASHTO Eq. (C4.7.4.3.2c-1) as

$$K = \frac{p_oL}{v_{s(\text{max})}} = \frac{\left(1.0 \; \dfrac{\text{kip}}{\text{ft}}\right)(200 \; \text{ft})}{(0.00548 \; \text{ft})\left(12 \; \dfrac{\text{in}}{\text{ft}}\right)}$$

$$= 3041 \; \text{kips/in} \quad (3040 \; \text{kips/in})$$

The total weight of the bridge superstructure and tributary substructure is given by AASHTO Eq. (C4.7.4.3.2c-2) as

$$W = \int w(x)dx = wL$$

$$= \left(10 \; \frac{\text{kips}}{\text{ft}}\right)(200 \; \text{ft})$$

$$= 2000 \; \text{kips}$$

The fundamental period of the bridge is given by AASHTO Eq. (C4.7.4.3.2c-3) as

$$T_m = 2\pi\sqrt{\frac{W}{gK}} = 0.32\sqrt{\frac{W}{K}}$$

$$= 0.32\ \frac{\text{sec}}{\text{in}^{-1}}\sqrt{\frac{2000\ \text{kips}}{3040\ \dfrac{\text{kips}}{\text{in}}}}$$

$$= 0.26\ \text{sec} > T_0$$

$$< T_S$$

The elastic seismic response coefficient is given by AASHTO Eq. (3.10.4.2-4) as

$$C_{sm} = S_{DS}$$

$$= 1.45$$

The uniform equivalent static seismic load is given by AASHTO Eq. (C4.7.4.3.2c-4) as

$$p_e = \frac{C_{sm}W}{L} = \frac{(1.45)(2000\ \text{kips})}{200\ \text{ft}}$$

$$= 14.5\ \text{kips/ft}$$

The maximum transverse displacement due to the equivalent seismic load is

$$v_{e(\max)} = \frac{p_e v_{s(\max)}}{p_o} = \frac{\left(14.5\ \dfrac{\text{kips}}{\text{ft}}\right)(0.00548\ \text{ft})}{1.0\ \dfrac{\text{kip}}{\text{ft}}}$$

$$= 0.079\ \text{ft}$$

The elastic transverse shear in the column is

$$V = v_{e(\max)}K_c$$

$$= (0.079)\left(12{,}000\ \frac{\text{kips}}{\text{ft}}\right)$$

$$= 948\ \text{kips}$$

The elastic transverse moment in the column is

$$M = \frac{V h_c}{2} = \frac{(948\ \text{kips})(30\ \text{ft})}{2}$$

$$= 14{,}220\ \text{ft-kips}$$

The Single-Mode Elastic Method

The fundamental period and the equivalent static force are obtained by using the technique detailed in AASHTO Sec. 4.7.4.3.2b. The method may be used for both transverse and longitudinal earthquake motions. The six stages in the procedure are as follows.

1. Calculate the static displacements $v_s(x)$ due to a uniform unit load p_o as shown in Fig. 8.15 for a

longitudinal load. The uniform load is resisted by the lateral stiffness of the central column, with the abutments assumed to provide no restraint.

Figure 8.15 *Longitudinal Displacement Due to Unit Longitudinal Load*

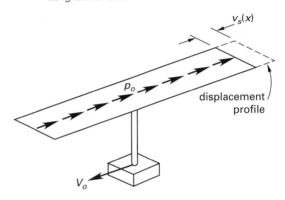

2. Calculate the factors α, β, and γ, which are given by

$$\alpha = \int v_s(x)\,dx \qquad \textit{[AASHTO C4.7.4.3.2b-1]}$$
$$\beta = \int w(x)v_s(x)\,dx \qquad \textit{[AASHTO C4.7.4.3.2b-2]}$$
$$\gamma = \int w(x)v_s^2(x)\,dx \qquad \textit{[AASHTO C4.7.4.3.2b-3]}$$

The limits of the integrals extend over the whole length of the bridge.

3. The fundamental period is given by

$$T_m = 2\pi\sqrt{\frac{\gamma}{p_o g\alpha}} \qquad \textit{[AASHTO C4.7.4.3.2b-4]}$$

4. The governing elastic seismic response coefficient, C_{sm}, is used to determine the elastic force in a member and is given by AASHTO Sec. 3.10.4.2.

5. The equivalent static seismic load is

$$p_e(x) = \frac{\beta C_{sm}w(x)v_s(x)}{\gamma} \qquad \textit{[AASHTO C4.7.4.3.2b-5]}$$

6. Apply $p_e(x)$ to the bridge as shown in Fig. 8.16 and determine the member forces due to the seismic load.

Figure 8.16 *Equivalent Static Seismic Load Applied to Bridge*

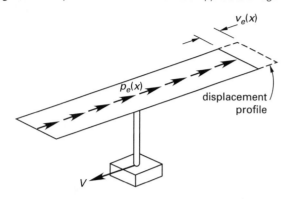

Example 8.40

Using the single-mode elastic method, determine the elastic seismic design moment in the column, in the longitudinal direction, for the bridge of Ex. 8.31.

Solution

The stiffness of the column, fixed at the top and bottom, is obtained from Ex. 8.39 as

$$K_c = 12{,}000 \text{ kips/ft}$$

Applying a uniform load of $p_o = 1.0$ kip/ft along the longitudinal axis of the bridge produces a longitudinal displacement of the superstructure of

$$v_s(x) = \frac{p_o L}{K_c} = \frac{\left(1.0 \ \dfrac{\text{kip}}{\text{ft}}\right)(200 \text{ ft})}{12{,}000 \ \dfrac{\text{kips}}{\text{ft}}}$$

$$= 0.0167 \text{ ft}$$

The factor α is given by AASHTO Eq. (C4.7.4.3.2b-1) as

$$\alpha = \int v_s(x)\,dx$$

$$= (0.0167 \text{ ft})(200 \text{ ft})$$

$$= 3.333 \text{ ft}^2$$

The factor β is given by AASHTO Eq. (C4.7.4.3.2b-2) as

$$\beta = \int w(x)v_s(x)\,dx$$

$$= \left(10 \ \frac{\text{kips}}{\text{ft}}\right)(3.333 \text{ ft}^2)$$

$$= 33.333 \text{ ft-kips}$$

The factor γ is given by AASHTO Eq. (C4.7.4.3.2b-3) as

$$\gamma = \int w(x)v_s^2(x)\,dx$$

$$= \left(10 \ \frac{\text{kips}}{\text{ft}}\right)(0.0167 \text{ ft})^2(200 \text{ ft})$$

$$= 0.557 \text{ ft}^2\text{-kip}$$

The fundamental period is given by AASHTO Eq. (C4.7.4.3.2b-4) as

$$T_m = 2\pi\sqrt{\frac{\gamma}{p_o g \alpha}}$$

$$= 2\pi\sqrt{\frac{0.557 \text{ ft}^2\text{-kip}}{\left(1.0 \ \dfrac{\text{kip}}{\text{ft}}\right)\left(32.2 \ \dfrac{\text{ft}}{\text{sec}^2}\right)(3.333 \text{ ft}^2)}}$$

$$= 0.45 \text{ sec} > T_0$$

$$< T_s$$

The elastic seismic response coefficient is given by AASHTO Eq. (3.10.4.2-4) as

$$C_{sm} = S_{DS}$$

$$= 1.45$$

The equivalent static seismic load is given by AASHTO Eq. (C4.7.4.3.2b-5).

$$p_e(x) = \frac{\beta C_{sm} w(x) v_s(x)}{\gamma}$$

$$= \frac{(33.333 \text{ ft-kips})(1.45)\left(10 \ \dfrac{\text{kips}}{\text{ft}}\right)(0.0167 \text{ ft})}{0.557 \text{ ft}^2\text{-kip}}$$

$$= 14.49 \text{ kips/ft}$$

The longitudinal displacement due to the equivalent seismic load is

$$v_e(x) = \frac{p_e(x) v_s(x)}{p_o}$$

$$= \frac{\left(14.49 \ \dfrac{\text{kips}}{\text{ft}}\right)(0.0167 \text{ ft})}{1.0 \ \dfrac{\text{kip}}{\text{ft}}}$$

$$= 0.242 \text{ ft}$$

The elastic shear in the column in the longitudinal direction is

$$V = v_e(x) K_c$$

$$= (0.242 \text{ ft})\left(12{,}000 \ \frac{\text{kips}}{\text{ft}}\right)$$

$$= 2904 \text{ kips}$$

The elastic moment in the column in the longitudinal direction is

$$M = \frac{V h_c}{2} = \frac{(2904 \text{ kips})(30 \text{ ft})}{2}$$

$$= 43{,}560 \text{ ft-kips}$$

Response Modification Factor

The seismic design force for a member is determined by dividing the elastic force by the response modification factor R. AASHTO Tables 3.10.7.1-1 and 3.10.7.1-2 lists the different structural systems and response modification factors. An abbreviated listing of response modification factors is provided in Tables 8.14 and 8.15.

Table 8.14 Response Modification Factors for Substructures

	importance category		
substructure	critical	essential	other
wall-type pier: strong axis	1.5	1.5	2.0
single column	1.5	2.0	3.0
multiple column bents	1.5	3.5	5.0

Table 8.15 Response Modification Factors for Connections

connection	all importance categories
superstructure to abutment	0.8
superstructure to column or pier	1.0
column or pier to foundation	1.0

Example 8.41

Determine the seismic design moment in the column, in the longitudinal and transverse direction, for the bridge of Ex. 8.31.

Solution

From Ex. 8.32, the importance category is "other." The response modification factor for a single column is given in Table 8.14 as
$$R = 3$$

The reduced design moment in the column in the longitudinal direction is

$$M_R = \frac{M}{R} = \frac{43{,}560 \text{ ft-kips}}{3} = 14{,}520 \text{ ft-kips}$$

The reduced transverse design moment in the column in the transverse direction is

$$M_R = \frac{M}{R} = \frac{14{,}220 \text{ ft-kips}}{3} = 4740 \text{ ft-kips}$$

Combination of Orthogonal Seismic Forces

AASHTO Sec. 3.10.8 requires the combination of orthogonal seismic forces to account for the directional uncertainty of earthquake motions and the simultaneous occurrence of earthquake forces in two perpendicular horizontal directions. Two load combinations are specified as follows.

- *load case 1:* 100% of the forces due to a seismic event in the longitudinal direction plus 30% of the forces due to a seismic event in the transverse direction

- *load case 2:* 100% of the forces due to a seismic event in the transverse direction plus 30% of the forces due to a seismic event in the longitudinal direction

Example 8.42

Determine the resultant seismic design moment in the column, due to the longitudinal and transverse forces, for the bridge of Ex. 8.31.

Solution

From Ex. 8.41, load case 1 governs, and the longitudinal moment is

$$M_x = (1.0)(14{,}520 \text{ ft-kips}) = 14{,}520 \text{ ft-kips}$$

The corresponding transverse moment is

$$M_y = (0.3)(4740 \text{ ft-kips}) = 1422 \text{ ft-kips}$$

For a circular column, the maximum resultant moment is given by

$$\begin{aligned} M_R &= \sqrt{(M_x)^2 + (M_y)^2} \\ &= \sqrt{(14{,}520 \text{ ft-kips})^2 + (1422 \text{ ft-kips})^2} \\ &= 14{,}589 \text{ ft-kips} \end{aligned}$$

Minimum Seat-Width Requirements

In accordance with AASHTO Sec. 4.7.4.4, minimum support lengths are required at the expansion ends of all girders as shown in Fig. 8.17. For seismic zone 1, with $A_s \geq 0.05$, the minimum support length in inches is given by

$$N = (8 + 0.02L + 0.08H)\left(1 + 0.000125S^2\right)$$

[AASHTO 4.7.4.4-1]

For seismic zone 1, with $A_s < 0.05$, the minimum support length is

$$N = (6 + 0.015L + 0.06H)(1 + 0.000125S^2)$$

For seismic zones 2, 3, and 4, the minimum support length is

$$N = (12 + 0.03L + 0.12H)\left(1 + 0.000125S^2\right)$$

The terms in these expressions are defined as follows.

L = length in feet of the bridge deck to the adjacent expansion joint or the end of the bridge deck

H = average height in feet of the columns

= 0 for a single-span bridge

S = angle of skew of the support in degrees measured from a line normal to the span

Figure 8.17 *Minimum Seat-Width Requirements*

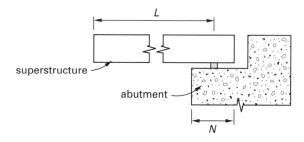

Example 8.43

Determine the minimum support length for the bridge of Ex. 8.31.

Solution

From Ex. 8.37, the seismic zone is 4.

From AASHTO Sec. 4.7.4.4, the minimum support length is given by

$$N = (12 + 0.03L + 0.12H)\left(1 + 0.000125S^2\right)$$
$$= \left(12 \text{ in} + \left(0.03\,\frac{\text{in}}{\text{ft}}\right)(200 \text{ ft}) + \left(0.12\,\frac{\text{in}}{\text{ft}}\right)(30 \text{ ft})\right)$$
$$\quad \times \left(1.0 + (0.000125)(0.0)^2\right)$$
$$= 21.6 \text{ in}$$

References

1. American Association of State Highway and Transportation Officials. *AASHTO LRFD Bridge Design Specifications*, 4th ed. 2007, with 2008 Interim Revisions.

2. American Institute of Steel Construction. *Moments, Shears, and Reactions: Continuous Highway Bridge Tables*. 1959.

3. Graudenz, H. *Bending Moment Coefficients in Continuous Beams*. Pitman. 1964.

4. Portland Cement Association. *Influence Lines Drawn as Deflection Curves*. 1948.

5. Williams, A. "The Determination of Influence Lines for Bridge Decks Monolithic with Their Piers." *Structural Engineer* (42). 1964, May.

6. Morice, P.B. and Little, G. *The Analysis of Right Bridge Decks Subjected to Abnormal Loading*. Cement and Concrete Association. 1973.

7. West, R. *Recommendations on the Use of Grillage Analysis for Slab and Pseudo-Slab Bridge Decks*. Cement and Concrete Association. 1973.

8. Loo, Y.C. and Cusens, A.R. "A Refined Finite Strip Method for the Analysis of Orthotropic Plates." *Proceedings Institution of Civil Engineers* (48). 1971, January.

9. Davis, J.D., Somerville, I.J., and Zienkiewicz, O.C. "Analysis of Various Types of Bridges by the Finite Element Method." *Proceedings of the Conference on Developments in Bridge Design and Construction, Cardiff, March 1971*. 1972.

10. Westergaard, H.M. "Computation of Stresses in Bridge Slabs Due to Wheel Loads." *Public Roads*. 1930, March.

11. Pucher, A. *Influence Surfaces of Elastic Plates*. Springer-Verlag. 1964.

12. Reynolds, C.E. and Steedman, J.C. *Reinforced Concrete Designers Handbook*. Cement and Concrete Association. 1981.

13. American Concrete Institute. *Building Code Requirements and Commentary for Structural Concrete, (ACI 318-05)*. 2005.

14. Hawkins, N.M., et al. *Simplified Shear Design of Structural Concrete Members*. NCHRP Report XXI, Transportation Research Board. 2005.

15. Federal Highway Administration. *Seismic Design and Retrofit Manual for Highway Bridges*. 1987.

16. Imbsen, R.A. "Seismic Design of Bridges." *Boston Society of Engineers, Fall Lecture Series*. BSCES. 1991.

17. Williams, A. *Structural Analysis in Theory and Practice*. Elsevier/International Code Council. 2009.

PRACTICE PROBLEMS

1. The reinforced concrete T-beam bridge shown in the following illustration is simply supported over a span of 40 ft. The deck has an overall width of 39 ft with five supporting beams, and has three 12 ft design lanes. The ratio of the modulus of elasticity of the beam and the deck slab is $n = 1.0$. The superimposed dead load on an interior beam due to surfacing is 0.25 kips/ft, and due to parapets is 0.2 kips/ft. For an interior beam, determine the bending moment produced by the permanent loads.

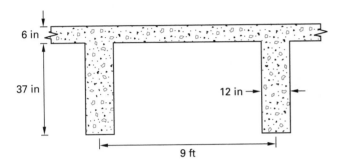

2. For the reinforced concrete T-beam bridge of Prob. 1, determine the maximum bending moment produced in an interior beam by the design lane load in combination with the design tandem.

For Prob. 3 through Prob. 5, assume that $M_C = 250$ ft-kips, $M_W = 50$ ft-kips, and $M_L = 614$ ft-kips.

3. For the reinforced concrete T-beam bridge of Prob. 1, what is the strength I limit state factored moment for design of a beam?

4. For the reinforced concrete T-beam bridge of Prob. 1, determine if the tensile reinforcement is satisfactory. The concrete strength is 4000 lbf/in², and the reinforcement consists of nine no. 9 grade 60 bars.

5. For the reinforced concrete T-beam bridge of Prob. 1, are the fatigue stress limits satisfactory? Ignore the effects of the 8 kip axle.

SOLUTIONS

1. The dead load acting on one beam due to the weight of the parapets, the weight of the deck slab, and the self-weight of the beam is

$$w_C = 0.20 \; \frac{\text{kips}}{\text{ft}}$$
$$+ \left(0.15 \; \frac{\text{kips}}{\text{ft}^3} \right) \left(\frac{(0.5 \text{ ft})(39 \text{ ft})}{5 \text{ beams}} + (1.0 \text{ ft})(3.08 \text{ ft}) \right)$$
$$= 1.25 \text{ kips/ft}$$

The bending moment produced in an interior beam at the center of the span by the parapets, deck slab, and beam self-weight is

$$M_C = \frac{w_C L^2}{8}$$
$$= \frac{\left(1.25 \; \frac{\text{kips}}{\text{ft}} \right) (40 \text{ ft})^2}{8}$$
$$= 250 \text{ ft-kips}$$

The bending moment produced in an interior beam at the center of the span by the surfacing is

$$M_W = \frac{w_W L^2}{8}$$
$$= \frac{\left(0.25 \; \frac{\text{kips}}{\text{ft}} \right) (40 \text{ ft})^2}{8}$$
$$= 50 \text{ ft-kips}$$

The total dead load bending moment produced in an interior beam at the center of the span is

$$M_D = M_C + M_W$$
$$= 250 \text{ ft-kips} + 50 \text{ ft-kips}$$
$$= 300 \text{ ft-kips}$$

2. The bending moment produced at the center of the span by the design lane load is

$$M_{LL} = \frac{w_{LL} L^2}{8}$$
$$= \frac{\left(0.64 \; \frac{\text{kips}}{\text{ft}} \right) (40 \text{ ft})^2}{8}$$
$$= 128 \text{ ft-kips}$$

As shown in the following illustration, the maximum moment due to the design tandem is produced under the lead axle of the design tandem when it is located 1 ft beyond the center of the span, and is given by

$$M_{DT} = \frac{(25 \text{ kips})(19 \text{ ft})(21 \text{ ft} + 17 \text{ ft})}{40 \text{ ft}}$$
$$= 451 \text{ ft-kips}$$

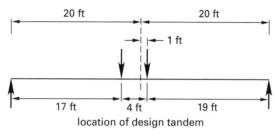

location of design tandem

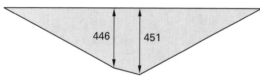

bending moment, ft-kips

The dynamic load allowance for the span moment is given by AASHTO Table 3.6.2.1-1 as

$$IM = 33\%$$

It is applied to the static axle loads of the design tandem and the dynamic factor to be applied is

$$I = 1 + \frac{IM}{100}$$
$$= 1 + \frac{33\%}{100}$$
$$= 1.33$$

The moment caused by the design tandem, including the dynamic load allowance, is

$$M = (1.33)(451 \text{ ft-kips})$$
$$= 600 \text{ ft-kips}$$

The combined moment produced by the design lane load and the design tandem, including the dynamic load allowance, is

$$M_{L+T} = 128 \text{ ft-kips} + 600 \text{ ft-kips}$$
$$= 728 \text{ ft-kips}$$

The ratio of the modulus of elasticity of the beam and the deck slab is given as

$$n = \frac{E_B}{E_D}$$
$$= 1.0$$

The moment of inertia of the beam is

$$I = \frac{bd^3}{12}$$
$$= \frac{(12 \text{ in})(37 \text{ in})^3}{12}$$
$$= 50,653 \text{ in}^4$$

The area of the beam is

$$A = bd$$
$$= (12 \text{ in})(37 \text{ in})$$
$$= 444 \text{ in}^2$$

The distance between the centers of gravity of the beam and the deck slab is

$$e_g = \frac{t_s + d}{2}$$
$$= \frac{6 \text{ in} + 37 \text{ in}}{2}$$
$$= 21.5 \text{ in}$$

The longitudinal stiffness parameter of the deck is defined by AASHTO Eq. 4.6.2.2.1-1 as

$$K_g = n(I + Ae_g^2)$$
$$= (1.0)\left(50,653 \text{ in}^4 + (444 \text{ in}^2)(21.5 \text{ in})^2\right)$$
$$= 255,892 \text{ in}^4 \quad \begin{bmatrix} \text{complies with AASHTO} \\ \text{Table 4.6.2.2b-1} \end{bmatrix}$$
$$> 10,000 \text{ in}^4$$
$$< 7,000,000 \text{ in}^4$$

The beam spacing is

$$S = 9 \text{ ft} \quad [\text{complies with AASHTO Table 4.6.2.2.2b-1}]$$
$$> 3.5 \text{ ft}$$
$$< 16.0 \text{ ft}$$

The deck slab thickness is

$$t_s = 6 \text{ in} \quad [\text{complies with AASHTO Table 4.6.2.2.2b-1}]$$
$$> 4.5 \text{ in}$$
$$< 12.0 \text{ in}$$

The superstructure span is

$$L = 40 \text{ ft} \quad [\text{complies with AASHTO Table 4.6.2.2.2b-1}]$$
$$> 20 \text{ ft}$$
$$< 240 \text{ ft}$$

The number of beams in the deck is

$N_b = 5$ [complies with AASHTO Table 4.6.2.2.2b-1]

Therefore, the configuration of the deck is in full conformity with the requirements of AASHTO Table 4.6.2.2.2b-1.

With one lane loaded, AASHTO Table 4.6.2.2.2b-1 gives the distribution factor for moment as

$$g_1 = 0.06 + \left(\frac{S}{14}\right)^{0.4} \left(\frac{S}{L}\right)^{0.3} \left(\frac{K_g}{12.0Lt_s^3}\right)^{0.1}$$

$$= 0.06 + \left(\frac{9\text{ ft}}{14}\right)^{0.4} \left(\frac{9\text{ ft}}{40\text{ ft}}\right)^{0.3}$$

$$\times \left(\frac{255{,}892\text{ in}^4}{(12.0)(40\text{ ft})(6\text{ in})^3}\right)^{0.1}$$

$$= 0.646$$

With two lanes loaded, as shown in the figure, AASHTO Table 4.6.2.2.2b-1 gives the distribution factor for moment as

$$g_m = 0.075 + \left(\frac{S}{9.5}\right)^{0.6} \left(\frac{S}{L}\right)^{0.2} \left(\frac{K_g}{12.0Lt_s^3}\right)^{0.1}$$

$$= 0.075 + \left(\frac{9\text{ ft}}{9.5}\right)^{0.6} \left(\frac{9\text{ ft}}{40\text{ ft}}\right)^{0.2}$$

$$\times \left(\frac{255{,}892\text{ in}^4}{(12.0)(40\text{ ft})(6\text{ in})^3}\right)^{0.1}$$

$$= 0.861 \quad \text{[governs]}$$

The live load moment for the design of an interior beam is

$$M_L = g_m M_{L+T}$$

$$= (0.861)(728\text{ ft-kips})$$

$$= 627\text{ ft-kips}$$

3. The relevant service level moments are

M_C = moment produced by the parapets, deck slab, and beam self-weight
  = 250 ft-kips
M_W = moment produced by the wearing surface
  = 50 ft-kips
M_L = moment produced by the design lane load and the design tandem, including the dynamic load allowance
  = 627 ft-kips

The factored design moment for the strength I limit state is given by AASHTO Eq. (3.4.1-1) and AASHTO Table 3.4.1-1 as

$$M_u = \eta_i(\gamma_p M_C + \gamma_p M_W + \gamma_L M_L)$$

$$= (1.0)(1.25M_C + 1.5M_W + 1.75M_L)$$

$$= (1.25)(250\text{ ft-kips}) + (1.5)(50\text{ ft-kips})$$

$$\quad + (1.75)(627\text{ ft-kips})$$

$$= 1485\text{ ft-kips}$$

4. The effective compression flange width is given by AASHTO Sec. 4.6.2.6.1 as

$$b = S$$

$$= (9\text{ ft})\left(12\ \frac{\text{in}}{\text{ft}}\right)$$

$$= 108\text{ in}$$

The height of the centroid of the tensile reinforcement is

$$\bar{c} = 2\text{ in} + (2.5)(1.125\text{ in})$$

$$= 4.81\text{ in}$$

The effective depth is

$$d = h - \bar{c}$$

$$= 43\text{ in} - 4.81\text{ in}$$

$$= 38.19\text{ in}$$

Assuming that the stress block lies within the flange, the required tension reinforcement is determined from the principles of AASHTO Sec. 5.7. The design moment factor is

$$K_u = \frac{M_u}{b_w d^2}$$

$$= \frac{(1485\text{ ft-kips})\left(12\ \frac{\text{in}}{\text{ft}}\right)\left(1000\ \frac{\text{lbf}}{\text{kip}}\right)}{(108\text{ in})(38.19\text{ in})^2}$$

$$= 113\text{ lbf/in}^2$$

$$\frac{K_u}{f_c'} = \frac{113\ \dfrac{\text{lbf}}{\text{in}^2}}{4000\ \dfrac{\text{lbf}}{\text{in}^2}}$$

$$= 0.0283$$

From Table A.1, in the appendix, the corresponding tension reinforcement index is

$$\omega = 0.0312$$

$$< 0.319\beta_1$$

$$= (0.319)(0.85)$$

$$= 0.271$$

Therefore, the section is tension controlled, and $\phi = 0.90$.

The required reinforcement ratio is

$$\rho = \frac{\omega f_c'}{f_y}$$

$$= \frac{(0.0312)\left(4000\ \dfrac{\text{lbf}}{\text{in}^2}\right)}{60{,}000\ \dfrac{\text{lbf}}{\text{in}^2}}$$

$$= 0.00208$$

The reinforcement area required is

$$A_s = \rho b d$$
$$= (0.00208)(108 \text{ in})(38.19 \text{ in})$$
$$= 8.58 \text{ in}^2$$

For nine no. 9 bars as shown in the following illustration, the reinforcement area provided is

$$A_s = 9 \text{ in}^2$$
$$> 8.58 \text{ in}^2 \quad \text{[satisfactory]}$$

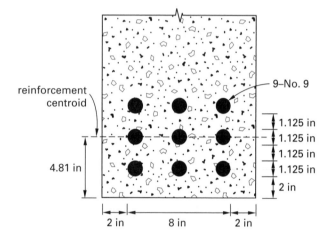

The stress block depth is

$$a = \frac{A_s f_y}{0.85 b f_c'}$$
$$= \frac{(9 \text{ in}^2)\left(60{,}000 \dfrac{\text{lbf}}{\text{in}^2}\right)}{(0.85)(108 \text{ in})\left(4000 \dfrac{\text{lbf}}{\text{in}^2}\right)}$$
$$= 1.47 \text{ in}$$
$$< h_f \quad \left[\begin{array}{l}\text{The stress block is contained}\\ \text{within the flange as assumed.}\end{array}\right]$$

5. The lever-arm for elastic design is conservatively obtained from Prob. 4 as

$$l_a = d - \frac{h_f}{2}$$
$$= 38.19 \text{ in} - \frac{6 \text{ in}}{2}$$
$$= 35.19 \text{ in}$$

Fatigue limits are determined using 75% of the stress produced by the design truck plus dynamic load allowance. The distance between the 32 kip axles of the design truck is fixed at 30 ft. Therefore, the maximum moment is developed at the location of the lead axle when this is positioned at the center of the span. The bending moment produced is

$$M_{DT} = \frac{WL}{4}$$
$$= \frac{(32 \text{ kips})(40 \text{ ft})}{4}$$
$$= 320 \text{ ft-kips}$$

The dynamic load allowance for the span moment is given by AASHTO Table 3.6.2.1-1 as

$$IM = 15\%$$

This is applied to the static axle loads of the design truck and the dynamic factor to be applied is

$$I = 1 + \frac{IM}{100}$$
$$= 1 + \frac{15\%}{100}$$
$$= 1.15$$

75% of the moment caused by the design truck, including the dynamic load allowance, is

$$M_f = (0.75)(1.15)M_{DT}$$
$$= (0.75)(1.15)(320 \text{ ft-kips})$$
$$= 276 \text{ ft-kips}$$

This is the maximum moment range producing fatigue. The corresponding maximum stress range is

$$f_f = \frac{M_f}{l_a A_s}$$
$$= \frac{(276 \text{ ft-kips})\left(12 \dfrac{\text{in}}{\text{ft}}\right)}{(35.19 \text{ in})(9 \text{ in}^2)}$$
$$= 10.46 \text{ kips/in}^2$$

The permanent dead load moment at the center of the span is derived in Prob. 1 as

$$M_D = 300 \text{ ft-kips}$$

This is the minimum moment at the center of the span. The corresponding minimum stress is

$$f_{\min} = \frac{M_D}{l_a A_s}$$
$$= \frac{(300 \text{ ft-kips})\left(12 \dfrac{\text{in}}{\text{ft}}\right)}{(35.19 \text{ in})(9 \text{ in}^2)}$$
$$= 11.37 \text{ kips/in}^2$$

The allowable range is given by AASHTO Eq. (5.5.3.2-1) as

$$
\begin{aligned}
f_{f(\text{all})} &= 23.4 - 0.33 f_{\min} \\
&= 23.4 - (0.33)\left(11.37\ \frac{\text{kips}}{\text{in}^2}\right) \\
&= 19.65\ \text{kips/in}^2 \\
&> f_f \quad [\text{satisfactory}]
\end{aligned}
$$

Appendices

Table A.1
Values of $M_u/f'_c bd^2$
For a tension-controlled section

ω	0.000	0.001	0.002	0.003	0.004	0.005	0.006	0.007	0.008	0.009
0	0.0000	0.0009	0.0018	0.0027	0.0036	0.0045	0.0054	0.0063	0.0072	0.0081
0.01	0.0089	0.0098	0.0107	0.0116	0.0125	0.0134	0.0143	0.0151	0.0160	0.0169
0.02	0.0178	0.0187	0.0195	0.0204	0.0213	0.0222	0.0230	0.0239	0.0248	0.0257
0.03	0.0265	0.0274	0.0283	0.0291	0.0300	0.0309	0.0317	0.0326	0.0334	0.0343
0.04	0.0352	0.0360	0.0369	0.0377	0.0386	0.0394	0.0403	0.0411	0.0420	0.0428
0.05	0.0437	0.0445	0.0454	0.0462	0.0471	0.0479	0.0487	0.0496	0.0504	0.0513
0.06	0.0521	0.0529	0.0538	0.0546	0.0554	0.0563	0.0571	0.0579	0.0588	0.0596
0.07	0.0604	0.0612	0.0621	0.0629	0.0637	0.0645	0.0653	0.0662	0.0670	0.0678
0.08	0.0686	0.0694	0.0702	0.0711	0.0719	0.0727	0.0735	0.0743	0.0751	0.0759
0.09	0.0767	0.0775	0.0783	0.0791	0.0799	0.0807	0.0815	0.0823	0.0831	0.0839
0.10	0.0847	0.0855	0.0863	0.0871	0.0879	0.0887	0.0895	0.0902	0.0910	0.0918
0.11	0.0926	0.0934	0.0942	0.0949	0.0957	0.0965	0.0973	0.0981	0.0988	0.0996
0.12	0.1004	0.1011	0.1019	0.1027	0.1035	0.1042	0.1050	0.1058	0.1065	0.1073
0.13	0.1081	0.1088	0.1096	0.1103	0.1111	0.1119	0.1126	0.1134	0.1141	0.1149
0.14	0.1156	0.1164	0.1171	0.1179	0.1186	0.1194	0.1201	0.1209	0.1216	0.1223
0.15	0.1231	0.1238	0.1246	0.1253	0.1260	0.1268	0.1275	0.1283	0.1290	0.1297
0.16	0.1304	0.1312	0.1319	0.1326	0.1334	0.1341	0.1348	0.1355	0.1363	0.1370
0.17	0.1377	0.1384	0.1391	0.1399	0.1406	0.1413	0.1420	0.1427	0.1434	0.1441
0.18	0.1448	0.1456	0.1463	0.1470	0.1477	0.1484	0.1491	0.1498	0.1505	0.1512
0.19	0.1519	0.1526	0.1533	0.1540	0.1547	0.1554	0.1561	0.1568	0.1574	0.1581
0.20	0.1588	0.1595	0.1602	0.1609	0.1616	0.1623	0.1629	0.1636	0.1643	0.1650
0.21	0.1657	0.1663	0.1670	0.1677	0.1684	0.1690	0.1697	0.1704	0.1710	0.1717
0.22	0.1724	0.1730	0.1737	0.1744	0.1750	0.1757	0.1764	0.1770	0.1777	0.1783
0.23	0.1790	0.1797	0.1803	0.1810	0.1816	0.1823	0.1829	0.1836	0.1842	0.1849
0.24	0.1855	0.1862	0.1868	0.1874	0.1881	0.1887	0.1894	0.1900	0.1906	0.1913
0.25	0.1919	0.1925	0.1932	0.1938	0.1944	0.1951	0.1957	0.1963	0.1970	0.1976
0.26	0.1982	0.1988	0.1995	0.2001	0.2007	0.2013	0.2019	0.2026	0.2032	0.2038
0.27	0.2044	0.2050	0.2056	0.2062	0.2069	0.2075	0.2081	0.2087	0.2093	0.2099
0.28	0.2105	0.2111	0.2117	0.2123	0.2129	0.2135	0.2141	0.2147	0.2153	0.2159
0.29	0.2165	0.2171	0.2177	0.2183	0.2188	0.2194	0.2200	0.2206	0.2212	0.2218
0.30	0.2224	0.2229	0.2235	0.2241	0.2247	0.2253	0.2258	0.2264	0.2270	0.2276
0.31	0.2281	0.2287	0.2293	0.2298	0.2304	0.2310	0.2315	0.2321	0.2327	0.2332
0.32	0.2338	0.2344	0.2349	0.2355	0.2360	0.2366	0.2371	0.2377	0.2382	0.2388
0.33	0.2393	0.2399	0.2404	0.2410	0.2415	0.2421	0.2426	0.2432	0.2437	0.2443
0.34	0.2448	0.2453	0.2459	0.2464	0.2470	0.2475	0.2480	0.2486	0.2491	0.2496
0.35	0.2501	0.2507	0.2512	0.2517	0.2523	0.2528	0.2533	0.2538	0.2543	0.2549
0.36	0.2554	0.2559	0.2564	0.2569	0.2575	0.2580	0.2585	0.2590	0.2595	0.2600
0.37	0.2605	0.2610	0.2615	0.2620	0.2625	0.2631	0.2636	0.2641	0.2646	0.2651
0.38	0.2656	0.2661	0.2665	0.2670	0.2675	0.2680	0.2685	0.2690	0.2695	0.2700
0.39	0.2705	0.2710	0.2715	0.2719	0.2724	0.2729	0.2734	0.2739	0.2743	0.2748

Table A.2
Values of the Neutral Axis Factor, k

ρn	0.000	0.001	0.002	0.003	0.004	0.005	0.006	0.007	0.008	0.009
0	0.0000	0.0437	0.0613	0.0745	0.0855	0.0951	0.1037	0.1115	0.1187	0.1255
0.01	0.1318	0.1377	0.1434	0.1488	0.1539	0.1589	0.1636	0.1682	0.1726	0.1769
0.02	0.1810	0.1850	0.1889	0.1927	0.1964	0.2000	0.2035	0.2069	0.2103	0.2136
0.03	0.2168	0.2199	0.2230	0.2260	0.2290	0.2319	0.2347	0.2375	0.2403	0.2430
0.04	0.2457	0.2483	0.2509	0.2534	0.2559	0.2584	0.2608	0.2632	0.2655	0.2679
0.05	0.2702	0.2724	0.2747	0.2769	0.2790	0.2812	0.2833	0.2854	0.2875	0.2895
0.06	0.2916	0.2936	0.2956	0.2975	0.2995	0.3014	0.3033	0.3051	0.3070	0.3088
0.07	0.3107	0.3125	0.3142	0.3160	0.3178	0.3195	0.3212	0.3229	0.3246	0.3263
0.08	0.3279	0.3296	0.3312	0.3328	0.3344	0.3360	0.3376	0.3391	0.3407	0.3422
0.09	0.3437	0.3452	0.3467	0.3482	0.3497	0.3511	0.3526	0.3540	0.3554	0.3569
0.10	0.3583	0.3597	0.3610	0.3624	0.3638	0.3651	0.3665	0.3678	0.3691	0.3705
0.11	0.3718	0.3731	0.3744	0.3756	0.3769	0.3782	0.3794	0.3807	0.3819	0.3832
0.12	0.3844	0.3856	0.3868	0.3880	0.3892	0.3904	0.3916	0.3927	0.3939	0.3951
0.13	0.3962	0.3974	0.3985	0.3996	0.4007	0.4019	0.4030	0.4041	0.4052	0.4063
0.14	0.4074	0.4084	0.4095	0.4106	0.4116	0.4127	0.4137	0.4148	0.4158	0.4169
0.15	0.4179	0.4189	0.4199	0.4209	0.4219	0.4229	0.4239	0.4249	0.4259	0.4269
0.16	0.4279	0.4288	0.4298	0.4308	0.4317	0.4327	0.4336	0.4346	0.4355	0.4364
0.17	0.4374	0.4383	0.4392	0.4401	0.4410	0.4419	0.4429	0.4437	0.4446	0.4455
0.18	0.4464	0.4473	0.4482	0.4491	0.4499	0.4508	0.4516	0.4525	0.4534	0.4542
0.19	0.4551	0.4559	0.4567	0.4576	0.4584	0.4592	0.4601	0.4609	0.4617	0.4625
0.20	0.4633	0.4641	0.4649	0.4657	0.4665	0.4673	0.4681	0.4689	0.4697	0.4705
0.21	0.4712	0.4720	0.4728	0.4736	0.4743	0.4751	0.4758	0.4766	0.4774	0.4781
0.22	0.4789	0.4796	0.4803	0.4811	0.4818	0.4825	0.4833	0.4840	0.4847	0.4855
0.23	0.4862	0.4869	0.4876	0.4883	0.4890	0.4897	0.4904	0.4911	0.4918	0.4925
0.24	0.4932	0.4939	0.4946	0.4953	0.4960	0.4966	0.4973	0.4980	0.4987	0.4993
0.25	0.5000	0.5007	0.5013	0.5020	0.5026	0.5033	0.5040	0.5046	0.5053	0.5059
0.26	0.5066	0.5072	0.5078	0.5085	0.5091	0.5097	0.5104	0.5110	0.5116	0.5123
0.27	0.5129	0.5135	0.5141	0.5147	0.5154	0.5160	0.5166	0.5172	0.5178	0.5184
0.28	0.5190	0.5196	0.5202	0.5208	0.5214	0.5220	0.5226	0.5232	0.5238	0.5243
0.29	0.5249	0.5255	0.5261	0.5267	0.5272	0.5278	0.5284	0.5290	0.5295	0.5301
0.30	0.5307	0.5312	0.5318	0.5323	0.5329	0.5335	0.5340	0.5346	0.5351	0.5357
0.31	0.5362	0.5368	0.5373	0.5379	0.5384	0.5389	0.5395	0.5400	0.5406	0.5411
0.32	0.5416	0.5422	0.5427	0.5432	0.5437	0.5443	0.5448	0.5453	0.5458	0.5464
0.33	0.5469	0.5474	0.5479	0.5484	0.5489	0.5494	0.5499	0.5505	0.5510	0.5515
0.34	0.5520	0.5525	0.5530	0.5535	0.5540	0.5545	0.5550	0.5554	0.5559	0.5564
0.35	0.5569	0.5574	0.5579	0.5584	0.5589	0.5593	0.5598	0.5603	0.5608	0.5613
0.36	0.5617	0.5622	0.5627	0.5632	0.5636	0.5641	0.5646	0.5650	0.5655	0.5660
0.37	0.5664	0.5669	0.5674	0.5678	0.5683	0.5687	0.5692	0.5696	0.5701	0.5705
0.38	0.5710	0.5714	0.5719	0.5723	0.5728	0.5732	0.5737	0.5741	0.5746	0.5750
0.39	0.5755	0.5759	0.5763	0.5768	0.5772	0.5776	0.5781	0.5785	0.5789	0.5794

Figure A.1
Interaction Diagram: Tied Circular Column
($f'_c = 4$ kips/in^2, $f_y = 60$ kips/in^2, $\gamma = 0.60$)

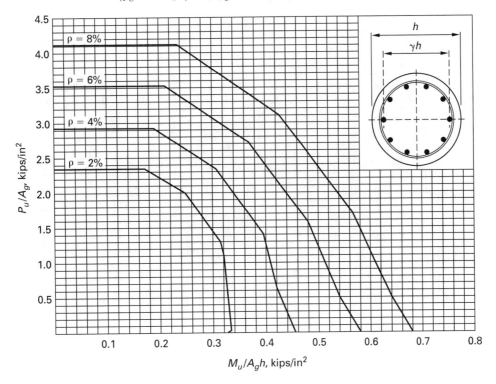

Figure A.2
Interaction Diagram: Tied Circular Column
($f'_c = 4$ kips/in^2, $f_y = 60$ kips/in^2, $\gamma = 0.75$)

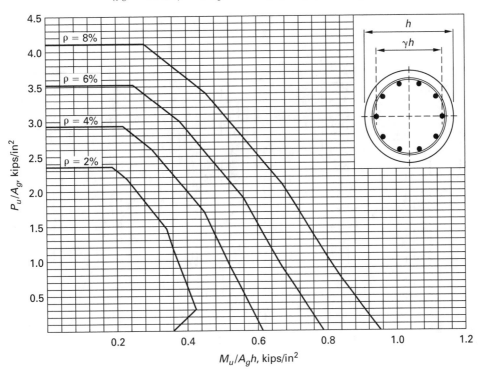

Figure A.3
Interaction Diagram: Tied Circular Column
($f'_c = 4$ kips/in², $f_y = 60$ kips/in², $\gamma = 0.90$)

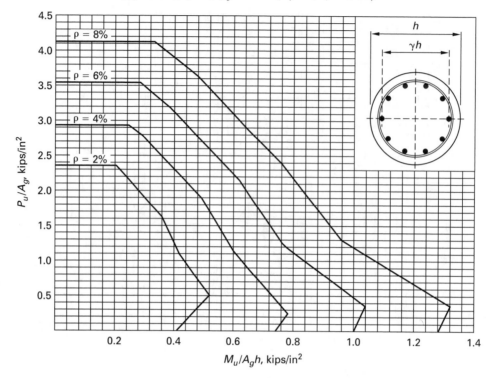

Figure A.4
Interaction Diagram: Tied Square Column
($f'_c = 4$ kips/in², $f_y = 60$ kips/in², $\gamma = 0.60$)

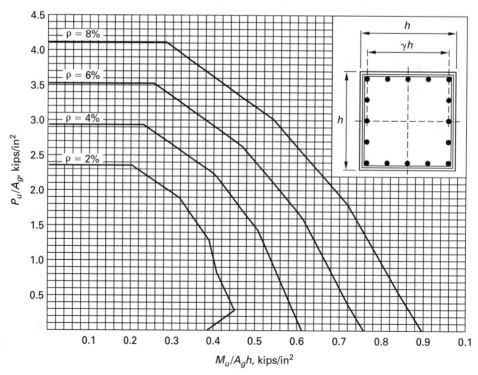

Figure A.5
Interaction Diagram: Tied Square Column
$(f'_c = 4 \text{ kips/in}^2, \ f_y = 60 \text{ kips/in}^2, \ \gamma = 0.75)$

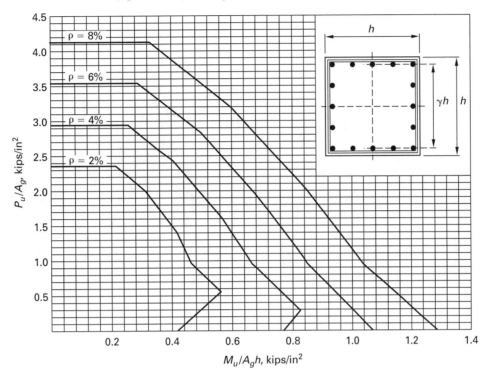

Figure A.6
Interaction Diagram: Tied Square Column
$(f'_c = 4 \text{ kips/in}^2, \ f_y = 60 \text{ kips/in}^2, \ \gamma = 0.90)$

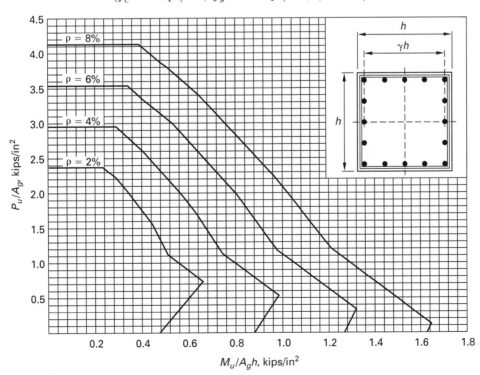

Index

Italicized page numbers represent sources of data in tables, figures, and appendices.

A

Abutment, rigid, 8-47
Acceleration
    adjusted response, 7-3
    coefficient, *8-47*
    earthquake spectral response, 7-2
    ground, effective peak, 7-3
    maximum considered, 7-2, 7-3
    parameters, design spectral response, 7-3
Accidental eccentricity, 6-7
Active earth pressure, total, 2-21
Adjusted modulus of elasticity, 5-3
Adjusted response
    accelerations, 7-3
    parameters, 8-46
Adjustment factor, 5-2–5-6, 5-18, 8-40–8-43
    for connection, *5-18*
Aggregate concrete factor, lightweight, 1-33
Alignment chart, 1-24, 4-11, 4-12, *4-13*
    effective length factor, *4-12*
    for *k*, *1-29*
Allowable axial load, 6-6
    maximum, 5-12
Allowable design value, 5-21
    in compression, 5-12
Allowable pressure, 2-12, 2-17
Allowable range of shear, 8-38
Allowable story drift, maximum, *7-24*
Allowable stress, 8-22, 8-23
    compressive, 6-1, 6-2, 6-6, 6-8
    design, 8-42
    range, 4-33
    shear, 6-5, 6-11, 6-12, 6-13
    tensile, 6-1
Allowance, dynamic load, 8-4, 8-5
Alternative tandem loading, 8-2, 8-3
Alternative tendon profile, 3-25, *3-26*
Ambient relative humidity, 3-20
Amplification factor
    acceleration based, 7-3
    deflection, 7-5
    long-period, 7-3
    short-period, 7-3
    sidesway, 4-16, 4-18, 4-19, 4-20
    velocity-based, 7-3
Analysis
    direct, 4-16, 4-18
    effective length method, 4-16, 4-22
    elastic, 6-1
    first-order, 4-16, 4-17, 4-19, 4-20, 4-21
    of procedure selection, 8-44, *8-44*
    procedure, seismic design, 8-43, *8-43*
    second-order, 4-16, 4-18
    serviceability limit state, 8-11
    simplified method, 4-16, 4-21
    strength limit state, 8-11
Anchor
    bar, *1-23*
    seating loss, 3-18

set, 3-2, 3-18
slip, 3-18
Anchorage
    length, *2-3*, 2-4
    post-tensioning, 3-2
Angle
    inclination, 1-18
    skew, 8-52
Applied load
    external, unfactored, 3-14
    factored, *2-13*
    factored column, 2-12
    net factored, 2-1
    service, *2-12*, *2-17*, *2-21*, 6-2
    superimposed, 3-23, 3-24
Applied moment, 6-8
    bending, 6-8
Applied shear force, 6-11, 8-36
    factored, 3-13, 8-20, 8-27
Area
    bearing, factor, 5-2
    closed stirrup, 3-15
    effective, 4-38
    effective net, 4-29–4-31
    elastic unit area, 4-35
    gross, 4-30
    horizontal reinforcement, minimum, 1-19
    longitudinal reinforcement, 1-25, 3-15
    reinforcement, 1-7, 1-8, 1-23, 1-24, 2-4,
        2-10, 2-15
    reinforcement, horizontal, *6-10*, *6-11*
    reinforcement, minimum, 2-8
    reinforcement, rectangular footing, *2-10*
    reinforcement, transformed, *6-8*, *6-9*
    reinforcement, vertical, *6-10*, *6-11*
    shear reinforcement, 6-5, 6-13
    shear reinforcement, minimum, 1-16,
        1-19, 3-13, 8-31
    stiffener, required, 4-43
    stirrups, minimum combined, 1-25
    tension reinforcement, 1-23
    transformed, of concrete slab, 8-38
    tributary, *1-45*
    vertical reinforcement, minimum, 1-19
    web, 4-8
Auxiliary reinforcement, 3-4, 3-7–3-9
Axial compression and flexure, combined,
    5-14, 5-15
Axial compression, in columns, 6-6
Axial compression load, concentric, 5-14
Axial load, 4-28, 4-29, 5-11, 6-7
    allowable, 6-6
    and flexure, combined, 4-22, 4-28
    capacity, short column, 1-26, 1-27
    factored, 1-31
    maximum allowable, 5-12
    stress, *6-8*
Axially loaded member, 4-12
Axial strength, 4-29
    design, 4-12, 4-13

Axis, neutral, 3-10
    depth factor, 1-15, 6-2
    factor, *A-2*
    plastic, 8-34

B

B-region, 1-19, *1-19*
Balanced
    design, 1-14
    strain, 1-5
Balancing load, 3-25, *3-26*
Bar, 1-27, 1-28
    anchor, *1-23*
    bent-up, 1-18
    bundled, 1-35, 1-37, 1-40
    diameter, 1-39
    framing, *1-23*
    Grade 60, 2-4
    hooked, in tension, 1-37
    hooked, tie spacing, 1-38
    inclined, 1-18
    inclined, beam, *1-18*
    longitudinal, 1-27
    straight, in compression, 1-36
    straight, in tension, 1-35
Base
    design, counterfort, 2-26
    plate, column, 4-23
    plate, on steel column, *2-7*, 2-7
    plate, thickness, required, 4-23
    pyramid, *2-9*
    shear distribution, 7-10, 7-18
    shear, seismic, 7-9
Basic design value, 8-40
Beam (*see also type*)
    and slab equivalent dimensions, *1-38*
    both ends continuous, *1-12*
    cantilever, *1-12*
    composite, 4-49, 4-50, 8-34, 8-35, *8-35*
    compression-controlled, 1-3, 1-4
    continuous, 1-38, 3-27, 3-28, 4-7, 6-2
    coped, 4-8
    deep, 1-19, *1-19*
    design requirements, 4-17
    doubly symmetric, nonhybrid, 4-33, 4-34
    edge, column strip moment distribution,
        1-43
    flanged, 1-8, *1-9*
    framing into column, 1-39
    framing into girder, 1-39
    mechanism, 4-26, 4-27
    noncomposite, 8-27
    notched, 5-9
    one end continuous, *1-12*
    rectangular, 5-9, 8-15
    rectangular, nominal flexural strength, 1-3
    reinforced concrete, 1-3, 1-7, 8-13
    simply supported, *1-12*, 6-2
    span, 5-6
    stability factor, 5-2, 5-5–5-7, 8-41, 8-43

Italicized page numbers represent sources of data in tables, figures, and appendices.

steel, 8-34, 8-35
strap, 2-16, *2-17*, 2-18, 2-19
tension-controlled, 1-3, 1-4
two-span, 3-27
web shear, 4-8
with compression reinforcement, *1-7*
with inclined bars, *1-18*
with inclined stirrups, *1-16*
W shape, rolled, 4-8
Bearing
  area factor, 5-2
  capacity, 2-8
  length, 8-43
  of bolts, 4-35
  on footing concrete, *2-9*
  plate, *1-23*, 4-9
  pressure, soil, 2-12, *2-17*, 2-18
  pressure, uniform, *2-12*
  reaction, factored, 4-48
  stiffener design, 4-48
  strength, 2-9
  strength, design, 4-27, 4-48
  strength, nominal, 4-34
  wall system, 7-5, *7-6*
Bearing-type bolt, 4-34, 4-35
Bending
  capacity, section, 5-9
  coefficient, 4-4, 4-5, 4-10
  coefficient, typical value, *4-5*
  design value, 5-3, 5-6–5-8
  design value, biaxial, 4-7, 6-4
  design value, reference, 5-2, 8-40, 8-41
  load, 5-14
  moment, 2-10, 2-25, 3-14, 4-7, 4-14, 8-4
  moment, applied, 6-8
  moment, diagram, 3-28
  moment, non-uniform, 1-27
  moment stress, *6-8*
  moment, uniform, 4-4, 4-5
  plastic moment of resistance, 8-40
  strength, nominal, 8-41
  stress (*see* Flexural strength)
Bent-up bar, 1-18
Biaxial bending, 4-7, 6-4
Block
  equivalent rectangular stress, depth of,
    1-3
  rectangular stress, 1-3, *1-4*, 1-8
  shear, 4-8
  stress, 8-25
  stress, depth, 4-50
Bolt
  bearing-type, 4-34, 4-35
  edge distance, 5-19, 5-20
  end distance, 5-19, 5-20
  group, eccentrically loaded, 4-35, 4-37
  in bearing, 4-34
  pitch, 4-33
  slip-critical, 4-34
  spacing, 5-19, 5-20
  spacing for full design values, *5-20*
Bolted connection, 4-3, 4-33, 4-35, *5-10*,
  5-19, *5-20*,
Bonded reinforcement, 8-23, 8-25
  auxiliary, 3-4
Bonded tendon, 3-8, 8-23, 8-25
Bond stress, 1-39
Bridge
  category, 8-43
  deck, 8-1
  deck, length, 8-52
  deck system, 8-5

essential, 8-45, *8-47*
irregular, 8-47, *8-47*
non-essential, 8-45
other, 8-45
regular, 8-47, *8-47*
single-span, 8-47
transverse stiffness, 8-47
type, 8-44
Buckling
  capacity, web, 4-48
  coefficient, Euler, 8-40
  design value, critical, 5-3, 8-40
  flange local, 4-44, 4-45
  inelastic, 4-44
  lateral-torsional, 4-44
  length coefficient, 5-11, *5-12*
  torsional, lateral, 4-3, 4-5
  web, sidesway, 4-48
Building
  configuration requirements, 7-10
  frame system, 7-5, *7-6*
  systems, 7-5
Built-up section, 4-13
Bundled bar, 1-31, 1-32, 1-36
Bundle, equivalent diameter, 1-31

C
Cable, concordant, 3-28
Cantilever
  beam, *1-12*
  moment, 2-26
  retaining wall, 2-20–2-23
  retaining wall with applied service loads,
    2-21
  slab, *1-12*
Cantilevered column structure, 7-5
Capacity (*see also* Strength)
  bearing, 2-9
  bending, section, 5-10
  design resistance, 8-6
  flexural, 1-38, 1-40
  moment, full plastic, 4-4
  reinforcement, 2-9
  shear, 1-38, 1-45
  shear, combined, 3-13, 8-29
  shear, design, 4-39
  shear, nominal, 3-12, 3-14, 4-45, 4-46,
    8-20, 8-28
  shear, punching, 2-6
  shear, section, 5-9
  torsional, nominal, 3-16
  web buckling, 4-48
Category
  C, seismic, 6-10, *6-10*
  D, seismic, 6-10, *6-10*, 6-11
  E, seismic, 6-10, *6-10*
  F, seismic, 6-10, *6-10*
  importance, 8-43, 8-44
  occupancy, 7-4
  seismic design, 6-10, *6-10*, *6-11*, 7-4, *7-5*
  stress, 4-24
Center of rotation, instantaneous, method,
  4-36, 4-40
Central column stiffness, 8-49
Chart
  alignment, 1-28, 4-11, 4-12
  alignment, effective length factor, *4-12*
  alignment, for *k*, *1-29*
Chord modulus of elasticity, 6-1
Circular column, concrete, 1-27
Class A splice, 1-40
Class A tension lap splice, 1-40

Class B tension lap splice, 1-40
Classification
  occupancy category, 7-4, *7-5*
  site, 7-2
  structural system, 7-5
Clear spacing between spirals, 1-27
Clear span, 1-19, *1-39*, 5-6, 6-2, 8-43
Closed stirrups, *1-23*, 1-24, *1-25*, 3-15
  (*see also* Closed ties)
Closed ties, 1-23
C_m factor, 4-15, 4-17, 4-19
Coefficient
  acceleration, 8-43
  bending, 4-4, 4-5, 4-10
  bending, typical value, *4-5*
  buckling length, 5-12, *5-12*
  elastic seismic response, 8-46, 8-48
  friction, 3-23, 8-32
  reduction, 4-18, 4-30, 4-31
  response modification, 7-5, *7-6*
  seismic response, 7-9
  shear, 4-38
  site, 7-3, *7-3*, 8-43
  stability, 7-13
  stud group, 4-51–4-53
  stud position, 4-51–4-53
  upper limit on building period, 7-7, 7-8,
    7-8
  web buckling, 4-46
  wobble, 8-32
Collapse
  mechanism, 4-27
  mechanism, partial, 4-24
Column
  axial compression, 6-6
  base plate, 4-23
  bent, single curvature, 1-26
  cantilevered, 7-5
  central, stiffness, 8-47, 8-49
  circular, concrete, 1-28
  composite, 4-15
  concrete, 1-27–1-34
  concrete, effective length, 1-28
  concrete, radius of gyration, 1-28
  concrete, rectangular, 1-28
  concrete, slenderness ratio, 1-27, 1-28,
    1-29, 1-30
  concrete, stiffness ratio, 1-28
  concrete, with spirals, 1-27, 1-30
  concrete, with ties, 1-27
  corner, *1-46*
  design requirement, 4-28
  dimension limitations, 6-7
  edge *1-46*
  effective length, 1-28, 4-14
  end, *1-39*
  end moment, 1-31, 1-34
  in rigid frame, 4-11
  interior, *1-39*
  isolated, rectangular footing, 2-10
  isolated, square footing, 2-5
  load, factored applied, 2-12
  load, table, 4-14
  long, concrete, 1-31, 1-33
  long, with sway, 1-33
  masonry, 6-6
  nonsway, 1-31, 1-32
  nonsway, concrete, 1-28
  parameter, 5-3
  reinforcement limitations, 6-7
  short, concrete, 1-30, 1-31
  short, design axial load capacity, 1-30

Italicized page numbers represent sources of data in tables, figures, and appendices.

short, lateral tie reinforcement, 1-30
short, spiral reinforcement, 1-30
slenderness ratio, 4-11, 4-13, 4-28
spirally reinforced, 1-41
stability factor, 5-3
steel, with base plate, 2-7
strip, 1-43, *1-43*, 1-44
strip, exterior negative moment
  distribution, *1-43*
strip, interior negative moment
  distribution, *1-43*
strip moment, 1-43
strip moments, distribution to edge beam,
  1-43
strip, positive moment distribution, *1-43*
tie, *1-28*
typical, *1-43*
Combination
  load, 8-51
  of orthogonal seismic forces, 8-51
Combined area, minimum, stirrups, 3-16
Combined compression and flexure, 6-8
  axial, 5-14, 5-15
Combined footing, 2-12–2-16
  with applied factored loads, *2-13*
  with applied service loads, *2-12*
Combined load, 7-13
  axial (compression) and flexure, 4-22,
    4-28
  lateral and withdrawal, 5-22–5-24
Combined mechanism, 4-26, 4-27
Combined shear capacity, 3-13, 8-20, 8-29
Combined stresses, 6-4, *6-8*
Compactness criteria, 4-3
Compact section, 4-4
Compatibility
  of displacements, 3-28
  strain, 3-10
Complete-penetration groove weld, 4-38
Composite
  action, 4-51
  beam, 4-49, 8-27
  column, 4-15
  construction, 3-21
  section, 3-21, 3-23, 3-24, 8-28
Compression
  and flexure, combined, 4-22, 4-28, 6-7
  axial, in columns, 6-6
  design, 5-11
  design value, allowable, 5-12
  design value, reference, 5-3
  load, axial, concentric, 5-14
  member, design, 4-10, 4-11
  member, laced, 4-14
  reinforcement, 1-7, 3-8
  reinforcement, beam, *1-7*
  test, 6-1
  zone, 8-35
  zone depth, 8-35
Compressive
  force, 3-10
  strength, 8-14
  strength, maximum, 6-1
  stress, 1-16, 3-13, 3-14, 3-20, 8-13, 8-29
  stress, allowable, 6-1, 6-2, 6-6, 6-8
  stress, at centroid of prestressing steel,
    8-32
  stress, calculated, 6-8
Concentrated load, 4-9, 4-10
Concentric axial compression load, 5-14

Concordant
  cable, 3-28
  profile, 3-28
Concrete
  at transfer, 8-32
  beam, reinforced, 1-3, 1-8
  beam, with compression reinforcement,
    1-7
  column, 1-27–1-34
  column, circular, 1-27
  column, effective length, 1-28
  column, long, 1-31–1-34
  column, long, with sway, 1-33
  column, nonsway, 1-28
  column, radius of gyration, 1-28
  column, rectangular, 1-28
  column, short, 1-30, 1-31
  column, short, lateral tie reinforcement,
    1-30
  column, short, spiral reinforcement, 1-30
  column, slenderness ratio, 1-28, 1-30,
    1-31, 1-33
  column, stiffness ratio, 1-28
  column, with spirals, 1-27, 1-30
  column, with ties, 1-27
  creep, 8-32
  design, reinforced, 8-14
  factor, lightweight aggregate, 1-37
  footing, bearing on, *2-9*
  modulus of elasticity, 8-16, 8-32
  post-tensioned, 3-2
  prestressed, 8-21
  section, gross, 8-16
  slab, 8-11
  slab, design, 8-11
  slab, transformed area, 8-37
  slab, transformed width, 8-37
  slab, width, effective, 8-35
  stress, 3-25, 8-32
  wall, *2-3*
Condition
  of use, 8-40
  of use, normal, 8-40
  restraint, 4-11
  service load, 8-5, 8-23
  ultimate load, 8-25
Confinement factor, 1-37
Connection
  adjustment factor, 5-18
  bolted, 4-30, 4-33, 5-10, 5-19
  common wire nail, 5-24
  design, 5-17–5-19
  eccentricity, 4-23
  lag screw, *5-18*, 5-21
  length, 4-24
  nail, 5-24, 5-25
  nailed, *5-18*
  rupture, 4-29, 4-31, 4-33
  screwed, *5-18*
  shear, 4-41, 8-29
  shear, double, 5-20
  shear plate, *5-18*, 5-22
  shear, single, 5-19, 5-20, 5-23
  spike, 5-24
  split ring, *5-18*, 5-22
  steel side plate, 5-22
  toe-nailed, 5-24, 5-25
  welded, 4-31, 4-32
  welded, design, 4-38
  wood screw, 5-23, 5-24
  wood side member, 5-22

Connector (*see* Connection)
  number required, 8-39
  shear, 4-50–4-53
  spacing, required, 8-38
Constant gravity, 8-44
Construction
  composite, 3-21
  diaphragm, 5-24
  non-propped, 3-24
  propped, 3-24
  shored, 4-53
  unshored, 4-53
Continuing reinforcement, 1-38
Continuous
  beam, 1-38, 3-28, 4-7, 6-2
  span, 8-2, *8-5*
Control
  cracking, 8-17, 8-18
  crack width, 1-10
  flexural cracking, 8-17
Coped beam, 4-8
Corbel, 1-23
  depth, 1-23
  details, *1-23*
Corner column, *1-41*
Correlation factor, 3-23
  concrete, 1-23
Counterfort
  moment reinforcement, *2-25*
  retaining wall, 2-25, *2-25*, 2-26
  tie reinforcement, *2-25*
Coupler, post-tensioning, 3-2
Cover factor, 1-37
Crack
  horizontal, *1-35*
  vertical, *1-35*
Cracked section, 6-8
  properties, *6-9*
Cracking, 1-22, 6-8, 8-17, 8-18
  flexural, 8-15, 8-17
  flexural, control, 8-17
  flexural-shear, 8-28
  flexure-shear, 3-14
  moment, 3-6, 3-14, 8-16, 8-17, 8-26
  moment requirement, 8-16
  web-shear, 3-14, 8-28
Crack width control, 1-9
Creep, 3-2, 6-7
  loss, 3-20
Crippling, web, 4-10
Criteria compactness, 4-3
Critical buckling design value, 5-3
Critical elastic moment, 4-5
Critical load, 8-2
  Euler, 1-32
Critical perimeter, 1-45, *1-45*, *1-46*, 2-5, 2-12
  punching shear, *2-6*
  reduction, *1-46*
Critical plane, *1-45*
Critical section, *2-6*, *2-7*
  flexural shear, 2-3, 2-7, 2-13
  flexure, 2-4, 2-7, 2-10, 2-13
  footing with steel base plate, *2-6*
  for flexure and shear, *2-3*
  for shear, 1-16, *1-16*, *1-45*, 3-13, 8-13, 8-29
  shear force, 2-13
Critical stress, 4-13, 4-35, 4-44, 4-45
Culvert, 8-5
Curb, 8-2
Curtailment, reinforcement, 1-38, *1-38*
Curvature factor, 5-5
Curve, stress-strain, 3-10

Italicized page numbers represent sources of data in tables, figures, and appendices.

Cutoff point, theoretical, *1-38*
Cycle factor, stress, 8-38

D

D-region, 1-19, *1-19*
Dead load, 1-12, 7-12, 7-18
  total, 7-6
  unfactored, 3-14
  unfactored superimposed, 3-14
Deck
  bridge, 8-1
  system, bridge, 8-5
  width, 8-1
Deep beam, 1-19, *1-19*
Deflection, 3-25, *3-26*, 4-38, 8-16
  amplification factor, 7-5, 7-12
  determination, 1-12
  final, 1-13
  floor, 1-12
  horizontal, at level x, 7-24
  limitation, 1-11
  live load, 1-11
  long-term, 1-12
  requirement, 8-6
  roof, flat, 1-11
  short-term, 1-12
  total, 1-11, 1-12
Deformation, 4-35
  elastic, 3-2
  inelastic, 7-12
  shape, 8-19
Deformed high-strength bar, 8-25
Depth
  effective, 1-16, 1-19, 2-7, 2-12, 6-11, 8-7
  effective, web, 1-10
  equivalent rectangular stress block, 8-25
  equivalent stress block, 8-14
  factor, neutral axis, 1-14, 6-2
  factor, penetration, *5-18*, 5-19, 5-23
  flange, T-beam, 8-15, 8-17
  footing, 2-4, 2-13
  minimum, superstructure, *8-16*
  of compression zone, 8-35
  of equivalent rectangular stress block, 1-3
  of penetration, 5-21–5-23
  of stress block, 8-35
  overall, 6-11
  slab, 8-35
  slab, equivalent, 4-50
  stress block, 4-50
  superstructure, minimum, *8-16*
  to centroid of prestressing steel, 8-28
  to neutral axis, 8-26
Depth-to-thickness ratio, 4-43
Derivation
  of K_{tr}, *1-35*
  transverse reinforcement index, 1-35
Design
  axial load capacity, short column, 1-30
  balanced, 1-14
  beam, 4-25
  bolted connection, 4-30, 4-33
  category, seismic, 7-4, *7-5*
  column, 4-28
  compression, 5-11
  compression member, 4-10, 4-11
  concrete slab, 8-11
  connection, 5-17–5-19
  counterfort, 2-26
  elastic, 8-38
  fatigue, 4-32
  flexural shear, 2-3, 2-7, 2-13

flexure, 1-39, 2-4, 2-7, 2-10, 2-15, 2-23,
  4-2, 6-1–6-4, 8-21, 8-33
  force, seismic, 8-51
  lane, 8-1
  load, factored, 8-12, *8-12*
  masonry column, 6-6
  method, elastic, 1-13, 6-2
  method, load factor, 8-11, 8-15
  method, mechanism, 4-26
  method, strength, 8-15, 8-34
  moment, strength, flange, 1-8
  plastic, 4-24, 4-25
  prestressed concrete, 8-21
  procedure selection, 8-44
  punching shear, 2-5, 2-12
  reinforced concrete, 8-14
  response spectrum, 8-46
  seismic, 7-1
  shear, 1-45, 3-11, 3-12, 4-8, 4-46, 5-8, 6-5,
    8-20, 8-27, 8-43
  shear and flexure, cantilever retaining
    wall, 2-23
  shear, capacity, 4-43
  shear, maximum, 6-5
  shear, strength, 4-8
  shear wall, 6-10
  spectral response acceleration
    maximum considered earthquake, 7-2
    parameters, 7-3
  stage, 3-1– 3-10
  stage, serviceability, 3-2, 3-5
  stage, strength, 3-2, 3-6
  stage, transfer, 3-2
  statical, 4-24
  steel, structural, 8-33
  strap beam for flexure, 2-20
  strap beam for shear, 2-18
  strength, 1-2, 4-2, 4-9, 4-10, 4-12
  strength, axial, 4-12, 4-14
  strength, bearing, 4-35, 4-48
  strength, shear, 4-34
  strength, tension, 4-34
  strength, weld, 4-38
  strip details, *1-43*
  tandem, 8-2
  tension, 5-16
  tension member, 4-29
  torsion, 3-12, 3-15
  traffic lanes, *8-2*
    tandem, 8-2
    truck, 8-2
  value, adjusted, 5-21
  value, base, 8-40
  value, bending, 5-5, 5-7
  value, bending, tabulated, 5-2
  value, compression, allowable, 5-12
  value, compression, reference, 5-3, 5-4
  value, critical buckling, 5-3
  value, full, bolt spacing, *5-20*
  value, lateral, *5-18*
  value, lateral, in side grain, 5-21, 5-23
  value, reference, 5-18, 5-22–5-24
  value, parallel to grain, *5-18*
  value, perpendicular to grain, *5-18*
  value, withdrawal, *5-18*
  value, withdrawal, in side grain, 5-22, 5-24
  web buckling capacity, 4-48
  welded connection, 4-38
Details, counterfort retaining wall, *2-25*
Determinate structure, statically, 4-24
Determination, shear, 8-7
Development length, 1-34–1-35, 1-40, 2-4

Diagram
  bending moment, 3-28
  force, *6-9*
  free moment, 4-16
  interaction, *A-3–A-5*
  moment, 3-27, 4-19
  shear force, *2-13*, 2-13
  strain, *6-9*
Diameter
  bar, 1-39
  equivalent, bundle, 1-35
  hole, specified, 4-30
  longitudinal bar, 1-27
  minimum, 1-25, 3-15
  spiral, minimum, 1-27
Diaphragm
  construction, 5-24
  factor, *5-18*, 5-19, 5-24
  horizontal, 7-11
  load, 7-11
Dimension
  equivalent, beam and slab, *1-42*
  lumber, 5-5
  column, limitations, 6-7
Direct design method, slab, 1-42–1-45
Directional uncertainty, 8-51
Displacement
  compatibility, 3-28
  inelastic, maximum, 8-47
  lateral, 7-12
  lateral, maximum, 8-47
  transverse, *8-47*
  virtual, 4-26, 4-27
Distance
  between centroids of tensile and
    compressive force, 8-36
  edge, 4-35
Distribution
  base shear, 7-10, 7-18
  column strip moments to edge beam, *1-43*
  exterior negative moment to column strip,
    1-43
  factors, *1-43*, 8-5, 8-6
  interior negative moment to column strip,
    1-43
  net pressure on footing, *2-2*
  of loads, 8-5, 8-6
  plastic stress, 4-37
  positive moment to column strip, *1-43*
  pressure, 2-1, 2-12, 2-17, 2-21
  reinforcement, 2-4
  strain, 6-8
  stress, 4-3
  vertical, base shear, 7-10, 7-18
  vertical, of seismic force, 7-10
Double shear
  connection, 5-20
  value, reference, 5-24
Drift
  simplified determination of, 7-18
  story, maximum allowable, 7-12, *7-12*
  story, 7-12
Dual system, 7-5, *7-6*
Duration factor, load, 5-2, 5-18, *5-18*
Dynamic
  analysis, 7-10
  factor, 8-5
  load allowance, 8-4

Italicized page numbers represent sources of data in tables, figures, and appendices.

E

Earth pressure, 2-25
  active, total, 2-21
  passive, total, 2-21
Earthquake spectral response acceleration,
  7-2
Eccentric shear stress, 2-6
Eccentrically loaded
  bolt group, 4-35, 4-37
  weld group, 4-40–4-42
Eccentric braced steel frame, 7-15
Eccentricity, 3-28, 4-31, 4-35, 4-36, 4-40,
  4-41, 6-6, 7-9, 7-19
  connection, 4-31
Edge
  beam, column strip moment distribution,
    1-43
  column, *1-46*
  distance, bolt, 5-20, 5-21
  distance, lag screw, 5-21
  distance, shear plate, 5-22
  distance, split ring, 5-22
  distance, wood screw, 5-23
Effect
  P-delta, 1-24, 1-26, 7-13
  second-order, 4-14, 4-20
Effective area, 4-38
Effective depth, 1-15, 1-17, 2-7, 2-12, 6-10,
  8-6
  web, 1-9
Effective flange width, 3-22
Effective length, 4-11–4-14, 4-16, 4-18, 5-11,
  8-41, *8-41*
  alignment chart, *4-13*
  column, 1-28
  factor, 4-48
  factor, *k*, *1-29*
  span, *6-2*
  typical values, *5-3*
Effective moment of inertia, 8-10
Effective net area, 4-29–4-31
Effective prestress, 8-25
Effective seismic weight, 7-6
Effective span length, 5-3, 5-6, 6-2
Effective thickness, 4-38
  throat, 4-38
Effective width, 4-50
  concrete slab, 8-35
Elastic analysis, 6-1
Elastic deformation, 3-2
Elastic design, 8-38
  method, 1-14, 6-2
Elastic force, 8-49, 8-51
Elastic instability, 4-13
Elasticity, modulus of, 5-2, 5-3, 6-1
  concrete, 8-16, 8-32, 8-40
  equivalent, 4-14
  prestressing steel, 8-33
Elastic lateral torsional buckling, 4-5
Elastic moment, critical, 4-5
Elastic seismic response coefficient, 8-46,
  8-48
Elastic shortening, 8-32
  loss, 3-19
Elastic unit area method, 4-36
Elastic vector analysis technique, 4-40, 4-41
Electrode, weld, nominal strength, 4-38
Element, lateral load resisting, 7-9
Embedment length, 1-34
End
  column, *1-34*
  condition, restraint, 4-11
  distance, bolt, 5-20, 5-21

distance, lag screw, 5-21
distance, shear plate, 5-22
distance, split ring, 5-22
distance, wood screw, 5-23
girder, *1-39*
grain factor, *5-18*, 5-19
moment, 1-31, 1-33
restraint, pinned, 8-47
Epoxy-coated reinforcement factor, 1-37
Equation
  equilibrium, 2-16, 2-17
  interaction, 4-7, 5-13, 5-23
Equilibrium
  check, static, 4-27
  equation, 2-17
Equivalent
  beam and slab dimensions, *1-38*
  diameter, bundle, 1-35
  effective length, 4-14
  height of fill, 2-21
  lateral force, 7-2, 7-6, 7-14
  lateral force procedure, 7-2
  number of bars, 1-9
  rectangular stress block, depth, 1-3
  slab depth, 4-50
  static force, 8-49
  stress block, depth, 8-15
  uniform static seismic load, 8-48
Essential
  bridge, 8-45, *8-47*
  facility, 7-4, *7-4*
Estimated losses, 8-33
Euler
  buckling coefficient, 8-40
  buckling strength, 4-11
  critical load, 1-32
Excess reinforcement, 3-7
  factor, 1-37
Exposure conditions, 8-17
Exterior
  beam, 8-8
  negative moment distribution to column
    strip, *1-43*
External applied load, unfactored, 3-14
External force, 6-8

F

Facility
  essential, 7-4, *7-4*
  hazardous, *7-4*
Factor
  adjustment, 5-2–5-8, 5-18
  adjustment, connection, *5-18*
  bar yield strength, 1-33
  beam stability, 5-2, 5-5–5-7, 8-41, 8-43
  bearing area, 5-2
  C_m, 4-15, 4-17, 4-19
  column stability, 5-3
  compression zone, 8-25
  concrete, lightweight aggregate, 1-33
  confinement, 1-37
  correction, concrete, 1-23
  correlation, 3-23
  cover, 1-37
  curvature, 5-5
  deck, 8-42
  deflection amplification, 7-5
  depth, neutral axis, 6-2
  diaphragm, *5-18*, 5-19, 5-24
  distribution, 8-6
  dynamic, 7-5
  effective length, 4-11, *4-12*, 4-48
  effective length, *k*, *1-28*

end grain, *5-18*
flat use, 5-5, 8-42
format conversion, 8-42
geometry, *5-18*, 5-19
group action, 5-18, *5-18*, 5-21, 5-22
importance, 7-4, *7-4*
incising, 5-5, 5-11, 8-42
lever-arm, 1-14, 6-2
load, 1-1, 4-1, 8-12, *8-12*
load, design method, 8-11
load duration, 5-2, 5-18, *5-18*
magnification, 1-32, 1-34, 4-10, 4-17, 4-19,
  4-21
metal side plate, *5-18*, 5-19
modification, 8-51
multiple presence, 8-6
neutral axis, *A-2*
neutral axis depth, 1-14
of safety, overturning, 2-21
of safety, sliding, 2-21
overstrength, 7-6
penetration, *5-18*, 5-19, 5-21–5-23
prestressing steel, 8-25
prestressing tendon type, 3-8
reduction, 1-2, 4-15, 4-17, 4-19, 8-40
reduction, strength, 8-1
redundancy, 7-19
reinforcement, epoxy-coated, 1-37
reinforcement, excess, 1-37
repetitive member, 5-4
resistance, 4-2, 4-38, 8-42
response modification, 8-51, *8-51*
shape, 4-3
site, 8-43, 8-45, *8-45*, 8-46
size, 5-5, 5-7, 8-41, 8-43
stability, beam, 8-41
stiffness reduction, 4-10, 4-19
strength reduction, 1-2, 8-1
stress cycle, 8-38
temperature, 5-2, *5-2*, 5-3, 5-6, *5-18*, 5-18
tie, 1-37
time-dependent, sustained load, 1-12
time-dependent, value, *1-12*
time effect, 8-41, *8-41*
toe-nail, *5-18*, 5-19, 5-25
volume, *5-2*, 5-5, 5-6, 5-8
wet service, 5-2, *5-2*, 5-6, 5-18, *5-18*, 8-41
Factored applied column load, 2-12
Factored applied load, net, 2-1
Factored applied shear force, 3-13
Factored axial load, 1-32
Factored bearing reaction, 4-49
Factored footing pressure, 2-12
Factored forces, 2-23
  footing, 2-18
  on strap footing, *2-19*
Factored load, 1-1, 2-12, 4-1, 4-24, 4-26
  applied, *2-13*
  total, 2-18
Factored moment, 1-3, 1-7, 1-16, 2-20, 3-12,
  8-13
  maximum, 1-3, 3-14
  total, 3-14
Factored pressure on footing, net, *2-13*
Factored shear force, 1-16, 1-38, 3-14, 3-22,
  3-23
Factored static moment, 1-43
  total, *1-44*
Factored torque, 1-24, 3-15
Factored torsion causing cracking, 1-25
Factors, distribution, *M_o*, *1-43*

Italicized page numbers represent sources of data in tables, figures, and appendices.

Failure
    fatigue, 4-32
    flexural, 3-7
    sudden tensile, 8-17, 8-27
Fastener (*see also* Connection)
    staggered, 5-18, *5-19*
Fatigue, 8-18, 8-19, 8-38
    constant, 4-33
    design, 4-32
    failure, 4-32
    stress, 8-38
    stress limits, 8-18
Fill, equivalent height, 2-21
Fillet weld, 4-39
Final deflection, 1-14
Final prestress, tendon, 3-10
Flange
    depth, T-beam, 8-15, 8-17
    design moment strength, 1-7
    local buckling, limit state, 4-45
    width, effective, 3-21
Flanged beam, 1-8
Flanged section
    torsion, *1-25*
    with tension reinforcement, 1-8, *1-9*
Flare groove weld, 4-38
Flat slab, 3-9
Flat use factor, 5-5, 8-42
Flexural capacity, 1-38
    nominal, 8-25
    shear, 1-45
Flexural cracking, 8-17
    control, 8-17
Flexural failure, 3-7
Flexural member, shear, 6-5
Flexural rigidity, 1-32
Flexural shear, 1-45, *1-45*, 2-3, *2-3*, 2-7, 2-13
    cracking, 3-14, 8-28
Flexural strength, 3-8, 3-9, 8-25
    beam, 1-3
    design, 1-7
    nominal, 3-7, 4-4–4-5, 4-6, 4-44, 4-51,
        8-25, 8-35
    using strain compatibility, 3-10
Flexural tension reinforcement, 8-18
Flexure, 2-4, 2-7, 2-10, 2-15, 2-23, 6-8
    and axial compression, combined, 5-14,
        5-15
    and axial load (compression), combined,
        4-22, 4-28
    and compression, combined, 6-8
    critical section, *2-3*, 2-4, *2-6*, 2-7, 2-8, 2-23
    design, 1-39, 4-2, 4-45, 4-53, 5-6, 6-1–6-4,
        8-14, 8-21, 8-33
    strap beam, 2-20
Floor
    live load, 7-6, 7-13
    load, 7-6
Footing
    combined, 2-12–2-16
    combined, with applied factored loads,
        2-13
    combined, with applied service loads, *2-12*
    concrete, bearing on, *2-9*
    depth, 2-4, 2-13
    factored forces, 2-18
    net factored pressure, *2-13*
    net pressure distribution, *2-2*
    pad, 2-17
    pressure, factored, 2-12
    rectangular, on isolated column, 2-10
    rectangular, reinforcement areas, *2-10*

shear strength, 2-3
square, on isolated column, 2-5
strap, 2-16, *2-19*
strap, with applied service loads, *2-17*
strip, 2-1, 2-7
    with steel base plate, *2-7*
Force (*see also* Load, *see also type*)
    axial, 4-29
    compressive, 3-10
    diagram, *6-9*
    elastic, 8-49, 8-51
    external, 6-9
    factored, 2-23
    factored, footing, 2-18
    factored, on strap footing, *2-19*
    frictional, 2-21
    horizontal, bolt, 4-36
    horizontal, weld, 4-40, 4-41
    internal, 6-8
    lateral, 7-2, 7-18
    lateral, equivalent, 7-2
    lateral, procedure, 7-2
    lateral, simplified procedure, 7-14
    minimum allowable, horizontal diaphragm,
        7-11, 7-12
    prestress, final, 3-14
    prestressing, 3-2, 3-18, 3-19, 3-27
    pre-tension, minimum, 4-34
    resultant, bolt, 4-36
    resultant, weld, 4-40, 4-41
    seismic design, 8-51
    seismic, vertical distribution, 7-10
    shear, 1-38, 3-14, 4-37, 6-10, 6-13
    shear, applied, 6-11
    shear, applied factored, 3-13
    shear, critical section, 2-13
    shear diagram, *2-13*, 2-13
    shear, factored, 1-16, 1-23, 1-38, 3-14,
        3-22, 3-23
    shear, for vehicle loads, 8-43
    shear, horizontal, 4-51
    shear, range, 8-38
    shear, total, 8-39
    shear, unfactored, 3-14
    static, equivalent, 8-49
    tendon jacking, 3-2
    tensile, 4-37
    tensile, corbel, 1-23
    tensile, total, 3-4, 3-9, 3-10
    transfer at column base, 2-8
    vertical, bolt, 4-35
    vertical, due to effects of vertical
        acceleration, 7-12
    vertical, weld, 4-40, 4-41
Format conversion factor, 8-42
Formula, Hankinson, 5-22, 5-23
Frame
    braced, 4-12, 4-28, 7-4, 7-12, *7-18*
    braced steel, eccentric, *7-6*
    building, 7-5, *7-6*
    indeterminate, three-degree, 4-18
    moment, 7-5, *7-6*
    moment-resisting, 7-5, *7-6*
    nonsway, 1-28, *1-29*, 1-30
    rigid, 4-27
    rigid, columns in, 4-12
    sway, *1-29*, 1-30, 4-13, 4-28, 4-29
    with rigid joints, 4-12
Framing bar, *1-23*
Framing system, structural, *7-6*
Free moment diagram, 4-24, 4-25

Friction
    coefficient, 3-23, 8-32
    loss, 3-2, 3-17, 8-32
Frictional force, 2-21
Full design value, bolt spacing, *5-20*
Full plasticity, 4-3
Full plastic moment capacity, 4-4
Fundamental period of vibration, 7-6, 7-7,
    8-44

G
General procedure response spectrum, 7-20
Geometry factor, *5-18*, 5-19
Girder
    end, *1-39*
    plate, 4-42
    with unstiffened web, 8-37
    with web stiffener, 4-12
Glued laminated member, 5-2, 5-5, 5-6, 5-7,
    5-9, *5-10*
Glued laminated timber, 8-41
Grade
    50 reinforcement, 6-1
    60 bar, 2-4
    60 reinforcement, 1-6, 1-12, 6-2
Gravity
    constant, 8-38
    load, 8-12
Groove weld
    complete-penetration, 4-38
    flare, 4-38
    partial-penetration, 4-38
Gross area, 4-29
Gross concrete section, 8-17
Gross moment of inertia, 8-17
Ground
    acceleration, 8-44
    acceleration, effective peak, 7-3
    coefficient, peak seismic, 8-43, 8-45, *8-45*
    motion parameter, 7-2
Group
    action factor, 5-17, *5-17*, 5-21–5-23
    bolt, eccentrically loaded, 4-35, 4-37
Gyration, radius, 1-28, 4-14

H
Hankinson formula, 5-20, 5-22
Hazardous structure, 7-4, *7-4*
Height
    of fill, equivalent, 2-21
    to diameter ratio, 8-38
High-strength bar, 8-25
Hinge, plastic, 4-24, 4-25, 4-26
History, time, method, *8-44*
HL-93 design load, 8-3
Hole diameter, specified, 4-30
Hooked bar
    in tension, 1-37
    tie spacing, 1-37
Horizontal crack, *1-35*
Horizontal deflection at level x, 7-12
Horizontal diaphragm, 7-11
Horizontal force
    bolt, 4-36
    weld, 4-40–4-42
Horizontal irregularity, 7-10
Horizontal reinforcement, *6-10*, *6-11*
    area, minimum, 1-19
Horizontal shear
    force, 4-51
    range, 8-38
    reinforcement, 6-11

Italicized page numbers represent sources of data in tables, figures, and appendices.

requirement, 3-23
   strength, nominal, 3-23
Horizontal span moment, 2-25
Horizontal tie, 2-26
h/r ratio, 6-6, 6-7

I
Impact with live load, 8-38
Importance
   category, 8-43
   factor, 7-4, *7-4*
Incising factor, *5-2*, 5-5, 5-11, 8-42
Inclined bar, 1-18
   beam, *1-18*
Inclined stirrups, beam, *1-16*
Independent mechanisms, number, 4-26
Indeterminate frame, three-degree, 4-26
Indeterminate structure, 3-27
Index
   reinforcement, 3-7
   stability, 1-25
   tension reinforcement, 1-3, *A-1*
   transverse reinforcement, 1-34
   transverse reinforcement, derivation, *1-35*
Ineffective length, *1-46*
Inelastic buckling, 4-46
Inelastic displacement, maximum, 8-47
Inelastic instability, 4-13
Inertia
   effective moment, 1-12
   moment of, 4-35, 4-36, 4-50, 8-17
   moment of, required, 4-38
Inflection point, 1-39
Influence line, *8-3*, 8-4, *8-4*, *8-8*
Instability
   elastic, 4-13
   inelastic, 4-13
Instantaneous center of rotation method,
   4-36, 4-40, 4-42
Integrity reinforcement, 1-31
Interaction
   diagram, *A-3–A-5*
   equation, 4-14, 5-14, 5-24
   expression, 4-7, 4-28
Interface
   roughened, 3-23
   smooth, 3-23
Interior
   beam, 8-6
   column, *1-34*
   negative moment distribution to column
      strip, *1-39*
Intermediate stiffener, 4-43, 4-44, 4-47
   design, 4-46
Internal force, 6-8
Inverted pendulum structure, 7-5
Irregular bridge, 8-47, *8-47*
Irregularity
   horizontal, 7-10
   vertical, 7-10
Isolated column
   rectangular footing, 2-10
   square footing, 2-5

J
Jacking stress, tendon, 8-22
Joint, rigid, in frame, 4-12
Joist, lumber, 5-1, 5-9

K
k, alignment chart, *1-29*
K factor, 4-11
K_{tr}, derivation, *1-35*

L
Laced compression member, 4-14
Lag
   screw, connection, *5-18*, 5-21
   screw, edge distance, 5-21
   screw, end distance, 5-21
   screw, spacing, 5-21
   shear, 4-31
Laminated member, glued, 5-2, 5-5–5-11
Laminated timber, glued, 8-41
Lane
   design, 8-1
   design traffic, *8-2*
   load, 8-2
Lap
   length, 1-40
   splice, 1-36, 1-40
   splice, *c_s* value, *1-40*
Lateral design value, *5-18*
   in side grain, 5-21, 5-23
Lateral displacement, 7-12
   maximum, 8-47
   static, 8-49
Lateral force, 7-1, 7-2
   equivalent, 7-2
   procedure, IBC, 7-1
   procedure, simplified, 7-14
   simplified vertical distribution, 7-18
   vertical distribution of, 7-10
Lateral load, 1-25
   and withdrawal, combined, 5-20, 5-22,
      5-23
Laterally unbraced length, 4-4, 4-5
Lateral stiffness, 8-47, 8-49
Lateral support, 4-4
   spacing, 6-2
Lateral tie, 6-7
   reinforcement, short concrete column,
      1-26
Lateral torsional buckling, 4-5
   limit state, 4-44
Length
   anchorage, *2-3*, 2-4
   bearing, 8-43
   bridge deck, 8-43
   buckling, coefficient, 5-12, *5-12*
   connection, 4-32
   development, 1-34–1-41, 2-4
   effective, 4-11, 4-14, 4-16, 4-18, 5-11,
      8-41, *8-41*
   effective, column, 1-28
   effective, factor, *1-29*
   effective, typical values, *5-4*
   embedment, 1-39
   factor, effective, *4-12*, 4-48
   ineffective, *1-46*
   lap, 1-41
   laterally unbraced, 4-4, 4-5
   span, effective, 5-3, 5-6, 6-2, *6-2*
   support, minimum, 8-43, 8-47
   thread, 5-22
   total, weld, 4-42
   unbraced, maximum, 4-25, 4-28
Lever-arm factor, 1-14, 6-2
Lever-rule method, 8-6
Lightweight aggregate concrete factor, 1-33
Limit
   service, state, 8-11
   state, extreme event, 8-12
   state, fatigue, 8-11
   state, flange local buckling, 4-45

state, lateral-torsional buckling, 4-45
   strength, state, 8-12
Limitation, deflection, 1-11
Line
   influence, 8-4
   of pressure, resultant, 3-28
Linear-elastic analysis, 1-32
Live load, 1-12, 8-2, 8-5, 8-23
   deflection, 1-12
   floor, 7-6, 7-13
   plus impact, 8-38
   roof, 7-6
   surcharge, 2-20, 2-21
Load (*see also* Force, *see also type*)
   applied, 4-39, 6-8
   applied, superimposed, 3-24
   axial, 4-28, 5-11, 6-6, 6-8
   axial, allowable, 6-6
   axial, factored, 1-32
   axial, maximum allowable, 5-12
   axial, stress, *6-9*
   balancing, 3-26, *3-26*
   balancing procedure, 3-26
   bending, 5-13
   capacity, axial, short column, 1-30, 1-31
   case 1, 8-51
   case 2, 8-51
   collapse, 4-20
   column, factored applied, 2-12
   combination, 6-1, 8-11, 8-51
   combined, 7-13
   concentrated, 4-9, 4-10
   concentric axial compression, 5-13
   conditions, service, *1-12*
   condition, ultimate, 8-25
   critical, 8-2
   dead, 1-12, 7-6
   dead, total, 7-6
   dead, unfactored, 3-14
   dead, unfactored superimposed, 3-14
   design, 8-2
   design, factored, 8-12
   design lane, 8-2, *8-2*
   design tandem, *8-1*, 8-2
   design truck, 8-2, *8-2*
   diaphragm, 7-11
   distribution, 8-5, 8-6
   duration factor, 5-1, 5-17, *5-17*
   dynamic, 8-4, 8-5
   Euler critical, 1-32
   factor, 1-1
   factor design, 4-36
   factored, 1-1, 2-12, 4-1, 4-24, 4-26
   factored, applied, *2-13*
   factored, total, 2-18
   floor, 7-6, 7-13
   HL-93 design, 8-3
   lane, 8-2, 8-3, 8-4, *8-4*
   lateral, 1-25
   live, 1-12, 8-2, 8-5, 8-25
   live, deflection, 1-12
   live, floor, 7-6, 7-13
   live, plus impact, 8-6, 8-38
   live, surcharge, 2-20, 2-21
   longitudinal, 8-47, *8-47*
   net factored applied, 2-1
   permanent, 8-5
   seismic, 7-18, 8-6, 8-45, 8-47
   service, 1-1, 1-9, 1-13, 2-11, 2-12, 3-5, 4-1,
      4-33, 6-1, 7-6, 8-3, 8-15
   service, applied, *2-12*, *2-17*, *2-21*, 6-2
   service, total, 2-17

Italicized page numbers represent sources of data in tables, figures, and appendices.

snow, 7-6, 7-13
sustained, 1-12
transfer, 2-8
transverse, 5-4, 8-45, *8-45*
ultimate, 8-25, 8-35
unfactored external applied, 3-14
uniform, 3-14, 8-2, 8-44, 8-47
uniform, method, 8-47
uniformly distributed, 8-2
uniform unit, 8-47
vehicle, shear force, 8-43
vertical, story, 1-29
working, 1-1, 4-1 (*see also* Load, service)
zero shear, *2-13*
Loaded bolt group, eccentrically, 4-35–4-37
Loading
  design tandem, 8-2, 8-3, *8-3*
  cycle, 4-25
  HL-93 design, 8-3
  transverse, 4-15
  vehicular, 8-2
Local web yielding, 4-9
Long column
  concrete, 1-31–1-33
  with sway, 1-33
Longitudinal
  bar, 1-27
  load, 8-49, *8-49*
  reinforcement, 1-18, 1-24, 1-27, 6-7
  skin reinforcement, 8-18
  torsional reinforcement, 3-15
Long-term deflection, 1-12
Loss
  anchor seating, 3-18
  creep, 3-20
  elastic shortening, 8-32
  estimated, 8-33
  friction, 3-2, 3-18, 8-32
  prestress, 3-17–3-21, 8-32, 8-33
  prestressing, 8-23
  relaxation, 3-33
  shortening, elastic, 3-19, 8-32
  shrinkage, 3-20
  time-dependent, 8-33
Low-relaxation steel, 8-25
Low-relaxation strand, 8-25, 8-33
Lumber
  joist, 5-1
  dimension, 5-1, 5-5
  machine graded dimension, 5-2
  sawn, 8-40, 8-41
  sawn, visually graded, 5-5, 5-6
  visually graded dimension, 5-2

M
Machine graded dimension lumber, 5-1
Magnification factor, 1-32, 1-34, 4-10, 4-17,
  4-19, 4-21
Magnified moment, 1-33, 1-34
Main reinforcement, 2-4
Masonry
  column design, 6-6
  shear stress, 6-5, 6-12
  wall, *2-3*
  wall, without shear reinforcement, 6-11
  wall, with shear reinforcement, 6-13
Maximum allowable axial load, 5-12
Maximum allowable story drift, 7-12, *7-12*
Maximum allowable stress, 8-23
Maximum allowable width-to-thickness ratio,
  4-47
Maximum compressive strength, 6-1

Maximum considered earthquake spectral
  response acceleration, 7-2
Maximum factored moment, 1-3, 3-14
Maximum inelastic displacement, 8-47
  negative, 2-15, 8-2
  on strap, *2-19*
  positive, 2-15
Maximum reinforcement ratio, 1-4
Maximum shear
  on strap, *2-19*
Maximum spacing
  closed stirrups, 3-16
  main reinforcement, 2-4
  transverse reinforcement, *1-36*
Maximum strain, 1-3, 3-2, 3-10
Maximum unbraced length, 4-25, 4-28
Mechanism
  beam, 4-26, 4-27
  combined, 4-26, 4-27
  design method, 4-26
  independent, number of, 4-26
  partial collapse, 4-24
  sway, 4-26, 4-27
Member
  compression, laced, 4-14
  flexural, 6-5, 8-17
  post-tensioned, 8-22, 8-33
  precast, 8-18
  pre-tensioned, 8-22, 8-33
Metal side plate factor, *5-18*, 5-19
Method
  elastic design, 1-13, 6-2
  lever-rule, 8-6
  load balancing, 3-25
  mechanism design, 4-26, 4-27
  shear-friction, 3-23
Middle strip, *1-43*
Minimum area
  combined, stirrups, 3-16
  horizontal reinforcement, 1-19
  shear reinforcement, 1-16, 1-18, 3-13,
    8-20, 8-31
  vertical reinforcement, 1-19
Minimum depth, superstructure, *8-16*
Minimum diameter, 1-25, 3-15
Minimum nominal reinforcement
  requirements, 6-10
Minimum pre-tension force, 4-34
Minimum ratio, reinforcement area to gross
  concrete area, 2-4
Minimum reinforcement
  area, 2-8
  ratio, 1-6
Minimum required moment strength, 1-3
Minimum seat-width requirement, 8-51
Minimum superstructure depth, *8-16*
Minimum support length, 8-47, 8-51
Modification factor, response, 8-44, 8-51
M_o distribution factors, *1-44*
Modular ratio, 3-21, 8-33, 8-38
Modulus of elasticity, 5-2, 5-3, 6-1
  concrete, 8-16, 8-32, 8-40
  modified, 4-13
  prestressing steel, 8-33
Modulus of rupture, 3-6, 8-17, 8-26
Moisture content, 5-2, 5-18, 8-41
Moment
  applied, 6-2, 6-8
  arm between centroids of tensile and
    compressive force, 8-34
  bending, 2-10, 2-25, 3-14, 4-7, 8-4
  bending, applied, 6-8

bending, non-uniform, 1-27
bending, stress, *6-8*
bending, uniform, 4-4, 4-5
cantilever, 2-25
capacity, full plastic, 4-4
causing flexural cracking, 8-28
column strip, 1-40
cracking, 3-6, 3-14, 8-16, 8-17, 8-26
critical elastic, 4-5
design strength, flange, 1-7
diagram, 3-28, 4-28
diagram, bending, 3-28
distribution, column strip to edge beam,
  1-43
end, 1-31, 1-33
factored, 1-3, 1-16, 2-20, 3-12, 8-13
factored, maximum, 3-14
factored, total, 3-14
frame, *7-6*
free, diagram, 4-24
magnification factor, 4-10, 4-17, 4-19, 4-21
magnified, 1-32, 1-33
maximum, 8-3, *8-3*, *8-4*
maximum, on strap, *2-19*
negative, 1-43
negative, distribution to column strip,
  1-43
negative, maximum, 2-15
negative, reinforcement, *1-39*
nonsway, 1-33
of inertia, 4-35, 4-36, 4-40, 4-41, 4-42, 8-6
of inertia, effective, 1-12, 8-16
of inertia, gross, 8-16
of inertia, required, 4-47
of resistance, plastic, 4-3
out-of-balance, 3-25
plastic, 4-11, 4-29, 4-43
plastic, required, 4-26
positive, 1-43, 8-3
positive, distribution to column strip, *1-43*
positive, maximum, 2-15, *8-3*
positive, reinforcement, *1-34*
primary, 1-28, 3-27, 4-10
primary, in a story, 7-13
reinforcement, counterfort, *2-25*
reinforcement, negative, 1-39
reinforcement, positive, 1-38
residual, 1-7, 1-9
resistance, ultimate, 3-6
resultant, 3-27, 3-28
secondary, 1-24, 1-26, 3-27, 4-10
secondary, in a story, 7-13
service, 1-14
span, horizontal, 2-25
statical, 8-38
static, total factored, 1-43, *1-44*
strength, 3-7
strength, minimum required, 1-3
sway, 1-33
unfactored, total, 3-14
Moment-resisting frame, 7-5, 7-6, 7-13
Multimode elastic method, *8-44*

N
Nail
  connection, 5-24, 5-25
  wire, 5-24
Nailed connection, *5-18*
Negative moment, 1-43
  distribution to column strip, *1-43*
  maximum, 2-15
  reinforcement, 1-39, *1-39*

Italicized page numbers represent sources of data in tables, figures, and appendices.

Negative reinforcement, 1-39
Net area, effective, 4-30–4-32
Net factored applied load, 2-1
Net factored pressure on footing, *2-13*
Net pressure, 2-1
    distribution on footing, *2-2*
    factored footing, 2-12
Neutral axis, 3-10, 8-17
    depth factor, 1-14, 6-2
    depth to, 8-25
    factor, *A-2*
    plastic, 8-34
Nodal zone, 1-20, *1-20*, 1-21, *1-21*
Nominal capacity
    shear (strength), 1-16, 3-12–3-14, 4-45,
        4-46, 8-20, 8-28
    torsional, 3-16
Nominal reinforcement requirements, shear
    wall, 6-10
Nominal strength, 1-2, 1-39, 4-2, 4-3,
    4-9–4-10
    beam, 1-3
    bearing, 4-35, 4-48
    flexural, 1-3, 3-7, 4-4, 4-5, 4-43, 4-44,
        4-51, 8-25
    flexural, beam, 1-3
    resistance, 8-42, 8-43
    shear, 1-15–1-19, 3-12, 4-8, 4-34, 4-46,
        8-20, 8-28
    shear, horizontal, 3-23
    tensile, 4-34
    weld electrode, 4-38
Noncomposite beam, 8-27
Non-essential bridge, 8-45
Nonprestressed reinforcement, 3-6
Nonprestressed tension reinforcement, 8-25
Non-propped construction, 3-24
Nonsway column, 1-31, 1-32
Nonsway frame, 1-28, *1-29*, 1-30
Nonsway moment, 1-33
Nonsway story, 1-29
Non-uniform bending moment, 1-32
Normal conditions of use, 8-40
Notched beam, 5-9, *5-10*
Notch, shear stress, 5-9
Number
    independent mechanisms, 4-26
    loading cycles, 4-32
    of design lanes, 8-1, 8-2

O
Occupancy, 7-4, *7-4*
    category, 7-4
    importance factor, 7-4, *7-4*
One-way shear, 1-45 (*see also* Flexural
    shear)
Orthogonal seismic forces, 8-51
Out-of-balance moment, 3-26
Overstrength factor, 7-16
Overturning, factor of safety, 2-21

P
Pad footing, 2-17
Parallel-to-grain design value, *5-18*
Parameter
    column, 5-3
    crack control, 8-11
    design spectral response acceleration,
        7-13
    ground motion, 7-12
    longitudinal stiffness, 8-6
    slenderness, 4-35

Partial collapse mechanism, 4-24
Partial-penetration groove weld, 4-38
Passive earth pressure, total, 2-21
P-delta effect, 1-28, 4-15, 7-13
Pendulum structure, inverted, 7-5
Penetration
    depth, 5-21–5-23
    factor, *5-18*, 5-19, 5-21–5-23
Percentage
    distribution, column strip moments to
        edge beam, *1-43*
    distribution, exterior negative moment to
        column strip, *1-43*
    distribution, interior negative moment to
        column strip, *1-43*
    distribution, positive moment to column
        strip, *1-43*
Performance zone, 8-44
    seismic category, 6-10, *6-10*
Perimeter, critical, 1-45, *1-45*, 2-5, 2-12
    punching shear, *2-6*
    reduction, *1-46*
Period
    corner, 8-44, 8-46, *8-46*
    of vibration, fundamental, 7-6, 7-7, 7-8,
        8-45–8-47
    reference, 8-44, 8-46, *8-46*
Permanent load, 8-5
Permissible stress, 1-14, 3-2, 3-4, 3-5
Perpendicular reinforcement, 6-12
Perpendicular to grain design value, *5-17*
Pinned end restraint, 8-47
Pitch, bolt, 4-33
Plane
    critical, *1-45*
    truss, 4-12
Plastic
    design, 4-24, 4-25
    hinge, 4-24, 4-25, 4-26
    moment, 4-11, 4-29, 4-43
    moment capacity, 4-4
    moment capacity in bending, 8-35
    moment of resistance, 4-3
    moment, required, 4-27
    neutral axis, 8-34
    stress distribution, 4-37
Plasticity, 4-3
Plate
    base, column, 4-23
    bearing, *1-23*, 4-9
    factor, metal side, *5-18*, 5-19
    girder, 4-42
    in tension, 4-29
Point of inflection, 1-39
Polar moment of inertia, 4-35, 4-36, 4-41
Positive moment, 1-43
    distribution to column strip, *1-43*
    maximum, 2-15
    reinforcement, 1-36, 1-38, *1-39*
Positive reinforcement, 1-38
Post-tensioned concrete, 3-2
Post-tensioned member, 8-22, 8-33
Post-tensioned tendon, 3-18–3-20
Post-tensioning anchorage, 3-2
Post-tensioning coupler, 3-2
Precast section, 3-23, 3-24
Pressure
    allowable, 2-12, 2-17
    bearing, soil, 2-12, *2-17*, 2-17
    bearing, uniform, *2-12*
    distribution, 2-1, 2-12, 2-17, 2-21
    distribution on footing, net, *2-2*

earth, 2-25
earth, total active, 2-21
earth, total passive, 2-21
factored, on footing, 2-13, *2-13*
line, resultant, 3-28
net, 2-1
soil, 2-1, 2-17, 2-23
surcharge, total, 2-21
uniform, soil, 2-12
Prestrain in tendon due to final prestress,
    3-10
Prestress
    effective, 8-25
    final, tendon, 3-10
    force, final, 3-14, 8-30
    loss, 3-17–3-21, 8-23, 8-32, 8-33
Prestressed
    concrete design, 8-21
    reinforcement, 3-8
Prestressing
    force, 3-2, 3-18, 3-19, 3-27
    losses, 8-23, 8-32, 8-33
    steel, 8-25
    steel factor, 8-25
    steel, modulus of elasticity, 8-33
    tendon, 3-7, 3-14, 3-19, 3-26
    tendon type factor, 3-8
Pre-tension force, minimum, 4-34
Pre-tensioned member, 8-33
Pre-tensioned tendon, 3-18–3-20
Primary moment, 1-28, 3-27, 4-15
    in a story, 7-13
Primary reinforcement, *1-23*
Prismatic beam, 4-26
Procedure
    analysis, 8-44
    load balancing, 3-25
    Rayleigh, 7-7
    simplified lateral force, 7-14
Profile
    concordant, 3-28
    tendon, alternative, 3-26, *3-26*
    type, soil, 7-2, *7-2*, 8-45, *8-45*
Property, section, 3-21
    transformed, 6-8
Propped construction, 3-24
Punching shear, 1-45, *1-45*, 1-46, 2-5, 2-12
    capacity, 2-6
    critical perimeter, *2-6*
    strength, 2-6
Pyramid base, *2-9*

R
Radius of gyration, 1-24, 4-13
Range
    of horizontal shear, 8-38
    of shear force, 8-38
    stress, 8-18
Rankine's theory, 2-21
Ratio
    horizontal tensile force to vertical force,
        1-23
    h/r, 6-6, 6-7
    modular, 3-21, 8-33, 8-38
    of height to diameter, 8-38
    reinforcement, 1-3, 1-6, 3-7
    reinforcement area to gross concrete area,
        minimum, 2-4
    reinforcement, maximum, 1-4
    reinforcement, minimum, 1-6
    slenderness, 4-8, 4-11, 4-13, 4-28, 5-3,
        5-12, 8-41

Italicized page numbers represent sources of data in tables, figures, and appendices.

slenderness, concrete column, 1-28, 1-30, 1-31, 1-33
span-to-depth, 3-9
stiffness, 4-11, 4-12
stiffness, concrete column, 1-24
tension reinforcement, 1-3
volume of spiral reinforcement to core volume, 1-27
web depth-to-thickness, 4-43
width-to-thickness, maximum, 4-47
Rayleigh procedure, 7-7
Reaction, support, 8-13, 8-29
Recommended minimum superstructure depth, *8-16*
Rectangular beam, 5-9, 8-15
nominal flexural strength, 1-3
Rectangular column
concrete, 1-27
tie size, 1-27
Rectangular footing
on isolated column, 2-10
reinforcement areas, *2-10*
Rectangular section, 8-25
torsion, *1-25*
Rectangular stress block, 1-3, *1-4*, 1-8
equivalent, depth of, 1-3, 8-25
Recurrence interval, 8-44
Reduction
coefficient, 4-18, 4-30, 4-31
critical perimeter, *1-45*
factor, 4-15, 4-19, 8-40
factor, strength, 1-2, 8-25
Redundancy factor, 7-9
Reference value
design, 5-18, 5-20, 5-22–5-24
design value, 8-40–8-43
double shear, 5-20, 5-24
periods, 8-46
rectangular beam, 1-3
Regular bridge, 8-47, *8-47*
Reinforced concrete
beam, 1-3, 1-7, 8-13
design, 8-14
Reinforcement
area, 1-7, 1-8, 1-23, 1-36, 2-4, 2-10, 2-15
area, minimum, 2-8
area, transformed, 6-8, *6-8*
areas, rectangular footing, *2-10*
auxiliary, 3-4, 3-7–3-9
bonded, 8-22, 8-23, 8-25
capacity, 2-8
column, limitations, 6-7
compression, 1-7, 3-8
compression, beam, *1-7*
continuing, 1-38
curtailment, 1-38, *1-38*
details, shear wall, *6-10*
development length, 1-34–1-37
distribution, 2-4
excess, 3-7
factor, epoxy-coated, 1-37
factor, excess, 1-37
Grade 40, 1-12
Grade 50, 6-1
Grade 60, 1-6, 1-12, 6-2
horizontal, *6-10*, 6-11
index, 1-3, 3-7, 8-15
index, tension, *A-1*
index, transverse, derivation, *1-35*
integrity, 1-36
lateral tie, short concrete column, 1-30
longitudinal, 1-18, 1-24, 1-27, 6-6

longitudinal torsional, 3-15
main, 2-4
maximum, 1-4
minimum, 1-6
moment, *2-25*
negative, 1-39
negative moment, 1-39, *1-39*
nonprestressed, 3-6
perpendicular, 6-13
positive, 1-39
positive moment, 1-36, 1-39, *1-39*
prestressed, 3-8
primary, *1-23*
ratio, 1-3, 1-6, 3-7
requirement, 1-27
requirement, nominal, shear wall, 6-10
shear, 1-16–1-18, 1-25, 3-12, 3-15, 6-5, 6-10, 6-11, 8-20, 8-28
shear, horizontal, 6-10
shear, vertical, 1-16
skin, 1-10, 8-18
skin, longitudinal, 8-18
spiral, 1-37
spiral, short concrete column, 1-30
splice length, 1-34
stress, 1-10, 1-14
temperature, 1-36
tension, 1-3, 1-6–1-8, *1-9*, *1-19*, 1-35, 6-1, 8-18
tension, flanged section, *1-9*
tension, flexural, 8-18
tension, nonprestressed, 8-25
tie, *2-25*
torsion, 1-24
transverse, index, 1-35
transverse, maximum spacing, *1-40*
vertical, 6-10, *6-10*, 6-11
Relative humidity, 3-20
Relaxation loss, 3-20
Repetitive member factor, 5-5
Required base plate thickness, 4-16
Required moment of inertia, 4-47
Required plastic moment, 4-27
Required reinforcement area, 2-4, 2-9, 2-15
rectangular footing, *2-10*
Required shear strength, 4-43
Required strength, 1-1, 4-1
Requirement
building configuration, 7-10
deflection, 8-16
horizontal shear, 3-22, 3-23
Residual moment, 1-7, 1-9
Residual stress, 5-5
Resistance
factor, 4-2, 8-42
moment, plastic, 4-3
moment, ultimate, 3-6
shear, 1-18
Response
coefficient, seismic, 7-9, 8-43, 8-45
modification factor, 8-40, 8-48, *8-48*
modification coefficient, 7-5, *7-6*
seismic, 8-43, 8-44
spectrum, 7-3, 7-8, *7-8*, 7-9
Restraint
end condition, 4-11
end, pinned, 8-47
support, 3-27, 3-28
Resultant force
bolt, 4-36
weld, 4-40–4-42
Resultant line of pressure, 3-28

Resultant moment, 3-28
Retaining wall
cantilever, 2-20–2-25
cantilever, with applied service loads, *2-21*
counterfort, 2-25, *2-25*, 2-26
Rigid abutment, 8-47
Rigid frame, 4-26
Rigidity, flexural, 1-27
Rolled section in tension, 4-31
Rolled W shape beam, 4-8
Rotation, instantaneous center of, method, 4-36, 4-40–4-42
Roughened interface, 3-22
Rupture
modulus, 3-6, 8-17, 8-26
tension, 4-29

S

Safety factor
overturning, 2-21
sliding, 2-21
Sawn lumber, 8-40–8-41
visually graded, 5-5, 5-6
Screw
lag, spacing, 5-21
wood, 5-23
wood, spacing, 5-23
Screwed connection, *5-18*
Seating loss, anchor, 3-18
Seat-width, minimum, 8-51
Secondary moment, 1-24, 1-26, 3-27
in a story, 7-13
Second-order effect, 4-15, 4-19, 4-28
Section
built-up, 4-13
compact, 4-4
composite, 3-21, 3-23, 3-24, 8-26, 8-38, *8-38*
compression-controlled, 1-3, 1-4
cracked, 6-8
cracked, properties, *6-9*
critical, *2-6*, *2-7*
critical, flexure, 2-4, 2-7, 2-10, 2-23
critical, flexure and shear, *2-3*
critical, flexural shear, 2-3, 2-7, 2-13
critical, for shear, *1-16*, *1-45*, 8-13, 8-29
critical, shear, 1-16, 3-13
critical, shear force, 2-13
flanged, 8-25
flanged, tension reinforcement, 1-8, *1-9*
flanged, torsion, *1-25*
gross concrete, 8-16
precast, 3-24
properties, 3-21
properties, transformed, 6-8
rectangular, 8-25
rectangular, torsion, *1-25*
rolled, in tension, 4-31
shear capacity, 5-9
tension-controlled, 1-3, 1-4
transformed, *1-12*
uncracked, 6-8
uncracked, properties, *6-8*
Seismic base shear, 7-9
Seismic category
C, 6-10, *6-10*
D, 6-10, *6-10*
E, 6-10, *6-10*
F, 6-10, *6-10*
Seismic design, 8-43
analysis procedure, *8-44*

Italicized page numbers represent sources of data in tables, figures, and appendices.

category, 7-4, *7-4*
force, 8-51
Seismic force
  orthogonal, 8-51
  vertical distribution, 7-10, 7-18
Seismic load, 7-18, 8-48, 8-49, *8-50*
Seismic performance category, 6-9, *6-9, 6-10*
Seismic performance zone, 8-47, *8-47*
Seismic response, 8-43, 8-44
  coefficient, 7-9, 8-43, 8-46, *8-46*
  coefficient, elastic, 8-46
Seismic weight, effective, 7-6
Selection, analysis procedure, 8-47, *8-47*
Service
  factor, wet, 5-2, 5-18, *5-18*, 8-41
  limit state, 8-14
  live load plus impact, 8-16
  load, 1-1, 1-9, 1-13, 2-12, 3-5, 4-1, 4-33,
    6-1, 8-15, 8-23
  load, applied, *2-12, 2-17, 2-21*, 6-2
  load condition, *1-12*, 8-15, 8-23
  load, total, 2-17
  moment, 1-14
Serviceability design stage, 3-2, 3-5
Set, anchor, 3-2, 3-18
Shape
  factor, 4-3
  W, 4-4, 4-8
Shear, 1-15–1-19, 1-25
  applied, 8-38
  base, 7-9, 7-10, 7-18
  base, distribution, 7-10
  beam web, 4-8
  block, 4-8
  cantilever retaining wall, 2-23
  capacity, 1-34, 1-40
  capacity, combined, 3-13, 8-20, 8-29
  capacity, design, 4-43, 4-46
  capacity, nominal, 3-12, 3-14, 4-46, 8-20,
    8-28
  capacity, punching, 2-5
  capacity, section, 5-9
  capacity, total nominal, 3-14
  coefficient, 4-47
  connection, 4-50, 8-37
  connection, double, 5-20
  connection, single, 5-19, 5-20, 5-23
  connector, 4-50–4-53
  critical section, 1-16, *1-16, 1-45, 2-3, 2-6,*
    3-13, 8-13, 8-29
  deep beams, 1-19–1-21
  design, 3-11, 3-12, 4-8, 4-46, 5-8, 6-5,
    8-20, 8-27, 8-28, 8-36
  design, maximum, 6-5
  design strength, 4-34
  design value, reference, 5-11
  determination, 8-7
  double, reference value, 5-24
  flexural, 1-45, *1-45*, 2-3, *2-3*, 2-7, 2-13
  force, 1-38, 3-14, 4-37, 6-10–6-13
  force, applied, 6-11
  force, applied factored, 3-13
  force, critical section, 2-13
  force diagram, *2-13*, 2-13
  force, factored, 1-15, 1-19, 1-34, 3-14,
    3-22, 3-23
  force for vehicle loads, 8-42
  force, horizontal, 4-51
  force, range, 8-38
  force, total, 8-39
  force, unfactored, 3-14
  friction reinforcement area, 1-19

horizontal, range, 8-38
in flexural member, 6-5
lag, 4-31
maximum, on strap, *2-19*
one-way, 1-45 (*see also* Shear, flexural)
plate connection, *5-18*, 5-22
plate edge distance, 5-22
plate end distance, 5-22
plate spacing, 5-22
punching, 1-45, *1-45*, 1-46, 2-5, 2-12
punching, critical perimeter, *2-6*
reinforcement, 1-16–1-18, 3-12, 3-13,
  3-15, 6-5, 6-10, 6-11, 8-20, 8-28
reinforcement, horizontal, 6-11
reinforcement, vertical, 1-16
requirement, horizontal, 3-23
resistance, 1-18
rupture, 4-8
seismic base, 7-9
span-to-depth ratio, 1-23
story, 1-29
strap beam, 2-18
strength, design, 4-8
strength, footing, 2-3
strength, horizontal, nominal, 3-22, 3-23
strength, nominal, 1-15–1-19, 3-12, 4-8,
  4-34, 4-46, 4-51, 8-13
strength, punching, 2-5
strength, required, 4-43
stress, 1-19, 4-8, 5-9, 6-13
stress, allowable, 6-5, 6-11, 6-13
stress, eccentric, 2-6
stress, masonry, 6-5, 6-13
two-way, 1-45 (*see also* Shear, punching)
wall, 6-10, *6-10*
wall design, 6-10
wall nominal reinforcement requirements,
  6-10
wave velocity, *8-45*
zero, *2-13*, 2-15, 4-18
Shear-friction method, 3-23
Shored construction, 4-53
Short column
  axial load capacity, 1-30, 1-31
  concrete, 1-30, 1-31
  lateral tie reinforcement, 1-30
  spiral reinforcement, 1-30
Shortening, elastic, 8-32, 8-33
Shortening loss, elastic, 3-19
Short-term deflection, 1-12
Shrinkage, 1-36, 3-2
  loss, 3-20
  strain, 3-20
Side grain, 5-21, 5-23
Side plate, steel, connection, 5-22
Sidesway, 4-11
  web buckling, 4-48
Simple span, *8-16*
Simplified determination, seismic base shear,
  7-17, *7-17*, 7-18
Simplified lateral force procedure, 7-14–7-16
Simplified vertical distribution, base shear,
  7-18
Simply supported beam, *1-12*, 6-2
Simply supported slab, *1-12*, 8-11
Single shear connection, 5-19, 5-20, 5-22,
  5-24
Single support, 1-35
Single-mode elastic method *8-44*, 8-49
Single-span bridge, 8-47
Site class, 8-44, 8-45
Site factor, 8-43, 8-45, *8-45*, 8-46

Site classification characteristics, 7-2, *7-2*
Site coefficient, 7-3, *7-3*
Size
  factor, 5-2, 5-5, 5-7, 8-41, 8-43
  tension reinforcement, 8-17
Skew, angle of, 8-52
Skin
  reinforcement, 1-10, 8-18
  reinforcement, longitudinal, 8-18
Slab
  and beam equivalent dimensions, *1-42*
  and wall, c_s value, *1-40*
  both ends continuous, *1-12*
  cantilever, *1-12*
  concrete, 8-9, 8-11
  concrete, design, 8-9
  concrete, transformed area, 8-38
  concrete, transformed width, 8-38
  depth, 8-35
  depth, equivalent, 4-50
  flat, 3-9
  non-staggered reinforcement, *1-40*
  one end continuous, *1-12*
  simply supported, *1-12*, 8-5
  staggered reinforcement, *1-40*
  systems, two-way, 1-43–1-47
  width, 8-9
  width, effective, 8-35
Slenderness ratio, 4-11, 4-13, 4-28, 5-2, 5-12,
  8-41
  column, 4-11, 4-13, 4-28
  concrete column, 1-28, 1-30, 1-31, 1-33
Sliding, factor of safety, 2-21
Slip anchor, 3-18
Slip-critical bolt, 4-34, 4-35
Smooth interface, 3-22
Snow load, 7-13
Soil
  bearing pressure, 2-12, *2-17*, 2-17
  pressure, 2-1, 2-16, 2-22
  pressure, uniform, 2-12
  profile, 8-45, *8-45*
  profile type, 7-2, *7-2*
Source, seismic, classification, 7-2
Spacing
  bolt, 5-19, 5-20, *5-20*
  clear, between spirals, 1-27
  closed stirrups, maximum, 1-25
  connector, 8-38
  lag screw, 5-19, 5-21
  lateral support, 6-2
  maximum, closed stirrups, 3-16
  maximum, main reinforcement, 2-4
  maximum, transverse reinforcement, *1-40*
  shear plate, 5-19, 5-22
  shear reinforcement, 6-5, 6-12
  split ring, 5-19, 5-22
  stirrup, 3-12, 8-29
  tension reinforcement, 8-17
  tie, 3-23
  tie, hooked bar, 1-37
  vertical, tie, 1-27
  wood screw, 5-19, 5-23
Span
  beam, 5-6
  clear, 1-19, *1-39*, 5-6, 6-2
  continuous, 8-2, 8-4
  effective, 5-3, 5-6, 6-2
  length, effective, *6-2*
  moment, horizontal, 2-25
  simple, *8-16*
  typical, *1-38*

Italicized page numbers represent sources of data in tables, figures, and appendices.

Span/depth ratio, 1-11, *1-11* (compare to Span-to-depth ratio)
Span-to-depth ratio, 3-9
   shear, 1-23
Specified hole diameter, 4-30
Spectral analysis
   multimode *8-44*
   single-mode, *8-44*, 8-47
Spectrum response, 7-3, 7-8, *7-8*, 7-9
Spike connection, 5-24
Spiral
   clear spacing, 1-27
   minimum diameter, 1-27
   reinforcement, 1-37
   reinforcement, concrete column, 1-27
   reinforcement, short concrete column, 1-30
Spirally reinforced column, 1-41
Splice
   bar in tension, 1-40
   Class A, 1-40
   compression bar, 1-41
   lap, 1-41
   lap, c_s value, *1-40*
   tension lap, 1-31
   tension lap, Class A, 1-40
   tension lap, Class B, 1-40
Split ring connection, 5-18, 5-22
Split ring edge distance, 5-22
Split ring end distance, 5-22
Split ring spacing, 5-22
Square footing on isolated column, 2-5
Stability
   coefficient, 7-13, 7-14
   factor, beam, 8-41
   factor, column, 5-3
   index, 1-29
Stage
   design, serviceability, 3-2, 3-5
   design strength, 3-2, 3-6
   design, transfer, 3-2
Staggered fastener, 5-18, *5-19*
Staggered reinforcement slab and wall, *1-40*
State service limit, 8-11
Static
   equilibrium check, 4-27
   force, equivalent, 8-49, *8-50*
   lateral displacement, 8-47
   lateral force procedure, 7-2
   moment, total factored, 1-43, *1-44*
   seismic load, equivalent, 8-48, *8-48*, 8-49, *8-50*
Statical design method, 4-24
Statical moment, 8-38
Statically determinate structure, 4-17
Statically indeterminate structure, 3-27
Steel
   column with base plate, 2-6, *2-6*, 2-7, *2-7*
   design, structural, 8-33
   factor, prestressing, 8-25
   low-relaxation, 8-25
   prestressing, 8-25
   prestressing, modulus of elasticity, 8-33
   side plate connection, 5-22
   stress-relieved, 8-25
   structural, 8-33
Stem design, counterfort, 2-25
Stiffener
   area, required, 4-47
   bearing, 4-48
   intermediate, 4-43, 4-44, 4-46, 4-47
   pair, 4-46
Stiffness
   central column, 8-49

lateral, 8-47, 8-49
ratio, 4-12, 4-13
ratio, concrete column, 1-24
reduction factor, 4-10, 4-19
transverse, bridge, 8-48
value, 4-12
Stirrup, 1-16, 1-18, *1-35*, 1-38
   closed, *1-23*, *1-25*, 3-15
   closed, maximum spacing, 1-25
   inclined, beam, *1-16*
   minimum combined area, 1-25
   spacing, 3-12, 8-29
Story
   drift, maximum allowable, 7-12, *7-12*
   drift, 7-24
   nonsway, 1-29
   primary moment, 7-13
   secondary moment, 7-13
   shear, 1-29
   total vertical load, 1-29
Straight bar
   in compression, 1-36
   in tension, 1-35
Straight-line theory, 1-14
Strain
   balanced, 1-6
   compatibility, 3-10
   diagram, *6-9*
   distribution, *1-4*
   maximum, 1-6, 3-2, 3-10
   shrinkage, 3-20
   ultimate, 1-4
Strand
   low-relaxation, 8-25
   post-tensioned, 8-33
   pre-tensioned, 8-33
   stress-relieved, 8-25
Strap
   beam, 2-16, *2-17*, 2-18, 2-19
   footing, 2-16, *2-19*
   footing with applied service loads, *2-17*
Strength (*see also* Capacity *and* Resistance)
   axial, 4-21
   axial, design, 4-14, 4-15
   bearing, 2-9
   bearing, design, 4-34, 4-39
   bearing, nominal, 4-35, 4-48
   compressive, maximum, 6-1
   design, 1-1, 4-2, 4-10, 4-13
   design, flexural, 1-3
   design method, 8-15, 8-34
   design moment, flange, 1-8
   design stage, 3-2, 3-6
   flexural, 3-8, 3-9
   flexural, nominal, 3-7, 4-5, 4-6, 4-43, 4-44, 8-35
   flexural, using strain compatibility, 3-10
   moment, 3-7
   moment, minimum required, 1-3
   nominal, 1-3, 1-39, 4-2, 4-10, 4-11
   nominal flexural, rectangular beam, 1-3
   nominal shear, 1-15–1-19
   nominal weld electrode, 4-30
   of welded stud shear connector, nominal, 8-40
   reduction factor, 1-2, 8-25, 8-37
   required, 1-1, 4-1
   shear, design, 4-8, 4-34
   shear, footing, 2-3
   shear, nominal, 3-12, 4-8, 4-34, 4-46, 8-20, 8-31, 8-37
   shear, nominal horizontal, 3-23

shear, punching, 2-5
shear, required, 4-34
tensile, 8-25
tensile, design, 4-34
tensile, nominal, 4-34
tension design, 4-34
theoretical ultimate, 1-2
ultimate, 8-38, 8-39
ultimate, theoretical, 4-2
weld design, 4-30
yield, 3-6
Stress
   allowable, 8-22, 8-23
   axial load, *6-8*
   bending moment, *6-8*
   block depth, 4-50, 8-35
   block, equivalent depth, 8-15
   block, equivalent rectangular, depth, 1-3, 8-25
   block, rectangular, 1-3, *1-4*, 1-8
   bond, 1-39
   category, 4-33
   combined, 6-4, *6-8*
   combined, compression and flexure, 5-14, 5-15
   compressive, 1-16, 3-13, 3-14, 3-19, 8-13, 8-29
   compressive, allowable, 6-1, 6-2, 6-6, 6-8
   compressive, at centroid of prestressing steel, 8-32
   compressive, calculated, 6-8
   concrete, 3-26
   critical, 4-13
   cycle factor, 8-38
   distribution, *1-4*, 4-3
   eccentric shear stress, 2-6
   fatigue, 8-38
   limits, fatigue, 8-18
   permissible, 1-14, 3-2, 3-4, 3-5
   range, 8-18
   range, actual, 4-33
   range threshold, 4-33
   reinforcement, 1-10, 1-14
   residual, 5-5
   reversal, 4-32
   shear, 1-19, 4-8, 5-9, 6-12, 8-31, 8-43
   shear, allowable, 6-5, 6-11, 6-13
   tendon, 3-2
   tensile, 1-14, 1-36, 3-4, 3-9, 3-14, 4-27, 6-2
   tensile, allowable, 6-1
   yield, 1-39, 4-14, 4-25
Stress-relieved steel, 8-22, 8-25
Stress-strain curve, 3-10
Stringer, *8-10*
Strip
   column, 1-43, *1-43*
   column, exterior negative moment distribution, *1-43*
   column, interior negative moment distribution, *1-43*
   column, moment distribution to edge beam, *1-43*
   column, positive moment distribution, *1-43*
   design, details, *1-43*
   footing, 2-1, 2-6, 2-7
   middle, *1-43*
Structural steel design, 8-33
Structural system, 7-5, *7-14*
   classification, 7-5
Structure
   low-rise, 7-14
   seismic load on, 7-18

Italicized page numbers represent sources of data in tables, figures, and appendices.

statically determinate, 4-17
statically indeterminate, 3-27
timber, 8-43
wood, 8-40
Strut-and-tie, 1-19–1-21, *1-20*, *1-21*
Stud, welded, shear connector, 4-51–4-53,
    8-40
Substructure, 8-44
Sudden tensile failure, 8-17, 8-27
Superimposed applied load, 3-24
Superimposed dead load, unfactored, 3-14
Superstructure, 8-5, 8-47
    depth, minimum, 8-16
    recommended minimum depth, *8-16*
Support
    lateral, 4-4
    lateral, spacing, 6-2
    length, minimum, 8-43, 8-47
    reaction, 8-13, 8-29
    restraint, 3-27, 3-28
    simple, 1-39
Surcharge
    live load, 2-20, 2-21
    pressure, total, 2-21
Surface
    faying, 4-34, 4-35, 4-40, 4-41
    wearing, 8-12
Sustained load, 1-12
Sway, 1-33
    frame, *1-29*, 1-30, *4-13*, 4-28, 4-29
    mechanism, 4-26, 4-27
    moment, 1-33
System
    bearing wall, 7-5, *7-6*
    bridge deck, 8-1, 8-5
    building frame, 7-5, *7-6*
    cantilevered column, 7-5, *7-6*
    dual, 7-5, *7-6*
    inverted pendulum structure, 7-5
    moment-resisting frame, 7-5, *7-6*
    slab, two-way, 1-42–1-47
    structural, 7-5, *7-6*, *7-17*

T
Tabulated design value
    bending, 8-40, 8-41
T-beam, 8-15, 8-17
    flange depth, 8-15, 8-17
Temperature
    factor, 5-2, *5-2*, 5-3, 5-6, 5-18, *5-18*
    reinforcement, 1-31
Tendon
    bonded, 3-8, 8-23
    jacking force, 3-2
    post-tensioned, 3-18–3-20
    prestrain, 3-10
    prestressing, 3-7, 3-14, 3-19, 3-25
    pre-tensioned, 3-18–3-20
    profile, alternative, 3-26, *3-26*
    stress, 3-2
    type, factor, 3-8
    unbonded, 3-9, 8-23
Tensile
    failure, sudden, 8-17, 8-27
    force, 4-34
    force, corbel, 1-23
    force, total, 3-4, 3-9, 3-10
    rupture at connection, 4-29
    strength, 8-25
    strength, design, factored load, 4-34
    strength, nominal, 4-34
    stress, 1-14, 1-40, 3-4, 3-9, 3-14, 6-2, 6-7

stress, allowable, 6-1
zone, 3-7
Tension
    design, 5-16
    design strength, 4-34
    face, 1-9
    field action, 4-44–4-47
    lap splice, 1-36, 1-40
    member design, 4-29
    plate, 4-29
    reinforcement, 1-3, 1-8, *1-9*, *1-19*, 1-35,
        6-1, 8-18
    reinforcement, flanged section, *1-8*
    reinforcement, flexural, 8-18
    reinforcement index, 1-3, *A-1*
    reinforcement, nonprestressed, 8-25
    reinforcement ratio, 1-3, 1-4, 1-6, 1-7
    rolled section, 4-31
    rupture, 4-8
Test, compression, 6-1
Theoretical ultimate strength, 1-2, 4-2
Theory
    Rankine's, 2-21
    straight-line, 1-14
    Westergaard's, 8-9
Thickness
    base plate, required, 4-23
    effective, 4-38
    throat, effective, 4-38
Thread length, 5-22
Three-degree indeterminate frame, 4-23
Throat thickness, effective, 4-38
Tie
    column, *1-27*
    factor, 1-37
    horizontal, 2-26
    lateral, 6-6
    reinforcement, counterfort, *2-25*
    size, rectangular column, 1-27
    spacing, 3-23
    spacing, hooked bar, 1-37
    vertical spacing, 1-27
Timber
    glued laminated, 8-41
    visually graded, 5-2
Time-dependent factor
    sustained load, 1-12
    value, *1-12*
Time effect factor, 8-41, *8-41*
Time history method, *8-44*
Toe-nailed connection, 5-24, 5-25
Toe-nail factor, *5-18*, 5-19, 5-25
Torque, factored, 1-24, 3-15
Torsion, 1-15, 1-24, 1-25
    design, 3-11, 3-15
    factored, causing cracking, 1-25
    flanged section, *1-25*
    rectangular section, *1-25*
    reinforcement, 1-24
Torsional buckling, lateral, 4-3, 4-5
Torsional capacity, nominal, 3-16
Torsional moment (*see* Torque)
Torsional reinforcement, longitudinal, 3-15
Total
    active earth pressure, 2-21
    dead load, 7-6, 7-13
    deflection, 1-11, 1-12
    factored load, 2-18
    factored moment, 1-39, 3-14
    factored static moment distribution
        factors, *1-43*
    length, weld, 4-33

nominal shear capacity, 3-14
passive earth pressure, 2-21
service load, 2-17
shear force, 8-39
surcharge pressure, 2-21
tensile force, 3-10
unfactored moment, 3-14
weight of bridge superstructure and
    tributary superstructure, 8-43
Traffic lane, *8-2* (*see also* Design lane)
Transfer
    concrete, 8-22
    conditions, 8-22
    design stage, 3-2
    force at column base, 2-8
    load, 2-8
Transformed area
    of concrete slab, 8-38
    reinforcement, 6-8, *6-8*
Transformed section, *1-11*
    properties, 3-21, 6-7
Transformed width of concrete slab, 8-38
Transverse load, 5-4, 8-47, *8-47*
Transverse reinforcement
    index, 1-34
    index derivation, *1-35*
    maximum spacing, *1-40*
Transverse stiffness, bridge, 8-47
Tributary area, *1-45*
Truss, plane, 4-12
Two-span beam, 3-27
Two-way shear, 1-45 (*see also* Punching
    shear)
Two-way slab systems, 1-42–1-47
Typical column, *1-43*
Typical span, *1-43*
Typical value
    bending coefficient, *4-6*
    effective length, *5-3*

U
Ultimate load, 8-25, 8-28, 8-39
    condition, 8-25
Ultimate moment of resistance, 3-6
Ultimate strength, 8-38, 8-39
    required, 4-1
    theoretical, 1-2, 4-2
Unbonded tendon, 3-9, 8-23
Unbraced length, 4-4–4-6
    maximum, 4-25, 4-28
Uncertainty, directional, 8-48
Uncracked section, 6-8
    properties, *6-8*
Unfactored dead load, 3-14
    superimposed, 3-14
Unfactored external applied load, 3-14
Unfactored moment, total, 3-14
Unfactored shear force, 3-14
Uniform bearing pressure, *2-12*
Uniform bending moment, 4-4, 4-5
Uniform load, 3-14, 8-44, 8-47
    static seismic, 8-48, *8-48*
    method, 8-44
    soil pressure, 2-12, 2-17
    unit, 8-45, 8-46
Uniformly distributed load, 8-2
Unit
    area method, elastic, 4-36
    load, uniform, 8-47
Unshored construction, 4-53

Italicized page numbers represent sources of data in tables, figures, and appendices.

V

Value
  of c_s, lap splice, *1-40*
  of time-dependent factor, *1-12*
  stiffness, 4-12
Vector analysis technique, elastic, 4-40, 4-41
Vehicle loads shear force, 8-43
Vehicular loading, 8-2
Vertical crack, *1-35*
Vertical distribution
  base shear, 7-10, 7-18
  of seismic force, 7-10, 7-18
Vertical force
  bolt, 4-35
  weld, 4-31
Vertical reinforcement, 6-10, *6-10*, 6-11
  area, minimum, 1-19
  shear, 1-14, 1-19
Vertical spacing, tie, 1-24
Vibration, 8-44
  fundamental period, 8-44–8-48
  period of, fundamental, 7-6, 7-7
Virtual displacement, 4-26, 4-27
Virtual work method, 8-47
Visually graded lumber
  dimension, 5-1
  sawn, 5-4, 5-6
Visually graded timber, 5-2
Volume factor, *5-2*, 5-5, 5-6, 5-8, 8-41, 8-42, 8-43

W

W shape, 4-4, 4-8
  beam, rolled, 4-8
Wall
  bearing, 7-5, *7-6*
  concrete, *2-3*
  design, shear, 6-10
  masonry, *2-3*, 6-10, 6-13
  masonry, with shear reinforcement, 6-13

masonry, without shear reinforcement, 6-11
  non-staggered reinforcement, *1-40*
  retaining, cantilever, 2-20–2-25
  retaining, cantilever, with applied service loads, *2-21*
  retaining, counterfort, 2-25, *2-25*, 2-26
  shear, *6-10*, 6-10
  staggered, *1-36*
  Wearing surface, 8-12
Web
  area, 4-8
  beam, shear, 4-8
  buckling capacity, design, 4-48
  buckling coefficient, 4-46
  buckling, sidesway, 4-48
  crippling, 4-10
  depth-to-thickness ratio, 4-43
  effective depth, 1-10
  shear cracking, 3-14, 8-28
  stiffener on girder, 4-12
  tear out, 4-8
  unstiffened, 8-37
  yielding, local, 4-9
Weight, effective seismic, 7-6
Weld
  design strength, 4-38
  electrode, nominal strength, 4-38
  fillet, 4-38, 4-39
  groove, complete-penetration, 4-38
  groove, flare, 4-38
  groove, partial-penetration, 4-38
  group, eccentrically loaded, 4-40, 4-41
  total length, 4-42
Welded connection, 4-30, 4-31
  design, 4-38
Welded stud shear connector, 8-40
Westergaard's theory, 8-9
Wet service factor, 5-2, *5-2*, 5-6, 5-18, *5-18*, 8-41, *8-41*
Wheel, 8-2

Width
  control, crack, 1-9
  deck, 8-1
  effective, 4-50
  effective, concrete slab, 8-35
  flange, effective, 3-21
  seat, minimum, 8-51
  transformed, of concrete slab, 8-38
Width-to-thickness ratio, maximum allowable, 4-47
Withdrawal
  and lateral load, combined, 5-20, 5-22, 5-23
  design value, *5-18*
  design value, in side grain, 5-20, 5-22, 5-23
Wobble coefficient, 8-32
Wood
  modulus of elasticity, 8-40
  screw, connection, 5-23
  screw, edge distance, 5-23
  screw, end distance, 5-23
  screw, spacing, 5-23
  side member connection, 5-22
  structure, 8-40
Work, virtual, method, 8-47
Working load, 1-1, 4-1 (*see also* Service load)

Y

Yield strength, 3-6, 4-28
Yield stress, 1-39, 4-14
Yielding, web, local, 4-9

Z

Zero shear, 2-15, 4-26
Zone
  compression, 8-35
  compression, depth, 8-35
  seismic performance, 8-44
  tensile, 3-7

Index of Codes

AASHTO
1.3.2, 8-11
1.3.3, 8-12
1.3.4, 8-12
1.3.5, 8-12
2.5.2.6.2, 8-16
3.6, 8-1, 8-2
3.6.1.1.1, 8-11
3.6.1.1.2, 8-3, 8-6, 8-8
3.6.1.2, 8-2
3.6.1.2.1, 8-3
3.6.1.3.1, 8-2, 8-3, 8-4, 8-5, 8-8, 8-19
3.6.1.4, 8-19
3.6.2, 8-4, 8-43
3.10.2, 8-43
3.10.2.1, 8-44
3.10.3.1, 8-44, 8-45
3.10.4, 8-44
3.10.6, 8-47
3.10.8, 8-51
4.4, 8-6
4.6.2.1, 8-9
4.6.2.2, 8-5
4.6.2.2.1, 8-5
4.6.2.3, 8-9, 8-11
4.6.2.6.1, 8-35
4.7.4.3, 8-47
4.7.4.3.1, 8-44
4.7.4.3.2c, 8-44
4.7.4.4, 8-51, 8-52
5.4.2.6, 8-17, 8-26
5.5.3, 8-18
5.5.4.2.1, 8-20, 8-25, 8-26, 8-29
5.6.3, 8-20
5.7, 8-15
5.7.2.2, 8-25
5.7.3.3.2, 8-17, 8-27
5.7.3.6.2, 8-16, 8-17
5.7.3.4, 8-17, 8-18
5.8.2.7, 8-20, 8-29
5.8.2.9, 8-13
5.8.3, 8-20
5.8.3.2, 8-14, 8-29
5.8.3.3, 8-20, 8-28
5.8.3.4.1, 8-20
5.9.3, 8-22, 8-23
5.9.4.1, 8-22
5.9.4.2, 8-23
5.9.5.2.2, 8-32
5.9.5.2.3a, 8-33
5.9.5.2.3b, 8-33
6.5.4.2, 8-35, 8-37, 8-40
6.10.7, 8-34
6.10.9.2, 8-37
6.10.10.4.1, 8-39
6.10.10.4.2, 8-39
8.4.4, 8-40
8.4.4.1, 8-40
8.4.4.9, 8-41
8.5.2.2, 8-42
8.6.2, 8-42

9.7.3.2, 8-9
C4.6.2.2.1, 8-6, 8-8
C5.4.2.4, 8-32
D6.1, 8-34, 8-35
App. D6, 8-34
Eq. (1.3.2.1-2), 8-12
Eq. (1.3.2.1-3), 8-12
Eq. (3.6.2.2-1), 8-5
Eq. (4.6.2.2.1-1), 8-6, 8-7
Eq. (4.6.2.2.1-2), 8-6
Eq. (4.6.2.3-1), 8-11
Eq. (4.6.2.3-2), 8-11
Eq. (4.7.4.4-1), 8-51
Eq. (5.5.3.2-1), 8-18
Eq. (5.7.3.1.1-1), 8-25
Eq. (5.7.3.1.1-4), 8-25
Eq. (5.7.3.2.1-1), 8-25
Eq. (5.7.3.2.2-1), 8-25
Eq. (5.7.3.3.2-1), 8-17
Eq. (5.7.3.4-1), 8-18
Eq. (5.8.2.5-1), 8-20
Eq. (5.8.2.9-1), 8-21, 8-29
Eq. (5.8.3.3-4), 8-20, 8-28
Eq. (5.8.3.4.3-2), 8-28
Eq. (6.10.7.1.1-1), 8-35
Eq. (6.10.7.1.2-2), 8-35
Eq. (6.10.9.2-1), 8-37
Eq. (6.10.9.2-2), 8-37
Eq. (6.10.10.1.2-1), 8-38
Eq. (6.10.10.1.2-3), 8-38
Eq. (6.10.10.2-1), 8-39
Eq. (6.10.10.2-2), 8-38
Eq. (6.10.10.4.3-1), 8-40
Eq. (8.4.4.1-1), 8-40
Eq. (8.4.4.2-2), 8-41
Eq. (8.4.4.2-3), 8-41
Eq. (8.6.1-1), 8-41
Eq. (8.6.2-1), 8-42
Eq. (8.6.2-3), 8-41
Eq. (8.6.2-4), 8-41
Eq. (8.6.2-5), 8-41
Eq. (8.7-1), 8-42
Eq. (8.7-2), 8-42
Eq. (C4.7.4.3.2b-1), 8-49
Eq. (C4.7.4.3.2b-2), 8-49
Eq. (C4.7.4.3.2b-3), 8-49
Eq. (C4.7.4.3.2b-4), 8-49
Eq. (C4.7.4.3.2b-5), 8-49
Eq. (C4.7.4.3.2c-1), 8-47
Eq. (C4.7.4.3.2c-2), 8-47
Eq. (C4.7.4.3.2c-3), 8-48
Eq. (C4.7.4.3.2c-4), 8-48
Eq. (C5.4.2.4-1), 8-16, 8-40
Eq. (C5.8.2.9-1), 8-20, 8-28
Fig. 3.10.2.1-1, 8-44
Fig. 3.10.2.1-21, 8-44
Table 2.5.2.6.3-1, 8-16
Table 3.4.1-1, 8-12, 8-13, 8-19
Table 3.4.1-2, 8-12
Table 3.6.1.1.2-1, 8-3, 8-6, 8-8
Table 3.6.2.1-1, 8-4, 8-5, 8-9, 8-54, 8-56

Table 3.10.7.1-1, 8-51
Table 3.10.7.1-2, 8-51
Table 4.6.2.1.3-1, 8-10
Table 4.6.2.2.1-1, 8-5, 8-6
Table 4.6.2.2.2b-1, 8-6
Table 4.6.2.2.3a-1, 8-9
Table 5.9.5.2.2b-1, 8-32
Table 5.9.5.3-1, 8-33
Table 8.4.1.1.4-1, 8-40
Table 8.4.1.1.4-2, 8-40
Table 8.4.1.2.3-1, 8-40
Table 8.4.1.2.3-2, 8-40
Table 8.4.4.2, 8-41
Table 8.4.4.3-1, 8-41
Table 8.4.4.3-2, 8-41
Table 8.4.4.5-1, 8-42
Table A4-1, 8-10
Table C5.7.3.1-1, 8-25
Table D6.1-1, 8-34

ACI
2.1, 3-7
7.10, 1-27
7.10.5, 1-27
7.10.5.1, 1-30
7.12, 1-5, 1-36
7.12.2, 2-4, 2-5, 2-8, 2-11, 2-15, 2-28
7.13, 1-36
8.3.3, 1-43, 2-25
8.5, 3-19, 3-20, 3-22
8.5.1, 1-32
8.10, 3-21
9.2, 1-1, 2-1, 2-2
9.2.4, 2-22, 2-25
9.3, 1-2
9.3.2.1, 1-4
9.3.2.2, 1-4
9.5.2.3, 1-12, 3-6
9.5.2.5, 1-12
10.17, 2-9
10.17.1, 2-8
10.2, 1-3, 1-19, 2-4
10.2.3, 1-3
10.2.7.3, 3-7, 3-8
10.3, 1-6, 2-8, 2-14
10.3.3, 1-4, 1-7
10.3.4, 1-4, 3-9
10.3.6.1, 1-30
10.3.6.2, 1-30
10.5, 1-6
10.5.3, 2-15
10.5.4, 2-4
10.6, 1-10
10.6.7, 1-10
10.9, 1-27
10.9.3, 1-27
10.11.4.1, 1-28
10.11.4.2, 1-29
10.11.5, 1-28, 1-31
10.12.1, 1-28
10.12.3, 1-32

10.13.2, 1-30
10.13.5, 1-33
10.17.1, 2-9
11.1.1, 3-13
11.1.3, 1-16, 3-13
11.1.3.1, 2-3
11.3, 2-3
11.4.2, 3-12
11.4.3.1, 3-14
11.4.3.2, 3-14
11.5.1.2, 1-18
11.5.4.3, 1-18
11.5.5, 3-12
11.5.6, 3-13
11.5.5.1, 1-16, 1-17, 1-19
11.5.5.3, 1-17, 1-18
11.5.6.1, 1-17
11.5.6.3, 1-16, 1-17
11.5.6.8, 3-12
11.5.7.2, 3-12
11.5.7.4, 1-16, 1-18
11.5.7.5, 1-18
11.5.7.9, 1-18, 3-12
11.6.1, 1-24, 3-15, 3-16, 11-26
11.6.2.2, 1-25, 3-15
11.6.3.6, 1-24, 3-15
11.6.3.7, 1-24, 3-15
11.6.5.2, 3-15
11.6.5.3, 1-24
11.6.6.1, 3-15
11.6.6.2, 1-25
11.7.1, 1-19
11.7.4, 3-23
11.7.4.3, 1-23, 3-23
11.8.1, 1-19
11.8.2, 1-19
11.8.3, 1-19
11.8.4, 1-19
11.8.5, 1-19
11.9, 1-23
11.9.3.2, 1-23
11.9.3.4, 1-23
11.9.4, 1-23
11.12.1.2, 1-45, 2-5, 2-6
11.13.1.1, 1-45
12.2.2, 1-35
12.2.5, 1-36
12.3.1, 2-9
12.3.2, 1-36, 2-9
12.3.3, 1-37
12.4.1, 1-35
12.4.2, 1-35
12.5.1, 1-37
12.5.2, 1-37
12.5.3, 1-37
12.10.3, 1-38
12.10.4, 1-38
12.10.5, 1-38
12.11, 1-36
12.11.1, 1-38
12.11.3, 1-39
12.12, 1-39, 1-40
12.14, 1-40
12.15, 1-36, 1-40
12.16.1, 1-41
12.16.2, 1-41
12.17.2, 1-41
13.3.8, 1-36
13.6.1, 1-42
13.6.3, 1-43
13.6.4, 1-43
13.6.7, 1-43
14.3, 2-28
15.4.2, 2-4

15.4.4.2, 2-9, 2-14
15.5.2, 2-3, 2-5
15.8, 2-8
15.8.2.1, 2-8, 2-9
17.5.3, 3-23
17.5.3.1, 3-23
17.5.3.2, 3-23
17.5.3.3, 3-23
17.5.2.4, 3-23
17.5.2.5, 3-23
17.6.1, 3-23
18.0, 3-7, 3-8
18.2, 3-2
18.4.1, 3-2–3-4, 3-30
18.4.2, 3-5, 3-6
18.5.2, 3-2
18.6.2.1, 3-17
18.7.2, 3-7–3-10
18.7.3, 3-6
18.8, 3-7
18.8.1, 3-9, 3-10, 3-31
18.8.2, 3-7
18.9, 3-9, 3-10
A.3, 1-13
App. A, 1-13
Eq. (9-4), 2-22
Eq. (9-7), 1-12
Eq. (9-8), 1-12, 1-13
Eq. (9-9), 1-13, 3-6
Eq. (9-10), 1-13
Eq. (9-11), 1-12, 1-14, 3-6
Eq. (10-1), 1-30
Eq. (10-2), 1-30
Eq. (10-4), 1-10
Eq. (10-5), 1-27
Eq. (10-6), 1-29
Eq. (10-7), 1-30, 1-32
Eq. (10-15), 1-33
Eq. (10-16), 1-33
Eq. (10-18), 1-33
Eq. (11-1), 1-16, 3-16
Eq. (11-2), 1-16, 3-14, 3-16
Eq. (11-3), 1-16, 1-18, 1-45, 2-4, 2-7, 2-14, 2-27
Eq. (11-5), 1-16
Eq. (11-9), 3-12, 3-13
Eq. (11-10), 3-14, 3-15
Eq. (11-11), 3-14, 3-15
Eq. (11-12), 3-14, 3-15
Eq. (11-13), 1-16, 1-35, 3-13, 3-16
Eq. (11-14), 3-13, 3-16
Eq. (11-15), 1-16, 3-12, 3-14, 3-16
Eq. (11-16), 1-16, 1-18
Eq. (11-17), 1-18
Eq. (11-21), 1-24
Eq. (11-22), 1-24, 3-15
Eq. (11-23), 1-25, 3-15
Eq. (11-24), 1-24, 3-15
Eq. (11-33), 1-46, 2-6
Eq. (11-34), 1-46
Eq. (11-35), 1-46, 2-6, 2-13, 2-27
Eq. (12-1), 1-35, 2-5
Eq. (12-5), 1-39
Eq. (13-3), 1-43
Eq. (18-1), 3-17
Eq. (18-2), 3-17
Eq. (18-3), 3-8, 3-31
Eq. (18-4), 3-9
Eq. (18-5), 3-9
Eq. (18-6), 3-9
Eq. (A-2), 1-20
Eq. (A-3), 1-20
Eq. (A-7), 1-21
Eq. (A-8), 1-20

Fig. R10.12.1, 1-28
A.1, 1-20
A.2.5, 1-20
A.3.2, 1-20
A.3.3, 1-19, 1-20
A.4.1, 1-20
A.4.3.2, 1-21
A.5.1, 1-21
A.5.2, 1-20
R1.1, 1-14
R10.12, 1-28
R11.4.3, 3-14
R11.6.3.10, 3-15
R15.8.1.2, 2-9
R18.4.1, 3-4
Table 9.5(a), 1-11
Table 9.5(b), 1-11, 1-13

ASCE
1.5.1, 7-4
2.4.1, 6-5, 6-9, 7-21
11.4.1, 7-2, 7-17
11.4.2, 7-2
11.4.3, 7-3
11.4.4, 7-3, 7-4
11.4.5, 7-8
11.6, 7-4
11.7, 7-4
12.2.1, 7-5
12.3.1.1, 7-2
12.3.4, 7-10, 7-19, 7-20
12.3.4.1, 7-19
12.3.4.2, 7-19, 7-20
12.7.2, 7-6
12.8.1, 7-12, 7-13
12.8.1.1, 7-9, 7-12, 7-21
12.8.2, 7-6, 7-7, 7-8
12.8.2.1, 7-6
12.8.3, 7-10
12.8.6, 7-12
12.8.6.2, 7-12
12.8.7, 7-12, 7-13
12.10.1.1, 7-11, 7-12, 7-21
12.11.2.1, 7-21
12.11.2, 7-21
12.14, 7-14
12.14.2, 7-18
12.14.3.1.1, 7-16
12.14.3.2.1, 7-16
12.14.3.1.1, 7-16
12.14.8.1, 7-17, 7-18
12.14.8.5, 7-18
17.5.4.2, 7-2
Eq. (12.8-1), 7-9, 7-10
Eq. (12.8-2), 7-9
Eq. (12.8-3), 7-9
Eq. (12.8-4), 7-9
Eq. (12.8-5), 7-9
Eq. (12.8-6), 7-9
Eq. (12.8-7), 7-7
Eq. (12.8-8), 7-7
Eq. (12.8-15), 7-5, 7-12
Eq. (12.10-1), 7-11, 7-19
Eq. (12.14-1), 7-14
Eq. (12.14-2), 7-15, 7-16
Eq. (12.14-11), 7-17
Eq. (12.14-12), 7-18
Fig. 11.4-1, 7-8
Fig. 22-15, 7-9
Fig. 22-20, 7-9
Table 1-1, 7-4
Table 11.4-1, 7-3, 7-17
Table 11.4-2, 7-3
Table 11.4-1, 7-4

Table 11.4-5, 7-9
Table 11.4-1, 7-4, 7-17
Table 11.4-6, 7-9
Table 11.4-7, 7-9
Table 11.6-1, 7-4, 7-5
Table 11.6-2, 7-4, 7-5
Table 11.3-1, 7-10
Table 11.3-2, 7-10
Table 11.6-1, 7-10
Table 12.2-1, 7-5, 7-6, 7-9, 7-13
Table 12.3-1, 7-10
Table 12.3-2, 7-10
Table 12.8-1, 7-7, 7-8
Table 12.12-1, 7-12
Table 12.14-1, 7-17, 7-18
Table 12.14-11, 7-18
Table 12.14-12, 7-18

BCRMS
1.8.2.1, 6-1
1.13.5.2.3, 6-16
1.14.2.2.1, 6-10
1.14.2.2.4, 6-10
1.14.2.2.5, 6-10, 6-12
1.14.5.3.3, 6-17
1.14.6.3, 6-10, 6-11, 6-12, 6-14
1.14.7.3, 6-10
2.1.1, 6-1
2.1.1.1, 6-1
2.1.2.1, 6-1
2.1.2.3, 6-4, 6-9, 6-10, 6-12
2.1.3.4, 6-1
2.1.6.1, 6-7
2.1.6.2, 6-7
2.1.6.3, 6-7
2.1.6.4, 6-7, 6-16
2.1.6.5, 6-7
2.2.3.1, 6-8
2.3, 6-3, 6-4, 6-11, 6-13, 6-18
2.3.2, 6-1–6-3, 6-11, 6-17, 6-18
2.3.2.1, 6-1, 6-2
2.3.2.2.1, 6-17
2.3.2.2.2, 6-7, 6-16
2.3.3, 6-3, 6-11, 6-17, 6-18
2.3.3.2.1, 6-6
2.3.3.2.2, 6-2, 6-8, 6-9
2.3.3.4, 6-2
2.3.3.4.2, 6-2
2.3.3.4.4, 6-2, 6-18
2.3.5.2.1, 6-5, 6-11
2.3.5.2.2, 6-5, 6-6, 6-11, 6-12, 6-18
2.3.5.2.2(b), 6-11
2.3.5.2.3, 6-5, 6-6, 6-18, 6-19
2.3.5.2.3(b), 6-13, 6-14
2.3.5.3, 6-5, 6-11–6-14
2.3.5.3.2, 6-14
2.3.5.5, 6-5
5.9.1.1, 6-6
5.9.1.2, 6-6
5.9.1.4, 6-6
5.9.1.6, 6-6
7.2.1.2, 6-6
Eq. (2-12), 6-8, 6-17
Eq. (2-13), 6-8
Eq. (2-17), 6-6, 6-7, 6-16
Eq. (2-18), 6-7
Eq. (2-19), 6-5, 6-6, 6-11, 6-12, 6-14, 6-19
Eq. (2-20), 6-5, 6-6, 6-18
Eq. (2-21), 6-11
Eq. (2-22), 6-11, 6-12
Eq. (2-23), 6-5, 6-6, 6-18
Eq. (2-24), 6-13
Eq. (2-25), 6-13, 6-14

Eq. (2-26), 6-5, 6-6, 6-13, 6-14, 6-19
Eq. (7-10), 6-18

BCRMS Commentary
2.3, 6-5
2.3.5.3, 6-11, 6-12, 6-14

IBC
1605.2.1, 4-1
2106.5.1, 6-10
2107.1, 6-1
Eq. (16.1), 4-1
Eq. (16.6), 4-1
Table 301.7(2), 7-4
Table 2306.3.1, 7-21

LRFD
A-1.2, 4-25
A-1.3, 4-7, 4-18, 4-25
A-1.4, 4-25
A-1.5.2, 4-29
A-1.6, 4-28
A-3, 4-33
A5.3, 4-2
A-7.1, 4-18
A-7.3, 4-19
C1.3a, 4-12
C-1.7, 4-28
C2, 1-30, 4-12, 4-20
C2.2, 4-16, 4-20, 4-21, 4-22
C-C2, 4-28
C-D3.3, 4-31, 4-32
D2, 4-29, 4-57
D3, 4-31, 4-55
D3.1, 4-30
D3.2, 4-9, 4-30
Eq. (A-1-1), 4-28
Eq. (A-1-2), 4-28
Eq. (A-1-7), 4-25, 4-26, 4-28
Eq. (A-3-1), 4-33
Eq. (A-3-2), 4-33
Eq. (A-7-2), 4-18, 4-19
Eq. (A-7-3), 4-18, 4-19
Eq. (C2-1), 4-15, 4-16, 4-18, 4-19
Eq. (C2-2), 4-15, 4-16, 4-17
Eq. (C2-3), 4-16, 4-17, 4-19
Eq. (C2-4), 4-15
Eq. (C2-5), 4-15
Eq. (C2-6), 4-16, 4-17, 4-19
Eq. (D3-1), 4-31, 4-57
Eq. (E3-2), 4-13
Eq. (E3-3), 4-13
Eq. (E3-4), 4-13
Eq. (F1-1), 4-5, 4-6
Eq. (F2-3), 4-5
Eq. (F2-6), 4-4
Eq. (F4-7), 4-44
Eq. (F5-1), 4-44
Eq. (F5-2), 4-44
Eq. (F5-3), 4-44
Eq. (F5-4), 4-44
Eq. (F5-5), 4-44
Eq. (F5-6), 4-44
Eq. (F13-3), 4-44
Eq. (F13-4), 4-44
Eq. (G2-1), 4-8, 4-44–4-46
Eq. (G2-4), 4-47
Eq. (G2-5), 4-47, 4-49
Eq. (G2-6), 4-47
Eq. (G3-3), 4-47, 4-49
Eq. (H1-1a), 4-22, 4-23, 4-28
Eq. (H1-1b), 4-7, 4-22, 4-28
Eq. (I3-1), 4-51
Eq. (I3-3), 4-51

Eq. (J2-4), 4-38
Eq. (J2-5), 4-38, 4-57
Eq. (J2-9), 4-39
Eq. (J3-2), 4-38
Eq. (J3-3a), 4-38
Eq. (J3-6a), 4-35
Eq. (J3-6b), 4-35
Eq. (J4-5), 4-8, 4-9, 4-57
Eq. (J10-2), 4-10
Eq. (J10-3), 4-9
Eq. (J10-4), 4-10
Eq. (J10-5), 4-10
F1, 4-5, 4-6, 4-45
F1.3, 4-7
F2.2, 4-4
F5, 4-43, 4-44
F5.3, 4-44, 4-45
F13.2, 4-43
G2.1, 4-8, 4-43, 4-46, 4-47, 4-48
G2.2, 4-43, 4-47
G3.3, 4-44
H1.1, 4-7, 4-22
I2, 4-15
I3.1, 4-49
I3.2, 4-51, 4-52
J2.2, 4-38, 4-39
J2.4, 4-38, 4-39, 4-48
J3.2, 4-34
J3.6, 4-34
J3.7, 4-34
J3.9, 4-34
J3.10(a), 4-35
J4.1, 4-30
J7, 4-48, 4-49
J8, 4-23
J10.2, 4-9
J10.3, 4-10
J10.4, 4-48
J10.8, 4-11, 4-48
Part 2, 4-22
Part 3, 4-3, 4-4, 4-13, 4-14, 4-45, 4-51, 4-53
Part 14, 4-23
Table 3-2, 4-4, 4-7, 4-55
Table 3-5, 4-8
Table 3-6, 4-24, 4-26, 4-29
Table 3-10, 4-5, 4-7
Table 3-16, 4-46
Table 3-17, 4-46
Table 3-19, 4-53, 4-56
Table 3-20, 4-50, 4-51
Table 3-21, 4-51, 4-53, 4-56
Table 3-23, 4-7
Table 4-1, 4-13, 4-14, 4-16, 4-23, 4-28
Table 4-14, 4-15
Table 4-22, 4-14, 4-29, 4-49, 4-55
Table 6-1, 4-22, 4-23
Table 6-2, 4-14
Table 7-1, 4-34, 4-36
Table 7-2, 4-34
Table 7-3, 4-34
Table 7-4, 4-34, 4-35
Table 7-5, 4-35
Table 7-6, 4-35
Table 7-7, 4-36
Table 7-8, 4-36
Table 7-14, 4-36
Table 8-4, 4-40, 4-42
Table 8-8, 4-40, 4-41
Table 8-11, 4-40
Table 9-4, 4-10
Table A-3.1, 4-33
Table B4.1, 4-3, 4-44, 4-47
Table C-C2.2, 4-12
Table D3.1, 4-31, 4-57

Table J2.1, 4-38
Table J2.2, 4-38
Table J2.4, 4-38, 4-39, 4-48
Table J2.5, 4-38
Table J3.1, 4-34
Table J3.2, 4-34, 4-37
Table J3.3, 4-9, 4-30

NDS
2.2, 5-18
2.2.1, 5-6
2.3.3, 5-3
3.2.3, 5-9
3.3.3, 5-2, 5-3, 5-6, 5-7, 5-9, 5-15
3.4.2, 5-9
3.4.3.2, 5-9
3.5.2, 5-28
3.7.1, 5-13, 5-29
3.7.1.4, 5-12
3.7.1.5, 5-13, 5-29
3.9.1, 5-16, 5-17
3.9.2, 5-14, 4-15
3.10.3, 5-27
3.10.4, 5-2, 5-27, 5-27
4.3.8, 5-5, 5-7, 5-11
4.3.9, 5-5
4.4.1, 5-3
5.3.6, 5-5
5.3.8, 5-5, 5-6
10.3, 5-18
10.3.3, 5-18
11.1.5, 5-25
11.3, 5-22, 5-23
11.4.1, 5-22, 5-23
11.4.2, 5-24
11.5, 5-20
11.5.1, 5-19, 5-27
11.5.2, 5-19
11.5.3, 5-19
11.5.4, 5-19, 5-25

12.2.4, 5-19
12.2.5, 5-23
12.3.2, 5-19
C11.1.5, 5-25
Eq. (3.3-5), 5-3, 5-8
Eq. (3.3-6), 5-2
Eq. (3.4-3), 5-9
Eq. (3.4-5), 5-11
Eq. (3.4-6), 5-9, 5-11
Eq. (3.4-7), 5-9
Eq. (3.7-1), 5-3
Eq. (3.9-1), 5-16
Eq. (3.9-2), 5-16
Eq. (3.10-2), 5-2
Eq. (4.3-1), 5-5
Eq. (5.3-1), 5-5
Eq. (5.3-2), 5-5
Eq. (C3.9-3), 5-14
Eq. (12.2-1), 5-23
Table 2.3.2, 5-27
Table 2.3.3, 5-3, 5-6
Table 3.3.3, 5-3, 5-18
Table 10.3.3, 5-19
Table 10.3.6A, 5-18, 5-19
Table 10.3.6B, 5-18, 5-22
Table 10.3.6C, 5-18
Table 10.3.6D, 5-18, 5-22
Table 11.2A, 5-22
Table 11.3.2A, 5-22
Table 11.5.1A, 5-21
Table 11.5.1B, 5-21
Table 11.5.1C, 5-21
Table 11.5.1D, 5-21
Table 11.5.1E, 5-22
Table 11A, 5-19, 20
Table 11B, 5-20
Table 11C, 5-20
Table 11D, 5-20
Table 11E, 5-20
Table 11F, 5-20

Table 11G, 5-20
Table 11H, 5-20
Table 11I, 5-20
Table 11J, 5-19, 5-21, 5-22
Table 11K, 5-19, 5-22
Table 11L, 5-19, 5-23
Table 11M, 5-19, 5-23, 5-24
Table 11N, 5-19, 5-24. 5-25
Table 11O, 5-19
Table 11P, 5-19, 5-24
Table 11Q, 5-19
Table 11R, 5-19
Table 11.2B, 5-23
Table 11.2C, 5-24
Table 12.2A, 5-22
Table 12.2B, 5-22, 5-23
Table 12.2.3, 5-19, 5-22
Table 12.3, 5-22, 5-23
Table 12.2.4, 5-22
Table C11.1.4.7, 5-23
Table C11.1.5.6, 5-24
Table G1, 5-12

NDS SUPP
Table 4A, 5-2, 5-5, 5-7, 5-11, 5-13, 5-14,
    4-15, 5-16, 5-27
Table 4B, 5-2, 5-5
Table 4C, 5-2, 5-5
Table 4D, 5-2
Table 4E, 5-2
Table 4F, 5-2
Table 5A, 5-2, 5-5, 5-6
Table 5B, 5-2, 5-5
Table 5C, 5-2, 5-5
Table 5D, 5-2, 5-5, 5-28

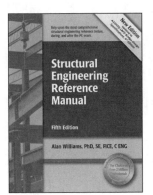
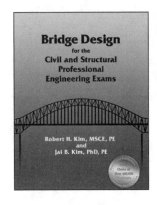
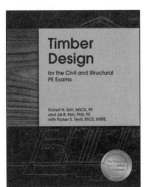

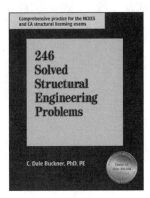